A FIRST COURSE IN VIBRATIONS AND WAVES

A First Course in Vibrations and Waves

Mohammad Samiullah

Truman State University

OXFORD
UNIVERSITY PRESS

Great Clarendon Street, Oxford, OX2 6DP,
United Kingdom

Oxford University Press is a department of the University of Oxford.
It furthers the University's objective of excellence in research, scholarship,
and education by publishing worldwide. Oxford is a registered trade mark of
Oxford University Press in the UK and in certain other countries

First Edition published in 2015

Impression: 1

Published in the United States of America by Oxford University Press
198 Madison Avenue, New York, NY 10016, United States of America

British Library Cataloguing in Publication Data
Data available

Library of Congress Control Number: 2015930647

ISBN 978–0–19–872978–5 (hbk.)
ISBN 978–0–19–872979–2 (pbk.)

Printed and bound by
CPI Group (UK) Ltd, Croydon, CR0 4YY

This book is dedicated to the memory of my sister Mrs. Rehana Sultana Ilyas whose wit and warmth brightened the lives she touched.

Preface

The study of vibrations and waves is central to physics and engineering. A course dealing with vibrations and waves often forms a bridge course between introductory physics courses and more advanced physics and engineering courses. This book is written for such a course, which I have taught regularly for a number of years at Truman. The book emphasizes the understanding of physics based on the fundamental principles expressed through mathematical equations.

While teaching the Vibs and Waves course I noticed that the mathematical maturity of students assumed by the existing textbooks was not realistic. Two textbooks that I tried in the course posed undue mathematical difficulties to students, even though students had already completed the Calculus sequence. To aid students I began to write subsidiary modules which they found to be very helpful. Some of them even relied on them completely rather than on the textbook. These modules have evolved into the present book. To make the book more complete and more useful to a broader audience I have expanded it by including some topics which I normally do not cover in my course due to a lack of time in a one-semester course.

The book follows the standard logical progression from simple harmonic motion to waves. It is organized into three parts. Part I contains a preliminary chapter that serves as a review of relevant ideas of mechanics and complex numbers. Although this chapter is just a review of the basics, I have found that most students need a review, especially, to learn or re-learn the language of equations of motion and exponential complex notation.

Part II is devoted to a detailed discussion of vibrations of mechanical systems. Chapters 2–4 are devoted to free harmonic motions in increasingly complicated systems. In Chapter 2, I present various systems which can be approximated by simple harmonic motion. They include the classic mass/spring system, the plane pendulum, the physical pendulum, fluids, and electric circuits. Both undamped and damped motion of free simple harmonic oscillators are presented here. In Chapter 3, we study the motion of two coupled oscillators. Here we learn various techniques for obtaining normal modes and the use of normal coordinates for understanding arbitrary motion. Chapter 4 extends the treatment of two coupled oscillators to N-bodies and continuous systems. The Fourier series is introduced in Chapter 4 along with expansion in the normal modes of a continuous system. Chapter 5 is devoted to driven oscillators. In this chapter we study driven oscillations and resonance in systems with one degree of freedom as well as many degrees of freedoms. The discussions of Chapter 5 are limited to closed systems in which the driven system occupies a finite space so that driving the system does not lead to the propagation of waves but rather to the excitement of standing waves.

Part III is concerned with waves. Here, the emphasis is on the discussion of common aspects of all types of waves. The basic language of traveling waves is more easily visualized in a one-dimensional situation of a transverse wave on a taut string. In Chapter 6 we make use of this example to study fundamental aspects of waves. Although the wave on a string is easy to visualize and serves to introduce the idea of waves well, the one-dimensional nature of the string limits its scope. In Chapter 7 we study waves in three dimensions, where we also introduce the fundamental nature of sound, electromagnetic, and matter waves.

We also learn about polarization of waves. In Chapter 8 we address the issue of the reaction of a wave with the boundary between different media. The reflection and transmission of waves is illustrated for waves on a string as well as the more complicated case of electromagnetic waves. Chapter 9 contains the important topic of interference of waves with applications in interferometry. The book concludes with a chapter on diffraction, where the application to diffraction gratings is presented.

The book includes many examples to illustrate main ideas. The exercises at the end of chapters are integral to the text. There are some simple exercises that help clarify concepts or familiarize students with important formulas and some challenging ones that help explore ideas further. Many exercises require the student to think of simplifying aspects and use analogies to solve them. The solutions to exercises are included in the Appendix. However, to get the full benefit of the exercises, a student should first try the exercises before consulting the solution.

Preliminary versions of this book were used as a textbook for my course and I would like to thank numerous students who have given feedback that has improved the book. My wife Huping deserves special thanks for encouraging me to convert my notes into a useful textbook and for giving me considerable free time to pursue this project.

M. Samiullah

Contents

Part I
Preliminaries

Review of Mechanics and Complex Algebra

1

Chapter Goals

In this chapter we will review the basics of mechanics at a little more of an abstract level than you have encountered in Introductory Physics courses. We will also review the algebra of complex numbers.

1.1 Review of Mechanics

1.1.1 Kinematics

Consider a point particle moving in space with respect to an observer O. The most basic quantity that describes the motion of the particle is the direct distance vector of the particle from O (see Fig. 1.1). We call this vector the position vector and denote it by $\vec{r}$,

$$\text{Position vector} = \vec{r}. \quad \text{(with respect to O)} \tag{1.1}$$

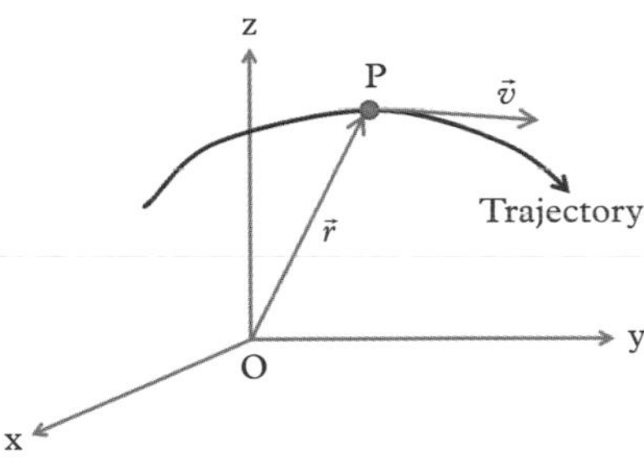

Fig. 1.1 *The trajectory, position and velocity of a particle P.*

We will denote vectors by arrows over their symbols. The magnitude of the vector will be denoted by the same symbol without the arrow. For instance, the symbol $\vec{r}$ will denote the **position vector** and the symbol r will denote just the distance.

The position vector may be a function of time if either its magnitude or the direction changes with time. Denoting t for the time variable we write this functional dependence as $\vec{r}(t)$. The derivative of this vector with respect to t gives the instantaneous rate at which the position of the particle changes with time. This quantity is called the **velocity** or the velocity vector, $\vec{v}$, with respect to the observer O,

$$\vec{v} = \frac{d\vec{r}}{dt}. \quad \text{(relative to the observer O)} \tag{1.2}$$

The velocity may also change with time if either its magnitude or direction changes and we can denote this by writing the velocity vector as a function of time t, that is, $\vec{v}(t)$. The rate of change of the velocity vector, as given by the derivative of $\vec{v}(t)$ with respect to time t is called the **acceleration** or the acceleration vector, $\vec{a}$,

$$\vec{a} = \frac{d\vec{v}}{dt}. \quad \text{(relative to the observer O)} \tag{1.3}$$

A First Course in Vibrations and Waves. First Edition. Mohammad Samiullah.

Note that the acceleration obtained by taking the time derivative of velocity is the acceleration with respect to the observer O since the velocity was with respect to O. Substituting $\vec{v}$ from Eq. 1.2 we can write acceleration as the second derivative of the position vector.

$$\vec{a} = \frac{d\vec{v}}{dt} = \frac{d^2\vec{r}}{dt^2}. \tag{1.4}$$

The acceleration vector can also change with time. The rate of change of acceleration is called jerk. In this book we will not study the rate at which acceleration of a body changes.

The study of the vector equations, Eqs. 1.1–1.3, is more easily done in the analytic picture of vectors by introducing convenient Cartesian coordinates. The vectors are decomposed into their x-, y-, and z-components. The components are denoted by attaching a subscript to the symbol of the vector without the arrow, e.g. v_x for the x-component of the velocity $\vec{v}$; we do not use this rule for the position vector $\vec{r}$, whose Cartesian components are simply the x-, y-, and z-coordinates. The components of the position, velocity, and acceleration vectors are given in Table 1.1.

The defining equations for the velocity and acceleration can be integrated. Thus, when we integrate Eq. 1.2 we obtain

$$\vec{r}(t_2) - \vec{r}(t_1) = \int_{t_1}^{t_2} \vec{v}(t)dt. \tag{1.5}$$

Integration of Eq. 1.3 gives

$$\vec{v}(t_2) - \vec{v}(t_1) = \int_{t_1}^{t_2} \vec{a}(t)dt. \tag{1.6}$$

Suppose $t_1 = 0$ and t_2 is some arbitrary time t, then we will have

$$\vec{r}(t) = \vec{r}_0 + \int_0^t \vec{v}(t')dt', \tag{1.7}$$

where I have replaced the dummy integration variable by t' so that we do not confuse the dummy integration variable with the t on the limit of integration, which is some definite instant. If the initial velocity is denoted by $\vec{v}_0$, then we get the following for the velocity at time t:

$$\vec{v}(t) = \vec{v}_0 + \int_0^t \vec{a}(t')dt'. \tag{1.8}$$

Table 1.1 *Position, Velocity, Acceleration.*

Property	x-component	y-component	z-component
$\vec{r}$	x	y	z
$\vec{v}$	$v_x = \frac{dx}{dt}$	$v_y = \frac{dy}{dt}$	$v_z = \frac{dz}{dt}$
$\vec{a}$	$a_x = \frac{dv_x}{dt}$	$a_y = \frac{dv_y}{dt}$	$a_z = \frac{dv_z}{dt}$

Example 1.1 Constant Acceleration

A particle has a constant acceleration $\vec{a}_0$. (a) What would be the velocity at instant t when the velocity at instant $t = 0$ is $\vec{v}_0$? (b) What would be the position at time t if the position at time $t = 0$ is $\vec{r}_0$?

Solution

(a) We use Eq. 1.8 with $\vec{a}(t') = \vec{a}_0$, constant in the integral on the right side to obtain

$$\vec{v}(t) = \vec{v}_0 + \int_0^t \vec{a}_0 dt' = \vec{v}_0 + \vec{a}_0 \int_0^t dt' = \vec{v}_0 + \vec{a}_0 t.$$

(b) Now, we use this velocity function in the integral in Eq. 1.7 to obtain

$$\vec{r}(t) = \vec{r}_0 + \int_0^t (\vec{v}_0 + \vec{a}_0 t)\, dt' = \vec{r}_0 + \vec{v}_0 t + \frac{1}{2}\vec{a}_0 t^2.$$

Example 1.2 Varying Acceleration

A particle has an acceleration given by $\cos(t)\hat{\imath}$, where $\hat{\imath}$ is a unit vector pointed towards the positive x-axis. The acceleration of this particle will point towards the positive x-axis at some time, as at $t = 0$, and towards the negative x-axis at some other times, such as at $t = \pi$. (a) What would be the velocity at instant t when the velocity at instant $t = 0$ is $\vec{v}_0$? (b) What would be the position at time t if the position at time $t = 0$ is $\vec{r}_0$?

Solution

(a) We use Eq. 1.8 with $\vec{a}(t') = \cos(t')\hat{\imath}$ in the integral on the right side to obtain

$$\vec{v}(t) = \vec{v}_0 + \int_0^t \cos(t')\hat{\imath} dt' = \vec{v}_0 + \sin(t)\hat{\imath}.$$

(b) Now, we use this velocity function in the integral in Eq. 1.7 to obtain

$$\vec{r}(t) = \vec{r}_0 + \int_0^t \left(\vec{v}_0 + \sin(t')\hat{\imath}\right) dt' = \vec{r}_0 + \vec{v}_0 t + [1 - \cos(t)]\,\hat{\imath}.$$

Example 1.3 Sinusoidally Varying Position

The position of a particle with respect to an observer O is given by $\vec{r}(t) = \cos(2\pi f t)\,\hat{\imath}$, where f is a constant. (a) What is the velocity of the particle at $t = \tau$? (b) What is the acceleration of the particle at $t = \tau$?

Solution

(a) We use the definition of velocity given in Eq. 1.2 to obtain

$$\vec{v}(t) = \frac{d\vec{r}}{dt} = -2\pi f \sin(2\pi f t)\,\hat{\imath}.$$

continued

Example 1.3 *continued*

Note that when we take the derivative of $\hat{\imath}$ with respect to t we get a zero since both the magnitude and direction of $\hat{\imath}$ are constants. Now, we can evaluate this for $t = \tau$ to obtain the velocity at time $t = \tau$,

$$\vec{v} = -2\pi f \sin(2\pi f\tau)\ \hat{\imath}.$$

(b) We use the definition of acceleration given in Eq. 1.3 to obtain

$$\vec{a}(t) = \frac{d\vec{v}}{dt} = -(2\pi f)^2 \cos(2\pi ft)\ \hat{\imath}.$$

Now, we can evaluate this for $t = \tau$ to obtain the acceleration at time $t = \tau$,

$$\vec{a} = -(2\pi f)^2 \cos(2\pi f\tau)\ \hat{\imath}.$$

1.1.2 Dynamics: Newton's Laws of Motion

The dynamics of the motion of a particle deals with the changes in the motion of the particle as a result of its interactions with other objects. The interaction is given by the forces between the particle and other objects. Newton's second law of motion states that a force on a particle causes the change in momentum of the particle. Suppose a force $\vec{F}$ is acting on a particle whose momentum at that instant is $\vec{p}$. Then, Newton's second law of motion states that the force will be proportional to the rate at which the momentum of the particle is changing at that instant. If the force is expressed in the unit Newton (N), momentum in kg.m/s, and time in s, then the relation is

$$\vec{F} = \frac{d\vec{p}}{dt}, \tag{1.9}$$

where the momentum $\vec{p}$ of a particle of mass m and velocity $\vec{v}$ is given by

$$\vec{p} = m\vec{v}. \tag{1.10}$$

Note that the force does not give the momentum of the particle but rather the rate at which the momentum is changing at that instant. Each force on the particle will cause its own rate of change of momentum. Suppose two forces $\vec{F}_1$ and $\vec{F}_2$ are acting on a particle of mass m during an interval t to $t + \Delta t$. Let the changes in momentum by the two forces be $\Delta\vec{p}_1$ and $\Delta\vec{p}_2$ respectively. Then the net change in momentum $\Delta\vec{p}$ will be

$$\vec{F}_1\Delta t + \vec{F}_2\Delta t = \Delta\vec{p}_1 + \Delta\vec{p}_2 = \Delta\vec{p}. \tag{1.11}$$

Dividing both sides by Δt and taking the limit $\Delta t \to 0$ we get

$$\vec{F}_1 + \vec{F}_2 = \frac{d\vec{p}}{dt}. \tag{1.12}$$

That is, the net force gives the net rate of change of the momentum of the particle. We can generalize this argument for any number of forces,

$$\vec{F}_1 + \vec{F}_2 + \cdots + \vec{F}_N = \frac{d\vec{p}_{\text{net}}}{dt}. \tag{1.13}$$

This is written compactly as

$$\vec{F}_{\text{net}} = \frac{d\vec{p}_{\text{net}}}{dt}. \tag{1.14}$$

Often we omit the subscript, net, from the symbols and keep that information implicit rather than explicit. For constant mass $d\vec{p}/dt = m\vec{a}$. Therefore, we get the following simpler equation when mass is constant,

$$\vec{F} = m\vec{a}. \quad \text{(constant mass)} \tag{1.15}$$

This law shows that if we know forces at each instant, then we can find the acceleration. Once we know the acceleration at all instants, we can predict the changes in velocity and position over time.

Note that when we measure the acceleration of a particle we measure the net acceleration and not the acceleration due to individual forces. Therefore, we only get the net force from a measurement of the acceleration. To learn about individual forces, we need to perform experiments that involve one force at a time.

Example 1.4 Free Fall

A particle is falling freely near the Earth. Assuming the weight of the particle to be the only force on the particle, find the distance the particle will fall in the duration $t = 0$ to $t = \tau$ if the initial speed of the particle were zero.

Solution

Let us use a Cartesian coordinate system with the y-axis pointed up and work with the y-component of Eq. 1.15. The net force on the particle has only the y-component and since the positive y-axis is pointed up, the y-component of the net force will be negative,

$$F_y = -mg.$$

Therefore from the y-component of Eq. 1.15 we obtain

$$-mg = ma_y.$$

This give the y-component of acceleration to be

$$a_y = -g.$$

Using this constant a_y in Eq. 1.8 we find the y-component of the velocity at instant t to be

$$v_y(t) = -gt, \text{ using } v_y(0) = 0.$$

Now, from Eq. 1.7 we obtain the change in the y-coordinate of the particle during time $t = 0$ to $t = \tau$ to be

$$y - y_0 = -\frac{1}{2}g\tau^2,$$

where y_0 is the y-coordinate at $t = 0$.

Example 1.5 Varying Force

A particle is subject to the following force, $\vec{F} = bt\hat{i}$, where b is constant and $\hat{i}$ is the unit vector pointed towards the positive x-axis. What would be the change in the velocity during the interval $t = 0$ to $t = \tau$?

Solution

Since the force is pointed along the x-axis, the acceleration will also point along the x-axis. This means that only the x-component of the velocity will change. We therefore work only with the x-components, with the x-component of the force being

$$F_x = b\,t.$$

The x-component of Eq. 1.15 gives

$$b\,t = m\,a_x.$$

Therefore, the x-component of acceleration will be

$$a_x = \frac{b}{m}\,t.$$

Using this a_x in Eq. 1.8 we find the x-component of the velocity at instant t to be

$$v_x(t) = v_{0x} + \frac{b}{2m}\,t^2.$$

At time $t = \tau$ this gives

$$v_x = v_{0x} + \frac{b}{2m}\,\tau^2.$$

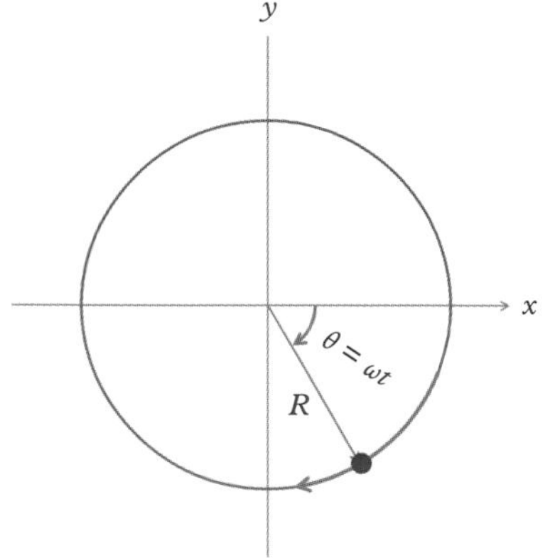

Fig. 1.2 *Example 1.6.*

Example 1.6 Circular Motion

A particle of mass m is observed to be moving in a circle in the xy-plane of a Cartesian coordinate system, as shown in Fig. 1.2. The x and y coordinates of the particle are given by

$$x(t) = R\cos(\omega t),$$
$$y(t) = -R\sin(\omega t),$$

where R and ω are constants. What are the magnitude and direction of the net force on the particle?

Solution

Let us work out the components of the acceleration by taking two time derivatives of $x(t)$ and $y(t)$. The z-component of the acceleration would be zero since the particle is moving only in the xy plane.

$$a_x = \frac{d^2x}{dt^2} = -R\omega^2\cos(\omega t),$$
$$a_y = \frac{d^2y}{dt^2} = R\omega^2\sin(\omega t),$$
$$a_z = 0.$$

Therefore, the components of the net force would be

$$F_x = -m R\omega^2 \cos(\omega t),$$
$$F_y = m R\omega^2 \sin(\omega t),$$
$$F_z = 0.$$

This force has a constant magnitude $mR\omega^2$ and the direction pointed towards the origin, which changes with time as the particle goes around in a circle.

Example 1.7 Position-dependent Force

A particle of mass m is subject to the following force:

$$\vec{F} = -kx\hat{i},$$

where k is a constant and x is the x-coordinate of the particle. Suppose that at time $t = 0$ the particle were at $x = A$ and its speed were zero, where will the particle be at $t = \tau$?

Solution

Since the force is pointed along the x-axis, the acceleration will also point along the x-axis. This means that only the x-component of the velocity will change. We therefore work only with x-components,

$$F_x = -k x.$$

The x-component of $\vec{F} = m\vec{a}$ gives

$$-k x = m a_x.$$

Solving for a_x we obtain

$$a_x = -\frac{k}{m} x. \tag{1.16}$$

Can we plug this a_x in Eq. 1.8 to find the change in the x-component of the velocity? We will get

$$v_x(t) = v_{0x} - \frac{k}{m} \int_0^t x(t')dt'.$$

It turns out that we are stuck at this point since we do not yet know $x(t)$. Another approach is needed when acceleration is not given as a function of time, but instead as a function of position or velocity. We will rewrite Eq. 1.16 as a differential equation by making use of the definition of acceleration in terms of the two derivatives of position. This turns Eq. 1.16 into a differential equation for the variable $x(t)$:

$$\frac{d^2x}{dt^2} = -\frac{k}{m} x. \tag{1.17}$$

The general solution of this differential equation is

$$x(t) = C_1 \cos(\omega t) + C_2 \sin(\omega t), \tag{1.18}$$

where $\omega = \sqrt{k/m}$, and C_1 and C_2 are constants to be determined from the given initial position and velocity. To make use of the zero velocity at $t = 0$ we take a time derivative of Eq. 1.18 and obtain the x-component of the velocity,

$$v_x(t) = \omega \left[-C_1 \sin(\omega t) + C_2 \cos(\omega t)\right]. \tag{1.19}$$

continued

Example 1.7 *continued*

Now we put $t = 0$ in Eqs. 1.18 and 1.19 to use the following conditions at $t = 0$:

$$x(0) = A \quad \text{and} \quad v_x(0) = 0.$$

A simple algebra shows that

$$C_1 = A, \; C_2 = 0.$$

Therefore the position of the particle at an arbitrary time t is given by

$$x(t) = A\cos(\omega t).$$

We can now set $t = \tau$ to obtain the position at that instant,

$$x(\tau) = A\cos(\omega\tau).$$

1.1.3 Work, Energy, and Power

Work and Potential Energy

To analyze the energy of a particle it is helpful to introduce the concept of work by a force. Suppose a particle is subject to a force $\vec{F}$ and moves from a position $\vec{r}$ at time t to the position $\vec{r} + d\vec{r}$ at time $t + dt$ (see Fig. 1.3). We say that the force $\vec{F}$ does work dW on the particle during the interval t to $t + dt$. The **work** dW is given by the dot product between the force and the displacement vectors,

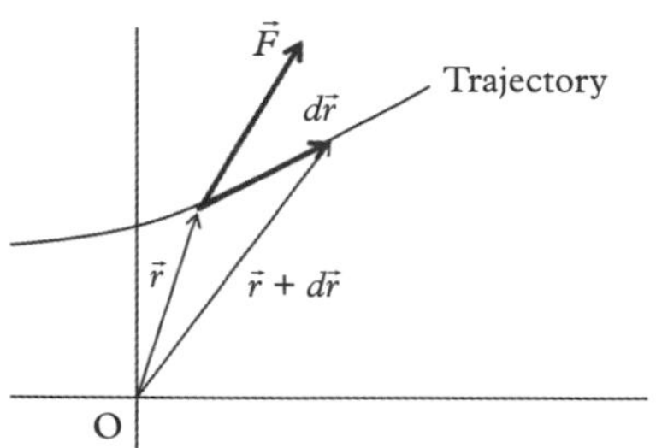

Fig. 1.3 *Work by force $\vec{F}$ during the displacement $d\vec{r}$ is $dW = \vec{F}\cdot d\vec{r}$.*

$$dW = \vec{F}\cdot d\vec{r}. \tag{1.20}$$

This can be written more explicitly in terms of the magnitude $|\vec{F}|$ of the force, the magnitude $|d\vec{r}|$ of the displacement, and the angle θ between them,

$$dW = \vec{F}\cdot d\vec{r} = |\vec{F}|\,|d\vec{r}|\cos\theta. \tag{1.21}$$

Note that the work of a force is zero if it is perpendicular to the motion of the particle, since the infinitesimal displacement $d\vec{r}$ will be in the direction of the motion given by the direction of the instantaneous velocity $\vec{v}$. For instance, magnetic force on a charged particle is always perpendicular to the velocity of the particle, therefore a magnetic force does not do any work on a moving charged particle. To obtain work by the force over a finite interval of time t_i to t_f, we just need to add up the contributions from work on each infinitesimal displacement. This is written formally as a line integral,

$$W_{if} = \int_i^f \vec{F}\cdot d\vec{r}. \tag{1.22}$$

This integral may depend on the path of the particle between its initial and final positions. If the integral depends on the path of the particle then we say that the force is a non-conservative force. If the integral does not depend on the path but just on the end points, we say that the force is a conservative force and write the result as

$$W_{if} = U_i - U_f, \tag{1.23}$$

where U is called the **potential energy** of the particle corresponding to that particular conservative force. It turns out that all fundamental forces in nature are conservative. The commonly occurring effective forces such as the weight of a particle and the force of a spring on a body are also conservative forces. Dissipative forces, such as friction and drag, are not conservative, and therefore, there is no potential energy concept associated with them.

The potential energy due to gravity, denoted by U_g, and that due to spring force, denoted by U_s, take the following forms:

$$U_g = mgh, \quad \text{(with reference } h = 0) \tag{1.24}$$

$$U_s = \frac{1}{2}k(\Delta l)^2, \quad \text{(with reference } \Delta l = 0), \tag{1.25}$$

where

$h =$ the height above a reference point, and

$\Delta l -$ the change in the length of the spring.

From Eqs. 1.22 and 1.23 we can derive an important relation between potential energy of conservative force and the force itself. Consider an infinitesimal process in which the displacement occurs from $\vec{r}$ to $\vec{r} + d\vec{r}$. The work done will be just $\vec{F} \cdot d\vec{r}$. Therefore, we can write the change in potential energy to be

$$U(\vec{r}) - U(\vec{r} + d\vec{r}) = \vec{F} \cdot d\vec{r}. \tag{1.26}$$

Considering the change in various directions, you can prove that

$$\vec{F} = -\vec{\nabla} U, \tag{1.27}$$

where $\vec{\nabla}$ is the vector operator with the following representation in the Cartesian coordinates:

$$\vec{\nabla} = \hat{i}\,\frac{\partial}{\partial x} + \hat{j}\,\frac{\partial}{\partial y} + \hat{k}\,\frac{\partial}{\partial z}. \tag{1.28}$$

Equation 1.27 is an important relation for finding force since, in many situations, it is easier to guess the potential energy than to account for all forces directly.

Kinetic Energy and Work–Energy Theorem

Suppose we do the line integral in Eq. 1.22 that defines the work on both sides of the equation for Newton's second law for a fixed mass particle, what will we get?

$$\int_i^f \vec{F} \cdot d\vec{r} = \int_i^f m\vec{a} \cdot d\vec{r}. \tag{1.29}$$

The left side is the work by all of the forces, which we will denote as W_{if}^{net}. Note that W_{if}^{net} is **not the work by the net force** but rather **net work by all forces**–although the two works are the same for a single-particle system, they may be different for multiparticle systems due to the work by the internal forces which is not included in the work by the net force. The right side looks complicated but can be simplified to

$$\text{Right side} = \int_i^f m\vec{a}\cdot d\vec{r} = \frac{1}{2}mv_f^2 - \frac{1}{2}mv_i^2,$$

where v_i and v_f are the speed of the particle at the initial and final instants respectively. This shows that the sum of the work done by all forces on a particle is equal to the change in a quantity of the form $\frac{1}{2}mv^2$. We call this quantity the **kinetic energy** (*KE*) or simply K,

$$K = \frac{1}{2}mv^2. \tag{1.30}$$

Therefore, Eq. 1.29 says that

$$W_{if}^{\text{net}} = K_f - K_i. \tag{1.31}$$

This important result is called the **Work–Energy theorem**. Suppose all forces on the particle are conservative forces, then we can replace the work on the left side of this equation by the negative of the change in the potential energy, as defined above,

$$U_i - U_f = K_f - K_i. \tag{1.32}$$

Rearranging the terms we get

$$K_f + U_f = K_i + U_i. \text{ (only conservative forces)} \tag{1.33}$$

We see that if only conservative forces act on the particle then the sum of the kinetic and potential energies does not change. We define **mechanical energy** E of the particle by this sum,

$$E = K + U. \tag{1.34}$$

Then, Eq. 1.33 is the statement of the conservation of energy for conservative systems where all forces are conservative.

Example 1.8 Energy Conservation

A block is attached to a spring and the other end of the spring is attached to a fixed support. The block is then placed on a frictionless horizontal surface and pulled along the line of the spring. After extending the spring by a length A, the block is released from rest. (a) What will be the speed of the block when it returns to a point where the spring is relaxed? (b) What will be the speed of the block when it returns to a point where the spring has an extension equal to $\frac{1}{2}A$?

Solution

(a) The forces on the blocks are gravity, normal, and the spring force. Since gravity and normal are perpendicular to the velocity of the block they will not do any work on the block. Since the spring force is a conservative force the energy of the block will be conserved. Figure 1.4 shows the two instants of interest. At the initial instant the energy is

$$E_i = K_i + U_{si} = 0 + \frac{1}{2}kA^2.$$

The energy at the final instant has no contribution from the potential energy since the spring is neither extended nor compressed,

$$E_f = K_f + U_{sf} = \frac{1}{2}mv_f^2 + 0.$$

Equating E_f to E_i we find

$$v_f = \left(\frac{k}{m}\right)^{1/2} A.$$

(b) In this part we keep the same t_i as in part (a) but change the situation at t_f to when $\Delta l = A/2$ for the spring and v for the speed at that instant. This gives

$$E_f = \frac{1}{2}mv^2 + \frac{1}{2}k\left(\frac{A}{2}\right)^2.$$

Now, equating E_f to E_i we find

$$v = \left(\frac{3k}{4m}\right)^{1/2} A.$$

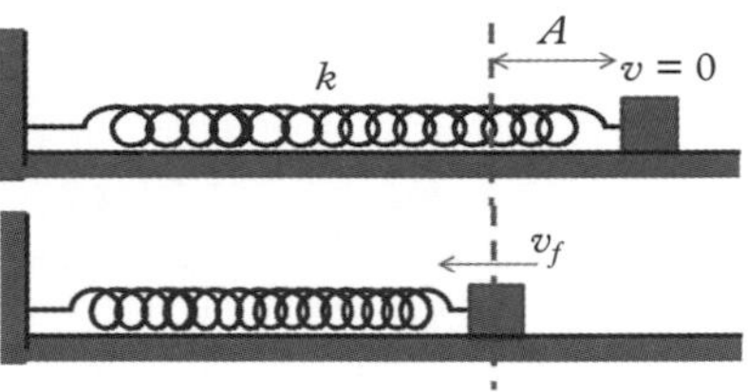

Fig. 1.4 *Example 1.8.*

Power

The rate at which a force does work is called its power P. Suppose a force $\vec{F}$ does work dW during the time interval from t to $t + dt$, then the instantaneous power is defined as

$$P(t) = \frac{dW}{dt}. \tag{1.35}$$

Writing the work in terms of force $\vec{F}$ and the displacement $d\vec{r}$ we obtain

$$\frac{dW}{dt} = \frac{\vec{F}\cdot d\vec{r}}{dt} = \vec{F}\cdot\frac{d\vec{r}}{dt} = \vec{F}\cdot\vec{v}. \tag{1.36}$$

Therefore, the **instantaneous power** can also be written as

$$P(t) = \vec{F}\cdot\vec{v}. \tag{1.37}$$

We will often be interested in the average power $\langle P \rangle$ during a finite interval t_i to t_f. The average power is obtained by time-averaging as follows:

$$\langle P \rangle = \frac{1}{t_f - t_i} \int_{t_i}^{t_f} P(t)\,dt.$$

Frequently we will be working with power $P(t)$ that is periodic in time with some period T. In that case, the average power will refer to the quantity obtained upon time-averaging $P(t)$ over one period of time,

$$\langle P \rangle = \frac{1}{T} \int_0^T P(t)\,dt. \quad \text{(periodic } P(t)\text{)} \tag{1.38}$$

Example 1.9 Instantaneous Power and Average Power

A block of mass m oscillates along the x-axis. The position of the block is given by $x(t) = A\cos\omega_0 t + B\sin\omega_0 t$, where A, B, and ω_0 are constants. The block is subject to a force $\vec{F}$ whose x-component is given by $F_x = F_0\cos\omega t$, where F_0 and ω are constants. Find the instantaneous power and average power over one cycle of the sinusoidally varying force.

Solution

The instantaneous power of the force would be

$$P(t) = \vec{F}\cdot\vec{v} = F_x v_x.$$

Therefore, we need the x-component of the velocity, which can be obtained from $x(t)$ by taking the time derivative,

$$v_x = \frac{dx}{dt} = -A\omega_0 \sin\omega_0 t + B\omega_0\cos\omega_0 t.$$

Therefore,

$$P(t) = -F_0 A\omega_0 \sin\omega_0 t \cos\omega t + F_0 B\omega_0 \cos\omega_0 t\cos\omega t.$$

To find the time average of this power over one cycle of the sinusoidally varying force we can integrate $P(t)$ over one time period of the force, which is $2\pi/\omega$,

$$P_{\text{av}} = \frac{1}{T}\int_0^T P(t)dt, \quad \left(T = \frac{2\pi}{\omega}\right).$$

This integral is messy for arbitrary ω and ω_0. The calculation simplifies in the case of $\omega = \omega_0$. In this case, the instantaneous power would be

$$P(t) = -\frac{1}{2}F_0 A\omega\sin(2\omega t) + F_0 B\omega\cos^2\omega t, \quad (\text{Case: } \omega = \omega_0).$$

Integrating this over one period gives

$$P_{\text{av}} = \frac{1}{2}F_0 B\omega, \quad (\text{Case: } \omega = \omega_0).$$

1.1.4 Conservation Laws

Conservation laws are very important in physics. There are three conservation laws in mechanics.

(1) The conservation of energy
(2) The conservation of momentum
(3) The conservation of angular momentum

We have already encountered the law of conservation of energy. Although, in this book, we will use only the conservation of energy, I will give a brief treatment of the two other conservation laws for the sake of completeness. For a more thorough treatment with applications of each of the conservation laws you should consult a book on mechanics. To find the origin and meaning of the law of conservation of momentum, recall Newton's second law in the original form,

$$\vec{F} = \frac{d\vec{p}}{dt}. \tag{1.39}$$

Now, if the left side is zero, then we immediately get

$$\text{if } \vec{F} = 0 \text{ then } \frac{d\vec{p}}{dt} = 0.$$

That is, if external force is absent or cancels out at any instant, then the rate of change of momentum at that instant will be zero. If this happens at each instant of an interval, $t_1 \leq t \leq t_2$, then momentum over that interval will not change. This conclusion is known as the **principle of conservation of momentum**. The fundamental basis of this principle can be traced to the fact that laws of mechanics are invariant with respect to translation in space.

For a system that has only one particle of constant mass, this law gives the same information as Newton's first law of motion. That is, in the absence of external force, the velocity of a particle will remain constant.

For a system that has more than one particle this law becomes very useful. Thus, in a collision of two particles, one particle can change the momentum of the other, but their combined momentum after the collision will be the same as before the collision,

$$\vec{p}_1 + \vec{p}_2 = \vec{p}_1' + \vec{p}_2'. \tag{1.40}$$

The conservation of angular momentum also follows from the second law of motion. The **angular momentum** $\vec{L}$ of a particle of momentum $\vec{p}$ and position $\vec{r}$ is defined by

$$\vec{L} = \vec{r} \times \vec{p}. \tag{1.41}$$

Let us take a derivative of both sides of this equation with respect to time t,

$$\frac{d\vec{L}}{dt} = \frac{d\vec{r}}{dt} \times \vec{p} + \vec{r} \times \frac{d\vec{p}}{dt} = \vec{v} \times \vec{p} + \vec{r} \times \vec{F}. \tag{1.42}$$

The first term on the right side will be zero since $\vec{v} \parallel \vec{p}$. Therefore,

$$\frac{d\vec{L}}{dt} = \vec{r} \times \vec{F}. \tag{1.43}$$

The quantity $\vec{r} \times \vec{F}$ is called the torque of the force $\vec{F}$ about the origin. Torque on a particle will be zero if force is radially directed, either towards the origin or away from the origin, for instance, the torque of the force of gravity on Earth about the Sun. The torque will also be zero if no force acts. Whenever torque is zero, the rate of change of angular momentum will be zero,

$$\text{if } \vec{r} \times \vec{F} = 0 \text{ then } \frac{d\vec{L}}{dt} = 0. \tag{1.44}$$

That is, if torque is zero at any instant, then the rate of change of angular momentum at that instant will be zero. If this happens at each instant of an interval, $t_1 \leq t \leq t_2$, then angular momentum over that interval will not change. This conclusion is known as the **principle of conservation of angular momentum.** The fundamental basis of this principle can be traced to symmetry of space under rotation.

1.2 Complex Numbers

Complex algebra is very useful in vibrations and waves. In this chapter we will review some basic aspects of calculations involving complex numbers. A complex number is a number that contains the number $\sqrt{-1}$, which is denoted by the letter i or the letter j. Multiples of i or j by real numbers are called **imaginary numbers**. Although no physical measurement gives $\sqrt{-1}$ or its multiples as an outcome, we use complex numbers in physics because they make calculations simpler and help with physical insights.

A **complex number** c can always be written as an ordinary sum of a real number and an imaginary number, $c = a + ib$, where a and b are real numbers. We can also think of a complex number c as an ordered pair of two real numbers, (a, b). The real numbers a and b that make up a complex number $c = a + ib$ are called **the real and imaginary parts** of the complex number, respectively. We often indicate them as follows:

$$c = a + i\,b, \quad a = \mathrm{Re}[c], \quad b = \mathrm{Im}[c]. \tag{1.45}$$

Two complex numbers that are related such that replacing all i in one number by $-i$ gives the other number are called the **complex conjugates** of each other. For instance, the complex conjugate of $3 + i4$ is $3 - i4$ and vice versa. We will denote the complex conjugate of a number by placing an asterisk next to it. For example, the complex conjugate of a complex number c will be denoted by c^*. Clearly, the complex conjugate of the complex conjugate will give back the original complex number,

$$(c^*)^* = c. \tag{1.46}$$

1.2.1 The Complex Plane

When you add two complex numbers, the real parts add independently from the imaginary parts, as the following calculation of the sum of two complex numbers $c_1 = (a_1 + ib_1)$ and $c_2 = (a_2 + ib_2)$ shows:

$$c = c_1 + c_2 = (a_1 + ib_1) + (a_2 + ib_2) = (a_1 + a_2) + i(b_1 + b_2). \tag{1.47}$$

The real part of the sum is the sum of the real parts of the summand, and similarly for the imaginary part of the sum. *This addition follows the same rule as the addition of components of vectors in a plane*. Therefore, it was realized very early on that a good way to visualize a complex number is to think of the real and imaginary parts as x- and y-components of a *fictitious vector* (Fig. 1.5). The fictitious vector representing the complex number $a + ib$ is from the origin to the point (a, b) in the xy-plane. Such a picture of a complex number is called its **Argand diagram**.

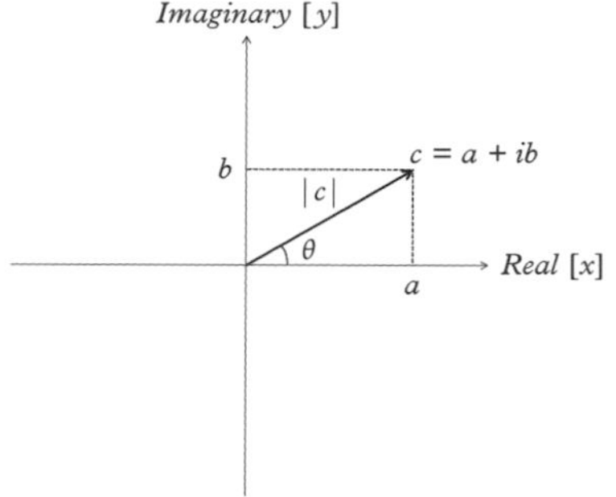

Fig. 1.5 *The complex plane and the Argand diagram. The complex number $c = a + ib$ is shown as a vector in the complex plane. The length of the vector is the absolute value of the complex number $|c|$ and the angle the vector makes with the positive x-axis is the argument of the complex number.*

Note that this xy-plane is an abstract plane and should not be confused with the real space plane. The abstract plane of complex numbers is called the **complex plane**. An arbitrary point in the xy-plane of the Argand diagram is usually denoted by the letter z,

$$z = x + iy. \tag{1.48}$$

Just like real two-dimensional planar space, we can also introduce polar coordinates in the complex plane. In polar coordinates, the point (x, y) is give by the radial distance r and the angle θ of the vector with the positive x-axis,

$$r = \sqrt{x^2 + y^2} \tag{1.49}$$

$$\theta = \begin{cases} \tan^{-1}(y/x) & \text{if } x \geq 0. \\ \tan^{-1}(y/x) + \pi & \text{if } x < 0. \end{cases} \tag{1.50}$$

The magnitude r of the vector representing a complex number is called the **amplitude or absolute value** of the complex number. We denote the absolute value of a complex number by placing vertical bars around the symbol for the complex number, e.g. $|z|$ for the complex number z. It is obvious from the construction that the absolute value of a complex number is a non-negative number. It can also be written in terms of the complex number z and its complex conjugate z^*,

$$|z| = \sqrt{z^* z} = \sqrt{x^2 + y^2}, \quad \text{for } z = x + iy. \tag{1.51}$$

The "direction" θ of the vector in the complex plane is also called the **argument or phase** of the complex number, usually denoted by Arg[z] for the complex number z,

$$\text{Arg}[z] = \theta = \tan^{-1}\left(\frac{y}{x}\right). \tag{1.52}$$

Thus, we have two equivalent representations of a complex number: rectangular, i.e. the Cartesian representation in terms of its real and imaginary parts (x, y), and the polar representation in terms of its absolute value and argument (r, θ). Both representations are useful. We will find that adding complex numbers is easiest in the rectangular form and multiplying or dividing is easiest in the polar form.

A Note on Arctangent and Quadrants

Beware of the complexities associated with the arctangent in the domain $[0, 2\pi]$, as given in Eq. 1.50. For instance, points $(1, 1)$ and $(-1, -1)$ will both give the same value of the angle when calculated by $\tan^{-1}(y/x)$ in a calculator, but one makes an angle of 45° counterclockwise from the positive x-axis, and the other makes an angle of $180 + 45 = 225°$ counterclockwise from the positive x-axis. Similarly for $(1, -1)$ and $(-1, 1)$. Therefore, you must be careful when interpreting the value you obtain by taking arctangent in a calculator; interpret your answer based on the quadrant in which the point (x, y) is located as indicated in the definition given in Eq. 1.50.

Multiple Values of θ

The argument θ of a complex number z refers to a direction in the complex plane. The direction angle in a two-dimensional plane of the physical space is restricted to the range $0 \le \theta < 2\pi$ or $0 \le \theta < 360°$. But, unlike the direction in space, the angle θ of a complex number can take any real value,

$$\text{Argument of complex number:} \quad -\infty < \theta < \infty.$$

Since any value $\theta = \theta_0$, where θ_0 is restricted to the 2π range, and the value $\theta = \theta_0 + 2n\pi$ with n any integer, positive or negative corresponds to the same direction in the complex plane, both will represent the same complex number,

$$\theta = \theta_0 \text{ and } \theta = \theta_0 + 2n\pi \text{ same complex number.}$$

Example 1.10 a Complex Number in Polar Form

Find the polar form of $1 - i$.

Solution

Here $x = 1$ and $y = -1$. Therefore,

$$r = \sqrt{1^2 + (-1)^2} = \sqrt{2},$$

and

$$\theta = \tan^{-1}(-1/1) + 2n\pi = \frac{3\pi}{4} + 2n\pi,$$

where $n = 0, \pm1, \pm2, \ldots$. Figure 1.6 shows the vector diagram of this complex number, where only the primary value of θ is shown.

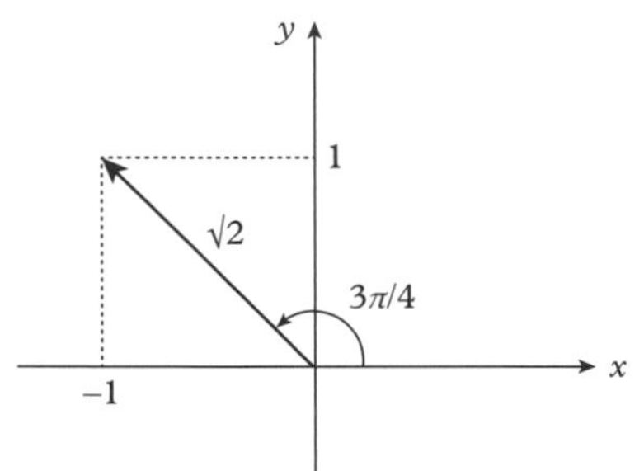

Fig. 1.6 *Example 1.10.*

1.2.2 The Exponential Form

Recall that the polar representation of a complex number consists of the absolute value r and the argument θ. By using the Argand diagram, we can easily see that the real part of the complex number is $(r\cos\theta)$ and the imaginary part is $(r\sin\theta)$. Euler showed that these trigonometric functions of the real and imaginary parts can be combined into a very useful exponential form,

$$z = r\,e^{i\theta}, \tag{1.53}$$

which can be shown to be identical to $(r\cos\theta) + i(r\sin\theta)$, or,

$$e^{i\theta} = \cos\theta + i\sin\theta. \tag{1.54}$$

You can prove this equation by expanding both sides in a Taylor series about $\theta = 0$. The complex conjugate of Eq. 1.54 gives another identity,

$$e^{-i\theta} = \cos\theta - i\sin\theta, \tag{1.55}$$

which can be used to write the trigonometric functions in terms of complex exponentials,

$$\cos(\theta) = \frac{e^{i\theta} + e^{-i\theta}}{2}, \quad \sin(\theta) = \frac{e^{i\theta} - e^{-i\theta}}{2i}. \tag{1.56}$$

Box 1.1 summarizes various relations between the rectangular and polar representations of complex numbers.

Box 1.1 *Rectangular and polar representations of complex numbers*

$$z = x + iy$$
$$r = \sqrt{x^2 + y^2}$$
$$\theta = \begin{cases} \tan^{-1}(y/x) & \text{if } x \geq 0. \\ \tan^{-1}(y/x) + \pi & \text{if } x < 0. \end{cases}$$
$$x = r\cos(\theta)$$
$$y = r\sin(\theta)$$
$$r\,e^{i\theta} = r\cos(\theta) + i\,r\sin(\theta)$$
$$e^{i\theta} = \cos(\theta) + i\sin(\theta)$$
$$e^{-i\theta} = \cos(\theta) - i\sin(\theta)$$
$$\cos(\theta) = \frac{e^{i\theta} + e^{-i\theta}}{2}$$
$$\sin(\theta) = \frac{e^{i\theta} - e^{-i\theta}}{2i}$$

1.2.3 Complex Algebra

The addition, subtraction, multiplication, and division of complex numbers follow the usual rules of algebra familiar from real numbers. We have already talked about addition of two complex numbers: the real parts add to the real parts, and the imaginary parts add to the imaginary parts. For example, the sum of $2 + i3$ and $4 + i5$ is equal to $(2 + 4) + i(3 + 5) = 6 + i8$. Subtraction works the same way, subtracting $7 + i3$ from $4 + i5$ gives $(4 - 7) + i(5 - 3) = -3 + i2$.

The multiplication of two complex numbers written in the rectangular form follows the usual rule of distribution,

$$(a_1 + ib_1)(a_2 + ib_2) = a_1a_2 + ia_1b_2 + ib_1a_2 + i^2b_1b_2, \tag{1.57}$$

which can be simplified by using $i^2 = -1$ and summing the two imaginary numbers,

$$(a_1 + ib_1)(a_2 + ib_2) = (a_1a_2 - b_1b_2) + i(a_1b_2 + b_1a_2), \tag{1.58}$$

which gives the real and imaginary parts of the product. For example, the product of the two numbers $2 + i3$ and $4 + i5$ gives $(2 \times 4 - 3 \times 5) + i(2 \times 5 + 3 \times 4) = -7 + i22$.

As announced above, the multiplication is more easily done in the polar form. The exponents simply add when two numbers with the same base are multiplied,

$$(r_1e^{i\theta_1})(r_1e^{i\theta_1}) = (r_1\ r_2)e^{i(\theta_1+\theta_2)}. \tag{1.59}$$

The absolute value of the product is the product of the absolute values of the two numbers and the argument of the product is the sum of the arguments. Note that any angle greater than 2π or less than -2π can be brought in the range $[0, 2\pi]$ by adding or subtracting an appropriate integral multiple of 2π.

The division of a complex number by another complex number is tricky. If the numerator and denominator are given in the rectangular form, we first try to convert the denominator to a real number by multiplying the numerator and denominator by the complex conjugate of the denominator,

$$\frac{z_1}{z_2} = \left(\frac{z_1}{z_2}\right)\left(\frac{z_2^*}{z_2^*}\right) = \frac{z_1\ z_2^*}{|z_2|^2}. \tag{1.60}$$

This trick makes the division into multiplication of two complex numbers z_1 and z_2^*. We illustrate this important case with an example.

Example 1.11 Division for Complex Numbers

Find the real and imaginary parts of $(2+i5)/(3-i4)$.

Solution

We follow the suggestion given in the text. The trick is to multiply both the numerator and the denominator by the complex conjugate of the denominator,

$$\begin{aligned}\frac{2+i5}{3-i4} &= \left(\frac{2+i5}{3-i4}\right)\left(\frac{3+i4}{3+i4}\right)\\ &= \frac{(2+i5)(3+i4)}{25}\\ &= \frac{1}{25}(-14+i23)\\ &= -\frac{14}{25}+i\frac{23}{25}.\end{aligned} \tag{1.61}$$

Therefore, the real part is $-14/25$ and the imaginary part $23/25$.

The division of a complex number by another is simpler if the numbers are given in the polar form. The amplitude of the resultant is obtained by dividing the amplitude of the numerator by the amplitude of the denominator; and the argument of the resultant is obtained by subtracting the argument of the denominator from that of the numerator,

$$\frac{z_1}{z_2} = \frac{r_1\, e^{i\theta_1}}{r_2\, e^{i\theta_2}} = \left(\frac{r_1}{r_2}\right)\, e^{i(\theta_1-\theta_2)}. \tag{1.62}$$

For example the division of $10e^{i3}$ by $2e^{i12}$ would give $5e^{-i9}$. Watch out for the common mistake here: do not divide both the amplitudes and the arguments; amplitudes divide but arguments subtract.

The exponential form is also very useful when calculating complex numbers raised to powers. Thus, a complex number z raised to a power p can be easily found to be

$$z^p = (re^{i\theta})^p = r^p e^{ip\theta}. \tag{1.63}$$

For instance, we can evaluate the $(3+i4)^5$ by first writing the complex number in the exponential form, and then raising the power by 5,

$$(3+i4)^5 = (5\, e^{i\tan^{-1}(4/3)})^5 = 875\, e^{i5\tan^{-1}(4/3)}.$$

On the other hand, exponentiating a real number by a complex number is done more easily by expressing the complex number in the rectangular form first,

$$a^z = a^{x+iy} = a^x\, a^{iy} = a^x\, e^{iy\ln(a)}. \tag{1.64}$$

Therefore, the complex number a^z with a real and $z = x+iy$ has magnitude a^x and argument $y\ln(a)$. While exponentiating as in Eq. 1.63, one must be mindful of the fact that θ is not restricted to $[0, 2\pi)$ but is multiple valued with $\theta = \theta_0 + 2n\pi$ with θ_0 in $[0, 2\pi)$ range as Example 1.12 illustrates.

Example 1.12 Complex Roots of 1

Find the values of z for which $z^6 = 1$. This problem finds application in systems with six-fold symmetry.

Solution

The exponential form of z is useful in solving this problem. Let $z = re^{i\theta}$ and write 1 as $1e^{i2n\pi}$ for n integer,

$$r^6 e^{i6\theta} = 1e^{i2n\pi}.$$

Therefore, $r = 1$ since $r > 0$, and

$$\theta = \frac{n}{3}\pi.$$

Now, we list the unique angles in the range 0 to 2π since any angle θ and $\theta \pm 2n\pi$ for n integer are identical points in the complex plane. The six solutions are placed symmetrically on the unit circle in the complex plane, as shown in Fig. 1.7 and listed in Table 1.2.

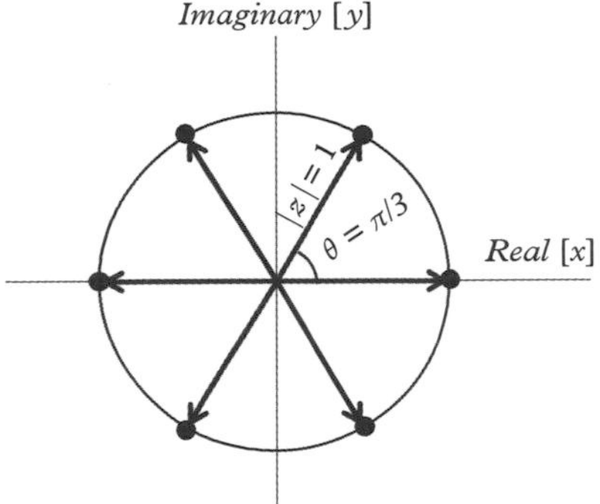

Fig. 1.7 *The Argand diagram of the complex solutions of $z^6 = 1$.*

Table 1.2 *The complex solutions of $z^6 = 1$.*

n	θ	Solution
0	0	1
1	$\frac{\pi}{3}$	$\frac{1}{2} + i\frac{\sqrt{3}}{2}$
2	$\frac{2\pi}{3}$	$-\frac{1}{2} + i\frac{\sqrt{3}}{2}$
3	π	-1
4	$\frac{4\pi}{3}$	$-\frac{1}{2} - i\frac{\sqrt{3}}{2}$
5	$\frac{5\pi}{3}$	$\frac{1}{2} - i\frac{\sqrt{3}}{2}$

1.2.4 Complex Exponential Function of Time

Complex exponentials of time appear frequently in physics. In this section, we will examine them graphically. As a concrete example, consider the following complex exponential function of time:

$$z(t) = e^{i\omega t}, \tag{1.65}$$

where ω is some constant with unit of 1/time so that the argument of the complex number is dimensionless. We have set amplitude to a value 1 to focus on the time behavior in the exponent only. The complex function $z(t)$ in Eq. 1.65 has the following rectangular expansion:

$$z(t) = e^{i\omega t} = \cos(\omega t) + i\sin(\omega t). \tag{1.66}$$

Note that due to the presence of i we cannot plot the complex function directly. Instead, we plot the real and imaginary parts of this complex function, which would be $\cos\omega t$ and $\sin\omega t$. A plot for these functions for $\omega = 1\ \text{sec}^{-1}$ is shown in Fig. 1.8.

Alternately, we can draw the arrow representation of the complex function $e^{i\omega t}$. At each instant the complex function will be a complex number. With the real and imaginary parts of the complex number we can draw the arrow in the Argand diagram. The arrow will point in some direction at instant t. The function $e^{i\omega t}$ is special in the sense that the length of the arrow is always 1. So, with time, the arrow changes direction only.

At $t = 0$ the vector is pointed along the x-axis since $z(0) = 1$. The vector then rotates counterclockwise with increasing time at a uniform angular speed ω such that the vector makes an angle ωt with the positive x-axis at time t, as shown in Fig. 1.9(a). This way of "plotting" the function $e^{i\omega t}$, is called the **rotating vector** or the **phasor** notation. The complex conjugate of $z(t)$, that is the function $e^{-i\omega t}$, will rotate clockwise with time as shown in Fig. 1.9(b).

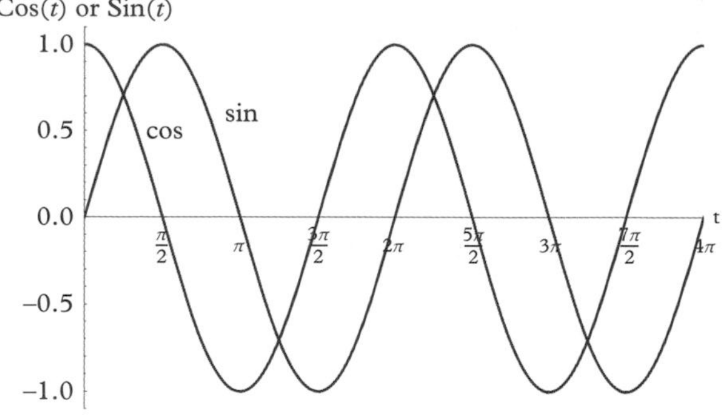

Fig. 1.8 *The real and imaginary parts of the complex function e^{it}.*

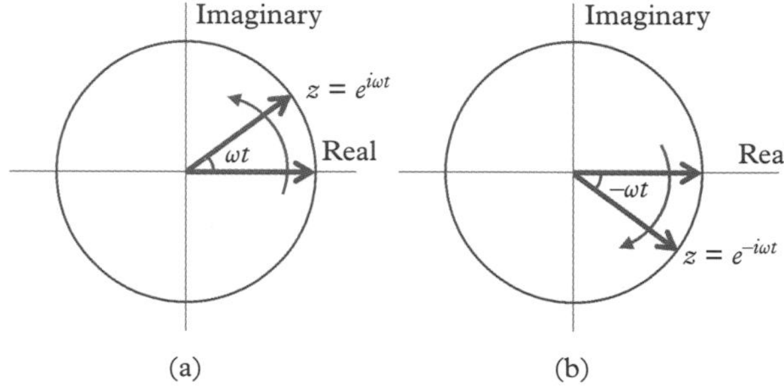

Fig. 1.9 *The rotating vector plot of the function (a) $e^{+i\omega t}$ and (b) $e^{-i\omega t}$. The magnitudes of these complex functions do not change with time, only the direction in which the vectors point changes uniformly in time.*

1.2.5 Vibrations and Complex Functions

The sinusoidal vibration of a one-dimensional oscillator of angular frequency ω with amplitude A is given by $x(t) = A\cos(\omega t + \phi)$, where ϕ is the phase constant. This vibration can be written in complex notation by using the Euler formula and replacing the cosine by an exponential function,

$$x(t) = A\cos(\omega t - \phi) = \text{Re}[Ae^{\pm i(\omega t - \phi)}], \tag{1.67}$$

where Re[·] stands for the real part of the complex number within the bracket. The choice of sign in the exponent of the complex number in Eq. 1.67 is arbitrary. Conventionally, in physics the negative sign is chosen. Therefore, we write Eq. 1.67 as follows:

$$x(t) = A\cos(\omega t - \phi) = \text{Re}[Ae^{-i(\omega t - \phi)}]. \tag{1.68}$$

It is customary to combine the amplitude and phase constant into a new complex amplitude D defined as

$$D \equiv Ae^{i\phi}. \tag{1.69}$$

Using the complex amplitude D we see that the sinusoidal vibration is written in a simple form,

$$x(t) = \text{Re}[De^{-i\omega t}]. \tag{1.70}$$

Note the complex amplitude D keeps track of two real constants of the motion, namely the amplitude A and the phase constant ϕ.

1.2.6 Adding Two Sinusoidal Vibrations—Beat Phenomenon

The beat phenomenon is often observed when two vibrations with similar frequencies overlap. For instance, when you play two notes that are close in frequency you can hear beats. In this section we will illustrate the power of the complex representation of sinusoidal vibrations by applying complex representation to the addition of two vibrations. For simplicity we will set the amplitudes of the two vibrations the same and the phase constants of two to zero. Therefore, consider two vibrations x_1 and x_2 that differ in frequency only,

$$x_1(t) = A\cos(\omega_1 t), \tag{1.71}$$

$$x_2(t) = A\cos(\omega_2 t). \tag{1.72}$$

Here ω_1 and ω_2 are angular frequencies. Let us denote the corresponding frequencies by f_1 and f_2,

$$\omega_1 = 2\pi f_1, \;\; \omega_2 = 2\pi f_2. \tag{1.73}$$

We wish to determine the sum of these vibrations,

$$x(t) = x_1(t) + x_2(t). \tag{1.74}$$

Note that complex exponentials are not the only way to perform the following calculations; as a matter of fact, the sum formula of two cosines works out just as easily with the following answer:

$$x(t) = A\cos\left(\frac{\omega_1 + \omega}{2}\, t\right)\cos\left(\frac{\omega_1 - \omega}{2}\, t\right). \tag{1.75}$$

We will use the complex exponential here as an illustration of the applications of complex exponentials. It turns out that we will find it useful to write the two angular frequencies in terms of their average ω_0 and their difference $\Delta\omega$,

$$\omega_1 = \omega_0 + \Delta\omega/2, \tag{1.76}$$

$$\omega_2 = \omega_0 - \Delta\omega/2. \tag{1.77}$$

Denote the average frequency by f_0 and the difference by Δf,

$$f_0 = \frac{f_1 + f_2}{2} = \frac{\omega_0}{2\pi}, \;\; \Delta f = f_1 - f_2 == \frac{\Delta\omega}{2\pi}. \tag{1.78}$$

Now, we first replace x_1 and x_2 by their corresponding complex forms, as suggested in Eq. 1.70, and then sum them. Then, we will extract the desired result $x(t)$ from the sum by separating the real part of the sum. Let us use the symbol z for the complex form of x,

$$\begin{aligned} z(t) &= z_1(t) + z_2(t) = Ae^{i\omega_1 t} + Ae^{i\omega_2 t} \\ &= Ae^{i\omega_0 t}\left[e^{i\Delta\omega t/2} + e^{-i\Delta\omega t/2}\right] = 2Ae^{i\omega_0 t}\cos\left(\frac{\Delta\omega}{2}t\right). \end{aligned} \tag{1.79}$$

Therefore, the physical vibration is

$$x(t) = \text{Re}[z(t)] = 2A\cos\left(\frac{\Delta\omega}{2}t\right)\cos(\omega_0 t). \tag{1.80}$$

Think of this result as

$$x(t) = B(t)\cos(\omega_0 t), \quad B(t) = 2A\cos\left(\frac{\Delta\omega}{2}t\right). \tag{1.81}$$

In this way of thinking about $x(t)$ we see that the sum of the two vibrations is another vibration at frequency the average of the two frequencies, but whose amplitude changes slowly with time if the difference in the frequencies is small. We say that the net vibration consists of oscillations at the average frequency ω_0 **modulated** by a slowly varying cosine, as shown in Fig. 1.10. The amplitude of the oscillations is not constant in time but rather varies between $-2A$ and $2A$. The oscillations with the largest amplitude occur at both the peak and trough of the slowly varying modulating function $\cos\left(\frac{\Delta\omega}{2}t\right)$, since intensity of sound is related to the square of this function. If the working frequencies f_1 and f_2 are in the audible range (20 Hz to 20 kHz), then both the peaks and the troughs of the slowly varying modulation correspond to the loudest sound; the sound is faintest when the modulation function is smallest. Thus, sound from two sources of nearby frequencies appear to go loud and soft periodically—the loud sounds are called **beats**. Since both the troughs and peaks of the function $\cos\left(\frac{\Delta\omega}{2}t\right)$ correspond to the loudest sound, there are two beats in one period of this function. Therefore, the beat period is half of the period of the modulating function $\cos\left(\frac{\Delta\omega}{2}t\right)$,

$$T_{\text{beat}} = \frac{1}{2}\left(\frac{2\pi}{\Delta\omega/2}\right) = \frac{2\pi}{\Delta\omega}. \tag{1.82}$$

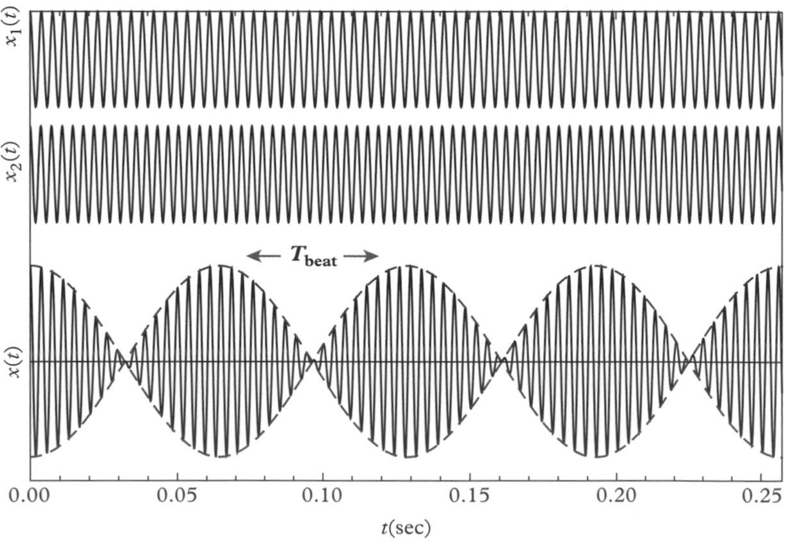

Fig. 1.10 *The beats as a result of superposition of two vibrations. The plot was made with $f_1 = 261.63$ Hz and $f_2 = 277.18$ Hz, corresponding to C_4 and $C_4^{\#}/D_4^{b}$ musical notes. The upper graph is the signal at frequency f_1, the middle at frequency f_2 and the bottom graph is the sum of the two. The peaks and troughs of the beats are separated by one beat period as shown.*

Writing this result in terms of frequency rather than angular frequency we get

$$T_{\text{beat}} = \frac{1}{\Delta f}. \tag{1.83}$$

Therefore, the beat frequency f_{beat} is the difference in frequencies of the two component vibrations,

$$f_{\text{beat}} = \Delta f = |f_1 - f_2|. \tag{1.84}$$

1.2.7 Complex Exponentials and Equations of Motion

Complex exponential functions sometimes make solving equations of motion an easier task. In this method of solving differential equations, the equation is replaced by a complex differential equation which is solved by using a complex exponential function of time. The use of a complex exponential of time converts the differential equation into an algebraic equation, thus reducing the complexity of the problem considerably. In this section we illustrate the use of complex exponential functions in solving equations of motion for harmonic oscillators.

Simple Harmonic Motion

Recall the equation of motion of a simple harmonic oscillator of angular frequency ω is given by

$$\frac{d^2x}{dt^2} = -\omega^2 x. \tag{1.85}$$

We start with complexifying the equation by replacing real x with complex z,

$$\frac{d^2z}{dt^2} = -\omega^2 z. \tag{1.86}$$

We now try an exponential solution, $z(t) = C\, e^{\alpha t}$, in Eq. 1.86, which results in an algebraic equation for α,

$$\alpha^2\; Ce^{i\alpha} = -\omega^2 Ce^{i\alpha}. \tag{1.87}$$

Since C is not zero, we can divide out $Ce^{i\alpha}$ from both sides, and solve for α,

$$\alpha = \pm i\omega. \tag{1.88}$$

Thus, a general solution of Eq. 1.86 is

$$z(t) = C_1 e^{i\omega t} + C_2 e^{-i\omega t}, \tag{1.89}$$

where C_1 and C_2 are arbitrary complex constants. Now, we can recover the original displacement from $z(t)$,

$$x(t) = Re[z] = Re[C_1 e^{i\omega t} + C_2 e^{-i\omega t}]. \tag{1.90}$$

How is this solution the same as $x(t) = A\cos(\omega t) + B\sin(\omega t)$, where A and B are real constants? Let us write C_1 and C_2 in their real and imaginary parts,

$$C_1 = C_{1R} + i\,C_{1I}, \tag{1.91}$$
$$C_2 = C_{2R} + i\,C_{2I}, \tag{1.92}$$

and expand the exponentials in Eq. 1.90,

$$\begin{aligned} x(t) &= Re[C_1 e^{i\omega t} + C_2 e^{-i\omega t}] \\ &= Re\,[(C_{1R} + i\,C_{1I})\,(\cos(\omega t) + i\sin(\omega t)) + (C_{2R} + i\,C_{2I})\,(\cos(\omega t) - i\sin(\omega t))] \\ &= (C_{1R} + C_{2R})\cos(\omega t) + (C_{2I} - C_{1I})\sin(\omega t). \end{aligned} \tag{1.93}$$

We can rename the arbitrary real constants $(C_{1R} + C_{2R})$ and $(C_{2I} - C_{1I})$ as A and B and write the solution as

$$x(t) = A\ \cos(\omega t) + B\ \sin(\omega t). \tag{1.94}$$

Steady State of Damped Driven Oscillator

Next, we illustrate the method of using complex exponentials for the steady state of a damped driven oscillator. A damped driven oscillator obeys the following differential equation:

$$\frac{d^2x}{dt^2} + \Gamma\frac{dx}{dt} + \omega_0^2 x = D_0\cos(\omega t), \tag{1.95}$$

where we are using the symbol ω for the driving frequency rather than ω_d. We replace the real amplitude x by the complex amplitude z and use $z = Ce^{\alpha t}$ in the differential equation, where C is constant. Furthermore, we replace $\cos(\omega t)$ by $e^{-i\omega t}$ to obtain the following equation:

$$\left(\alpha^2 + \Gamma\alpha + \omega_0^2\right) C\,e^{\alpha t} = D_0 e^{-i\omega t}. \tag{1.96}$$

From the time dependence in the exponentials, we deduce that

$$\alpha = -i\omega. \tag{1.97}$$

Putting this back in Eq. 1.96, we solve for the amplitude C,

$$C = \frac{D_0}{-\omega^2 - i\Gamma\omega + \omega_0^2}. \tag{1.98}$$

Therefore, the complex displacement is found to be

$$z(t) = \frac{D_0}{-\omega^2 - i\Gamma\omega + \omega_0^2}\, e^{-i\omega t}. \tag{1.99}$$

The real displacement is then deduced from $z(t)$ by extracting its real part,

$$\begin{aligned} x(t) &= Re[z(t)] \\ &= Re\left[\frac{D_0}{-\omega^2 - i\Gamma\omega + \omega_0^2} e^{-i\omega t}\right] \\ &= Re\left[\frac{D_0}{\sqrt{(\omega_0^2 - \omega^2)^2 + (\Gamma\omega)^2}} e^{i\delta} e^{-i\omega t}\right], \end{aligned}$$

where

$$\delta = \tan^{-1}\left[\frac{\Gamma\omega}{\omega_0^2 - \omega^2}\right].$$

Therefore, the real displacement is

$$x(t) = \frac{D_0}{\sqrt{(\omega_0^2 - \omega^2)^2 + (\Gamma\omega)^2}} \cos(\omega t - \delta).$$

This answer can also be obtained without resorting to the complex method but the calculation is more messy.

Summary of Steps in Solving Equations of Motion

Below is the list of steps involved in the method.

(1) Replace the displacement $x(t)$ by a complex displacement $z(t)$.

(2) Replace sinusoidal functions of time by an exponential function of time as follows:

(a) $\cos[\theta(t)] \implies e^{-i\theta(t)}$

(b) $\sin[\theta(t)] = \cos(\theta(t) - \pi/2) \implies e^{-i(\theta(t)-\pi/2)}$

Note. If there are cosines or sines in the equations that are not functions of time, we do not replace them; these cosines and sines are left in their original form.

(3) Assume exponential form for z: $z(t) = C\, e^{\alpha t}$, where C and α can be complex. With this assumption C and α are the unknowns in the problem. If we can figure out C and α, we have the solution.

(4) Insert the assumed solution form for z into the differential equation. This transforms the differential equation into an algebraic equation for C and α. Solve for C and α.

(5) If the differential equation is homogeneous in x, then C cancels out and becomes arbitrary, and the equation can be solved for the allowed values of α that must be used in $e^{\alpha t}$ to construct the general solution for $z(t)$.

(6) Each α value gives its own $e^{\alpha t}$. Their arbitrary sums are also solutions. Therefore, we multiply $e^{\alpha t}$ for each allowed α with an arbitrary complex constant and sum them to obtain the general solution. For instance, if there are two allowed values of α, say α_1 and α_2, the solution may look like $z(t) = C_1 e^{\alpha_1 t} + C_2 e^{\alpha_1 t}$, where C_1 and C_2 are complex constants.

(7) Recover the original displacement from $z(t)$ by extracting its real part: $x(t) = Re[z(t)]$.

(8) Rewrite or combine the arbitrary constants where possible so that $x(t)$ looks the simplest or as desired.

Example 1.13 Solving a Differential Equation

Find the steady state solution of the following equation:

$$\frac{d^2x}{dt^2} + \frac{9}{4}\frac{dx}{dt} + 25x = 5\cos(4t).$$

Solution

As explained in the text, we will replace the cosine by a complex exponential and x by a complex amplitude z. This would give us the following equation to solve:

$$\frac{d^2z}{dt^2} + \frac{9}{4}\frac{dz}{dt} + 25z = 5\,e^{-i4t}.$$

Now, we let

$$z = z_0 e^{-i4t}, \tag{1.100}$$

where z_0 is a complex number independent of t. After canceling out e^{-i4t} from every term we get

$$-16z_0 - i9z_0 + 25z_0 = 5.$$

Therefore,

$$z_0 = \frac{5}{9 - i9}.$$

This can be written in the exponential notation as

$$z_0 = \frac{5}{9\sqrt{2}}\,e^{i\frac{\pi}{4} + i2n\pi},$$

where n is any integer. Using this in Eq. 1.100 gives the following for z:

$$z = \frac{5}{9\sqrt{2}}\,e^{-i\left(4t - \frac{\pi}{4} - 2n\pi\right)}.$$

The real part of this gives x,

$$x(t) = \frac{5}{9\sqrt{2}}\cos(4t - \pi/4),$$

where I have used $\cos(\theta + 2n\pi) = \cos\theta$.

EXERCISES

(1.1) A particle moves on the x-axis of a Cartesian coordinate system so that its position varies as $x(t) = b_1 + b_2\,t^2 + b_3\cos(b_4\,t)$, where b_1, b_2, b_3, and b_4 are constants. Find the velocity and acceleration of the particle at $t = \tau$.

(1.2) A particle moves on the x-axis of a Cartesian coordinate system so that the x-component of its acceleration is given by $a_x = b_1 + b_2\,t^2 + b_3\cos(b_4\,t)$, where b_1, b_2, b_3, and b_4 are constants. At $t = 0$ the particle was at rest at $x = 0$. Find the position and velocity of the particle at $t = \tau$.

(1.3) A charged particle of charge q and mass m is placed in an electric field $\vec{E}(x, y, z, t) = \vec{E}_0$, where $\vec{E}_0$ is a constant vector. At $t = 0$ the particle was released at rest at the origin. Find the position and velocity of the particle at $t = \tau$.

(1.4) A particle of mass m moves along the x-axis of a Cartesian coordinate system. The particle is subject to the following force: $\vec{F} = b \cos(x)\, \hat{x}$, where $\hat{x}$ is the unit vector pointed towards the $+x$-axis, and x is the position of the particle at time t. Find the work done by the force when the particle moves a distance D from the origin on the positive x-axis.

(1.5) A particle of mass m is subject to a force whose magnitude changes with time but the direction is constant. With the $+x$-direction the same as the direction of the force, the force is given by $\vec{F} = c\, t\, \hat{x}$, where $\hat{x}$ is the unit vector pointed towards the $+x$-axis. At $t = 0$ the particle was released at rest at the origin.
(a) Find the position and velocity of the particle at $t = \tau$.
(b) What is the instantaneous power of the force?
(c) Find the work done by the force when the particle moves a distance D from the origin.

(1.6) A particle of mass m moves along the x-axis of a Cartesian coordinate system. The particle is subject to the following force: $\vec{F} = mg\,[1 - e^{-ct}]\hat{x}$, where $\hat{x}$ is the unit vector pointed towards the $+x$-axis, and g and c are constants. At $t = 0$ the particle was released at rest at $x = 0$.
(a) Find the change in the velocity of the particle during the time interval from $t = t_1$ to t_2.
(b) Find the displacement of the particle during the time interval from $t = t_1$ to t_2.
(c) Find the average power of the force during the time interval from $t = t_1$ to t_2.

(1.7) A rocket in space with an average mass burn rate α moves towards the positive x-axis with instantaneous velocity v_x given by

$$v_x(t) = u \ln \left[\frac{M}{M - \alpha\, t}\right],$$

where u is the constant speed of the burnt fuel with respect to the rocket and M is the initial mass of the rocket. What is the displacement of the rocket during the interval from $t = t_1$ to t_2?

(1.8) Two blocks of mass m each are connected by a spring of spring constant k and unstretched length l, and placed on a frictionless table. While holding one block the other block is pulled in the direction that stretches the spring and both blocks are released. Set up a coordinate system so that the spring is along the x-axis. Let x_1 and x_2 be the x-coordinates of the two blocks at instant t. Deduce the equations of motion of the two blocks. [You do not need to solve them.]

(1.9) A block is attached to two springs with spring constants k_1 and k_2, and the other ends of the springs are attached to two rigid supports so that the springs and the block are aligned in a straight line. Let the origin of the coordinates be at the point where the block is in equilibrium and the positive x-axis points towards one of the supports. The block is pulled along the x-axis and let go. Afterwards, the block moves along the x-axis only. Figure 1.11 shows the block at an arbitrary instant. Find the x-component of the equation of motion of the block.

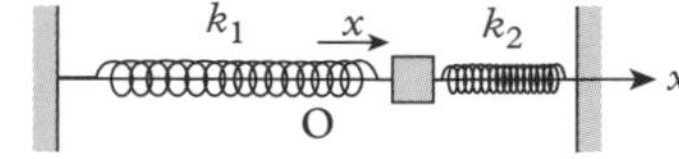

Fig. 1.11 *Exercise 1.9.*

(1.10) A block is attached to two springs with spring constants k_1 and k_2, and the other ends of the springs are attached to two rigid supports so that the springs and the block are aligned in a straight line. Let the block be placed on a frictionless surface so that it can move in a plane which we will take to be the xy-plane of a Cartesian

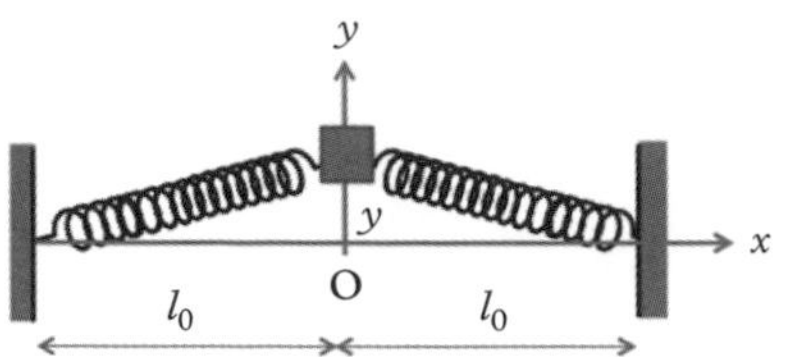

Fig. 1.12 *Exercise 1.10.*

coordinate system. Let the origin of the coordinates be at the point where the block is in equilibrium and the positive x-axis points towards one of the supports. The block is pulled along the y-axis and let go. Afterwards, the block moves along the y-axis. Figure 1.12 shows the block at an arbitrary instant. Find the y-component of the equation of motion of the block.

(1.11) Consider the problem above in Ex. 1.10.

(a) What is net potential energy stored in the springs when the block is at an arbitrary y on the y-axis?

(b) Use the relation between potential energy and force to find the y-component of the force on the block from the expression for the potential energy.

(c) Compare your answer for the force here to the expression for the force obtained in Ex. 1.10.

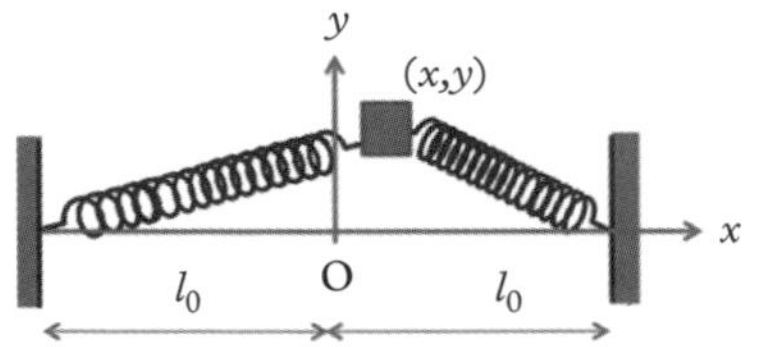

Fig. 1.13 *Exercise 1.12.*

(1.12) A block is attached to two springs with spring constants k_1 and k_2, and the other ends of the springs are attached to two rigid supports so that the springs and the block are aligned in a straight line. Let the block be placed on a frictionless surface so that it can move in a plane which we will take to be the xy-plane of a Cartesian coordinate system. Let the origin of the coordinates be at the point where the block is in equilibrium and the positive x-axis points towards one of the supports. Figure 1.13 shows the block at an arbitrary instant.

(a) Find the equation of motion of the x-coordinate of the block.

(b) Find the equation of motion of the y-coordinate of the block.

(1.13) Consider the problem above in Ex. 1.12.

(a) What is the net potential energy stored in the springs when the block is at an arbitrary (x, y)?

(b) Use the relation between potential energy and force to find the x- and y-components of the force on the block from the expression for the potential energy.

(c) Compare your answer for the force here to the expression for the force obtained in Ex. 1.12.

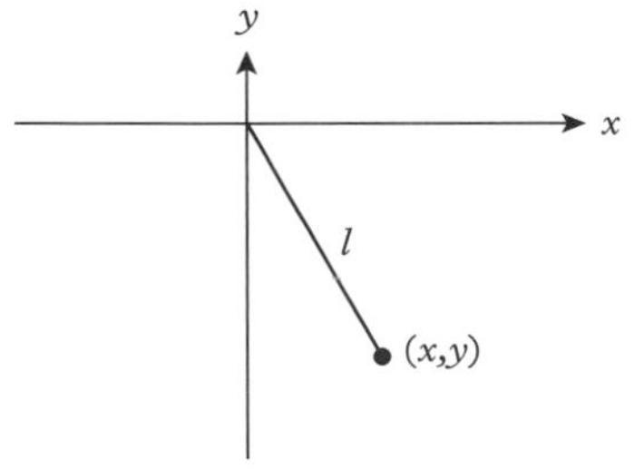

Fig. 1.14 *Exercise 1.14.*

(1.14) A pendulum of length l is hanging from a fixed support as shown in Fig. 1.14. Use a Cartesian coordinate system with the origin at the support, the positive x-axis to the right, and the positive y-axis vertically up. Let the pendulum move in the xy-plane and the position of the bob at an arbitrary time t be given by (x, y), as shown in Fig. 1.14.

Show that, for small oscillations, the equation of motion for the x-coordinate of the bob can be written as

$$\frac{d^2x}{dt^2} = -\frac{g}{l}\,x,$$

where g is the acceleration due to gravity.

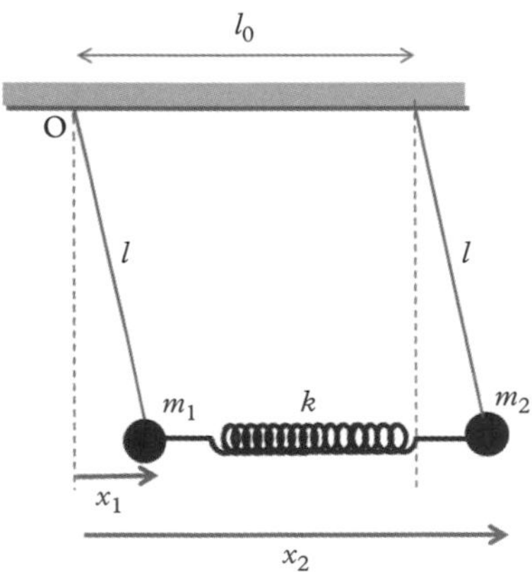

Fig. 1.15 *Exercise 1.15.*

(1.15) Two pendula of length l are hanging from a fixed support, as shown in Fig. 1.15. The two bobs are connected by a spring with spring constant k. The spring is in the relaxed state when the bobs are hanging vertically. Use a Cartesian coordinate system with the origin at the support of the pendulum on the left, the positive x-axis to the right, and positive y-axis vertically up. Let the pendula move in the xy-plane and the positions of the bobs at an arbitrary time t be given by (x_1, y_1) and (x_2, y_2). For small oscillations, you can assume that the y-coordinates do not change much with time.

(a) Find the x-component of the equations of motion of the x-coordinates x_1 and x_2 in the small oscillation approximation.

(b) Suppose $x'_2 = x_2 - l_0$. The coordinates x_1 and x'_2 refer to the displacement of the bobs from their equilibrium positions. What are the equations of motions of x_1 and x'_2?

(1.16) Express the following complex numbers in the polar form and show them on an Argand diagram:
(a) $1 + i$
(b) -5
(c) $-1 - i$
(d) $-i2$
(e) $\sqrt{3} + i2$
(f) $2 + i\sqrt{3}$

(1.17) Express the following complex numbers in the rectangular form and show them on an Argand diagram:
(a) $(1, 0)$
(b) $(1, \pi/2)$
(c) $(1, \pi)$
(d) $(1, 3\pi/2)$
(e) $(1, 2\pi)$
(f) $(2, \pi/3)$

(1.18) Give the polar and rectangular forms for the results:
(a) $(1 + i\sqrt{3})(1 + i\sqrt{3})$
(b) $(1 + i\sqrt{3})(1 - i\sqrt{3})$
(c) $(-1 + i\sqrt{3})(1 + i\sqrt{3})$
(d) $(1 + i)/(1 - i)$
(e) $(1, 0°)(2, 30°)$
(f) $(2, 90°)(3, -30°)$
(g) $(4, 3\pi/2)(2, -3\pi/2)$
(h) $(4, \pi)/(2, \pi/2)$
(i) $(6, 3\pi/2)/(2, -3\pi/2)$
(j) $(5, \pi/6)/(2, \pi/6)$

(1.19) (a) A complex number $z_1 = r_1 e^{i\theta_1}$ is multiplied by another complex number $z_2 = e^{i\theta_2}$. Draw z_1 in an Argand diagram and show what happens to the vector upon multiplication if (i) $\theta_2 > 0$ and (ii) $\theta_2 < 0$.
(b) A complex number $z = re^{i\theta}$ is multiplied by a real number a. Draw z in an Argand diagram and show what happens to the vector upon multiplication if (i) $a > 0$ and (ii) $a < 0$.

(1.20) If $z = a + ib$, what is the rectangular form of $1/z$?

(1.21) Prove that $(\cos\theta + i\sin\theta)^n = \cos n\theta + i\sin n\theta$, where n is an integer.

(1.22) Prove that if $z_1 = r(\cos\alpha + i\sin\alpha)$ and $z_2 = s(\cos\beta + i\sin\beta)$, then
(a) $z_1 z_2 = rs\,[\cos(\alpha + \beta) + i\sin(\alpha + \beta)]$, and
(b) $z_1/z_2 = (r/s)\,[\cos(\alpha - \beta) + i\sin(\alpha - \beta)]$.

(1.23) Let *Roots* be the set of all n^{th} roots of 1 with n an integer. Show that
(a) if $z_1 \in$ *Roots* and $z_2 \in$ *Roots*, then $z_1 z_2 \in$ *Roots*, and
(b) If $z \in$ *Roots*, then $1/z \in$ *Roots*.

(1.24) If $z = 1 + i\sqrt{3}$, find the rectangular and polar forms of z^{29} and z^{1000} and show them on an Argand diagram.

(1.25) Find all the solutions of
(a) $z = 0$,
(b) $z^4 = 1$,
(c) $z^4 = i$.

(1.26) Find the polar and rectangular forms of the following complex numbers and show them in an Argand diagram:
(a) $\sqrt{i}$,
(b) i^i.
Beware of multiple answers.

(1.27) With $z = x + iy$, express $\sqrt{z}$ in the polar and Cartesian forms.

(1.28) Prove the following identities:
(a) $\cos\theta = \frac{1}{2}\left(e^{i\theta} + e^{-i\theta}\right)$,
(b) $\sin\theta = \frac{1}{2i}\left(e^{i\theta} - e^{-i\theta}\right)$.

(1.29) Using the complex exponential representations of cosine and sine prove the following trigonometric identities:
(a) $\cos A + \cos B = 2\cos\left(\frac{A+B}{2}\right)\cos\left(\frac{A-B}{2}\right)$,
(b) $\cos A - \cos B = 2\sin\left(\frac{A+B}{2}\right)\sin\left(\frac{B-A}{2}\right)$,
(c) $\sin A + \sin B = 2\sin\left(\frac{A+B}{2}\right)\cos\left(\frac{A-B}{2}\right)$,
(d) $\sin A - \sin B = 2\sin\left(\frac{A-B}{2}\right)\cos\left(\frac{A+B}{2}\right)$.

(1.30) Two sound waves of frequencies 240 Hz and 260 Hz simultaneously vibrate your ear drum.
(a) What is the frequency of the sound you would hear?
(b) What would be the beat frequency you would hear?
(c) How many oscillations of main sound will occur in the duration of one beat?

(1.31) (a) Add the three vibrations $\psi_1(t) = \cos(\omega t)$, $\psi_2(t) = \cos(1.1\,\omega t)$, and $\psi_3(t) = \cos(1.2\,\omega t)$.
(b) Would the combined vibration have beats? If so, what would be the beat period?

(1.32) (a) Add $N+1$ vibrations $\psi_0(t) = \cos(\omega t)$, $\psi_1(t) = \cos(\omega t + \epsilon t)$, $\psi_2(t) = \cos(\omega t + 2\epsilon t), \ldots, \psi_N(t) = \cos(\omega t + N\epsilon t)$. Hint: The algebra using complex numbers is easier.
(b) Would the combined vibration have beats? If so, what would be the beat period?
(c) Plot the result for $N = 10$, $\omega = 2\pi$, $\epsilon = 0.1\,\omega$, from $t = 0$ to $t = 5$. Give an interpretation of your plot.

(1.33) Solve the following equations of motion by complex exponential method:
(a) $\frac{d^2x}{dt^2} = -25x$,
(b) $\frac{d^2x}{dt^2} = -25x - 6\frac{dx}{dt}$,
(c) $\frac{d^2x}{dt^2} = -25x - \frac{dx}{dt} + 3\cos(20\pi t)$,
(d) $\frac{d^2x}{dt^2} = -25x - \frac{dx}{dt} + 3\sin(20\pi t)$.

(1.34) A charged particle of charge q and mass m is placed in a magnetic field $\vec{B}(x, y, z, t) = B_0\,\hat{z}$, where B_0 is constant and $\hat{z}$ is a unit vector pointed towards the positive z-axis. At $t = 0$ the particle was released at the origin with velocity v_0 pointed towards the $+x$-axis.
(a) Find the position and velocity of the particle at an arbitrary time t.
(b) Show that the path of the particle is a circle and find the location of the center of the circle.

Part II
Vibrations

Free Oscillations—One Degree of Freedom

2

Chapter Goals

This chapter deals with fundamental aspects of vibrations by focusing attention on a simple system of a mass attached to a spring. You will learn to set up and solve the equation of motion of the ideal one-dimensional simple harmonic oscillator. Other one-dimensional systems will also be discussed. You will also learn how to incorporate damping in oscillating systems.

2.1 Basic Characteristics of an Oscillatory Motion

Oscillatory motion is ubiquitous in nature. You can find it in the swaying of branches of trees, rocking of boats, the back and forth motion of pendulums, vibrations of atoms in molecules, vibrations of a guitar string, current in an LC circuit, to name just a few. In this chapter, we will study in detail the oscillatory motion of a mass attached to an ideal spring that moves in only one dimension. The displacement of the mass from the equilibrium point is called the **dynamical variable** of the system.

The one-dimensional motion of a mass/spring system is an important example of a class of motion called **simple harmonic motion**. A simple harmonic motion is characterized by a single frequency. Suppose $x(t)$ is the dynamical variable of a system that executes a simple harmonic motion about the equilibrium value $x = 0$ with frequency f, then $x(t)$ can always be written as

$$x(t) = A\cos(2\pi ft - \phi), \tag{2.1}$$

where A is the largest displacement of x from the equilibrium, called the amplitude of the motion, and ϕ is called the phase constant. The x-component of the velocity of the oscillator will be

$$v = \frac{dx}{dt} = -2\pi f\sin(2\pi ft - \phi). \tag{2.2}$$

The phase constant depends on the choice of the zero of time. For instance, if the zero of time is chosen so that at $t = 0$, the oscillator is at rest, then $\phi = 0$.

$$\text{If } v = 0 \text{ at } t = 0, \text{ then } \phi = 0, \text{ making } x(t) = A\cos(2\pi ft). \tag{2.3}$$

A First Course in Vibrations and Waves. First Edition. Mohammad Samiullah.

We can also write the displacement of a simple harmonic motion using both sine and cosine functions as

$$x(t) = C_1 \cos(2\pi ft) + C_2 \sin(2\pi ft), \tag{2.4}$$

where $C_1 = A \cos\phi$ and $C_2 = A \sin\phi$.

The **frequency** of the motion refers to the number of cycles executed by the oscillatory motion in a unit time. Therefore, frequency f is just the inverse of the time period T,

$$f = \frac{1}{T}. \tag{2.5}$$

If the time period is expressed in seconds per cycle, the frequency would be cycles per second, which is also called **Hertz (Hz)**, named after the German physicist, Heinrich Rudolf Hertz (1857–1894), who was the first person to produce and study radio waves in the 1880s. Often in our calculations we will use angular frequency ω rather the frequency f,

$$\omega = 2\pi f. \tag{2.6}$$

Using the angular frequency makes the formula for the displacement of a harmonic oscillator simpler since it hides away the factor of 2π,

$$x(t) = A \cos(\omega t - \phi). \tag{2.7}$$

Although f is the frequency, which counts the number of cycles of motion per unit time, and ω is radians per unit time, we will often call ω itself frequency and hope that the meaning will be clear from the context.

Example 2.1 Vibrations of a Tuning Fork

Tuning forks are special devices that vibrate at a particular frequency, when struck. They are often used to tune pianos and other musical instruments. Consider a tuning fork that vibrates at 256 Hz, what is the time period of its vibrations?

Solution

The period and frequency of an oscillation are the inverse of each other. Therefore,

$$T = \frac{1}{f} = \frac{1}{265\ \text{Hz}} = 3.91 \times 10^{-3}\ \text{s} = 3.91\ \text{ms}.$$

2.2 Stable Equilibrium and Restoring Force

The concept of mechanical equilibrium is important for understanding oscillatory motion. An object is said to be in **mechanical equilibrium** if no net force acts on any point of the object. For instance, if a marble is sitting at rest at the bottom of a bowl, the normal force on the marble from the bowl balances the weight of the marble. The marble at rest at the bottom of the bowl is said to be in a mechanical equilibrium, and the bottom of the bowl is referred to as the **equilibrium position** of the marble (Fig. 2.1).

Similarly, a boulder at the top of a hill is also in a mechanical equilibrium since the net force on the boulder is zero when it is at rest at the top of the hill. However, these two examples of mechanical equilibrium differ drastically when the object under study is displaced from the equilibrium position. When the marble is released at rest from a point anywhere in the bowl except at the bottom, the marble tends to return to the bottom, while when the boulder is let go similarly at rest from a point other than the equilibrium point, it runs away from the equilibrium. We say that the marble has a **stable equilibrium** at the bottom of the bowl and the boulder an **unstable equilibrium** at the top of the hill.

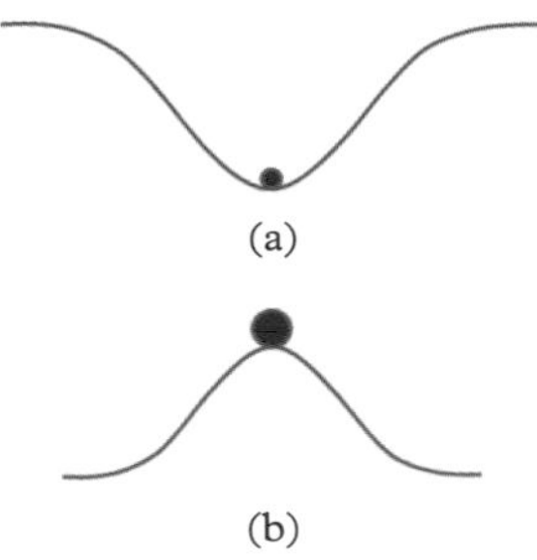

Fig. 2.1 *(a) A marble at rest at the bottom of a bowl—stable mechanical equilibrium. (b) A boulder at rest at the top of a hill—unstable mechanical equilibrium.*

Oscillatory motion is possible only near a stable equilibrium. The fundamental reason for the oscillatory motion can be traced to the presence of a **restoring force** when a system is near a stable equilibrium point. *When the system is at a stable equilibrium there is no net force on it, but when it is displaced from the equilibrium a restoring force acts on the system* (Fig. 2.2). The direction of the restoring force is always towards the equilibrium point. The restoring force causes the system to slow down when moving away from the equilibrium and speed up when moving towards the equilibrium point. As a result, when the system returns to the equilibrium point, it overshoots and goes past the equilibrium point due to the motional inertia. Once the system is on the other side of the equilibrium, the restoring force acts again but this time in the opposite direction. In this way, a back and forth or oscillatory motion takes place when a system is released at a point that is away from a stable equilibrium point.

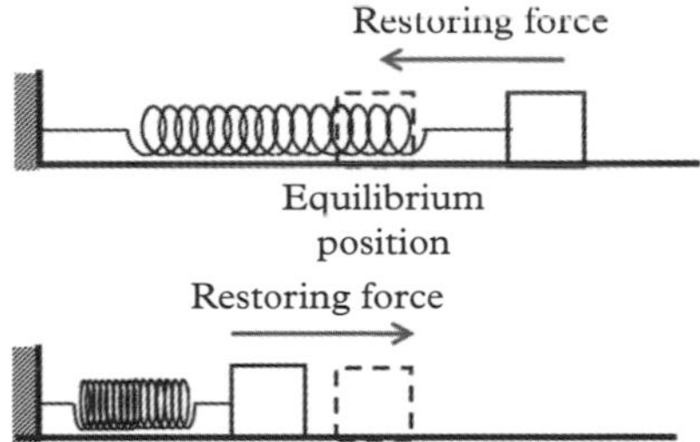

Fig. 2.2 *The restoring force from the spring on the mass. The restoring force changes both in magnitude and direction since its magnitude depends on the amount of displacement from the equilibrium and its direction is always towards the equilibrium position.*

2.3 Free Oscillations of a Mass/Spring System

The simplest type of restoring force occurs when the magnitude of the restoring force is proportional to one power of the displacement from the equilibrium point. This type of force is called **Hooke's law force** or **linear force**. The force by a spring is an example of a Hookean force,

$$F_{\text{restore}} \sim \text{displacement from equilibrium.}$$

To be concrete, consider a block of mass m attached to a spring with spring constant k. To complete the system, place the block on a frictionless table, and attach the other end of the spring to a fixed support, as shown in Fig. 2.3. When you pull the block a little distance from the equilibrium position and let go, the block begins to oscillate, accompanied by the contraction and expansion of the spring. Note that the net force on the block is in the horizontal direction, since here, the vertical forces, which are the gravity and normal forces on the block, cancel out. Let us choose the coordinate system such that the origin is located at the equilibrium position and the positive x-axis is pointed in the direction which leads to the extension of the spring, as shown in Fig. 2.3. With the choice of this origin, the extension or contraction of the spring is equal to the absolute value of the x-coordinate of the block. Therefore, the x-component of the spring force on the block is equal to the negative of the product of the spring constant k and the x-coordinate of the block,

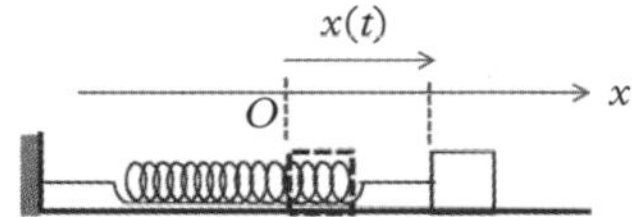

Fig. 2.3 *The oscillations of a mass attached to a spring. The origin of the x-axis will be taken at the equilibrium shown with the dashed position of the block when the spring is in the relaxed state.*

$$F_x = -k\,x. \tag{2.8}$$

Let us verify that the negative sign is correct here. We see that when $x > 0$, the spring would be stretched, therefore, the spring force would be pointed towards the negative x-axis and the x-component of the force, F_x will be negative. Similarly, when $x < 0$, the spring would be compressed, and therefore, the spring force would point to the right, giving $F_x > 0$ for $x < 0$, which is correctly given by Eq. 2.8. Thus, on both sides, the restoring force is pointed towards the equilibrium at the origin, as expected for a restoring force.

Newton's second law then gives us the following x-component of the equation of motion for the block:

$$m\, a_x = -k\, x, \tag{2.9}$$

where a_x is the x-component of the acceleration of the block. We see that the acceleration of the block is proportional to the displacement of the mass. Therefore, *acceleration is not constant in this system: it depends on where the block is at the time.* To solve Eq. 2.9, we find it helpful to write the acceleration as the second derivative of position with respect to time, so that the dependence of x on time is more clearly displayed,

$$m\, \frac{d^2x}{dt^2} = -k\, x. \tag{2.10}$$

Thus, the equation of motion of the block is a second-order ordinary differential equation. The solution of this equation will give us the position of the block at a particular time. When we specify the initial position and velocity we can obtain a unique solution of Eq. 2.10. The solution, written as position as a function of time, $x(t)$, gives us the position of the block at all times.

2.3.1 Solving the Equation of Motion

It is helpful to combine the parameters m and k in Eq. 2.10 into one parameter ω given by

$$\omega = \sqrt{\frac{k}{m}}. \tag{2.11}$$

The equation of motion, Eq. 2.10, can now written more compactly as

$$\frac{d^2x}{dt^2} = -\omega^2\, x. \tag{2.12}$$

The square on ω ensures that k/m would be positive since both k and m are positive. We will see below that $|\omega|/2\pi$ is the frequency of the oscillator and ω itself is the **angular frequency**.

Note that Eq. 2.12 shows that the displacement from equilibrium $x(t)$ is a function whose second derivative gives us back the function itself, multiplied by a negative constant. You may know from calculus that there are two functions with this property: sine and cosine,

$$\frac{d^2 \cos(\omega t)}{dt^2} = -\omega^2 \cos(\omega t), \tag{2.13}$$

$$\frac{d^2 \sin(\omega t)}{dt^2} = -\omega^2 \sin(\omega t). \tag{2.14}$$

Since Eq. 2.12 has only one power of x in both terms of the equation, any linear combination of sine and cosine with constant coefficients will also work, as you can easily verify for the following function $f(t)$ where C_1 and C_2 are constants:

$$f(t) = C_1 \cos(\omega t) + C_2 \sin(\omega t)$$
$$\frac{d^2}{dt^2} f(t) = -\omega^2 f(t). \tag{2.15}$$

Thus, a **general solution** of Eq. 2.12 can be written as

$$x(t) = C_1 \cos(\omega t) + C_2 \sin(\omega t), \tag{2.16}$$

where C_1 and C_2 are arbitrary constants. Different values of these constants correspond to different initial positions and velocities of the block, i.e. different ways the motion of the oscillator could be started. The solution given in Eq. 2.16 can also be written in the following form:

$$x(t) = A \cos(\omega t - \phi), \tag{2.17}$$

where A and ϕ are constants, which are related to constants C_1 and C_2,

$$A = \sqrt{C_1^2 + C_2^2}, \quad \tan\phi = C_2/C_1. \tag{2.18}$$

In the form of Eq. 2.17, the solution explicitly displays the range of displacement: the displacement x will be between $-A$ and A, since the cosine will be between -1 and $+1$. Thus, $|x| = A$ is the maximum displacement from the equilibrium on either side, as shown in Fig. 2.4. The constant A is called the **amplitude** and ϕ the **phase constant**, whose physical meaning will be discussed below.

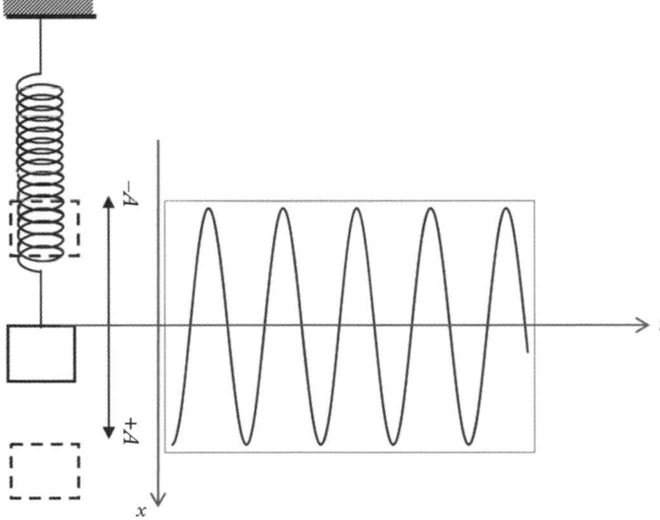

Fig. 2.4 *Plot of x versus t for $\phi = 0$ shows that the amplitude A is the largest displacement from the equilibrium in either direction.*

2.3.2 Specifying Initial Position and Velocity

The initial position and velocity of the block determine the amplitude and phase constant of the motion as we now show. Let the initial position and velocity be denoted by x_0 and v_0 respectively,

$$x(0) = x_0; \quad v(0) = v_0. \tag{2.19}$$

By setting $t = 0$ in Eq. 2.17 we obtain the following equation for x_0:

$$x_0 = A \cos\phi. \tag{2.20}$$

To use the initial condition on the velocity, we first take the derivative of x with respect to t and then set $t = 0$. This gives the following equation for v_0:

$$v_0 = v(0) = \left.\frac{dx}{dt}\right|_{t=0} = A\,\omega \sin\phi. \tag{2.21}$$

Equations 2.20 and 2.21 can be solved for A and ϕ in terms of x_0 and v_0,

$$A = \sqrt{x_0^2 + (v_0/\omega)^2}; \quad \phi = \tan^{-1}(v_0/\omega x_0). \tag{2.22}$$

2.3.3 Physical Meaning of ω

The solution $x(t)$ in Eq. 2.16 or Eq. 2.17 is periodic in time since it describes the periodic motion of the block attached to the spring. Let T be the time to complete one cycle of motion. We expect both position $x(t)$ and velocity $v(t)$ to be periodic functions of the variable t with period T,

$$x(t+T) = x(t) \text{ and } v(t+T) = v(t). \tag{2.23}$$

Demanding the periodicity in $x(t)$ given in Eq. 2.17, we find that ω is related to the period,

$$A\cos(\omega t + \omega T - \phi) = A\cos(\omega t - \phi), \implies \omega T = 2n\pi, \quad n = 0, \pm 1, \pm 2, \cdots, \tag{2.24}$$

Clearly $n = 0$ is not the solution since that would mean time period $T = 0$, which is not the case here. The $n = 1$ case refers to a situation after one time period, and $n = -1$, to a situation one time period earlier. The constant ω is therefore 2π over the time period, or 2π times frequency f,

$$\omega = \frac{2\pi}{T} = 2\pi f. \tag{2.25}$$

Analogy to Circular Motion

Some aspects of a particle moving in a circle provide useful tools for visualizing simple harmonic motion in another way. If a particle is moving in a circle of radius r about the origin and it is at an angle θ at time t, then x- and y-coordinates of the position are given by $r\cos(\theta)$ and $r\sin(\theta)$ respectively.

We found above that simple harmonic motion is given by a cosine function of time. Therefore, we can represent a simple harmonic motion by the x-component of the motion of a fictitious particle moving uniformly in a circle, as shown in Fig. 2.5. In this picture of

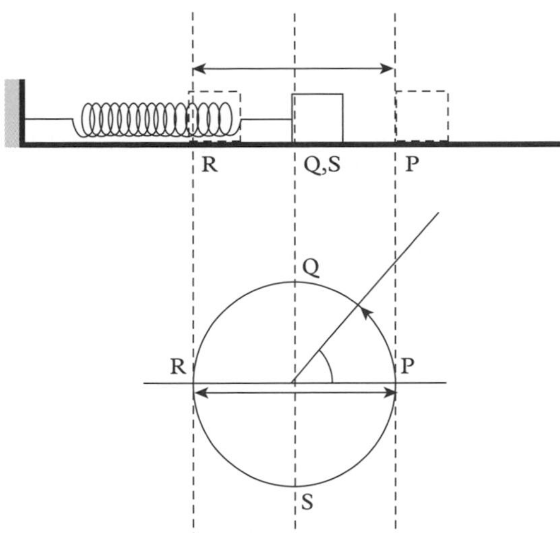

Fig. 2.5 *The cyclic process of oscillatory motion of the block represented as a point moving on a circle. The angular frequency ω corresponds to the angle in radians covered by the point on the circle per unit time; so that the angle covered in time Δt is $\omega\Delta t$. Here P, Q, R, S are successive positions of the block as it executes simple harmonic motion.*

a simple harmonic motion, ω refers to the angular speed of the fictitious particle. Since the fictitious particle moves uniformly, its angular displacement $\Delta\theta$ in time interval Δt would be given by

$$\Delta\theta = \omega\Delta t. \tag{2.26}$$

2.3.4 Physical Meaning of Phase Constant, ϕ

The phase constant ϕ is related to the relative position of the block in its cycle if the cycle is represented by a rotation by 2π radians of a fictitious particle in a circular motion, as explained above. To get a feel for the physical information contained in the phase constant, we examine two oscillators with $\phi_1 = 0$ and $\phi_2 = \pi/2$ radians respectively, as shown in Fig. 2.6. Notice that the second oscillator is always ahead of the first by a quarter of a cycle:

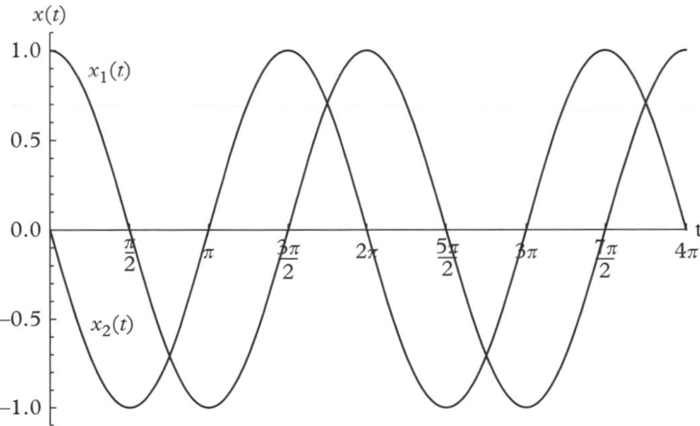

Fig. 2.6 *Oscillators started in two different ways such that they have the same amplitude but different phase constants. We plot their position in time for $\omega = 1$ rad/sec and the following initial conditions: (1) $x = A = 1$ m, $v = 0$, and (2) $x = 0, v = -A\omega = -1$ m/s. The relative phase difference is $\pi/2$ radians or 1/4 cycle.*

for instance, $x = A$ is reached by oscillator 2 before oscillator 1 in any cycle. In terms of 2π radians in one cycle, this corresponds to a phase difference of $\pi/2$ radians. Thus, the phase constant ϕ represents a measure of time in the cycle of the oscillator as measured in terms of the angle of the fictitious particle moving in a circle. You may practice drawing displacements of two oscillators of the same frequency and amplitude but differing in phase by π radians or $\pi/3$ radians.

2.4 Energy of a Simple Harmonic Oscillator

The energy of a simple harmonic oscillator consists of the kinetic energy of the moving block and the potential energy in the extended or compressed spring. The kinetic energy of the block is given by

$$KE = \frac{1}{2}mv^2, \tag{2.27}$$

and the potential energy stored in a spring with spring constant k and deformation (either extension or compression) x is given by

$$PE = \frac{1}{2}kx^2, \tag{2.28}$$

with the reference for the potential energy set at zero when the spring is in the relaxed state. Now, since the position and velocity of a harmonic oscillator oscillate between two values sinusoidally, the kinetic and potential energies of a simple harmonic oscillator would change with time. However, the total energy E of the oscillator remains constant since the spring force is a conservative force,

$$E = KE + PE = \frac{1}{2}mv^2 + \frac{1}{2}kx^2 = \text{constant}. \tag{2.29}$$

This statement of the conservation of energy can be taken as an alternative way to study the motion of a system that executes a simple harmonic motion. In the expression for the energy of a simple harmonic oscillator, there are two terms: an inertial term that is quadratic in velocity and an elastic restoring term that is quadratic in displacement. We can use Eq. 2.29 to deduce the frequency of the oscillator. Note that the ratio of the constant $(k/2)$ multiplying the square of the displacement and the constant $(m/2)$ multiplying the square of the velocity is equal to the square of the angular frequency,

$$\omega^2 = \frac{\text{Coeff. of } x^2 \text{ in PE}}{\text{Coeff. of } v^2 \text{ in KE}} = \frac{k}{m}. \tag{2.30}$$

When the block approaches the extreme points of the motion, the displacement from equilibrium $|x|$ approaches the maximum and the speed approaches zero. Thus, at the turning points, there is no kinetic energy and all the energy is in the potential energy. When the block approaches the equilibrium point the displacement x goes to zero, and hence, the potential energy goes to zero. When the block is at the equilibrium point, all of its energy would be in the kinetic energy. Therefore, the speed of the block would be greatest when it is passing the equilibrium point. We can rewrite Eq. 2.29 to include these observations,

$$E = \frac{1}{2}mv^2 + \frac{1}{2}kx^2 = \frac{1}{2}kA^2 = \frac{1}{2}mv_{max}^2, \tag{2.31}$$

where v_{max} is the speed when the block passes the equilibrium point.

Example 2.2 Conservation of energy of a harmonic oscillator

A block of mass 0.6 kg is attached to a spring with spring constant 180 N/m and negligible mass compared to the mass of the block. The block is placed on a frictionless horizontal table, pulled horizontally 3 cm from its equilibrium position, and let go from rest. Evaluate the speed of the block at the time (a) it crosses the equilibrium position, and (b) it is 1.5 cm from the equilibrium.

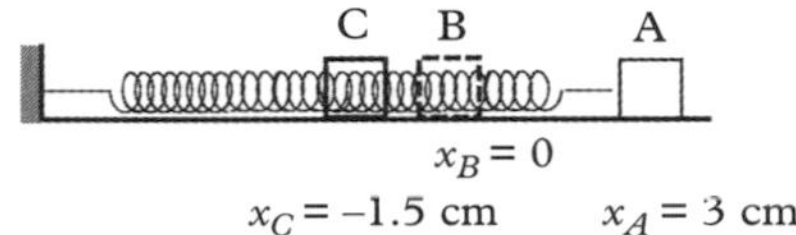

Fig. 2.7 *Example 2.2.*

Solution

(a) Since no friction acts on the block, the energy is conserved. Let us denote the original position as point A, the equilibrium as point B and 1.5 cm from the equilibrium as point C (see Fig. 2.7). When the block is at point A it is not moving, therefore all energy is the potential energy stored in the spring,

$$E_A = \frac{1}{2}kx_A^2 = \frac{1}{2}(180\ \text{N/m})(0.03\ \text{m})^2 = 0.081\ \text{J}.$$

When the block is at point B, the spring is neither stretched not compressed, therefore there is no potential energy and the entire energy is contained in the kinetic energy,

$$E_B = \frac{1}{2}mv_B^2 = \frac{1}{2}(0.6\ \text{kg})v_B^2.$$

Equating the energy at B to the energy at A we find the speed of the block when it is moving past the equilibrium point B,

$$0.3\ v_B^2 = 0.081 \Rightarrow v_B = 0.52\ \text{m/s}.$$

(b) When the block is at point C, which could be on either side of the equilibrium a distance of 1.5 cm away from it, the spring is either compressed or stretched. Therefore, there will be potential energy stored in the spring. But, the stored potential energy is less than the starting energy, hence, the rest of the energy will be in the form of kinetic energy of the block,

$$E_C = \frac{1}{2}mv_C^2 + \frac{1}{2}kx_C^2 = 0.3\ v_C^2 + 0.02\ \text{J}.$$

Equating E_C to E_A, and solving for v_C we find the speed when the block is 1.5 cm from the equilibrium to be

$$0.3\ v_C^2 + 0.02 = 0.081 \Rightarrow v_C = 0.45\ \text{m/s}.$$

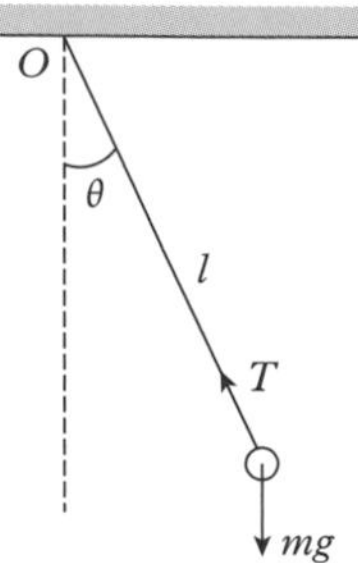

Fig. 2.8 *A plane pendulum oscillates in a plane, rotating about an axis through the point of suspension O and perpendicular to the plane of the pendulum motion. The angle θ that the cord makes with the vertical line gives the displacement of the pendulum from a stable equilibrium.*

2.5 Other Examples of Simple Harmonic Motion

Simple harmonic motion appears in a wide variety of physical settings. The fundamental requirement for a motion to be a simple harmonic motion is that when the system is displaced from a stable equilibrium, the restoring force is proportional to the displacement. In this section, we study some commonly encountered systems that exhibit simple harmonic motion if the displacement from the equilibrium is small enough.

2.5.1 Plane Pendulum

A pendulum consists of a bob of mass m suspended from a light, inextensible cord of length l. If the physical dimension of the bob is much smaller than the length of the cord, we can treat the bob as a point mass. A pendulum has a stable equilibrium position when the mass is hanging vertically down from the suspension point. The displacement of the bob from the equilibrium is given in terms of angle θ that the cord makes with the vertical line, as shown in Fig. 2.8. The displacement angle is positive for a counterclockwise change in the angle and negative for a clockwise change in the angle. Because we wish to study the angular displacement, it is more convenient to treat the pendulum problem as a rotation problem—the rotation of the point mass m about an axis through the point of suspension O and perpendicular to the plane of oscillation. Therefore, the pendulum bob obeys the following equation:

$$\tau = I\frac{d^2\theta}{dt^2}, \tag{2.32}$$

where τ is the torque and I the moment of inertia.

There are only two forces acting on the mass, the weight of the mass and the tension of the string. Since the force of tension goes through the point of suspension O, it does not exert any torque about an axis through O. The torque about O comes only from the weight which we can easily find to be

$$\tau = -mgl\sin\theta, \tag{2.33}$$

where the minus sign ensures that a counterclockwise sense torque will be positive since the counterclockwise rotation has been taken as positive angle here. The moment of inertia of the bob about the axis is

$$I = ml^2. \tag{2.34}$$

Therefore, the equation of motion of a pendulum is

$$\frac{d^2\theta}{dt^2} = -\frac{g}{l}\sin\theta. \tag{2.35}$$

Here, the acceleration of the angular displacement is not proportional to the angular displacement. Instead, the acceleration is proportional to the sine of the displacement. Since the acceleration of the dynamical variable is not proportional to one power of the dynamical variable (θ), a simple pendulum will not execute a simple harmonic motion. **Here is a system that oscillates but its motion is not a simple harmonic motion**. However, we note

Table 2.1 *Linear approximation of sine.*

Angle (deg)	Angle (rad)	Sin(angle)	Percentage difference
0	0	0	0.0
5	0.087266463	0.087155743	0.1
15	0.261799388	0.258819045	1.0
20	0.34906585	0.342020143	1.8
25	0.436332313	0.422618262	2.9
30	0.523598776	0.5	4.3
35	0.610865238	0.573576436	5.9
40	0.698131701	0.64278761	7.9
45	0.785398163	0.707106781	19.9

that for small angles, the sine of the angle can be approximated by the angle itself when the angle is expressed in radians,

$$\sin(\theta) \approx \theta \text{ (small angles in radians).} \tag{2.36}$$

How good is this approximation? Just to give you an idea, the difference between $\sin(\theta)$ and θ for $\theta < 15°$ is less than 1%, as illustrated in Table 2.1. The last column in the table gives the percentage difference,

$$\text{Percentage difference} = \frac{|\sin(\theta) - \theta(\text{rad})|}{\sin(\theta)}.$$

With the small angle approximation given in Eq. 2.36, the equation of motion of a pendulum, Eq. 2.35, good for small angles of oscillations, can be written as

$$\frac{d^2\theta}{dt^2} = -\frac{g}{l}\theta, \quad (\theta \text{ in radians}) \tag{2.37}$$

where we have replaced the approximation sign ($\approx$) with the equals sign ($=$). The modified equation of motion, Eq. 2.37 shows clearly that, if the oscillations are kept to small angles, the acceleration (two-time derivatives of θ) is proportional to the displacement (θ) and is opposed to the displacement. This is the same situation as the mass/spring system. Therefore, for small-angle displacements, a pendulum will execute a simple harmonic motion similar to that of the mass/spring system. Comparing the approximate equation of motion for a pendulum with the equation for a block attached to a spring, we immediately discover the following correspondence in symbols:

$$x \Longleftrightarrow \theta \tag{2.38}$$

$$(k/m) \Longleftrightarrow (g/l). \tag{2.39}$$

Exploiting this analogy, we can easily deduce the formula for the angular frequency of the plane pendulum:

$$\omega = \sqrt{\frac{g}{l}}, \tag{2.40}$$

which gives the period of the pendulum to be

$$T = \frac{2\pi}{\omega} = 2\pi\sqrt{\frac{l}{g}}. \tag{2.41}$$

The formula for the period shows that the period of the small oscillations of the pendulum does not depend on either the mass of the pendulum bob or the amplitude of oscillation, but only on the length of the pendulum cord and the acceleration due to gravity. Galileo appears to be the first person who noticed this aspect of pendulum motion, when he made the observation that different chandeliers of equal length had the same period regardless of their amplitude of swing or weight. The solution of the equation of motion (Eq. 2.37) for the small-angle pendulum can be written by analogy to the equation of motion of the mass/spring system,

$$\theta(t) = A\cos(\omega t - \phi). \tag{2.42}$$

The Energy Picture

The energy of a pendulum can also be written such that it contains two terms, one having the square of the velocity and the other the square of the displacement, similarly to that of the simple harmonic oscillator. Let x and y denote the x- and y-coordinates of the pendulum bob at an arbitrary time t with respect to the axes shown in Fig. 2.9. Then, in the small angle approximation, we have

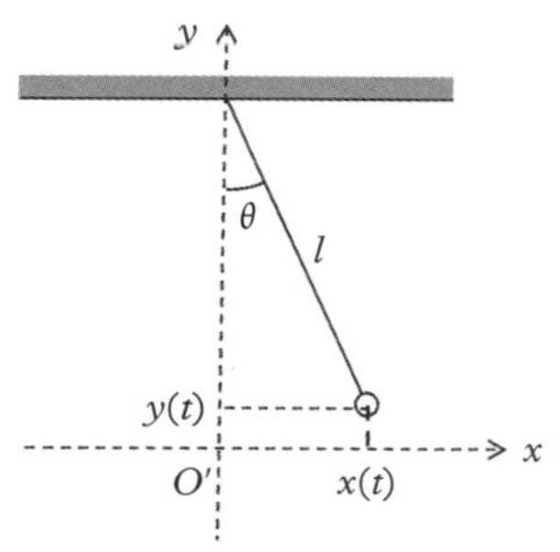

Fig. 2.9 *The x and y motions of the plane pendulum.*

$$x = l\sin\theta \approx l\theta, \tag{2.43}$$

$$y = l - l\cos\theta \approx \frac{1}{2}l\theta^2, \tag{2.44}$$

where we have dropped terms which are cubic or higher powers in θ. The velocities are

$$v_x = \frac{dx}{dt} = l\frac{d\theta}{dt}, \tag{2.45}$$

$$v_y = \frac{dy}{dt} = l\theta\frac{d\theta}{dt}. \tag{2.46}$$

Using the components of the velocity, we find the kinetic energy of the bob to be

$$KE = \frac{1}{2}\left(v_x^2 + v_y^2\right) \approx \frac{1}{2}ml^2\left(\frac{d\theta}{dt}\right)^2, \tag{2.47}$$

where we have dropped the term with four powers involving θ and $d\theta/dt$. The potential energy is

$$PE = mgy = mg\frac{l}{2}\theta^2. \tag{2.48}$$

Therefore, the energy of the pendulum in small angle approximation is

$$E = \frac{1}{2}ml^2\left(\frac{d\theta}{dt}\right)^2 + \frac{1}{2}mgl\theta^2, \tag{2.49}$$

which has the required form for a simple harmonic motion. Hence, we can determine the angular frequency also from the ratio of the coefficients of the restoring and inertial terms,

$$\omega^2 = \frac{\text{Coefficient of } \theta^2}{\text{Coefficient of } (d\theta/dt)^2} = \frac{\frac{1}{2}mgl}{\frac{1}{2}ml^2}, \implies \omega = \sqrt{\frac{g}{l}}, \tag{2.50}$$

which is identical to the expression we obtained from an application of the second law of motion as given in Eq. 2.40.

2.5.2 Torsion Pendulum

A torsion pendulum consists of a solid body, usually a dumbbell or a bar, suspended by a torsion wire from a fixed support, as shown in Fig. 2.10. A torsion wire is essentially a wire that could be twisted easily about its axis. The twisting of the wire applies a restoring torque on the supported solid that tends to bring the solid back to the configuration when the wire was not twisted, i.e. to the equilibrium. The restoring torque τ on the bar about the vertical axis is proportional to the angle of twist for small angles,

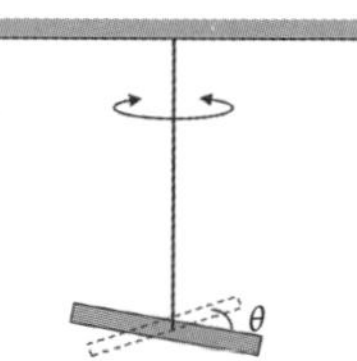

Fig. 2.10 *The torsion pendulum.*

$$\tau = -\kappa\theta, \tag{2.51}$$

where κ (read: kappa) is the torsional constant of the wire, which is analogous to the spring constant. The rotational motion of the bar is then given by the following equation of motion:

$$I\frac{d^2\theta}{dt^2} = -\kappa\theta, \tag{2.52}$$

where I is the moment of inertia of the bar. This equation is analogous to the equation for a plane pendulum in small angle approximation. Hence, the angular frequency ω of the oscillating motion of the torsion pendulum is given by

$$\omega = \sqrt{\frac{\kappa}{I}}. \tag{2.53}$$

Torsion pendulums are often used for time-keeping purposes, e.g. in the balance wheel of a mechanical watch. A torsion pendulum is also used in the Cavendish experiment for determining the value of Newton's gravitational constant.

Example 2.3 Torsion Constant of a Wire

A thin copper wire is tied to the center of a rod of mass 100 g and length 20 cm and hung from a fixed platform. When the copper wire is twisted and let go, the rod rotates about the wire with a time period of 10 seconds. What is the value of the torsion constant of the copper wire? Assume the moment of inertia of the wire itself about the axis of rotation to be small compared to that of the rod.

continued

Example 2.3 *continued*

Solution

From the given time period, we will find the angular frequency, and from the given geometry and masses, we will find the moment of inertia. Then we will use the formula for angular frequency to find the torsion constant of the wire. Angular frequency:

$$\omega = \frac{2\pi}{T} = \frac{2\pi}{10\text{ sec}} = 0.63\text{ rad/sec.}$$

Moment of inertia:

$$I = \frac{1}{12} m_{\text{rod}} l_{\text{rod}}^2 = 3.3 \times 10^{-4}\text{ kg.m}^2.$$

Therefore,

$$\kappa = I\omega^2 = 1.3 \times 10^{-4}\text{ N.m.}$$

2.5.3 Physical Pendulum

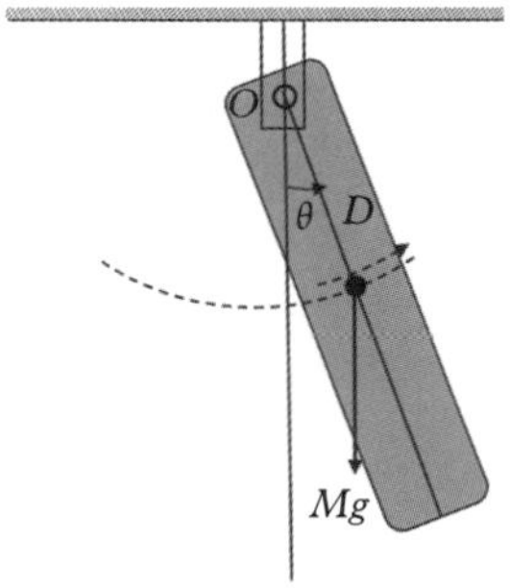

A rigid body hung from a post swings just like a pendulum. Such oscillating bodies are called physical pendulums. Almost anything can be a physical pendulum. An illustration is given in Fig. 2.11. The torque responsible for the oscillations comes from gravity acting at the center of mass of the body.

Let M be the mass and D the distance between the axis of rotation and the center of mass. From the torque due to gravity we obtain the following equation of motion:

$$I\frac{d^2\theta}{dt^2} = -MgD\sin\theta, \tag{2.54}$$

where I is the moment of inertia about an axis through O and perpendicular to the plane of the drawing. The mathematical situation here is identical to that of the plane pendulum. Once again, we restrict our studies to the domain of small angles and use the small-angle approximation. For small-angle oscillations, using $\sin\theta \approx \theta$ in Eq. 2.54 yields the following approximate equation of motion of a physical pendulum:

$$\frac{d^2\theta}{dt^2} = -\frac{MgD}{I}\theta. \tag{2.55}$$

Now, by analogy with the plane pendulum, we find that the angular frequency of oscillation of a physical pendulum would be given by

$$\omega = \sqrt{\frac{MgD}{I}}. \tag{2.56}$$

Fig. 2.11 *A physical pendulum.*

2.5.4 Oscillations of Freely Floating Objects

Buoyancy can provide restoring force to a freely floating object in a fluid. Recall that the Archimedes principle states that the force of buoyancy is pointed up and is equal to the weight of the displaced fluid. A consequence of the Archimedes principle is that a compact object whose density is less than that of the fluid will float in the fluid. Even an object with density more than that of the fluid can be made to float in the fluid if the object can be shaped such that the displaced fluid has more weight than the weight of the object itself. This is the principle by which a boat made of steel floats in water even though the density of steel is approximately eight times that of water: the boat is shaped such that the weight of displaced fluid is more than the weight of the boat itself.

Consider a compact object of density ρ floating in a fluid of density ρ_0, as shown in Fig. 2.12. For simplicity, let the object be shaped in the form of a rectangular box of cross-sectional area A and height h. Let h_0 be the height of the submerged part of the box from the bottom when it is in equilibrium, i.e. when the weight of the box is balanced by the force of buoyancy,

$$mg = F_B \implies Ah\rho g = Ah_0\rho_0 g \text{ (Equilibrium).} \tag{2.57}$$

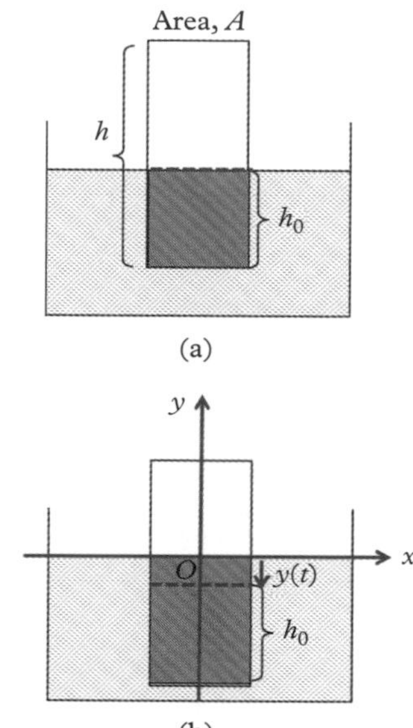

Fig. 2.12 *The oscillation of a floating object. (a) The box is partially submerged and in equilibrium. (b) The box is not in equilibrium since the force of buoyancy on the box is not equal to the weight of the box.*

Now, we push the box into the fluid by an amount B so that the submerged part has a height $h_0 + B$. When we let go of the box from this position, we find that the forces on the box are no longer balanced, since the force of buoyancy on the box will be greater than the weight of the box. As a result the box will accelerate upward. Once the box has risen so that the submerged part is less than h_0, the weight will be greater than the buoyancy, and the box will then accelerate downward. Thus, we see that the difference of the weight and the buoyancy provides a restoring force on the box.

Let us see if the motion will be a simple harmonic motion by working out the equation of motion. To find the equation of motion we will consider an arbitrary instant when the forces on the box are not balanced. Such an instant can be either when the height of the submerged part is greater than h_0 or less than h_0:

$$\text{If the submerged height} > h_0, \quad F_B > mg, \tag{2.58}$$

$$\text{If the submerged height} < h_0, \quad F_B < mg. \tag{2.59}$$

For our calculations we will point the y-axis upward with origin at the level of the fluid, assumed to remain at the same height regardless of the degree of submersion of the box in the fluid, which will be true if the fluid container is large and the box small. Let us mark the place on the box that is at the height of the fluid when it is in equilibrium; we will use this mark to follow the motion of the box. Let y be the coordinate of the mark on the box at an arbitrary time, as indicated in Fig. 2.12(b). The y-component of the equation of motion of the box is found to be

$$ma_y = F_y \implies Ah\rho\frac{d^2y}{dt^2} = -Ah\rho g + A(h_0 - y)\rho_0 g. \tag{2.60}$$

Using the equilibrium condition in Eq. 2.57, we simplify Eq. 2.60 and find that

$$\frac{d^2y}{dt^2} = -\frac{\rho_0 g}{\rho h}y. \tag{2.61}$$

Since the acceleration in y is proportional to y with a negative proportionality constant, the box will oscillate in a simple harmonic motion with angular frequency, given by

$$\omega = \sqrt{\frac{\rho_0 g}{\rho h}}. \tag{2.62}$$

2.5.5 Electromagnetic Oscillations in LC Circuits

The voltages and currents in an electric circuit containing a capacitor and an inductor (an LC circuit) oscillate just like the displacement of a simple harmonic oscillator. Since current and voltage in a circuit are easier to measure than the displacement of a block, an LC circuit provides a convenient practical device for studying simple harmonic motion.

To study the oscillation of an LC circuit consider, connecting a charged capacitor of capacitance C and initial charge q_0 to an inductor of inductance L at $t = 0$, as shown in Fig. 2.13.

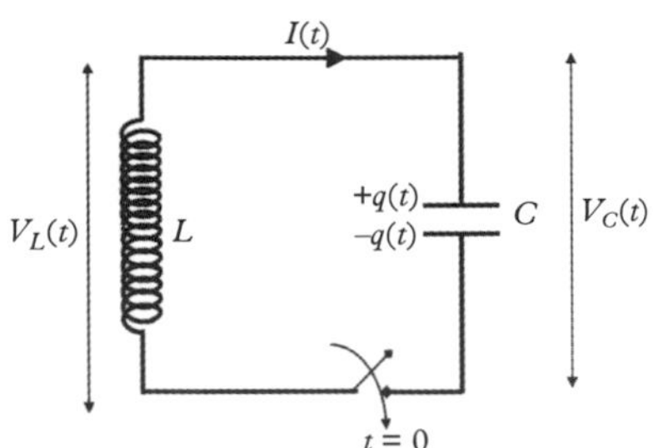

Fig. 2.13 *An oscillating LC circuit. The circuit is started with a charged capacitor. Thereafter the current in the circuit oscillates sinusoidally. The current $I(t)$ is shown at a time t when the charges are building up on the capacitor.*

Let V_0 be the potential difference across the capacitor at time $t = 0$. For the sake of concreteness we will write the equation of motion of the circuit by examining the circuit at a time when charge on the capacitor is building up. At this instant we will take the direction of the current to be towards the positive plate of the capacitor. This convention is convenient since it gives the following relation between the current into the capacitor and the rate at which charge builds up:

$$I = \frac{dq}{dt}. \tag{2.63}$$

Note that at $t = 0$ the current would point away from the positive plate since the capacitor would be discharging. Therefore, at $t = 0$ the current will be negative, which is consistent with $dq/dt < 0$ at $t = 0$. The voltage drops across the capacitor and inductor at the instant t are given by

$$V_C(t) = \frac{q(t)}{C}, \tag{2.64}$$

$$V_L(t) = L\frac{dI}{dt}. \tag{2.65}$$

The Kirchhoff's loop equation for the EMF around the circuit yields the following equation:

$$\frac{q}{C} + L\frac{dI}{dt} = 0. \tag{2.66}$$

Taking a time derivative and using Eq. 2.63 yields the following equation of motion for the current in the circuit:

$$\frac{d^2I}{dt^2} + \frac{1}{LC}I = 0. \tag{2.67}$$

This equation for the current I is analogous to the equation of motion for a simple harmonic oscillator,

$$\frac{d^2x}{dt^2} + \omega^2 x = 0. \tag{2.68}$$

The oscillating mass/spring system and the oscillating LC system give rise to the mechanical/electrical analogies which are summarized in Table 2.2.

Following this analogy the solution $I(t)$ of Eq. 2.67 will have the same form as the solution of Eq. 2.68. In particular, the current in the circuit will oscillate, given by the simple harmonic formula,

$$I(t) = I_0 \cos(\omega t - \phi), \tag{2.69}$$

Table 2.2 *Analogy between mechanical and electrical quantities.*

Mechanical system	Electrical system
Mass, m	Inductance, L
Spring constant, k	Inverse capacitance, $1/C$
Damping constant, b	Resistance, R
Displacement from equilibrium, x	Charge on capacitor, q
Velocity, v	Current, I
Kinetic energy, $1/2\ mv^2$	Magnetic energy, $1/2\ LI^2$
Potential energy, $1/2\ kx^2$	Electric energy, $1/2\ q^2/C$

with the angular frequency given by

$$\omega = \frac{1}{\sqrt{LC}}. \tag{2.70}$$

2.6 Simple Harmonic Motion Near Potential Minima

From our discussion in this chapter, you know that a restoring force that is proportional to the displacement from the equilibrium gives rise to a simple harmonic motion. Now, a conservative force $\vec{F}$ is related to potential energy U as follows:

$$\vec{F} = -\left(\frac{\partial U}{\partial x}\hat{i} + \frac{\partial U}{\partial y}\hat{j} + \frac{\partial U}{\partial z}\hat{k}\right), \tag{2.71}$$

where $\hat{i}$, $\hat{j}$, and $\hat{k}$ are unit vectors pointed towards the positive x-, y- and z-axes respectively. If the motion is only along the x-axis we work with the x-component only. Then, the x-component of the conservative force is related to the x-derivative of the potential energy, which is a function of x,

$$F_x = -\frac{dU}{dx}. \tag{2.72}$$

Therefore, a potential energy that is quadratic in x will give the linear restoring force appropriate for a simple harmonic motion. This is obviously the case with the potential energy in an ideal spring. In general, consider a potential energy function $U(x)$ that has a minimum at $x = x_0$, then we can make a Taylor series about $x = x_0$,

$$U(x) = U(x_0) + \left(\frac{dU}{dx}\right)_{x=x_0}(x-x_0) + \frac{1}{2!}\left(\frac{d^2U}{dx^2}\right)_{x=x_0}(x-x_0)^2 + \cdots. \tag{2.73}$$

Since $U(x)$ has a minimum at $x = x_0$, the first derivative is zero there, and the leading non-constant term is quadratic in the displacement from the equilibrium, $x - x_0$,

$$U(x) = U(x_0) + \frac{1}{2!}\left(\frac{d^2U}{dx^2}\right)_{x=x_0}(x - x_0)^2 + \cdots . \tag{2.74}$$

The value of the second derivative of the potential energy function for $x = x_0$ is a constant. Denote this constant by k,

$$k \equiv \left(\frac{d^2U}{dx^2}\right)_{x=x_0}. \tag{2.75}$$

Choosing the potential energy to be zero at the equilibrium, and placing the origin at the equilibrium point, we find that near a potential energy minimum, the leading behavior of the potential energy function is quadratic,

$$U(x) = \frac{1}{2}kx^2 + \cdots . \tag{2.76}$$

The quadratic potential energy will dominate the behavior near the potential energy minimum unless the potential energy function is so flat at the minimum that even the second derivative is zero. The quadratic potential energy function gives a linear restoring force of Hooke's law and leads to simple harmonic motion,

$$F(x) = -\frac{dU}{dx} = -kx + \text{ higher powers in } x. \tag{2.77}$$

Figure 2.14 illustrates how a quartic potential function can be approximated by a quadratic potential function for motion near a potential minimum. The double-well potential energy in the figure is

$$U(x) = (x + 1)^2(x - 1)^2, \tag{2.78}$$

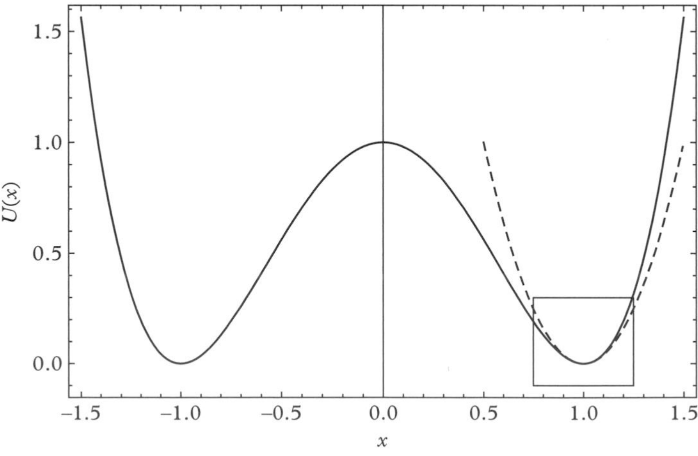

Fig. 2.14 *The quartic potential energy, $U(x) = (x + 1)^2(x - 1)^2$, shown as solid line, can be approximated by a quadratic potential energy, $U(x) = 4(x - 1)^2$, shown as a dashed line, near its minimum at $x = 1$. Similarly, near the minimum at $x = -1$, the potential energy function can be approximated by the quadratic function, $U(x) = 4(x + 1)^2$. As a result, the motion near $x = 1$ and $x = -1$ will be simple harmonic.*

which has minima at $x = -1$ and $x = 1$, with a local maximum at $x = 0$. The motion close to the minimum $x = 1$ can be approximated by the potential energy function,

$$U(x) = 4(x-1)^2, \tag{2.79}$$

as shown by the dashed curve. Similarly, the motion close to $x = -1$ can be approximated by the potential energy function,

$$U(x) = 4(x+1)^2. \tag{2.80}$$

These potential energy functions will give rise to linear restoring forces near $x = \pm 1$. Therefore, if the displacements from the minima are not large, we can ignore the original potential and pretend that the potential is quadratic.

The plane pendulum uses this approximation near the potential minimum, which we have seen previously in the observation that a plane pendulum is not a simple harmonic oscillator unless the angle of oscillation is small. We can understand this from the perspective of a quadratic potential energy function. The potential energy of a pendulum when it is displaced at angle θ is

$$U = mgl(1 - \cos\theta). \tag{2.81}$$

This is clearly not a quadratic function of the dynamical variable θ. Now, expanding $\cos(\theta)$ for small θ, we find

$$\cos(\theta) = 1 - \frac{1}{2!}\theta^2 + \frac{1}{4!}\theta^4 + \cdots. \tag{2.82}$$

If we keep only the leading term, namely, 1, we lose all physics information associated with θ. Therefore, we will keep two terms in this expansion. This gives the following expression for the potential energy near $\theta = 0$:

$$U = \frac{mgl}{2}\theta^2, \tag{2.83}$$

which is quadratic in the dynamical variable θ. Hence, for small angles we expect simple harmonic motion for the pendulum.

Example 2.4 Motion Through the Center of Earth

Imagine drilling a hole through the center of the earth. When you drop a ball in the hole, it will execute a simple harmonic motion if the density of the earth is assumed to be uniform. What will be the frequency of oscillation?

Solution

We will use a result from Newton's law of gravitation that for spherical masses such as the earth, the force on the ball will depend on only the mass from the ball's current location to the center of the earth. Let us examine the expression of the force at some arbitrary

continued

Example 2.4 *continued*

time when the ball is at a distance r from the center of the earth. The mass of the earth inside the radius r is

$$M = \frac{4}{3}\pi r^3 \rho,$$

where ρ is the density of Earth. Therefore, the force on the ball at this instant will be

$$F = -\frac{Gm}{r^2}\frac{4}{3}\pi r^3 \rho = -\left(\frac{4}{3}\pi \rho G m\right) r,$$

where minus is placed to indicate that the force is pointed towards $r = 0$ or the center of the earth. This force is a linear restoring force in the displacement variable r. Therefore, the constant k here is

$$k = \frac{4}{3}\pi \rho G m,$$

and the angular frequency of oscillation will be

$$\omega = \sqrt{\frac{4}{3}\pi \rho G}.$$

It is interesting to note that the frequency of oscillation does not depend on the properties of the ball. All objects dropped in the hole will oscillate at the same frequency.

2.7 Damping of Oscillations

So far we have studied the motion of ideal simple harmonic oscillators. We have found that the amplitude of oscillations of an ideal system never decreases with time. However, we know that real physical systems, such as sound in air or a pendulum in a grandfather clock, lose energy due to friction and other viscous forces. Therefore, the amplitude of the oscillations of a real system would decrease with time unless fresh energy were continually fed into the system. We say that the system is damped. In this section we will study the effect of damping without any additional energy supplied to the system. Systems which receive additional energy are called driven systems. We will study driven oscillators in a later chapter.

To study the damping phenomenon we subject our mass/spring system to a viscous force. Any viscous force will damp the motion, but a particularly simple case of a viscous force whose magnitude is proportional to the velocity of the oscillator has many applications and will be discussed here,

$$\vec{F}_{\text{visc}} = -b\,\vec{v} \quad (b > 0), \tag{2.84}$$

where b is the proportionality constant that depends on the viscous medium and the geometry of the oscillating body. The minus sign makes sure that viscous force is pointed in the opposite direction to the velocity, thus slowing the object. Viscous forces of this type act on objects moving in a fluid if their speeds are not too great. For instance, the viscous force on a sphere of radius R moving in a fluid of viscosity η is given by Stoke's force,

$$\vec{F}_{\text{visc}} = -6\pi \eta R\,\vec{v}. \tag{2.85}$$

Pictorially, it is customary to represent the damping force by attaching a dash pot to the oscillating block, as shown in Fig. 2.15. The dashpot has a piston that moves in a fluid. The net force on the oscillator now would be a vector sum of the restoring force by the spring and the viscous force.

Once again, we place the origin at the equilibrium position, and point the $+x$-axis in the direction that corresponds to the extension of the spring. Then the x-component of the second law of motion gives the following equation of motion:

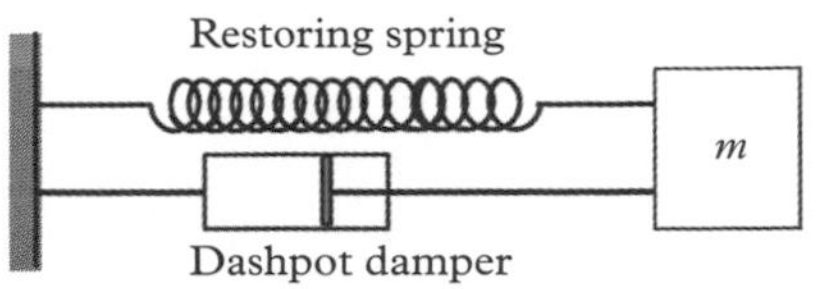

Fig. 2.15 *A harmonic oscillator with a dashpot. The dashpot has a piston that moves through a liquid providing a velocity-dependent damping force.*

$$m\frac{d^2x}{dt^2} = -kx - b\,\frac{dx}{dt}. \quad (2.86)$$

Dividing both sides by m reduces the number of parameters in this equation just as it did for the undamped case. We introduce new constants, Γ and ω_0, as follows to denote the resulting ratios,

$$\omega_0 = \sqrt{k/m}, \;\; \Gamma = b/m. \quad (2.87)$$

The damping in this system is characterized by the parameter Γ called the **damping parameter** or **damping constant**. Note also that we have introduced a different symbol ω_0 for $\sqrt{k/m}$ here than what we did when we studied the undamped oscillator, since the actual frequency of the damped oscillator will be different from the undamped case, as we will see below. The frequency ω_0 is the frequency of the system when the damping is absent. With these new parameters, Eq. 2.86 can be written as

$$\frac{d^2x}{dt^2} + \Gamma\,\frac{dx}{dt} + \omega_0^2 x = 0. \quad (2.88)$$

You should notice that the reduction of parameters from three (m, b, k) to two (Γ, ω_0) is as a result of reparameterization. Thus, oscillators with different (m, b, k) that have the same (Γ, ω_0) will have similar dynamics.

2.7.1 Solving the Equation of Motion

You already know how to solve the simpler equation when $\Gamma = 0$ in Eq. 2.88. Therefore, the trick is to make a variable change so that we obtain a new equation without the velocity term. The following change of variable achieves that objective. Let

$$x(t) = e^{-\frac{1}{2}\Gamma t}u(t), \quad (2.89)$$

where $u(t)$ is a variable of time t that will be determined from a simpler equation, as we now derive. Using this form for $x(t)$ in Eq. 2.88, we obtain the following equation for $u(t)$:

$$\frac{d^2u}{dt^2} + \left(\omega_0^2 - \frac{1}{4}\Gamma^2\right)u = 0. \quad (2.90)$$

The solution of Eq. 2.90 depends on whether the multiplier of u is positive, zero, or negative, i.e. on the relative values of ω_0 and $\frac{1}{2}\Gamma$. The three cases are given different names since they correspond to quite different physical behaviors,

$$\text{Under-damped:} \quad \omega_0 > \tfrac{1}{2}\Gamma$$
$$\text{Critically damped:} \quad \omega_0 = \tfrac{1}{2}\Gamma \tag{2.91}$$
$$\text{Over-damped:} \quad \omega_0 < \tfrac{1}{2}\Gamma.$$

Under-damped Oscillator

Since $\omega_0 > \frac{1}{2}\Gamma$ for an under-damped oscillator, let us replace the positive quantity in parenthesis in Eq. 2.90 by another symbol ω_1^2,

$$\omega_1^2 = \omega_0^2 - \frac{1}{4}\Gamma^2 > 0. \tag{2.92}$$

This changes Eq. 2.90 to the equation for a simple harmonic oscillator of natural frequency ω_1,

$$\frac{d^2u}{dt^2} = -\omega_1^2 u. \tag{2.93}$$

Therefore, it is easy to write down the solution for $u(t)$,

$$u(t) = C_1 \cos(\omega_1 t) + C_2 \sin(\omega_1 t), \text{ or, } A \cos(\omega_1 t - \phi), \tag{2.94}$$

where C_1 and C_2 or A and ϕ are arbitrary constants that are fixed by the initial conditions on $u(t)$, or, equivalently, on the initial conditions on $x(t)$. Putting this $u(t)$ in Eq. 2.89 we find that the displacement of the under-damped oscillator is given by

$$x(t) = e^{-\frac{1}{2}\Gamma t}\left[C_1 \cos(\omega_1 t) + C_2 \sin(\omega_1 t)\right], \tag{2.95}$$

or, equivalently, as

$$x(t) = \left[A\, e^{-\frac{1}{2}\Gamma t}\right] \cos(\omega_1 t - \phi). \tag{2.96}$$

The under-damped oscillator oscillates at the angular frequency of $\omega_1 = \sqrt{\omega_0^2 - \frac{1}{4}\Gamma^2}$ with decreasing amplitude, given within the square brackets, $[\cdots]$. Note that the presence of damping decreases the frequency of oscillation from the natural frequency ω_0 to frequency ω_1.

Critically Damped System

A critically damped system has $\omega_0 = \frac{1}{2}\Gamma$. Putting this condition in Eq. 2.90 shows that $u(t)$ obeys a different differential equation than the under-damped oscillator,

$$\frac{d^2u}{dt^2} = 0. \tag{2.97}$$

The general solution of this equation is

$$u(t) = C_1 + C_2 t, \tag{2.98}$$

where C_1 and C_2 are arbitrary constants. Therefore, the displacement of a critically damped system is given by

$$x(t) = e^{-\frac{1}{2}\Gamma t}\left(C_1 + C_2 t\right), \tag{2.99}$$

which shows that the critically damped system is not an oscillator at all, although it is commonly referred to as a "critically damped oscillator". Even though the factor $(C_1 + C_2 t)$ will grow with time for a positive C_2, the exponentially decreasing factor $e^{-\frac{1}{2}\Gamma t}$ will always force the product $e^{-\frac{1}{2}\Gamma t}(C_1 + C_2 t)$ to zero over a long enough time.

Over-damped System

Since $\omega_0 < \frac{1}{2}\Gamma$ for an over-damped oscillator, let us replace the negative quantity in parenthesis in Eq. 2.90 by another symbol, α^2, to obtain

$$\frac{d^2u}{dt^2} = \alpha^2 u, \tag{2.100}$$

where α is given by

$$\alpha = \sqrt{\frac{1}{4}\Gamma^2 - \omega_0^2}. \tag{2.101}$$

Here we take only the positive radical for α since the negative α does not contain any new information. The general solution of Eq. 2.100 is

$$u(t) = C_1 e^{-\alpha t} + C_2 e^{+\alpha t}. \tag{2.102}$$

Thus, the displacement of an over-damped oscillator is given by

$$x(t) = e^{-\frac{1}{2}\Gamma t}\left[C_1 e^{-\alpha t} + C_2 e^{+\alpha t}\right]. \tag{2.103}$$

Since $\frac{1}{2}\Gamma > \alpha$, as is evident from Eq. 2.101, the exponentially decaying multiplier outside of the bracket in Eq. 2.103 dominates, even for the exponentially increasing part $e^{\alpha t}$ inside the bracket, and the overall behavior is an exponentially decaying displacement. As a result, the over-damped "oscillator" will not oscillate, and therefore, the word "oscillator" should not be used when we refer to this case. The only case that has oscillations is the under-damped case.

Initial Conditions

The constants C_1 and C_2 in the three solutions given above are determined by the known position and velocity at some particular time, usually the initial time, as we show below. Let the initial position and velocity be x_0 and v_0 respectively,

$$x(0) = x_0, \tag{2.104}$$

$$v(0) = v_0. \tag{2.105}$$

I will work out a complete calculation for the under-damped case and leave the calculations for the critically damped and over-damped cases to the student. The under-damped oscillator has the following displacement, given in Eq. 2.95:

$$x(t) = e^{-\frac{1}{2}\Gamma t}\left[C_1\ \cos(\omega_1 t) + C_2\ \sin(\omega_1 t)\right]. \tag{2.106}$$

The time derivative of the displacement will give the velocity, more appropriately the x-component of the velocity,

$$v(t) = \frac{dx}{dt} = -\frac{1}{2}\Gamma x(t) + \omega_1 e^{-\frac{1}{2}\Gamma t}\left[-C_1\ \sin(\omega_1 t) + C_2\ \cos(\omega_1 t)\right]. \tag{2.107}$$

Setting $t = 0$ in Eqs. 2.106 and 2.107 and using the initial conditions in Eqs. 2.104 and 2.105 we obtain

$$x_0 = C_1,\ \ v_0 = -\frac{1}{2}\Gamma x_0 + \omega_1 C_2. \tag{2.108}$$

Therefore, in the case of the under-damped oscillator:

$$\textbf{Under-damped:}\quad C_1 = x_0,\ \ C_2 = \frac{v_0 + \frac{1}{2}\Gamma x_0}{\sqrt{\omega_0^2 - \frac{1}{4}\Gamma^2}}, \tag{2.109}$$

where we have replaced ω_1 by its definition in terms of more fundamental parameters of the oscillator. Similar calculations for critically damped and over-damped cases give the following expressions for C_1 and C_2 which a student should verify:

$$\textbf{Critically damped:}\quad C_1 = x_0,\ \ C_2 = v_0 + \frac{1}{2}\Gamma x_0. \tag{2.110}$$

$$\textbf{Over-damped:}\quad C_1 = \frac{1}{2\alpha}\left[\left(\alpha - \frac{1}{2}\Gamma\right)x_0 - v_0\right],\ \ C_2 = \frac{1}{2\alpha}\left[\left(\alpha + \frac{1}{2}\Gamma\right)x_0 + v_0\right]. \tag{2.111}$$

Plotting Solution

Figure 2.16 displays plots of x versus t for the three types of damped motion, starting with the same initial conditions, $x(0) = 1$, $v(0) = 0$. The under-damped case is the only one that oscillates with time. The critically damped system damps more smoothly than the over-damped case. In practical use of damping, as in shock absorbers for cars, you would want the car to relax gently whenever it hits a bump in the road. Therefore, shock absorbers are built with the characteristics of critical damping. However, with wear and tear

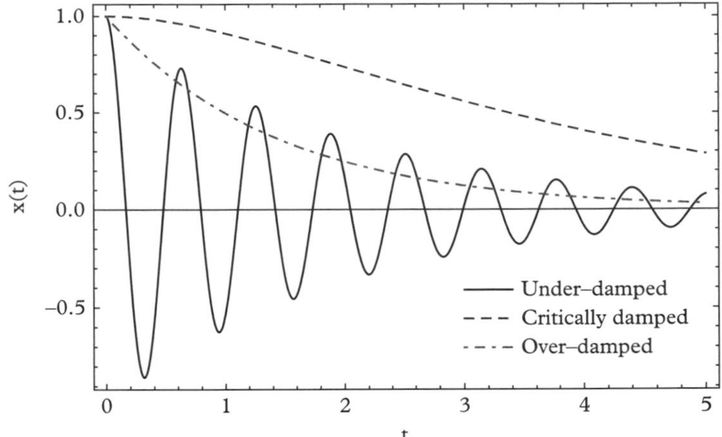

Fig. 2.16 *The displacement as a function of time for the under-damped, over-damped, and critically damped cases. Only the under-damped case is an oscillator. To plot the figures, the following values were used: Underdamped case, $\Gamma = 2$ rad/sec, $\omega_0 = \sqrt{101}$ rad/sec; Critically damped case, $\Gamma = 2$ rad/sec; Overdamped case, $\Gamma = 2$ rad/sec, $\omega_0 = \sqrt{0.96}$ rad/sec. The initial conditions for all cases were $x(0) = 1$ cm, $v(0) = 0$.*

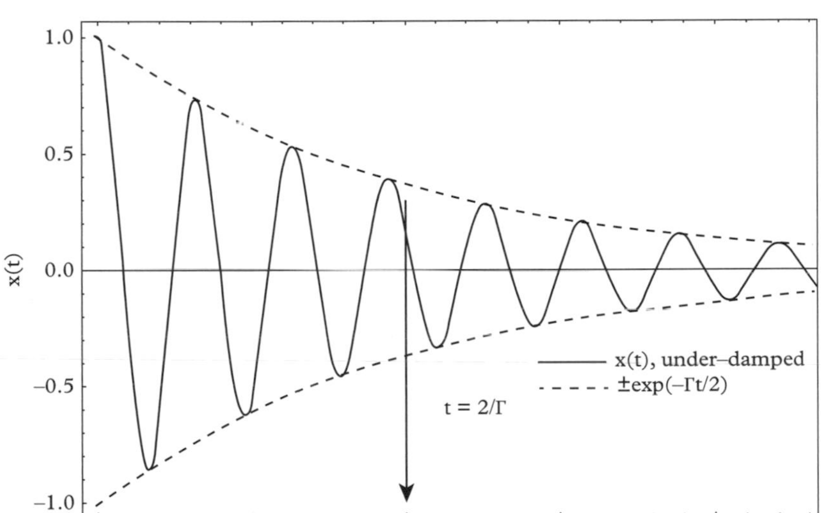

Fig. 2.17 *The dynamics of an under-damped oscillator shows oscillations with decreasing amplitude. Here $A = 1$ cm, Γ = 2 rad/sec, and $\omega_0 = \sqrt{101}$ rad/sec.*

the damping gives way to an under-damping and you find cars bounce up and down a few times whenever they hit a bump in the road, as if they were under-damped oscillators.

The displacement of an under-damped oscillator provides a physical meaning for the damping parameter Γ. We see from the graph in Fig. 2.17 that it takes a time of $2/\Gamma$ for the envelope of the oscillations to decrease by e^{-1} of its original value, where e is the Euler's number with value $e = 2.71828\cdots$. The time to relax to $1/e$ of the original amplitude is called the **time constant** of the oscillator, which is denoted by the Greek letter τ,

$$\tau = \frac{2}{\Gamma}. \tag{2.112}$$

The larger the damping constant, the shorter the time constant, and consequently, the faster the damping.

2.7.2 Dissipation of Energy and the Quality of an Oscillator

Under-damped oscillators oscillate with decreasing amplitude and eventually come to rest. A better oscillator would swing for many more cycles than a poorer oscillator before losing its energy. For an arbitrary damped oscillator, energy is a complicated function of time. But for a lightly damped oscillator, i.e. an oscillator that oscillates many times before running out of energy, the energy decreases exponentially. We will define a lightly damped oscillator by the following condition between the damping constant (Γ) and the natural frequency (ω_0) of the oscillator:

$$\Gamma \ll \omega_0 \quad \text{(Lightly damped oscillator)}. \tag{2.113}$$

Since a lightly damped oscillator is an under-damped oscillator the displacement $x(t)$ would be given by

$$x(t) = Ae^{-\frac{1}{2}\Gamma t}\cos(\omega_1 t - \phi), \tag{2.114}$$

where we leave A and ϕ unspecified. The velocity is obtained by taking a time derivative of x,

$$v(t) = A\,e^{-\frac{1}{2}\Gamma t}\left[-\frac{1}{2}\Gamma\,\cos(\omega_1 t - \phi) - \omega_1\,\sin(\omega_1 t - \phi)\right]. \tag{2.115}$$

Now, the energy of the oscillator at instant t is

$$E(t) = \frac{1}{2}mv^2 + \frac{1}{2}kx^2, \tag{2.116}$$

Using x from Eq. 2.114 and v from Eq. 2.115 we get a very messy expression, which can be simplified by using $\Gamma \ll \omega_0$ of the lightly damped oscillator condition. After some algebra, you can show that

$$E(t) \approx \frac{1}{2}m\omega_0^2 A^2 e^{-\Gamma t}. \tag{2.117}$$

Put another way,

$$E(t) = E_0 e^{-\Gamma t}, \tag{2.118}$$

where E_0 stands for the energy at $t = 0$.

That is, the energy decreases exponentially with time with rate constant $1/\Gamma$. The larger the value of Γ, the faster the energy dissipation. We define a time constant τ_E of the dissipation of energy by asking: how much time would it take for the energy to drop to e^{-1} of its value, i.e. before about two-thirds of the energy had dissipated? From Eq. 2.118, the answer is that

$$\tau_E = \frac{1}{\Gamma}. \tag{2.119}$$

Note that the energy of the oscillator does not decrease at the same rate as the envelope of the amplitude: it takes a time equal to $1/\Gamma$ for the energy to decrease by $1/e$, while it takes twice as long, $2/\Gamma$, for the amplitude of the oscillator to decrease by $1/e$. The reason for the energy decreasing twice as fast as the amplitude is that energy is proportional to the square of the amplitude. The same effect shows up in the rate at which sound fades compared to the rate at which a tuning fork loses vibrations. Since the intensity of sound generated by the vibrating tuning fork is related to the energy, the intensity of the sound will decrease faster than the amplitude of oscillations of the tuning fork itself.

The time τ_E for the oscillator to lose about two-thirds of its energy can be compared to the time period of the oscillations to get a sense of number of oscillations the oscillator will execute before a significant fraction of the energy has dissipated. For a lightly damped oscillator, the natural frequency ω_0 is very close to the actual frequency ω_1, therefore we will use ω_0 to get the time for one oscillation. From ω_0 we get the time T for one oscillation to be

$$T = \frac{2\pi}{\omega_0}. \tag{2.120}$$

Therefore, the number N of oscillations in the time τ_E will be

$$N = \frac{\tau_E}{T} = \frac{\omega_0}{2\pi\Gamma}. \tag{2.121}$$

The larger is N the better the oscillator. The physically meaningful part of N is the ratio of ω_0 to Γ, which is called the **Quality** or Q factor of an oscillator,

$$Q = \frac{\omega_0}{\Gamma}. \tag{2.122}$$

The larger the value of Q the better the oscillator. Good oscillators, such as tuning forks and guitar strings, have Q values in the thousands. Laser cavities have much higher Q values, exceeding 10^7. Of course, the undamped oscillator has zero Γ, and hence infinite Q. There is no Q for critically damped and over-damped systems since they do not oscillate.

Example 2.5 An Under-damped Harmonic Oscillator

A copper block of mass 1.5 kg is attached to a spring of stiffness 450 N/m, and hung from a platform above a beaker that contains a thick liquid so that the block oscillates entirely in the liquid with a damping constant of 3.0 kg/s (see Fig. 2.18). (a) How many oscillations will the block make before its amplitude drops by 90%? (b) What is the quality of this oscillator?

Solution

(a) We will first calculate the oscillation frequency ω_1 and the damping constant Γ for the damped oscillator. Since the peak of the displacements in successive cycles drops as $e^{-\frac{1}{2}\Gamma t}$, we can find the time for a 90% drop in amplitude from the value of Γ. We will obtain the required number of oscillations by dividing the time

continued

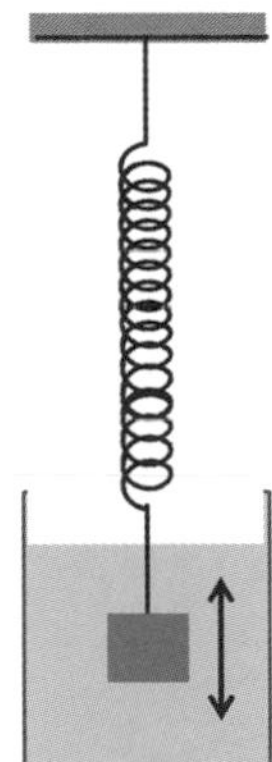

Fig. 2.18 *Example 2.5.*

Example 2.5 *continued*

required for the 90% drop in one time period,

$$\Gamma = \frac{b}{m} = \frac{3.0 \text{ kg/s}}{1.5 \text{ kg}} = 2.0 \text{ rad/sec},$$

$$\omega_0 = \sqrt{\frac{k}{m}} = \sqrt{\frac{450 \text{ N/s}}{1.5 \text{ kg}}} = 17.32 \text{ rad/sec}.$$

Therefore, the angular frequency of oscillation is

$$\omega_1 = \sqrt{\omega_0^2 - \frac{1}{4}\Gamma^2} \approx 17.3 \text{ rad/sec},$$

which gives the time period for oscillations to be

$$T = \frac{2\pi}{\omega_1} \approx 0.363 \text{ sec}.$$

Now, we use the decay of the amplitude envelope to find the time for the amplitude to drop by 90%,

$$\frac{\text{Amplitude left}}{\text{Original amplitude}} = 0.1.$$

Therefore,

$$e^{-\frac{1}{2}\Gamma t} = 0.1 \longrightarrow t - 2.3 \text{ sec}.$$

Hence, the number of cycles in which the amplitude drops by 90% will be

$$\text{Number of cycles} = \frac{t}{T} = \frac{2.3 \text{ sec}}{0.362 \text{ sec}} = 6.4 \text{ cycles}.$$

(b) The Q factor of the oscillator is

$$Q = \frac{\omega_0}{\Gamma} = \frac{17.32}{2} = 8.66.$$

2.8 The Damped AC Circuit

We have discussed the pure LC circuit without any resistance or source. The current in such a circuit can be started using a charged capacitor. When the circuit is closed, the capacitor starts to discharge. With an increasing flow of current through the inductor, the induction effect of L gives rise to a back EMF and eventually, current switches direction, which charges up the capacitor again. Thus, a pure LC circuit is just like an oscillator, oscillating between charging and discharging of the capacitor.

Without the resistance in the circuit, the LC circuit is an ideal oscillator. The resistance in the circuit causes the dissipation of energy. Thus, the RLC circuit is a damped oscillator.

The equation of motion of the RLC circuit can be obtained by voltage loop law as we did for the pure LC circuit. Consider a series RLC circuit without a power source, as displayed in Fig. 2.19.

The equation of motion for the charge on the positive plate of the capacitor $q(t)$ can be easily shown to be

$$L\frac{d^2q}{dt^2} + R\frac{dq}{dt} + \frac{1}{C}q = 0. \tag{2.123}$$

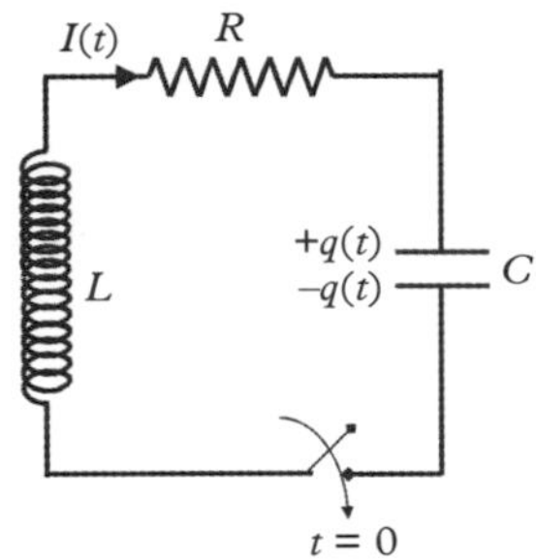

Fig. 2.19 *An RLC circuit started at t = 0 with a charged capacitor.*

Dividing out by L we get a two-parameter equation, which in analogy to the damped oscillator we label as Γ and ω_0,

$$\frac{d^2q}{dt^2} + \Gamma\frac{dq}{dt} + \omega_0^2 q = 0, \tag{2.124}$$
$$\Gamma = \frac{R}{L}, \quad \omega_0^2 = \frac{1}{LC}.$$

The underdamped oscillator will have

$$\frac{\Gamma}{2} < \omega_0.$$

Therefore, we need the following for the oscillator to be an under-damped oscillator:

$$\frac{R}{2L} < \frac{1}{\sqrt{LC}} \implies R < 2\sqrt{L/C}.$$

The charge on the capacitor will oscillate in time according to

$$q(t) = q_0 e^{-\Gamma t/2}\cos(\omega_1 t),$$

where q_0 is the charge at $t = 0$ and $\omega_1 = \sqrt{\omega_0^2 - \Gamma^2/4}$. The current in the circuit will be

$$I(t) = \frac{dq}{dt}.$$

For a lightly damped oscillator we need even more stringent conditions on Γ and ω_0,

$$\Gamma \ll \omega_0 \implies R \ll 2\sqrt{L/C}.$$

The Q of the oscillator will be

$$Q = \frac{\omega_0}{\Gamma} = \frac{1}{R}\sqrt{\frac{L}{C}} = \frac{\sqrt{LC}}{RC}. \tag{2.125}$$

Example 2.6 Q of an Electrical Oscillator

We need an electrical oscillator at frequency 20 kHz with a Q of 10,000. We have 750 pF capacitance in the circuit. What should be the values of the inductance and resistance in the circuit?

Solution

From the frequency requirement and C we can figure out L. Then by using the Q we can find the R needed,

$$f = \frac{1}{2\pi\sqrt{LC}} \implies L = \frac{1}{4\pi^2 f^2 C}.$$

Putting in the numerical values we get $L = 0.084$ H. Now, the requirement of Q gives

$$Q = \frac{1}{R}\sqrt{\frac{L}{C}} \implies R = \frac{1}{Q}\sqrt{\frac{L}{C}}.$$

Putting in the numbers we get $R = 1.06\ \Omega$.

EXERCISES

(2.1) A harmonic oscillator of mass 200 g is attached to a spring with spring constant 100 N/m. Find the angular frequency, frequency, and time period.

(2.2) Two oscillators of mass 200 g and 400 g oscillate with the same frequency. They move such that the displacement from their equilibrium positions are given by the following two functions x_1 and x_2 respectively:

$$x_1 = 2\ \cos(t)$$
$$x_2 = 2\ \cos(t + \pi)$$

where t is in seconds and x_1 and x_2 in cm. Note the arguments of cosines are in radians.

(a) What are the angular frequencies, amplitudes, and phase constants of the two oscillators?

(b) What are their positions and velocities at the initial time $t = 0$?

(c) Plot the positions of the two oscillators versus time on the same graph, and interpret which oscillator is ahead, and by how much.

(d) What are the kinetic and potential energies of the two oscillators at $t = 0$?

(2.3) An oscillator of frequency 20 Hz starts 3 cm from the equilibrium with an initial velocity of 10 cm/s pointed towards the equilibrium point. Find $x(t)$.

(2.4) Find C_1 and C_2 in terms of A and ϕ when $x(t) = C_1\cos(t) + C_2\sin(t)$ is used instead of $x(t) = A\cos(t - \phi)$.

(2.5) Find C_1 and C_2 in terms of x_0, v_0, and ω when $x(t) = C_1\cos(\omega t) + C_2\sin(\omega t)$, where $x_0 = x(0)$ and $v_0 = v(0)$.

(2.6) Consider a simple harmonic oscillator of mass m, amplitude A, and frequency f.

(a) What fraction of its energy is in the potential energy when the oscillator's displacement is half the amplitude? Give an expression for the potential energy in terms of the parameters given above when the oscillator is at this point.

(b) What fraction of its energy is in the kinetic energy when the oscillator's displacement is half the amplitude? Give an expression for the kinetic energy in terms of the parameters given above when the oscillator is at this point.
(c) What is the lowest speed of the oscillator and where in the oscillation does the oscillator have the lowest speed? Why?
(d) What is the highest speed of the oscillator and where in the oscillation does the oscillator have the highest speed? Why?

(2.7) A disk of mass m and radius R is suspended from a thin string of torsion constant κ. The string passes through the center of the disk and is perpendicular to it. When the disk is rotated by a small angle about the equilibrium, there is a restoring torque due to the twist in the string, which tends to bring the disk back to the equilibrium. Set up an equation for the rotational motion of the disk, and find the frequency of small oscillations about the equilibrium.

(2.8) A block of mass 5 kg is hung from a spring whose original length is 50 cm and spring constant 1.0 N/cm. As a result of the weight of the block, the spring now has a different length at equilibrium. The spring is attached to a ceiling which is at a height 2 m from the ground. The block is then hit with a hammer giving it an instantaneous velocity of 40 cm/s pointed downward. You may assume that, at this instant, the block is more or less still at the equilibrium position although it is now moving.
(a) Where is the equilibrium position of the block with respect to the floor?
(b) What is the potential energy of the block at the time it was hit by the hammer?
(c) What is the kinetic energy of the block after being hit by the hammer?
(d) How much work did the hammer do?
(e) What is the frequency of the oscillations of the block?
(f) What is the phase constant of the motion of the block?
(g) Write a function $y(t)$ for the y-coordinate of the block, stating clearly your choice for the origin.

(2.9) Two springs with spring constants k_1 and k_2 are attached on the two opposite sides of a block of mass m and the free ends of the springs are attached to two fixed supports such that the springs are taut and stretched. The block rests on a frictionless, flat horizontal surface and can move along the line of the two springs. When the block is pulled a little along the length of the springs from the equilibrium position and let go, as shown in Fig. 2.20, it executes a simple harmonic motion whose frequency depends on the spring constants of the two springs and the mass of the block.
(a) By looking at forces on the block at an arbitrary point in time, find the equation of motion of the block.
(b) From the equation of motion, show that the angular frequency of oscillation is given by $\omega = \sqrt{\omega_1^2 + \omega_2^2}$, where $\omega_1^2 = k_1/m$ and $\omega_2^2 = k_2/m$.
(c) Show that the system of two springs connected in parallel to an object has an effective spring constant equal to $k_{\text{eff}} = k_1 + k_2$.
(d) Deduce the formula for the frequency by examining the expression for the energy of the system.

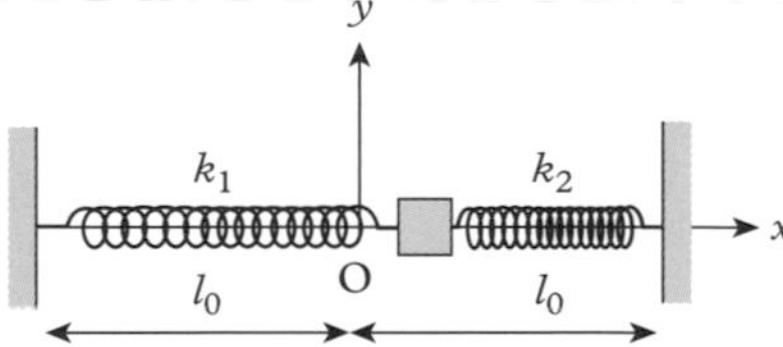

Fig. 2.20 *Exercise 2.9.*

(2.10) A block of mass m is attached to two springs with spring constants k_1 and k_2 that are glued together and the free end is fixed to a rigid support, as shown in Fig. 2.21. The block can move freely on a frictionless flat horizontal surface.
(a) Let l_1 and l_2 be the equilibrium lengths of the springs. Let x_1 and x_2 be the x-coordinates of the glue and the block. By examining the forces on the block deduce the equation of motion of the block.

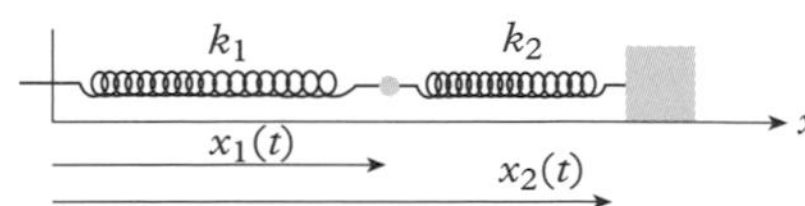

Fig. 2.21 *Exercise 2.10.*

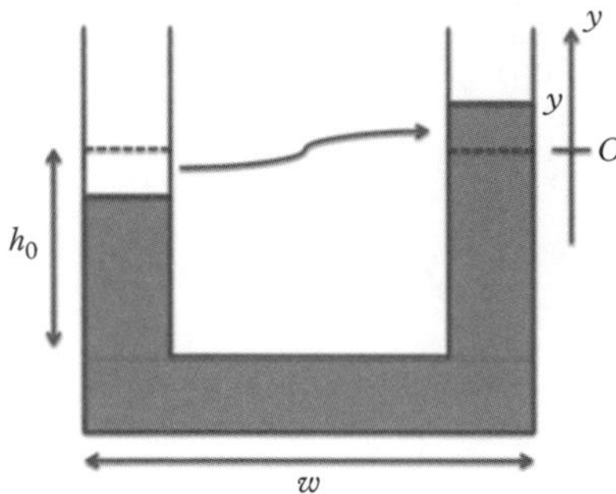

Fig. 2.22 *Exercise 2.11.*

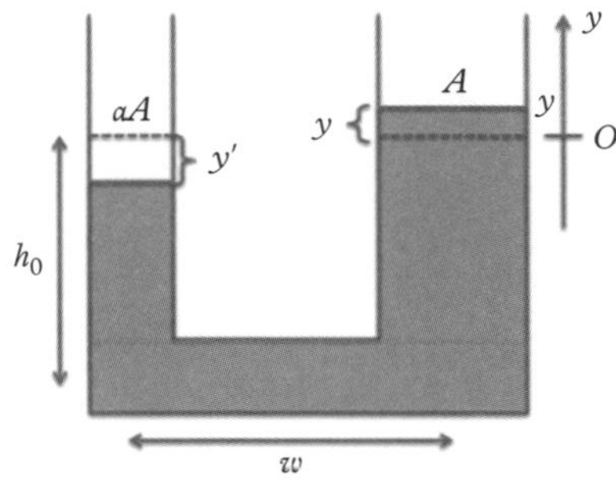

Fig. 2.23 *Exercise 2.12.*

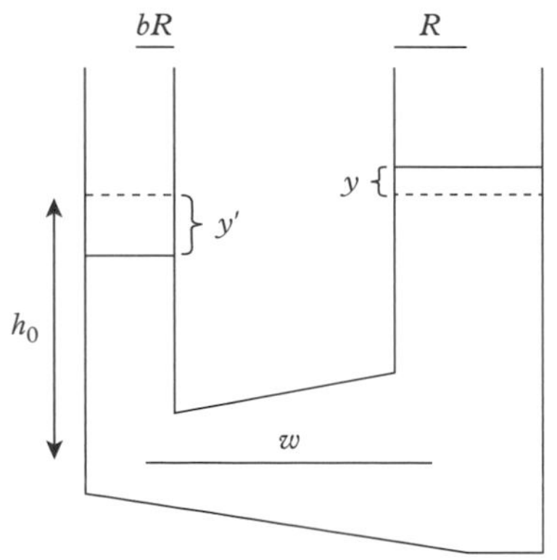

Fig. 2.24 *Exercise 2.12.*

(b) Write the equation of motion of the glue at the junction of the two springs and simplify the equation by setting the mass of the glue to zero.

(c) Combine the two equations to show that the frequency of oscillation of the block is given by $\omega = \frac{\omega_1\omega_2}{\sqrt{\omega_1^2+\omega_2^2}}$, where $\omega_1^2 = k_1/m$ and $\omega_2^2 = k_2/m$.

(d) Show that the system of two springs connected in series to an object has an effective spring constant that obeys $\frac{1}{k_{\text{eff}}} = \frac{1}{k_1} + \frac{1}{k_2}$.

(e) Deduce the formula for the frequency by examining the expression for the energy of the system.

(2.11) A U-tube of uniform cross-section of area A is filled by a liquid of density ρ up to a height h_0 on both sides (see Fig. 2.22). Let w be the distance between the necks. The liquid on one side is pushed in by a small height Δh and let go. After that the liquid executes a simple harmonic motion about the equilibrium level h_0. For this problem ignore the effects of resistance and viscosity.

(a) Find an expression for the potential energy of water at an arbitrary time when the displacement from the equilibrium height is $y < h_0$. Express your answer in terms of y and use the equilibrium as the zero reference for the potential energy.

(b) Find an expression of the kinetic energy at an arbitrary time when the displacement from the equilibrium height is $y < h_0$. Note that all liquid in the U-tube will be moving. That is, water in a total length $2h_0 + w$ of the tube will be moving at the same speed. Express your answer in terms of dy/dt, the velocity of the top at one side of the U-tube.

(c) Find the frequency of oscillation.

(2.12) Suppose the areas of cross-section of the two arms in problem (2.11) are different. Let the right side be area A and the left side area αA with $\alpha > 0$, as shown in Fig. 2.23. Let the area of cross-section of the joining arm be αA and length w. The U-tube is then filled to the same height h_0 as above. The fluid is then set in oscillation. What will be the frequency of oscillations now? Note: The incompressibility of the fluid means that the speed of flow in the two arms of different area will be inversely proportional to the area.

(2.13) Investigate problem (2.12) in the case of a U-tube with different cross-sections in the two arms and the radius of the joining tube varying from one end to the other, as shown in Fig. 2.24. For simplicity, let the two arms be circular in cross-section with radii R and bR. Suppose the radius at different points on the joining arm of length w vary from R at one end to bR at the other end linearly with distance, i.e. the radius r of the joining tube at a distance x from the end with radius bR will be a function of x given by

$$r(x) = bR + \frac{(1-b)R}{w}\,x,$$

where x is the horizontal coordinate from left.

If the tubes on the two arms are filled to the same height h_0 and set oscillating, prove that the angular frequency of oscillations is

$$\omega = \sqrt{\frac{(1+b^2)g}{bw + (1+b^2)h_0}}.$$

Note: the velocity of flow will change in the joining arm, and therefore, the kinetic energy will now involve an integration.

(2.14) (Adapted from A.P. French, *Vibrations and Waves*, MIT, 1971.) **Sloshing mode in a tank.** You may have noticed that it is hard to carry a pan of water without spilling. What happens is that, as you carry the pan, you inadvertently shake the water and excite modes of oscillation in the water. The lowest mode of oscillation is called the sloshing mode. In this mode water just wobbles back and forth. In this problem you will be able to find a formula for the frequency of this mode by a simple analysis.

Suppose you have water in a rectangular tank of base $L \times b$ and the height of water in the tank is H, as shown in Fig. 2.25. To study sloshing in the xy-plane we will ignore the z-axis which is along the side of dimension b. Let (X, Y) denote the CM at an arbitrary time t when the height on the right side wall is y_w above the equilibrium level, as shown in the figure. Assume the potential energy to be zero when the water surface is flat. Note: the actual shape of the water surface will be a sine or cosine function, but here, for simplicity, we assume it to be flat.

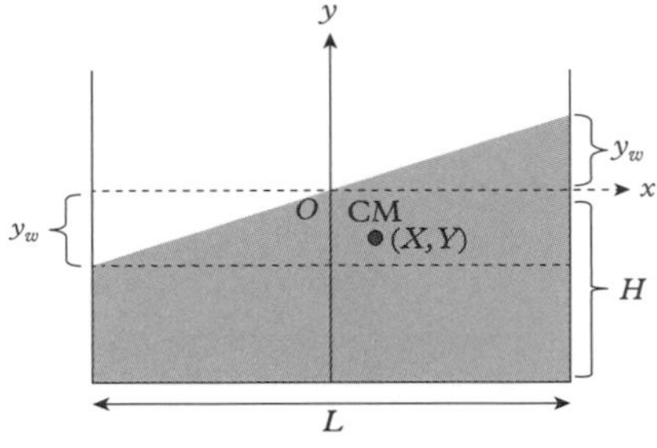

Fig. 2.25 *Exercise 2.14.*

(a) Find the x- and y-coordinates of the CM in terms of the height y_w and show that the CM moves in a parabola.

(b) Show that the potential energy is given by $U = \frac{1}{6}\rho bLgy_w^2$.

(c) You can find the horizontal velocity of the fluid at any point in the tank by finding how the height of the fluid changes within a small slice of the tank at a particular x during the interval t and $t + \Delta t$. Consider a slice between $x = x$ and $x = x + \Delta x$, as in Fig. 2.26.

Let the x-component of the velocity at x be v and pointed into the slice, and let the velocity at $x + \Delta x$ be $v + dv$ and pointed away from the slice. The height $H + y$ of the slice will change by Δy during an interval Δt due to more liquid flowing into the slice than leaving the slice. Show the following relation:

$$\frac{\Delta v}{\Delta x} = -\frac{1}{H}\frac{\Delta y}{\Delta t}.$$

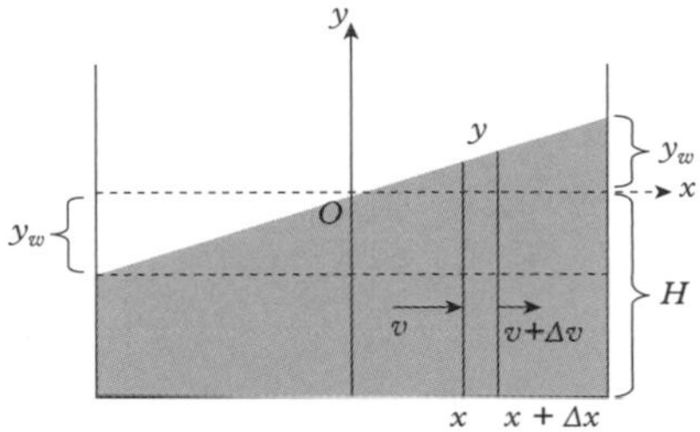

Fig. 2.26 *Exercise 2.14.*

In the infinitesimal limit this equation will become the following differential equation:

$$\frac{dv}{dx} = -\frac{1}{H}\frac{dy}{dt}.$$

(d) Let dy/dt at the right wall, i.e. when $x = L/2$, be denoted by dy_w/dt. Find dy/dt in the equation in terms of x of the slice and speed of the movement at the edge dy_w/dt.

(e) Show that

$$v(x) = v(0) - \frac{x^2}{HL}\frac{dy_w}{dt}.$$

With $v(L/2) = 0$ we get

$$v(0) = \frac{L}{4H}\frac{dy_w}{dt}.$$

(f) Write an expression for the kinetic energy of the fluid between x and $x + \Delta x$.

(g) Show that the kinetic energy of the fluid at an arbitrary instant will be

$$K = \frac{1}{60}\frac{\rho bL^3}{H}\left(\frac{dy_w}{dt}\right)^2.$$

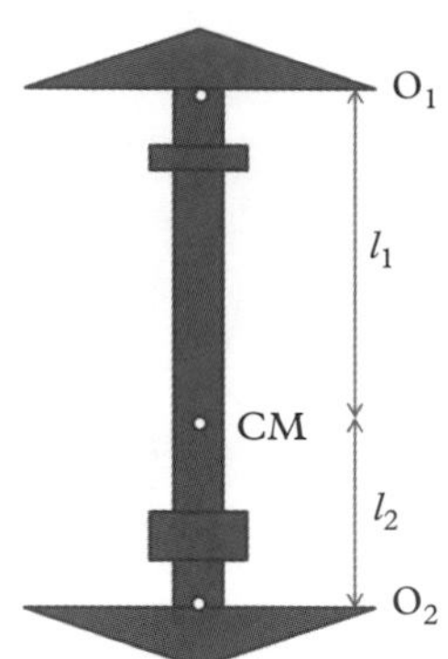

Fig. 2.27 *Exercise 2.15.*

(h) Based on the expression of the total energy, deduce the frequency of oscillations.

Note: An exact calculation will involve the water wave on the surface of the tank, which gives a frequency of $f = \sqrt{gH}/2L$.

(2.15) The **Kater's pendulum** consists of a long metallic rod R with a slidable weight on the rod, as shown in Fig. 2.27. The pendulum has two knife edges at two ends that are used to suspend the pendulum from a support, with a hole over which the knife edge rests. By adjusting the pivot points and the mass distribution over the rod, it is possible to obtain a configuration such that the time periods of oscillations about the two suspension points O_1 and O_2 are equal. Let k be the radius of gyration of the pendulum and l_1 and l_2 be the distances from the center of gravity to the suspensions O_1 and O_1 respectively. Let T_1 and T_2 be the time periods of oscillations when Kater's pendulum is hung from knife edges O_1 and O_2 respectively.

(a) Find the time periods T_1 and T_2 in terms of l_1, k, g, and l_2.

(b) Prove that when the time periods are equal, the period is given by $T = 2\pi\sqrt{\frac{L}{g}}$, where $L = l_1 + l_2$, the distance between the two knife edges.

(2.16) The potential energy of a particle of mass m moving along the x-axis is given as $U(x) = ax(x-1)^2$, where a is a constant.

(a) Find the location of the minimum of the potential energy function.

(b) Find the angular frequency of oscillation about the minimum of the potential.

(d) Are there any restrictions on the displacement for the particle to oscillate about the minimum? Explain.

(2.17) A block of mass m is attached to a spring with spring constant k. When the block moves there is a drag force on the block whose magnitude is equal to bv, where b is a constant and v is the velocity of the block. Decide what type of damping is in the following systems:

(a) $m = 0.2$ kg, $k = 10$ N/m, $b = 6$ kg/s,

(b) $m = 0.2$ kg, $k = 20$ N/m, $b = 8$ kg/s,

(c) $m = 0.8$ kg, $k = 32$ N/m, $b = 8$ kg/s,

(d) $m = 0.8$ kg, $k = 80$ N/m, $b = 16$ kg/s.

(2.18) If $x(0) = 1$, and $v_x(0) = 0$ for the systems in problem (2.17), find $x(t)$ for each block.

(2.19) Find the Q factor for the following underdamped oscillators. Here x is in cm and t in sec:

(a) $x(t) = 2\exp(-0.1t)\cos(2\pi t)$,

(b) $x(t) = 2\exp(-0.02t)\cos(t)$,

(c) $x(t) = 2\exp(-0.25t)\cos(200\pi t)$,

(d) $x(t) = 2\exp(-0.015t)\cos\left(2000\pi t + \frac{\pi}{2}\right)$.

(2.20) The position of a 250 g lightly damped oscillator is given by the following function of time, $x(t) = 2\exp(-0.1t)\cos(2\pi t)$, where t is in seconds and x in meters.

(a) Plot x versus t.

(b) How long does it take for the envelope of oscillations to drop by $\frac{1}{e}$?

(c) How long does it take for the envelope to drop by a factor $\frac{1}{e^2}$?

(d) What is the Q factor of the oscillator?

(e) What is the rate in J/s at which the energy of the oscillator is dissipated at $t = 0$?

(f) What is the frequency of oscillations?

(g) How many oscillations will the oscillator make before 90% of the energy is dissipated?

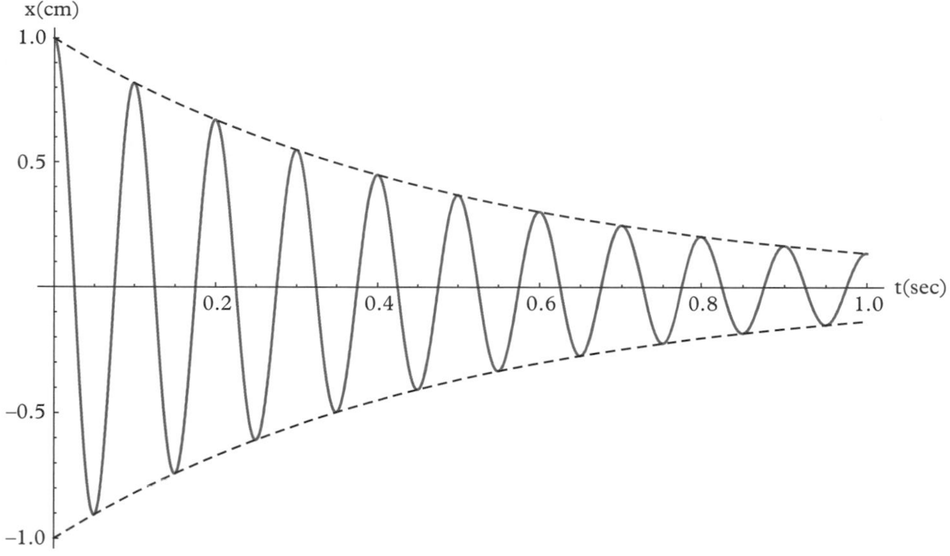

Fig. 2.28 *Exercise 2.22.*

(2.21) The displacements of two oscillators are given by $x_1 = 2\cos(2\pi t)$ and $x_2 = 2e^{-0.5t}\cos(2\pi t)$.
   (a) Plot x_1 on the x-axis and dx_1/dt on the y-axis for $t = [0, 5]$.
   (b) Plot x_2 on the x-axis and dx_2/dt on the y-axis for $t = [0, 5]$.
   (c) Compare the plots.

(2.22) Consider the graph shown in Fig. 2.28 of the displacement x of an oscillator subject to a viscous damping force, where the dashed line is the exponentially decaying envelope. You will deduce various properties of the oscillator from this "simulation" data.
   (a) What is the approximate value of the period of oscillations?
   (b) What is the approximate value of Γ of the oscillator?
   (c) What is the Q of the oscillator? State any approximations you may have made to get the numerical value of Q.
   (d) How much time would elapse for the energy to decay to a value that is $\frac{1}{e}$ of the initial value? State why is this value not equal to the time it takes for the amplitude of the oscillations of x to decay to $\frac{1}{e}$ of the initial value?

(2.23) A rectangular block of mass m is suspended from a support by a spring with spring constant k. The block with area of cross-section A at the bottom is then partially submerged in a fluid of density ρ, as shown in Fig. 2.29. The block is in equilibrium when a height h of the block is under the fluid. When the block is pushed into the fluid further by a small amount and released, it is found to oscillate.
   (a) Assuming no loss of energy in the drag or friction, find the period of oscillation.
   (b) If the drag force on the block is given by $-bv$, where v is the vertical component of the velocity and b is a constant, what would be the frequency of oscillations in the case of under-damped oscillations?

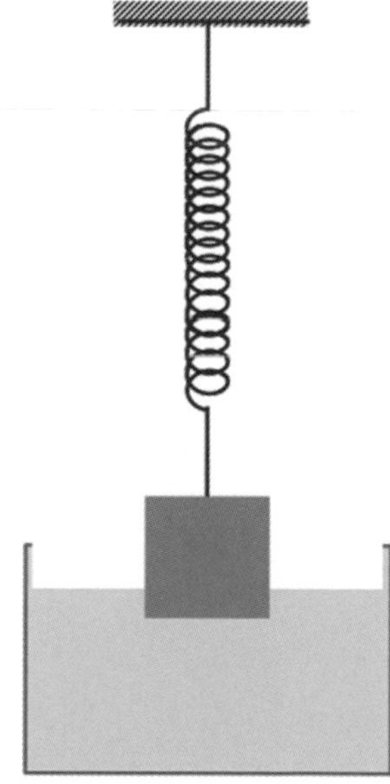

Fig. 2.29 *Exercise 2.23.*

3 Coupled Oscillations—Two Degrees of Freedom

Chapter Goals

In this chapter we will start our study of the motion of coupled systems with simple systems in which only two bodies are coupled. You will learn how to find normal modes and how to use the normal modes to study arbitrary motion.

In the last chapter we laid the foundation for studying oscillations of a body with a single degree of freedom. We found that there is a natural frequency at which the body oscillates. This fundamental observation that there are natural frequencies, also called normal mode frequencies, at which a body oscillates is also true for more realistic systems that have many degrees of freedom.

For instance, when you pluck a guitar string just right, the string can be made to vibrate at a particular note or frequency. They are normal modes of the plucked string. The configuration of the string for some of the lower frequency modes are illustrated in Fig. 3.1. An arbitrary vibration of a guitar string is more complicated than these normal modes but we will find that they can be understood as a superposition of these normal modes.

Fig. 3.1 *Some normal modes of guitar strings. In each normal mode every particle of the string vibrates up and down with the same frequency.*

The analysis of oscillations in a coupled system of two or more bodies relies on the method of normal modes. In this chapter we will use simple examples in which we can explore the discovery and usage of normal modes for understanding the complete dynamics of such systems. A good understanding of normal modes in two-variable systems will help us generalize the analysis to more complicated systems, which we will take up in the next chapter.

3.1 Linear Systems and Normal Modes

Many important physical systems involve two or more variables whose dynamics are coupled in the sense that the dynamics of one variable affects the dynamics of the others. Figure 3.2 illustrates some coupled systems in which two oscillators, each with one dynamical variable, are coupled.

The equations of motion of the dynamical variables in all three systems shown in the figure have the following linear form:

$$\frac{d^2x_A}{dt^2} = a\,x_A + b\,x_B, \tag{3.1}$$

$$\frac{d^2x_B}{dt^2} = c\,x_A + d\,x_B, \tag{3.2}$$

A First Course in Vibrations and Waves. First Edition. Mohammad Samiullah.

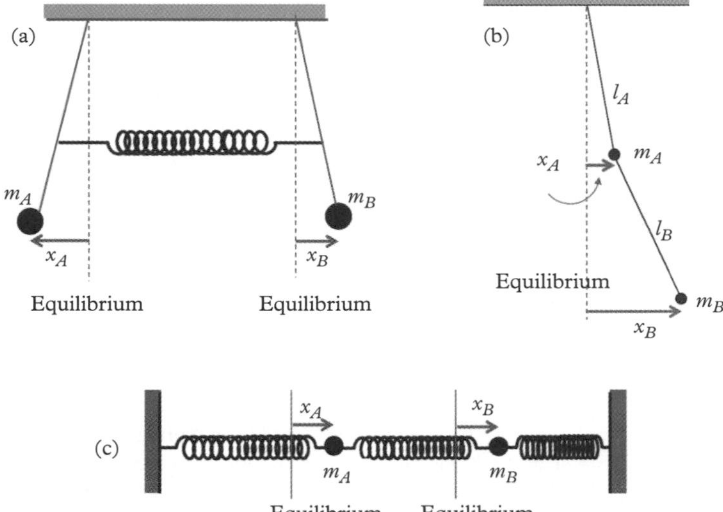

Fig. 3.2 *Examples of coupled systems with two degrees of freedom. (a) Two pendulums coupled by a spring, (b) two pendulums coupled by a suspension point of one pendulum controlled by the motion of the other pendulum, (c) two mass/spring systems coupled by a spring between them.*

for some constants a, b, c, and d. Here b and c arise due to the coupling of x_A and x_B. Note that $a < 0$ and $d < 0$ so that in the absence of coupling the two system are independently oscillatory. These systems are called **linear systems** since every term in their dynamical equations has only one power of a dynamical variable.

The solution of the equations of motion of linear systems can be written as a superposition of special solutions, called **normal modes** or simply **modes**. A system has as many normal modes as there are degrees of freedom in the system. For each system displayed in Fig. 3.2 there are two normal modes. It is often possible to guess the normal modes by utilizing fundamental characteristics of normal modes which I will illustrate in the following section for the coupled pendulums system in Fig. 3.2(a). But for now, I will list the special features of normal modes.

(1) In a normal mode all oscillators oscillate with the same frequency, which is one of the **normal mode frequencies** of the system.

(2) In a normal mode the ratios of the amplitudes of different oscillators remain constant with time.

(3) In a normal mode the motions of different oscillators have either the same phase or the opposite phase. Suppose you have two normal modes in the system with amplitudes x_A and x_B. Then, in one of the normal modes both x_A and x_B will be positive or both negative, and in the other normal mode one will be positive and the other negative.

(4) There are as many normal modes as there are degrees of freedom in the system.

3.2 Two Coupled Pendulums

As an example of coupled systems, consider identical pendulums of mass m and length l coupled by a spring with spring constant k, as in Fig. 3.3. The displacement of a pendulum

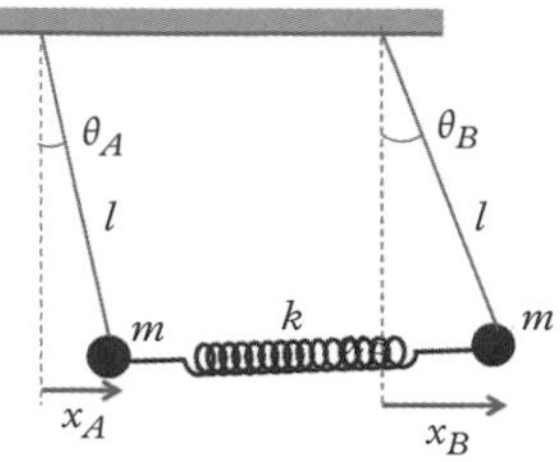

Fig. 3.3 *Small oscillations of coupled pendulums. Separately, the two pendulums oscillate at their natural frequencies. When the two pendulums are coupled by attaching the coupling spring between them, the "natural frequencies" would now be the normal mode frequencies which we will work out in this section. In small angle approximation, we will ignore the vertical displacements of the pendulums.*

is usually studied by the angle that the string makes with the vertical, but here, we will use the x-coordinate of the bobs instead since this is more directly related to the change in the length of the coupling spring. You may wish to check that, for small angles, the x-coordinate and the angle θ of a pendulum are related by the arclength-angle formula

$$x = l\theta.$$

We will denote the displacements of the two bobs by x_A and x_B. In this section we will first work out the normal modes and then use them to obtain the general solution in detail so that we will have a good understanding of the various concepts involved and how to work out the calculations.

Recall from the last chapter that when a pendulum is not coupled it will oscillate at the free pendulum frequency given by

$$\omega_0 = \sqrt{g/l}. \tag{3.3}$$

In the system of Fig. 3.3, the two pendulums are coupled by the spring. When the spring is not there, each will oscillate at this natural frequency ω_0. With the coupling spring between the pendulums we expect that the motion of one bob will affect the motion of the other. For brevity in writing formulas and equations below, we will introduce another frequency-like parameter associated with the coupling,

$$\omega_c = \sqrt{k/m}. \tag{3.4}$$

Note the notation for frequencies here—we are not using ω_0 for $\sqrt{k/m}$, but for $\sqrt{g/l}$, since the latter is the natural frequency of the uncoupled system. We will use this general guide for notation when deciding which frequency we denote by ω_0.

It turns out that, if you start the pendulums in special ways, which will be described more fully below, then the two pendulums oscillate at the same frequency. There are two special frequencies in this system. In one of the two modes, the two bobs move together in the same direction, and in the other, the two move in opposite directions. Now, it is possible to look for these *special motions* using a brute force systematic procedure starting from the equations of motion of the two pendulums, but it is also instructive to see if we can guess these modes based on intuition and symmetry. We will do the systematic procedure after we first have some practice with the guessing.

3.2.1 Guessing the Normal Modes

Sometimes it is possible to guess the normal modes if there are enough symmetries in the system. Based on our guess, we can work out not only the motion of each of the bobs in the modes but also their corresponding frequencies.

Lower Frequency Mode

Suppose you pull the two masses by an equal amount in the same direction and then let go of them from rest, as shown in Fig. 3.4. Since you don't stretch or compress the spring, the two pendulums should move independently thereafter. That is, we should see the two masses move by the same amount and in phase with each other, as illustrated in the figure, with the connecting spring remaining undisturbed.

The frequency of the mode is just the frequency of either of the two pendulums without the spring attachment. We will label the modes by the mode number and denote the mode frequencies by a subscript, e.g. ω_1 for the (angular) frequency for mode 1, and ω_2 for

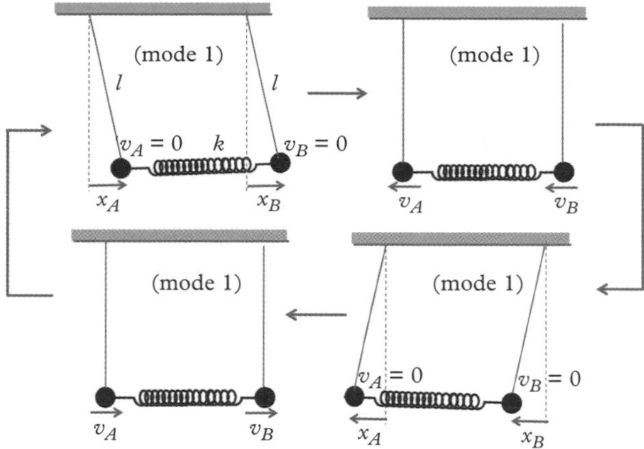

Fig. 3.4 *The lower-frequency mode of two coupled pendulums with equal mass and length. When the pendulum bobs are displaced to the same amount in the same direction and let go from rest, the two bobs oscillate with the same frequency and phase.*

mode 2. For the mode just described, we have the frequency equal to the natural frequency when not coupled,

$$\text{Frequency of mode 1: } \omega_1 = \omega_0 = \sqrt{g/l}.$$

Additionally, we will denote the displacements of the two bobs when they are moving in a particular mode by attaching the mode number to their symbols, e.g. x_{A1} and x_{B1} for the displacements of bobs A and B, respectively, when they are in mode 1. Let the initial displacements of the two masses be C, then the initial conditions for Fig. 3.4 would be

$$x_A(0) = C, \;\; x_B(0) = C, \;\; \left(\frac{dx_A}{dt}\right)_{t=0} = 0, \;\; \left(\frac{dx_B}{dt}\right)_{t=0} = 0.$$

For this initial condition, the displacements of the two bobs would all be cosines,

$$x_{A1}(t) = C\cos(\omega_1 t), \;\; x_{B1}(t) = C\cos(\omega_1 t). \tag{3.5}$$

Can you figure out why sines are missing? [Hint: check the initial velocity.] These expressions for the displacements give the correct phases of the two pendulums: they move in sync with each other and oscillate with the same frequency.

Upper Frequency Mode

The second normal mode of the coupled pendulum system can be excited if you pull the two masses to equal distance from the equilibrium but in the opposite direction and then let go from rest, as shown in Fig. 3.5. In this mode the masses oscillate with the same frequency but in opposite directions with respect to each other.

We can figure out the frequency of this mode by examining the equation of motion of either of the two masses, since they both oscillate at the same frequency. For our work we will choose bob B, the mass to the right. You are encouraged to carry out a similar calculation on bob A. The forces on B are the tension, weight, and spring force, as shown in Fig. 3.6.

The spring force will be proportional to the change in the length of the spring, which is given by $|x_B - x_A|$. In this mode we have $x_B = -x_A$.

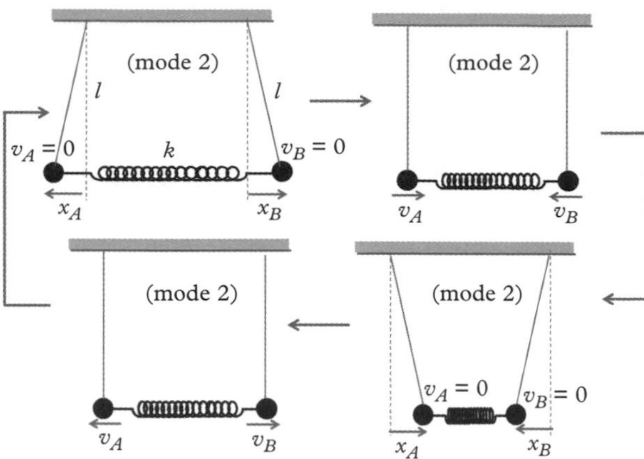

Fig. 3.5 *The upper frequency mode of two coupled pendulums with equal mass and equal length. When the pendulum bobs are displaced to the same amount in the opposite direction and let go from rest, the two bobs oscillate with the same frequency and opposite phase.*

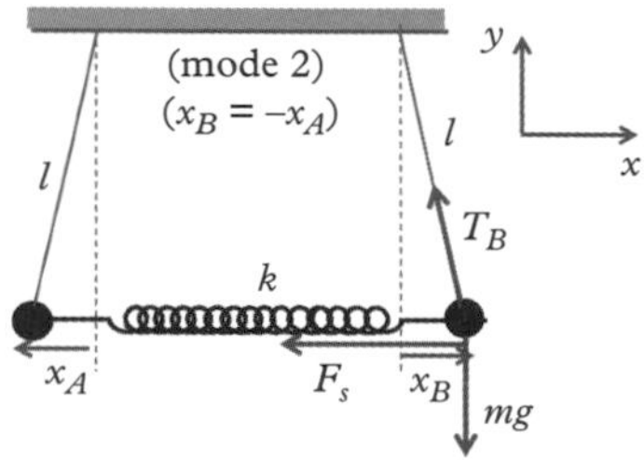

Fig. 3.6 *The forces on the mass on the right are the tension in the string, the weight of the bob, and the spring force. At the instant shown in the figure, the spring is stretched by $2x_B$, therefore the magnitude of the spring force is $2kx_B$.*

That means that the change in the length of the spring at any instant is $2|x_B|$. Therefore, the magnitude of the spring force would be $2k|x_B|$. Now, putting in the sign so that when the spring is stretched, the bob B is pulled towards the negative x-axis, we will get the following for the x-component of the spring force on this bob:

$$F_{sx} = -2kx_B. \tag{3.6}$$

In the small angle approximation the vertical component of the tension is approximately equal to the magnitude of the tension. In this approximation, we also ignore the vertical component of the acceleration. Thus, in the small angle approximation, we get the magnitude of tension equal to the weight of the bob,

$$T_B \approx mg.$$

Now, the x-component of the tension equals

$$T_{Bx} = -T_B\,\frac{x_B}{l} = -\frac{mg}{l}\,x_B. \tag{3.7}$$

Using Eqs. 3.6 and 3.7 in $\vec{F} = m\vec{a}$ for this bob, we get the x-component of equation of motion to be

$$m\frac{d^2x_B}{dt^2} = -2kx_B - \frac{mg}{l}\,x_B.$$

We now divide both sides by m and write the constants in terms of ω_0 and ω_c, previously defined. Then, the equation of motion can be written as

$$\frac{d^2x_B}{dt^2} = -\left(\omega_0^2 + 2\,\omega_c^2\right)\,x_B.$$

This equation is the same as the equation of a harmonic oscillator with frequency $\sqrt{\omega_0^2 + 2\,\omega_c^2}$. Therefore, we obtain the frequency of the second mode, ω_2,

Frequency of mode 2: $\omega_2 = \sqrt{\omega_0^2 + 2\,\omega_c^2}$, which we can write down in terms of the original physical parameters l and m, as

$$\text{Frequency of mode 2:} \quad \omega_2 = \sqrt{\frac{g}{l} + 2\,\frac{k}{m}}.$$

The displacements of the two bobs when moving in this mode, x_{A2}, and x_{B2} that conform to the condition $x_A(0) = -D$, $x_B(0) = +D$, and velocities zero at $t = 0$ can also be obtained readily with the following result:

$$x_{A2}(t) = -D\cos(\omega_2 t), \quad x_{B2}(t) = +D\cos(\omega_2 t). \tag{3.8}$$

Here, we gave a different initial amplitude for mode 2 than for mode 1 since they are independent experiments. With these x_A and x_B the two pendulums move in opposite directions and oscillate at frequency ω_2. The following is a summary of the two normal modes we found for the coupled pendulums with equal mass and the same length.

$$\text{Mode 1: Frequency, } \omega_1 = \omega_0, \tag{3.9}$$

$$\text{Displacements, } x_{A1} = C\cos(\omega_1 t), \quad x_{B1} = C\cos(\omega_1 t), \tag{3.10}$$

$$\text{Initial conditions: } x_{A1}(0) = C = x_{B1}(0), \quad v_{A1}(0) = 0 = v_{B1}(0), \tag{3.11}$$

$$\text{Mode 2: Frequency, } \omega_2 = \sqrt{\omega_0^2 + 2\omega_c^2}, \tag{3.12}$$

$$\text{Displacements, } x_{A2} = D\cos(\omega_2 t), \quad x_{B2} = -D\cos(\omega_2 t), \tag{3.13}$$

$$\text{Initial conditions: } x_{A2}(0) = -D, \; x_{B2}(0) = D, \; v_{A2}(0) = 0 = v_{B2}(0). \tag{3.14}$$

3.2.2 General Motion Using Normal Coordinates

The equations of motion of the two pendulums can be readily obtained by examining the forces on the bobs. First, note that the difference $|x_B - x_A|$ gives the change in the length of the spring, therefore, the spring force will have magnitude $k|x_B - x_A|$. Making use of the directions of the spring force and the x-component of the tension in the strings, we obtain the following equations of motion:

$$m\frac{d^2x_A}{dt^2} = -m\omega_0^2 x_A - m\omega_c^2(x_A - x_B), \tag{3.15}$$

$$m\frac{d^2x_B}{dt^2} = -m\omega_0^2 x_B - m\omega_c^2(x_B - x_A), \tag{3.16}$$

where $\omega_0^2 = g/l$ and $\omega_c^2 = k/m$, as previously introduced. Dividing by m and rearranging terms we get

$$\frac{d^2x_A}{dt^2} = -(\omega_0^2 + \omega_c^2)x_A + \omega_c^2 x_B, \tag{3.17}$$

$$\frac{d^2x_B}{dt^2} = -(\omega_0^2 + \omega_c^2)x_B + \omega_c^2 x_A. \tag{3.18}$$

It is hard to solve these two interdependent equations without some special trick. The normal mode provides just such a trick. Let us construct two new variables $q_1(t)$ and $q_2(t)$ from weighing x_A and x_B by the ratio of the coefficients in the two normal modes,

$$q_1 = x_A + (\text{ratio of amplitudes of } x_B \text{ to } x_A \text{ in mode 1})\ x_B, \tag{3.19}$$

$$q_2 = x_A + (\text{ratio of amplitudes of } x_B \text{ to } x_A \text{ in mode 2})\ x_B. \tag{3.20}$$

In mode 1, the amplitudes are in a 1 : 1 ratio and in mode 2 the amplitudes are 1 : –1. Therefore, we introduce composite variables $q_1(t)$ and $q_2(t)$ here by

$$q_1 = x_A + x_B, \tag{3.21}$$

$$q_2 = x_A - x_B. \tag{3.22}$$

The variables q_1 and q_2 are called **normal coordinates**. Note that particular expression of q_1 and q_2 in terms of x_A and x_B will depend upon the details of the given problem since the coefficients come from the normal modes. Warning: You cannot use Eqs. 3.21 and 3.22 as a general definition of q_1 and q_2; for each system, you will have to work out the ratios of amplitudes in the modes first, and then use Eqs. 3.19 and 3.20 to obtain q_1 and q_2 for that system.

Now, we take two derivatives of q_1 and q_2 and replace the second derivatives of x_A and x_B using the equation of motion, Eqs. 3.17 and 3.18,

$$\begin{aligned}\frac{d^2q_1}{dt^2} &= \frac{d^2x_A}{dt^2} + \frac{d^2x_B}{dt^2}\\ &= -\left(\omega_0^2 + \omega_c^2\right) x_A + \omega_c^2 x_B - (\omega_0^2 + \omega_c^2)x_B + \omega_c^2 x_A\\ &= -\omega_0^2\,(x_A + x_B) = -\omega_0^2 q_1,\end{aligned} \tag{3.23}$$

$$\begin{aligned}\frac{d^2q_2}{dt^2} &= \frac{d^2x_A}{dt^2} - \frac{d^2x_B}{dt^2}\\ &= -\left(\omega_0^2 + \omega_c^2\right) x_A + \omega_c^2 x_B + (\omega_0^2 + \omega_c^2)x_B - \omega_c^2 x_A\\ &= -(\omega_0^2 + 2\omega_c^2)\,(x_A - x_B) = -(\omega_0^2 + 2\omega_c^2)q_2.\end{aligned} \tag{3.24}$$

This is amazing; how various terms have conspired and given us one equation for q_1 and another for q_2! Unlike the equations of motion for x_A and x_B, which are inter-dependent, the equations for q_1 and q_2 are independent. We say that while x_A and x_B were coupled, q_1 and q_2 are **decoupled**. The variable q_1 obeys a harmonic oscillator equation of motion with the frequency of mode 1 and the variable q_2 of mode 2,

$$\frac{d^2q_1}{dt^2} = -\omega_1^2 q_1, \tag{3.25}$$

$$\frac{d^2q_2}{dt^2} = -\omega_2^2 q_2. \tag{3.26}$$

Now, you can appreciate why the variables q_1 and q_2 are called normal coordinates: these variables oscillate at the normal mode frequencies. Since q_1 and q_2 are not coupled, a change in one does not affect the other. The general solutions of Eqs. 3.25 and 3.26 are easy to write down from our studies is the last chapter,

$$q_1(t) = C_1\cos(\omega_1 t) + S_1\sin(\omega_1 t),\ \text{or},\ A_1\cos(\omega_1 t - \phi_1), \tag{3.27}$$

$$q_2(t) = C_2\cos(\omega_2 t) + S_2\sin(\omega_2 t),\ \text{or},\ A_2\cos(\omega_2 t - \phi_2). \tag{3.28}$$

Here the coefficients C_1, S_1, C_2, and S_2 (or, alternatively A_1, ϕ_1, A_2, ϕ_2) are determined by the initial conditions on the oscillators. For instance,

$$\text{If velocities are zero at } t = 0: \quad S_1 = 0, \quad S_2 = 0.$$

Then, the solution will have only the cosine terms,

$$q_1(t) = C_1 \cos(\omega_1 t),$$

$$q_2(t) = C_2 \cos(\omega_2 t).$$

To appreciate the use of q_1 and q_2, note that you can set up any problem in x_A and x_B or q_1 and q_2. But the calculations using q_1 and q_2 will be far easier since we can work on them independently. Once we have found q_1 and q_2, we can obtain x_A and x_B by inverting Eqs. 3.21 and 3.22,

$$x_A = \frac{q_1 + q_2}{2}, \tag{3.29}$$

$$x_B = \frac{q_1 - q_2}{2}. \tag{3.30}$$

Using the general solutions of q_1 and q_2, the original coordinates x_A and have the following general solutions:

$$x_A = \frac{1}{2}\left[C_1 \cos(\omega_1 t) + S_1 \sin(\omega_1 t) + C_2 \cos(\omega_2 t) + S_2 \sin(\omega_2 t)\right], \tag{3.31}$$

$$x_B = \frac{1}{2}\left[C_1 \cos(\omega_1 t) + S_1 \sin(\omega_1 t) - C_2 \cos(\omega_2 t) - S_2 \sin(\omega_2 t)\right]. \tag{3.32}$$

Alternatively, the general solution could be written in terms of the amplitudes and phases of each mode as

$$x_A = A_1 \cos(\omega_1 t - \phi_1) + A_2 \cos(\omega_2 t - \phi_2), \tag{3.33}$$

$$x_B = A_1 \cos(\omega_1 t - \phi_1) - A_2 \cos(\omega_2 t - \phi_2). \tag{3.34}$$

In this form, the constants A_1, ϕ_1, A_2, ϕ_2 are determined from the initial conditions. The sign multiplying A_2 in the second equation has come from the mode amplitude ratios. In a general motion of degrees of freedom without any symmetry you would get the mode ratios B/A having more factors than just $+1$ or -1. The general solution will use these ratios and will be given by

$$x_A = A_1 \cos(\omega_1 t - \phi_1) + A_2 \cos(\omega_2 t - \phi_2), \tag{3.35}$$

$$x_B = \left[\frac{B_1}{A_1}\right]_{\text{mode 1}} A_1 \cos(\omega_1 t - \phi_1) + \left[\frac{B_2}{A_2}\right]_{\text{mode 2}} A_2 \cos(\omega_2 t - \phi_2). \tag{3.36}$$

Here B_1/A_1 and B_2/A_2 are already known from the normal modes and the only things that need fixing from initial conditions are the constants in the first equation.

Example 3.1 Motion Started With Zero Velocity

A two-coupled pendulum system has normal frequencies 2 rad/s and 3 rad/s. The pendulums are started with the following initial conditions:

$$x_A(0) = 1,\ \ x_B(0) = \frac{1}{2},\ \ v_A(0) = 0,\ \ v_B(0) = 0.$$

Find $x_A(t)$ and $x_B(t)$.

Solution

Since the initial conditions do not correspond to any of the normal modes, let us convert these conditions to conditions on normal coordinates. We get

$$q_1(0) = x_A(0) + x_B(0) = 1 + \frac{1}{2} = \frac{3}{2},\ \ q_2(0) = x_A(0) - x_B(0) = 1 - \frac{1}{2} = \frac{1}{2},$$
$$\left(\frac{dq_1}{dt}\right)_{t=0} = v_A(0) + v_B(0) = 0,\ \ \left(\frac{dq_2}{dt}\right)_{t=0} = v_A(0) - v_B(0) = 0.$$

The solution of the normal coordinates for zero velocities for these coordinates has only cosines with an argument corresponding to the normal mode frequencies,

$$q_1(t) = C_1 \cos(2t),\ \ q_2(t) = C_2 \cos(3t).$$

Now, using the initial conditions we obtain the following conditions on the coefficients:

$$C_1 = \frac{3}{2},\ \ C_2 = \frac{1}{2}.$$

Therefore, we have the normal coordinates with all the initial conditions satisfied,

$$q_1(t) = \frac{3}{2}\cos(2t),\ \ q_2(t) = \frac{1}{2}\cos(3t).$$

It is trivial to get back x_A and x_B from these normal coordinates,

$$x_A = \frac{q_1 + q_2}{2} = \frac{3}{4}\cos(2t) + \frac{1}{4}\cos(3t),$$
$$x_B = \frac{q_1 - q_2}{2} = \frac{3}{4}\cos(2t) - \frac{1}{4}\cos(3t).$$

Example 3.2 Motion Started With Non-zero Velocity

A two-coupled pendulum system has normal frequencies 2 rad/s and 3 rad/s. The pendulums are started with the following initial conditions:

$$x_A(0) = 1,\ \ x_B(0) = 0,\ \ v_A(0) = 0,\ \ v_B(0) = 2.$$

We will suppress units in our calculations. Find $x_A(t)$ and $x_B(t)$.

Solution

Just as in the last example, the given initial condition does not correspond to an initial condition that leads to motion in only one of the modes. Therefore, we will proceed by using normal coordinates. We start with obtaining the initial conditions for q_1 and q_2

based on the initial conditions on x_A and x_B given in the problem statement,

$$q_1(0) = x_A(0) + x_B(0) = 1, \quad q_2(0) = x_A(0) - x_B(0) = 1,$$

$$\left(\frac{dq_1}{dt}\right)_{t=0} = v_A(0) + v_B(0) = 2, \quad \left(\frac{dq_2}{dt}\right)_{t=0} = v_A(0) - v_B(0) = -2.$$

Now, both q_1 and q_2 have initial velocities. Therefore, we will have both cosine and sine in their solution,

$$q_1(t) = C_1\cos(2t) + S_1\sin(2t), \quad q_2(t) = C_2\cos(3t) + S_2\sin(3t).$$

Now, using the initial conditions on q_1 and q_2 we obtain the following conditions on the coefficients:

$$1 = C_1, \quad 1 = C_2, \quad 2 = 2S_1, \quad -2 = 3S_2.$$

Therefore, we have the normal coordinates with all the initial conditions satisfied,

$$q_1(t) = \cos(2t) + \sin(2t), \quad q_2(t) = \cos(3t) - \frac{2}{3}\sin(3t).$$

It is trivial to get back x_A and x_B from these normal coordinates,

$$x_A = \frac{q_1 + q_2}{2} = \frac{1}{2}\left[\cos(2t) + \sin(2t) + \cos(3t) - \frac{2}{3}\sin(3t)\right],$$

$$x_B = \frac{q_1 - q_2}{2} = \frac{1}{2}\left[\cos(2t) + \sin(2t) - \cos(3t) + \frac{2}{3}\sin(3t)\right].$$

3.3 Systematic Method for Normal Modes

The problem of coupled pendulums has illustrated the central role played by normal modes when we want to study the dynamics. The problem is stated in the physical variables such as x_A and x_B of the two coupled pendulums we have studied above. In a coupled system, this leads to inter-dependent equations which are difficult to solve directly. We have solved the problem of coupled oscillators indirectly by first casting the problem in terms of the normal coordinates and then extracting the information about the original coordinates from the solution in terms of normal coordinates. To implement the normal coordinate route we need information about normal modes of the system.

That means the crucial first step in solving coupled systems is to figure out the normal modes. We need both the frequencies of the normal modes and the relative amplitudes and signs for the displacements in each mode. In Examples 3.1 and 3.2 we guessed the normal modes using our intuition based on the symmetry in the systems. That is not always possible. Therefore, we seek a more systematic approach. I will illustrate the systematic approach by working out the normal modes and normal coordinates of a classic system called the double pendulum.

3.3.1 The Double Pendulum

A double pendulum has one pendulum hanging from another pendulum, as shown in Fig. 3.7. The figure also shows the free-body diagrams of the two bobs.

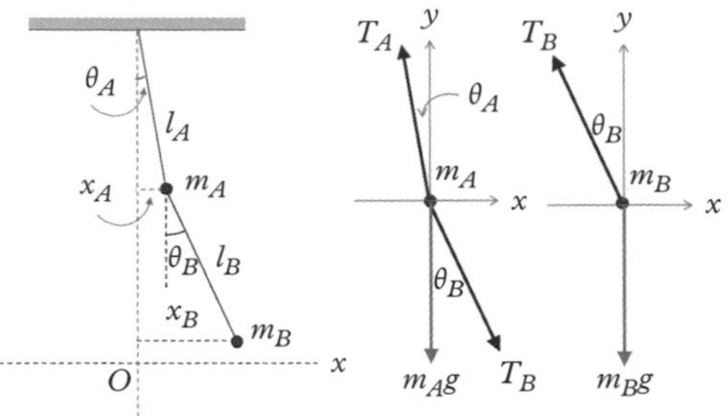

Fig. 3.7 *The double pendulum. The second pendulum is tied to the bob of the first pendulum. The free-body diagram of the forces is shown on the right side, where T_A and T_B are the tensions in the two strings. In the text, we use $m_A = m_B = m$, $l_A = l_B = l$, and the small-angle approximation, $\sin(\theta) \approx \theta$.*

Before we even do the problem, you should try to guess at least one normal mode of this problem. [Hint: in one mode they move together and in the other they move in opposite directions.] You can check whether your guess was right when I work out the normal modes below.

For simplicity, let the motion of the pendulum bobs be restricted to the same plane, their masses be equal, $m_A = m_B = m$, and the length of the pendulums also be equal, $l_A = l_B = l$. We assume small angles for the pendulums, and work out the x- and y-components of Newton's equations of motion of the two bobs. Let x_A and x_B be the x-coordinates of the two bobs, and y_A and y_B be their y-coordinates at some arbitrary time t.

The free-body diagrams of the forces on the two bobs, as given in Fig. 3.7, can be used to deduce their equations of motion. We find that the x- and y-components of the equation of motion for the bob A in the small-angle approximation are given as follows:

$$m\frac{d^2x_A}{dt^2} = -T_A\frac{x_A}{l} + T_B\frac{(x_B - x_A)}{l}, \tag{3.37}$$

$$m\frac{d^2y_A}{dt^2} = T_A - mg - T_B. \tag{3.38}$$

Similarly, x- and y-components of the equation of motion for the bob B in the small-angle approximation are found to be

$$m\frac{d^2x_B}{dt^2} = -T_B\frac{(x_B - x_A)}{l}, \tag{3.39}$$

$$m\frac{d^2y_B}{dt^2} = T_B - mg. \tag{3.40}$$

In the small-angle approximation, the y-displacements are considerably smaller than the x-displacements, and we can make the following assertions:

$$\frac{d^2y_A}{dt^2} \ll g, \tag{3.41}$$

$$\frac{d^2y_B}{dt^2} \ll g. \tag{3.42}$$

That is, we can neglect the y-acceleration of the masses compared to the acceleration due to gravity. With this assumption, we find that the tensions in the strings simplify:

$$T_A = 2mg, \tag{3.43}$$

$$T_B = mg. \tag{3.44}$$

Using these relations you can arrive at the following equations of motion in the small oscillation approximation:

$$\frac{d^2x_A}{dt^2} = -3\omega_0^2 x_A + \omega_0^2 x_B, \tag{3.45}$$

$$\frac{d^2x_B}{dt^2} = \omega_0^2 x_A - \omega_0^2 x_B, \tag{3.46}$$

where

$$\omega_0 = \sqrt{l/g}.$$

At this point in the solution we are not concerned with satisfying any initial conditions, but rather, we wish to determine only the frequencies of the normal modes and the amplitude ratios of the displacements in each mode. Once we have this information, we can build the most general solution in terms of the normal coordinates and satisfy any initial condition of the system that we like, as previously explained.

To start the process, we note the following fundamental characteristic of a normal mode:

> In a normal mode all particles oscillate at the same frequency.

Therefore, in a normal mode both x_A and x_B will oscillate with some (as of this point unknown) frequency ω. They may oscillate with different phase and different amplitude, but their frequency will be same. We then seek a solution of Eqs. 3.45 and 3.46 that has the following form:

$$x_A = A\cos(\omega t), \tag{3.47}$$

$$x_B = B\cos(\omega t). \tag{3.48}$$

A more general calculation that works, even for damped systems, starts with a complex exponential time part,

$$x_A = Ae^{i\omega t}, \tag{3.49}$$

$$x_B = Be^{i\omega t}. \tag{3.50}$$

Since we will be working here with undamped cases, we will use the cosine and do our calculations with real functions only. The question now is: are there A, B, ω for which Eqs. 3.45 and 3.46 can be satisfied? Let us plug these assumed forms of x_A and x_B into Eqs. 3.45 and 3.46 and see what happens. After canceling out the common factor of $\cos(\omega t)$ we find

$$-\omega^2 A = -3\omega_0^2 A + \omega_0^2 B, \tag{3.51}$$

$$-\omega^2 B = \omega_0^2 A - \omega_0^2 B. \tag{3.52}$$

Note that if A is zero, then B must be zero, and vice versa, which would be a trivial solution of these algebraic equations. We are not after this solution and will ignore this possibility.

Now, note that we have two equations in three unknowns, A, B, ω. Therefore, we cannot get all three of them. However, we really do not need all three to construct the normal coordinates. We just need ω and B/A for each mode, which we can get from these equations. Dividing by A and rearranging, we get two different expressions for B/A, one from each equation,

$$\frac{B}{A} = \frac{3\omega_0^2 - \omega^2}{\omega_0^2}, \tag{3.53}$$

$$\frac{B}{A} = \frac{\omega_0^2}{\omega_0^2 - \omega^2}. \tag{3.54}$$

For these two equations to be consistent, the right sides must be equal to each other. This gives a condition on the unknown ω,

$$\frac{3\omega_0^2 - \omega^2}{\omega_0^2} = \frac{\omega_0^2}{\omega_0^2 - \omega^2}.$$

This gives

$$\left(\omega^2/\omega_0^2\right)^2 - 4\left(\omega^2/\omega_0^2\right) + 2 = 0,$$

which has two solutions,

$$\omega_1^2 = \omega_0^2\left(2 - \sqrt{2}\right),$$

$$\omega_2^2 = \omega_0^2\left(2 + \sqrt{2}\right).$$

Taking square roots will again result in two solutions of each, but only the positive root is the physical solution since ω_1 and ω_2 are positive frequencies. Thus, we find that ω in the assumed solution of Eqs. 3.47 and 3.48 has two possibilities—these are the two normal mode frequencies,

$$\omega_1 = \omega_0\sqrt{2 - \sqrt{2}}, \tag{3.55}$$

$$\omega_2 = \omega_0\sqrt{2 + \sqrt{2}}. \tag{3.56}$$

How about B/A? Equations 3.53 and 3.54 tell us that B/A depends upon ω. Therefore, for each allowed ω, that is, for each normal mode frequency, we will get a particular B/A, which we will label with the mode number. Thus,

$$\text{When in mode 1, } \omega = \omega_1, \left(\frac{B}{A}\right)_1 = \frac{3\omega_0^2 - \omega_1^2}{\omega_0^2} = 1 + \sqrt{2}, \tag{3.57}$$

$$\text{When in mode 2, } \omega = \omega_2, \left(\frac{B}{A}\right)_2 = \frac{3\omega_0^2 - \omega_2^2}{\omega_0^2} = 1 - \sqrt{2}. \tag{3.58}$$

Clearly, one simple way to start mode 1 motion will be to pull the two pendulum bobs to the right, with x_B being $1 + \sqrt{2}$ times larger than x_A, and releasing them from rest. To start mode 2 motion we will pull them in opposite directions by different amounts: if x_A is one unit of distance to the right from vertical, then x_B will be $\sqrt{2} - 1$ unit to the left of the vertical, and then releasing them at rest. The modes of the double pendulum are shown in Fig. 3.8.

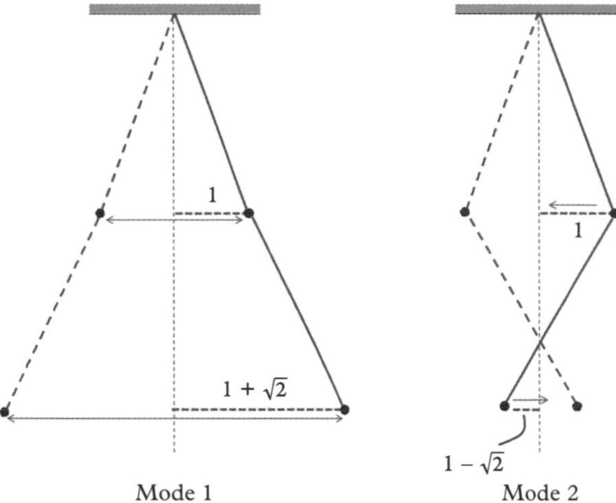

Fig. 3.8 *The modes of the double pendulum.*

3.3.2 Summary of Steps for Obtaining Normal Modes

The following steps summarize the general approach to solving coupled systems.

(1) Identify the dynamical variables. These are the variables that are appropriate for following the displacement of the system from equilibrium.

(2) Deduce the equations of motion for the dynamical variables.

(3) Search for the normal modes by assuming that every variable oscillates with the same frequency and same phase constant (modulo π radians), even though we do not know the specific values of the common frequency or the relative phases or amplitudes at this point of the calculation. We do this by assuming a form of the normal mode solution,

$$x_A(t) = A\cos(\omega t), \tag{3.59}$$

$$x_B(t) = B\cos(\omega t), \tag{3.60}$$

where the relative sign of the constants A and B could be different, i.e. A and B both positive or both negative, or, A positive and B negative, or vice versa.

For damped motions the damping term involves only one derivative of x with respect to t. This results in both cosine and sine in the equation and you cannot cancel out cosines from every term. Therefore, when dealing with damped systems, it is better to work with the complex representation in which cosine is replaced by a complex exponential,

$$x_A(t) = A\,e^{i(\omega t)},\quad x_B(t) = B\,e^{i(\omega t)}.$$

Since we will study the undamped motion here we will stay with the cosine representation.

(4) We input the assumed solution form Eqs. 3.59 and 3.60 into the equations of motion and cancel out the common factor $\cos(\omega t)$. This gives two equations in three unknowns, A, B, and ω.

(5) These equations are solved for possible values of ω. These are the normal mode frequencies.

(6) For each allowed value of ω, we find the ratio of the amplitudes B/A. If B/A is positive, then the two masses move in the same direction, and if B/A is negative, then they move in the opposite direction.

Example 3.3 Normal Modes and Normal Coordinates

A coupled system of two oscillators has the following equations of motion:

$$\frac{d^2x_A}{dt^2} = -9x_A + x_B, \qquad \frac{d^2x_B}{dt^2} = -9x_B + x_A.$$

(a) Find the normal modes. (b) Write normal coordinates and show that the equations of motion of normal coordinates are decoupled.

Solution

(a) We start with assuming the following form for x_A and x_B:

$$x_A = A\cos(\omega t), \quad x_B = B\cos(\omega t).$$

When we use these in the given equations of motion we get the following equation, after canceling out the common factor of $\cos(\omega t)$:

$$-\omega^2 A = -9A + B, \qquad -\omega^2 B = -9B + A.$$

This gives two expressions for B/A, which must be equal to each other,

$$\frac{B}{A} = -\omega^2 + 9 = \frac{1}{-\omega^2 + 9}. \tag{3.61}$$

Therefore,

$$-\omega^2 + 9 = \pm 1.$$

This gives the following for the squares of the two normal frequencies:

$$\omega_1^2 = 8, \ \omega_2^2 = 10.$$

Keeping the positive root only, we obtain the frequencies of the normal modes,

$$\omega_1 = 2\sqrt{2}, \ \omega_2 = \sqrt{10}.$$

From Eq. 3.61 we get the amplitude ratios corresponding to the two modes,

$$(B/A)_1 = -\omega_1^2 + 9 = 1, \ (B/A)_2 = -\omega_2^2 + 9 = -1.$$

(b) Using B/A of each mode we have the following normal coordinates:

$$q_1 = x_A + x_B, \ q_2 = x_A - x_B.$$

Taking two derivatives and using the equations of motion gives the following:

$$\frac{d^2q_1}{dt^2} = \frac{d^2x_A}{dt^2} + \frac{d^2x_B}{dt^2} = -9x_A + x_B - 9x_B + x_A$$

$$= -8(x_A + x_B) = -8q_1,$$

$$\frac{d^2q_2}{dt^2} = \frac{d^2x_A}{dt^2} - \frac{d^2x_B}{dt^2} = -9x_A + x_B + 9x_B - x_A$$

$$= -10(x_A - x_B) = -10q_2.$$

Recapping, we find the equations for q_1 and q_2 are decoupled,

$$\frac{d^2q_1}{dt^2} = -8q_1, \quad \frac{d^2q_2}{dt^2} = -10q_2.$$

3.4 Matrix Methods

We have seen that normal coordinates play a pivot role in solving the equations of motion of a coupled system. The way we got the normal coordinates from normal modes was not explained. I claimed that if you superpose the regular coordinates with appropriate weighting coefficients you would get normal coordinates in which equations decouple. Matrix methods provide another way of obtaining normal coordinates and normal mode frequencies.

3.4.1 Eigenvectors and Eigenvalues

To illustrate the matrix method for finding the normal coordinates let us work out the example of the coupled pendulum once again. Recall that the equations of motion of the dynamical variables x_A and x_B for this system are

$$\frac{d^2x_A}{dt^2} = -\left(\omega_0^2 + \omega_c^2\right) x_A + \omega_c^2 x_B, \tag{3.62}$$

$$\frac{d^2x_B}{dt^2} = \omega_c^2 x_A - \left(\omega_0^2 + \omega_c^2\right) x_B. \tag{3.63}$$

To deduce the normal coordinates, we demand that both x_A and x_B oscillate with the same frequency ω in a normal mode and assume the following form for x_A and x_B when in a normal mode:

$$\left.\begin{aligned} x_A &= C_1 \cos(\omega t), \\ x_B &= C_2 \cos(\omega t). \end{aligned}\right\} \text{(when in a normal mode)} \tag{3.64}$$

This assumed form of the solution transforms the equations of motion into two algebraic equations,

$$\left(\omega_0^2 + \omega_c{}^2\right) C_1 - \omega_c^2 C_2 = \omega^2 C_1, \tag{3.65}$$

$$-\omega_c^2 C_1 + \left(\omega_0^2 + \omega_c{}^2\right) C_2 = \omega^2 C_2. \tag{3.66}$$

We can write these equations in the matrix notation as

$$\begin{bmatrix} \omega_0^2 + \omega_c{}^2 & -\omega_c^2 \\ -\omega_c^2 & \omega_0^2 + \omega_c{}^2 \end{bmatrix} \begin{bmatrix} C_1 \\ C_2 \end{bmatrix} = \omega^2 \begin{bmatrix} C_1 \\ C_2 \end{bmatrix}. \tag{3.67}$$

We usually write the square matrix on the left as A, the column matrix of C's as vector $\vec{x}$, and ω^2 as λ,

$$A\,\vec{x} = \lambda\,\vec{x}. \tag{3.68}$$

The allowed values of λ are called the **eigenvalues** of the matrix A and the corresponding $\vec{x}$ the **eigenvectors** of A. This equation is also called the eigenvalue/eigenvector equation. Here, the eigenvalues give the square of the normal mode frequencies and the eigenvectors give the coefficients C_1 and C_2 in a normal coordinate. Therefore eigenvalues are also called **eigenfrequencies** and eigenvectors are referred to as **eigenmodes**.

Bringing $\lambda\vec{x}$ to the left side we obtain the following equation:

$$(A - \lambda I)\ \vec{x} = 0, \tag{3.69}$$

where I is a 2×2 unit matrix that has 1 along the diagonal and 0 for the off-diagonal entries. For a non trivial solution of this equation, i.e. where both C_1 and C_2 are not zero or the vector $\vec{x}$ is not a null vector, the determinant of the square matrix $A - \lambda I$ must be zero. This gives the following equation called the **characteristic equation**:

$$|A - \lambda I| = 0, \tag{3.70}$$

which yields the following explicit expression for the example above:

$$\begin{vmatrix} \omega_0^2 + \omega_c{}^2 - \omega^2 & -\omega_c^2 \\ -\omega_c^2 & \omega_0^2 + \omega_c{}^2 - \omega^2 \end{vmatrix} = 0 \implies \left(\omega_0^2 + \omega_c{}^2 - \omega^2\right)^2 = \left(\omega_c^2\right)^2. \tag{3.71}$$

Solving this equation for ω^2 we find the following equation for ω^2:

$$\omega^2 = \omega_0^2 + \omega_c{}^2 \mp \omega_c^2. \tag{3.72}$$

The minus and plus signs correspond to the two eigenvalues which give the two normal mode frequencies ω_1 and ω_2,

$$\omega_1 = \sqrt{\omega_0^2 + \omega_c{}^2 - \omega_c^2} = \omega_0, \tag{3.73}$$

$$\omega_2 = \sqrt{\omega_0^2 + \omega_c{}^2 + \omega_c^2} = \sqrt{\omega_0^2 + 2\omega_c{}^2}. \tag{3.74}$$

Eigenvectors and Ratio of Amplitudes

We can now work out the eigenvector for each eigenvalue by replacing ω in Eq. 3.67 by that particular eigenvalue and solving for the C_2/C_1 ratio. Note that we cannot determine both

C_1 and C_2 since the equations of motion are homogeneous in x_A and x_B; we can only get their ratio. The column matrix with entries C_1 and C_2 corresponding to each eigenvalue represents the eigenvector that goes with that eigenmode,

$$\text{Eigenvector}: \begin{bmatrix} C_1 \\ C_2 \end{bmatrix}. \tag{3.75}$$

Let us now work out eigenvectors for each eigenmode.

(1) When we set $\omega = \omega_1$ in Eq. 3.67 we get

$$\begin{bmatrix} \omega_0^2 + \omega_c^2 & -\omega_c^2 \\ -\omega_c^2 & \omega_0^2 + \omega_c^2 \end{bmatrix} \begin{bmatrix} C_1 \\ C_2 \end{bmatrix} = \omega_1^2 \begin{bmatrix} C_1 \\ C_2 \end{bmatrix}, \tag{3.76}$$

which gives us the following equations:

$$\begin{aligned} \left(\omega_0^2 + \omega_c^2\right) C_1 - \omega_c^2 C_2 &= \omega_0^2 C_1. \\ -\omega_c^2 C_1 + \left(\omega_0^2 + \omega_c^2\right) C_2 &= \omega_1^2 C_2 \end{aligned} \tag{3.77}$$

They are identical and you can use either to deduce:

$$\frac{C_2}{C_1} = 1. \tag{3.78}$$

Thus, if $C_1 = 1$, then $C_2 = 1$ also. As a column matrix this eigenvector takes the following representation:

$$\text{Eigenvector \# 1}: \begin{bmatrix} 1 \\ 1 \end{bmatrix}. \tag{3.79}$$

Often we write this column vector so that its length is 1. This is called **normalizing** the eigenvector. Recall that the length of a column vector is obtained by multiplying the vector with its transpose. Thus, the square of the length l of a vector with components a and b is

$$l^2 = \begin{bmatrix} a & b \end{bmatrix} \begin{bmatrix} a \\ b \end{bmatrix} = a^2 + b^2.$$

Therefore, the normalized eigenvector corresponding to the eigenvector in Eq. 3.79 will be

$$\text{Normalized Eigenvector \# 1}: \begin{bmatrix} 1/\sqrt{2} \\ 1/\sqrt{2} \end{bmatrix}. \tag{3.80}$$

The eigenvector can be used to write the normal coordinate q_1 in terms of x_A and x_B. We can get q_1 up to an overall factor, which we will set to 1 for simplicity,

$$q_1 = x_A + \left[(C_2/C_1)_{\text{this mode}}\right] x_B = x_A + x_B. \tag{3.81}$$

This is the same formula we have obtained above.

(2) A similar calculation from putting $\omega = \omega_2$ in Eq. 3.67 gives the eigenvector corresponding to the eigenvalue ω_2^2.

$$\text{Normalized Eigenvector \# 2 :} \begin{bmatrix} 1/\sqrt{2} \\ -1/\sqrt{2} \end{bmatrix} \tag{3.82}$$

This eigenvector gives the normal coordinate q_2 in terms of x_A and x_B.

$$q_2 = x_A + [(C_2/C_1)_{\text{this mode}}]\, x_B = x_A - x_B. \tag{3.83}$$

This is the same formula we have obtained above.

Example 3.4 Eigenvalues and Eigenvectors

Find the eigenvalues and eigenvectors of the following matrix:

$$\begin{bmatrix} 1 & 1 \\ 1 & 0 \end{bmatrix}.$$

Solution

The eigenvalue/eigenvector equation we try to solve is

$$\begin{bmatrix} 1 & 1 \\ 1 & 0 \end{bmatrix}\begin{bmatrix} C_1 \\ C_2 \end{bmatrix} = \lambda \begin{bmatrix} C_1 \\ C_2 \end{bmatrix}. \tag{3.84}$$

First we determine the eigenvalues from the characteristic equation, $Det[A - \lambda \mathbf{I}] = \mathbf{0}$. In the case of the given matrix A this becomes

$$\begin{vmatrix} 1-\lambda & 1 \\ 1 & -\lambda \end{vmatrix} = 0.$$

This gives

$$\lambda^2 - \lambda - 1 = 0.$$

Solving for λ gives the following values for the two eigenvalues:

$$\lambda_1 = \frac{1}{2} - \frac{\sqrt{5}}{2}, \quad \lambda_2 = \frac{1}{2} + \frac{\sqrt{5}}{2}.$$

Now, we work out the eigenvectors corresponding to the two eigenvalues by going back to Eq. (3.84).

Eigenvectors: We set $\lambda = \lambda_1$ in Eq. 3.84 and solve for the ratio C_2/C_1,

$$\begin{bmatrix} 1 & 1 \\ 1 & 0 \end{bmatrix}\begin{bmatrix} C_1 \\ C_2 \end{bmatrix} = \lambda_1 \begin{bmatrix} C_1 \\ C_2 \end{bmatrix}.$$

We get two equations from this equation, both of which will give us the same C_2/C_1,

$$C_1 + C_2 = \lambda_1 C_1,$$
$$C_1 = \lambda_1 C_2.$$

The amplitude ratio is $C_2/C_1 = 1/\lambda_1$. Therefore, the eigenvector 1 will be

$$\text{Eigenvector \#1:} \quad \begin{bmatrix} 1 \\ \frac{2}{1-\sqrt{5}} \end{bmatrix}.$$

A similar calculation using $\lambda = \lambda_2$ in Eq. 3.84 gives the following for eigenvector #2:

$$\text{Eigenvector \#2:} \quad \begin{bmatrix} 1 \\ \frac{2}{1+\sqrt{5}} \end{bmatrix}.$$

3.5 Longitudinal Vibration Modes

Longitudinal vibrations are of particular interest in sound and vibrations of solids. The particles of air vibrate along the line of the flow of sound when sound passes through air. In this section we will study a simple two-mass system connected by springs in which the two masses vibrate along the line of the springs. In analogy to sound we will also call these vibrations longitudinal vibrations. In a later section we will study the motion of two masses when they are displaced perpendicularly to the line of the springs. That motion will be called transverse vibration.

The system we have in mind here is shown in Fig. 3.9, with the motion restricted to the straight line joining the masses. You can think of this system as two mass/spring systems, one with mass m_A and spring k_A and the other mass m_B and spring k_B, coupled by the coupling spring k_c.

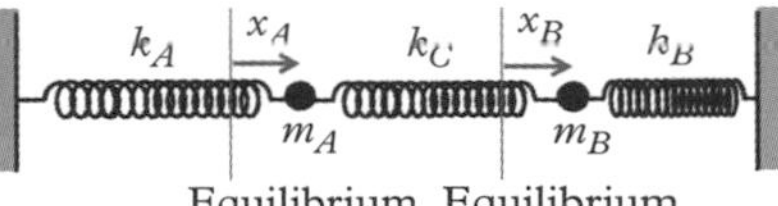

Fig. 3.9 *Two masses attached to three springs. The dynamical variables are the displacements of the two masses from their corresponding equilibrium positions. For longitudinal oscillations, these variables are the x_A and x_B shown in the figure. In the text we work with the system in which $m_A = m_B = m$ and $k_A = k_B = k$.*

To simplify the treatment, we will consider the case when $m_A = m_B = m$ and the spring constants of the springs connected to the supports are equal, $k_A = k_B = k$, while we keep the spring constant k_c of the coupling spring unrestricted. You should verify that the equations of motion of the two masses in terms of the displacement of each mass from its equilibrium position will be

$$m\frac{d^2x_A}{dt^2} = -kx_A + k_c(x_B - x_A), \tag{3.85}$$

$$m\frac{d^2x_B}{dt^2} = -kx_B - k_c(x_B - x_A). \tag{3.86}$$

Let us introduce the following frequency parameters:

$$\omega_0^2 = \frac{k}{m}, \quad \omega_c^2 = \frac{k_c}{m}. \tag{3.87}$$

The frequency ω_0 is the natural frequency of the separate oscillators and ω_c arises due to the coupling. Dividing Eqs. 3.85 and 3.86 by m, we can rewrite the equations of motion using these frequencies,

$$\frac{d^2x_A}{dt^2} = -\left(\omega_0^2 + \omega_c^2\right) x_A + \omega_c^2 x_B, \tag{3.88}$$

$$\frac{d^2x_B}{dt^2} = -\left(\omega_0^2 + \omega_c^2\right) x_B + \omega_c^2 x_A. \tag{3.89}$$

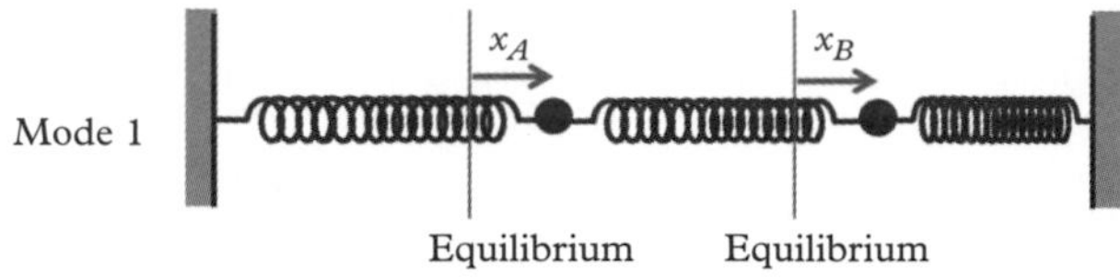

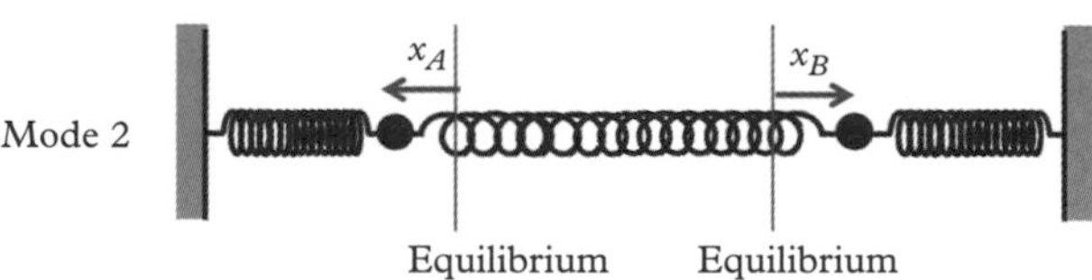

Fig. 3.10 *Illustration of the two normal modes. (a) In mode 1, the masses move in tandem so that the middle spring in unaffected. (b) In mode 2, the masses move in opposite directions so that the center of mass remains fixed.*

These equations are identical to the equations of coupled pendulums, namely Eqs. 3.17 and 3.18, previously studied. We can just copy the answer from that problem. Therefore, the normal mode frequencies and normal coordinates of our current problem will be

$$\text{Mode 1: } \omega_1 = \omega_0, \quad q_1 = x_A + x_B \tag{3.90}$$

$$\text{Mode 2: } \omega_2 = \sqrt{\omega_0^2 + 2\omega_c^2}, \quad q_2 = x_A - x_B. \tag{3.91}$$

We can display the normal modes by drawing arrows for the displacements. In mode 1, the two masses will move an equal distance in the same direction and in mode 2 they will move equal distances in the opposite direction, as shown in Fig. 3.10.

3.6 Transverse Vibrations

Transverse oscillations of coupled masses are easier to visualize and are important in understanding many physical phenomena, such as vibration of strings, ocean waves, and electromagnetic waves. *Here, by transverse vibrations we mean the vibrations in a direction perpendicular to the line of the springs.*

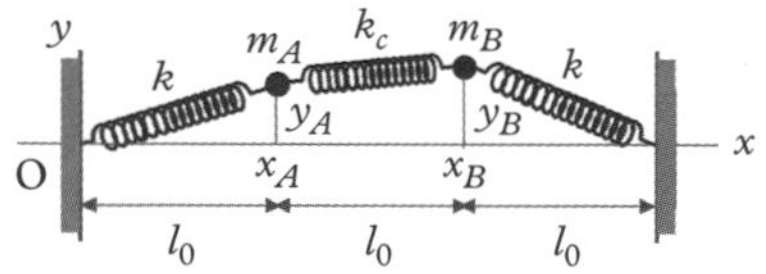

Fig. 3.11 *Two masses coupled with a spring and connected to supports with springs. The displacement of masses m_A and m_B in a transverse direction from their equilibrium positions are denoted by y_A and y_B. At equilibrium, the length of each spring is l, which is larger than the original length of the springs l_0. The x-coordinates are $x_A = l_0, x_B = 2l_0$.*

To be concrete, let us consider again the case of two masses joined by a spring. The masses are then connected to the supports by two additional springs, one on each side, as shown in Fig. 3.11. When the springs are connected to the masses and the support, each spring is stretched to length l_0 so that the distance between the supports is $3l_0$. Let the original unstretched lengths of the three springs be l_{01}, l_{02}, and $l_{03} = l_{01}$. Let k_c be the spring constant of the spring between the masses and k the spring constant of the spring connected to the supports. Let (x_A, y_A) and (x_B, y_B) be the coordinates of the two masses with the origin at the left support.

For simplicity, we will consider the equal-mass case, $m_A = m_B = m$, as before. *It can be shown that if the springs are stretched significantly compared to the transverse displacement, the transverse motion has a linear restoring force, which leads to a simple harmonic motion in the perpendicular direction.* Therefore, we will implement the following assumptions to ensure that the y force is linear in y as shown below:

$$|y_A| \ll |x_A - l_{01}|,$$

$$|y_B - y_A| \ll |x_B - x_A - l_{02}|,$$

$$|y_B| \ll |l_{01} + l_{02} + l_{03} - x_B|.$$

With these assumptions, the y-component of the equations of motion of the two masses are given by the following equations:

$$m\frac{d^2y_A}{dt^2} = -k_{\text{eff}}\, y_A + k^c_{\text{eff}}\, (y_B - y_A), \tag{3.92}$$

$$m\frac{d^2y_B}{dt^2} = -k_{\text{eff}}\, y_B - k^c_{\text{eff}}\, (y_B - y_A), \tag{3.93}$$

where the effective spring constants are

$$k_{\text{eff}} = k\left(\frac{l_0 - l_{01}}{l_0}\right), \tag{3.94}$$

$$k^c_{\text{eff}} = k_c\left(\frac{l_0 - l_{02}}{l_0}\right). \tag{3.95}$$

We leave the derivation of these equations as an exercise for the student. Note that $k(l_0 - l_{01})$ is the tension in the spring connected to the supports and $k_c(l_0 - l_{02})$ the tension in the spring connecting the two masses. The equations for the y-coordinates, Eqs. 3.92 and 3.93, are identical to the equations of motion for the longitudinal motion, i.e. the motion along the line of the springs as given in Eqs. 3.88 and 3.89, if we identify the following frequency-type parameters here to the corresponding ω_0 and ω_c in the longitudinal case,

$$\omega_0 = \sqrt{\frac{k_{\text{eff}}}{m}}, \tag{3.96}$$

$$\omega_c = \sqrt{\frac{k^c_{\text{eff}}}{m}}. \tag{3.97}$$

Using these frequency-type parameters, the equations of motion for the y-coordinates become

$$\frac{d^2y_A}{dt^2} = -\omega_0^2 y_A + \omega_c^2\,(y_B - y_A)\,, \tag{3.98}$$

$$\frac{d^2y_B}{dt^2} = -\omega_0^2 y_B - \omega_c^2\,(y_B - y_A)\,. \tag{3.99}$$

These equations have the same form as the equations for the longitudinal vibrations and the coupled pendulum problems. Therefore, we can just copy the normal mode frequencies and the normal coordinates,

$$\text{Mode 1: } \omega_1 = \omega_0, \quad q_1 = x_A + x_B \tag{3.100}$$

$$\text{Mode 2: } \omega_2 = \sqrt{\omega_0^2 + 2\omega_c^2}, \quad q_2 = x_A - x_B. \tag{3.101}$$

These modes are shown in Fig. 3.12. The lower frequency mode is symmetric about the equilibrium line, while the higher frequency mode has one place in its pattern where it crosses the zero displacement line, called a **node**. The particle of the spring at the node remains at rest while other particles of the mass/spring system oscillate up and down with the frequency ω_2 of the corresponding mode.

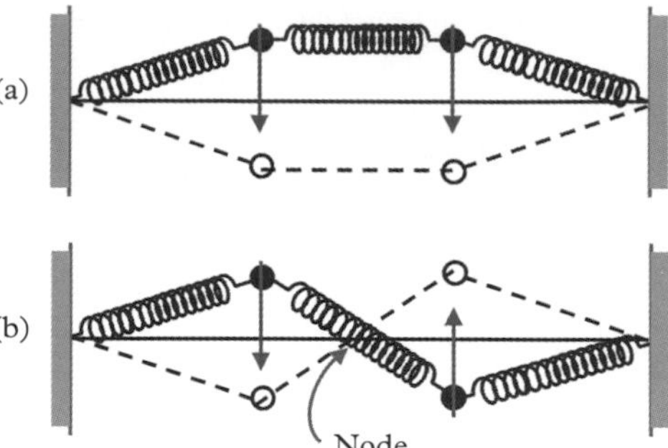

Fig. 3.12 *The transverse modes of two masses coupled with a spring and connected to supports with springs. (a) In the lower frequency mode, the masses move together in the same direction with the same amplitude. (b) In the higher frequency mode, the masses move in opposite directions with a node, that is, $y = 0$, in the line connecting the two masses. The point at the node remains at rest during the oscillations.*

3.7 Energy of Coupled Systems and Normal Coordinates

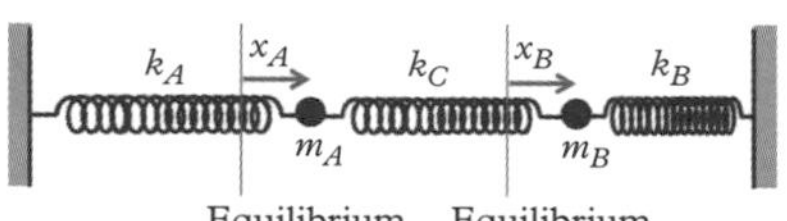

Fig. 3.13 *Two masses attached to three springs. Here we consider the case: $m_A = m_B = m$, $k_1 = k_2 = k$.*

The energy of coupled systems separates nicely into a sum of the energy in each normal mode. To see that, let us look again at the energy of the longitudinal mode problem of two masses and three springs, previously studied. Figure 3.13 is displayed as a quick reference.

The energy of the system will be the sum of the kinetic energies of the two blocks and the potential energies in the springs,

$$E = \frac{1}{2}m\left(\frac{dx_A}{dt}\right)^2 + \frac{1}{2}m\left(\frac{dx_B}{dt}\right)^2 + \frac{1}{2}kx_A^2 + \frac{1}{2}kx_B^2 + \frac{1}{2}k\left(x_B - x_A\right)^2. \tag{3.102}$$

The normal coordinates q_1 and q_2 of this system are related to the coordinates x_A and x_B as follows:

$$q_1 = x_A + x_B, \quad q_2 = x_A - x_B.$$

Therefore,

$$x_A = \frac{q_1 + q_2}{2}, \quad x_B = \frac{q_1 - q_2}{2}.$$

Replacing x_A and x_B in Eq. 3.102 and writing $k = m\omega_0^2$ and $k_c = m\omega_c^2$ gives

$$\begin{aligned} E = &\frac{1}{2}m\,\frac{1}{4}\left[\left(\frac{dq_1}{dt}\right)^2 + \left(\frac{dq_2}{dt}\right)^2 + 2\left(\frac{dq_1}{dt}\right)\left(\frac{dq_2}{dt}\right)\right] \\ &+ \frac{1}{2}m\,\frac{1}{4}\left[\left(\frac{dq_1}{dt}\right)^2 + \left(\frac{dq_2}{dt}\right)^2 - 2\left(\frac{dq_1}{dt}\right)\left(\frac{dq_2}{dt}\right)\right] \\ &+ \frac{1}{2}m\omega_0^2\,\frac{1}{4}\left(q_1^2 + q_2^2 + 2q_1q_2\right) + \frac{1}{2}m\omega_0^2\,\frac{1}{4}\left(q_1^2 + q_2^2 - 2q_1q_2\right) + \frac{1}{2}m\omega_c^2q_2^2 \\ = &\frac{1}{2}\,\frac{m}{2}\left(\frac{dq_1}{dt}\right)^2 + \frac{1}{2}\,\frac{m}{2}\left(\frac{dq_2}{dt}\right)^2 \\ &+ \frac{1}{2}\,\frac{m}{2}\omega_0^2q_1^2 + \frac{1}{2}\,\frac{m}{2}\left(\omega_0^2 + 2\omega_c^2\right)q_2^2. \end{aligned} \tag{3.103}$$

Let us replace ω_0^2 by ω_1^2 and $\omega_0^2 + 2\omega_c^2$ by ω_2^2, where ω_1 and ω_2 are the two normal frequencies. We will also replace $m/2$ by μ, the reduced mass of the two masses in the system. With these replacements and by grouping the terms by modes we see that the energy of the system can be written as

$$E = \left[\frac{1}{2}\mu\left(\frac{dq_1}{dt}\right)^2 + \frac{1}{2}\mu\omega_1^2 q_1^2\right] + \left[\frac{1}{2}\mu\left(\frac{dq_2}{dt}\right)^2 + \frac{1}{2}\mu\omega_2^2 q_2^2\right]. \tag{3.104}$$

Thus, the total energy of the system is the sum of the energies in the two modes,

$$E = E_1 + E_2. \tag{3.105}$$

Since the modes are independent of each other, the energies in each mode are independently conserved and so is their sum.

3.8 Coupled Electrical Oscillators

A coupled electrical circuit is shown in Fig. 3.14. By using Kirchhoff's rules we can get the equation of motion of two independent currents in the circuit. We will use the convention that the current into the positive plate of a capacitor corresponds to positive current. This will give us the following relations between current and charge on the positive plates of capacitors in the circuit:

$$I_A = \frac{dq_A}{dt}, \quad I_B = \frac{dq_B}{dt}, \quad I_C = \frac{dq_C}{dt}. \tag{3.106}$$

Now, we have the following equation from the current conservation at point (a) in the circuit:

$$I_A + I_B + I_C = 0. \tag{3.107}$$

In the loop a-b-c-a we get

$$-L_A\frac{dI_A}{dt} - \frac{1}{C_A}q_A + \frac{1}{C_C}q_C = 0. \tag{3.108}$$

In the loop a-d-c-a we get

$$-L_B\frac{dI_B}{dt} - \frac{1}{C_B}q_B + \frac{1}{C_C}q_C = 0. \tag{3.109}$$

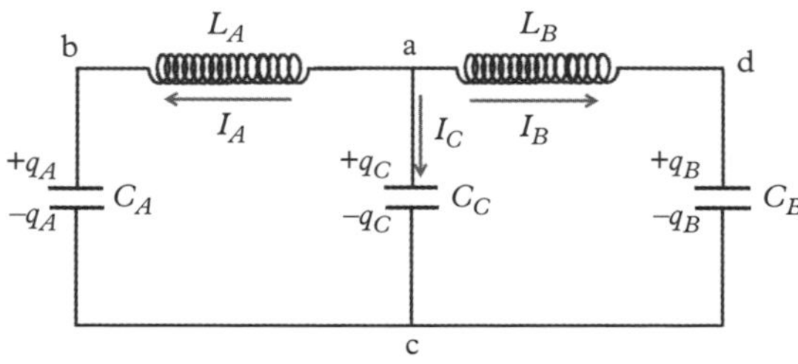

Fig. 3.14 *Coupled electrical oscillators with two degrees of freedom.*

Now, we take time derivatives of Eqs. 3.108 and 3.109, replace the time derivatives of the charges by appropriate currents, and then use the current law in Eq. 3.107 to eliminate I_C and arrive at the following equations for I_B and I_B:

$$L_A\frac{d^2I_A}{dt^2} = -\left(\frac{1}{C_A}+\frac{1}{C_C}\right)I_A + \frac{1}{C_C}I_B, \tag{3.110}$$

$$L_B\frac{d^2I_B}{dt^2} = -\left(\frac{1}{C_B}+\frac{1}{C_C}\right)I_B + \frac{1}{C_C}I_A. \tag{3.111}$$

Now, we divide the first equation by L_A and the second by L_B and introduce the following frequency-like parameters by their squares:

$$\omega_{aa}^2 = \left(\frac{1}{L_AC_A}+\frac{1}{L_AC_C}\right),\quad \omega_{ab}^2 = \frac{1}{L_AC_C}, \tag{3.112}$$

$$\omega_{bb}^2 = \left(\frac{1}{L_BC_A}+\frac{1}{L_BC_C}\right),\quad \omega_{ba}^2 = \frac{1}{L_BC_C}. \tag{3.113}$$

In terms of these parameters equations of motion take on a more compact form,

$$\frac{d^2I_A}{dt^2} = -\omega_{aa}^2I_A + \omega_{ab}^2I_B, \tag{3.114}$$

$$\frac{d^2I_B}{dt^2} = -\omega_{bb}^2I_B + \omega_{ba}^2I_A. \tag{3.115}$$

To look for normal modes, we assume the following forms for the two currents. (Since there is no resistance in the circuit, we can use the real functions.):

$$I_A = A\cos(\omega t),\quad I_B = B\cos(\omega t).$$

This gives the following equations for coefficients A and B and the unknown mode frequency ω:

$$(\omega_{aa}^2 - \omega^2)A = \omega_{ab}^2B, \tag{3.116}$$

$$(\omega_{bb}^2 - \omega^2)B = \omega_{ba}^2A. \tag{3.117}$$

Setting B/A from these two equations equal to each other gives the equation for ω,

$$(\omega^2)^2 - \left(\omega_{aa}^2 + \omega_{bb}^2\right)(\omega^2) + \left(\omega_{aa}^2\omega_{bb}^2 - \omega_{ab}^2\omega_{ba}^2\right) = 0.$$

The solution of this equation gives the two mode frequencies,

$$\omega_1 = \sqrt{\frac{1}{2}\left[\omega_{aa}^2 + \omega_{bb}^2 - \sqrt{\left(\omega_{aa}^2 - \omega_{bb}^2\right)^2 + 4\omega_{ab}^2\omega_{ba}^2}\right]}, \tag{3.118}$$

$$\omega_2 = \sqrt{\frac{1}{2}\left[\omega_{aa}^2 + \omega_{bb}^2 + \sqrt{\left(\omega_{aa}^2 - \omega_{bb}^2\right)^2 + 4\omega_{ab}^2\omega_{ba}^2}\right]}. \tag{3.119}$$

The B/A for each mode is obtained as before by using ω_1 and ω_2 in Eq. 3.116 or 3.117,

$$(B/A)_1 = \frac{1}{2\omega_{ab}^2}\left[\omega_{aa}^2 - \omega_{bb}^2 + \sqrt{\left(\omega_{aa}^2 - \omega_{bb}^2\right)^2 + 4\omega_{ab}^2\omega_{ba}^2}\right], \tag{3.120}$$

$$(B/A)_2 = \frac{1}{2\omega_{ab}^2}\left[\omega_{aa}^2 - \omega_{bb}^2 - \sqrt{\left(\omega_{aa}^2 - \omega_{bb}^2\right)^2 + 4\omega_{ab}^2\omega_{ba}^2}\right]. \tag{3.121}$$

The above formulas are complicated because we have different L and C in the two coupled circuits. If we work with a symmetric situation with $L_A = L_B \equiv L$ and $C_A = C_B \equiv C$ we will find that

$$\omega_{aa} = \omega_{bb} \text{ and } \omega_{ab} = \omega_{ba}.$$

In the symmetric case,

$$\omega_1 = \sqrt{\omega_{aa}^2 - \omega_{ab}^2}, \quad \omega_1 = \sqrt{\omega_{aa}^2 + \omega_{ab}^2}$$
$$(B/A)_1 = +1, \quad (B/A)_2 = -1.$$

3.9 Damped Coupled Systems

So far our coupled systems had no dissipative forces such as friction on them. In this section, we examine the solution of damped systems. From the damped single oscillator system we know that there are three types of solutions—under-damped, over-damped, and critically damped. Here we will be looking at only the under-damped case. To be concrete, consider two blocks of the same mass m attached to two supports by identical springs of spring constant $k_1 = k_2 = k$, as shown in Fig. 3.15. Let the blocks be further connected to each other by one spring with spring constant k_c.

Now, suppose the two masses have a drag force on them that is proportional to the speed and opposite in direction to that of the velocity, with the same damping constant Γ. This gives the following equations of motion for displacements x_A and x_B of the two masses from their respective equilibria:

$$\frac{d^2x_A}{dt^2} = -\left(\omega_0^2 + \omega_c^2\right)x_A - \Gamma\frac{dx_A}{dt} + \omega_c^2 x_B, \tag{3.122}$$

$$\frac{d^2x_B}{dt^2} = -\left(\omega_0^2 + \omega_c^2\right)x_B - \Gamma\frac{dx_B}{dt} + \omega_c^2 x_A, \tag{3.123}$$

where I have used $\omega_0^2 = k/m$ and $\omega_c = k_c/m$ as before. We wish to solve these equations for a given $x_A(0)$, $\dot{x}_1(0)$, $x_B(0)$, and $\dot{x}_2(0)$. The following change of variable helps here. Let

$$x_A = e^{-\frac{\Gamma}{2}t}u_A(t), \tag{3.124}$$

$$x_B = e^{-\frac{\Gamma}{2}t}u_B(t). \tag{3.125}$$

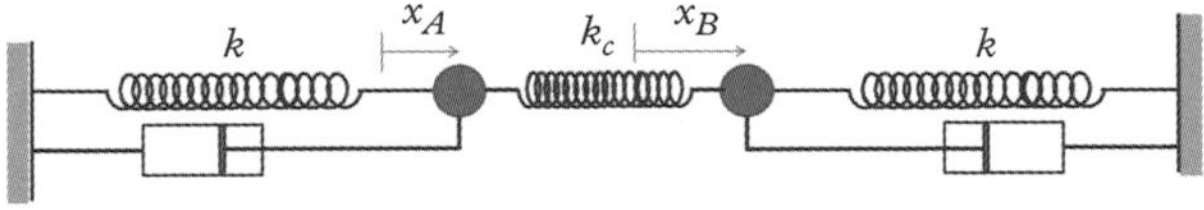

Fig. 3.15 *Two masses coupled with a spring and coupled to supports with springs. The displacement of the masses from their equilibrium positions are denoted by x_A and x_B. The oscillations of the masses are damped through by viscous damping through coupling to dashpots.*

With this change the equations of motion become

$$\frac{d^2u_A}{dt^2} = -\left(\omega_0^2 - \Gamma^2 + \omega_c^2\right) u_A + \omega_c^2 u_B, \tag{3.126}$$

$$\frac{d^2u_B}{dt^2} = -\left(\omega_0^2 - \Gamma^2 + \omega_c^2\right) u_B + \omega_c^2 u_A. \tag{3.127}$$

We notice that the velocity term has disappeared and these equations are similar to the equations without the damping term. Now, we can think of u_A and u_B as coupled oscillators without damping and borrow the results of similar systems we have solved above. In the present case, the normal frequencies and mode amplitude ratios for the modes will be

$$\omega_1 = \sqrt{\omega_0^2 - \Gamma^2}, \quad \text{mode ratio} \;\; C_2/C_1 = 1, \tag{3.128}$$

$$\omega_2^2 = \sqrt{\omega_0^2 + 2\omega_c^2 - \Gamma^2}, \quad \text{mode ratio} \;\; C_2/C_1 = -1. \tag{3.129}$$

Therefore, the general solution of these u-equations will be

$$u_A(t) = A\cos\left(\omega_1 t - \phi_1\right) + B\sin\left(\omega_2 t - \phi_2\right), \tag{3.130}$$

$$u_B(t) = A\cos\left(\omega_1 t - \phi_1\right) - B\sin\left(\omega_2 t - \phi_2\right). \tag{3.131}$$

Now, we can use these solutions in Eqs. 3.124 and 3.125 to get the general solution for the original variables,

$$x_A = e^{-\frac{\Gamma}{2}t}\left[A\cos\left(\omega_1 t - \phi_1\right) + B\sin\left(\omega_2 t - \phi_2\right)\right], \tag{3.132}$$

$$x_B = e^{-\frac{\Gamma}{2}t}\left[A\cos\left(\omega_1 t - \phi_1\right) - B\sin\left(\omega_2 t - \phi_2\right)\right]. \tag{3.133}$$

As usual, the four unknown constants here, A, B, ϕ_1, and ϕ_2, can be determined from initial conditions. The calculation is left as an exercise for the student.

EXERCISES

(3.1) Two identical pendulums A and B are coupled by a spring with parameters $\omega_0 = \sqrt{g/l} = 2$ rad/sec and $\omega_c = \sqrt{k/m} = 1$ rad/sec. The small oscillations of this system are studied by the horizontal displacements x_A and x_B of the two bobs in the small-angle approximation. Suppose the initial conditions at $t = 0$ are $x_A(0) = 0$, $v_A(0) = 0$, $x_B(0) = 1$ cm, $v_B(0) = 0$. Find the positions of the two pendulum bobs at an arbitrary instant $t > 0$. Assume small angles.

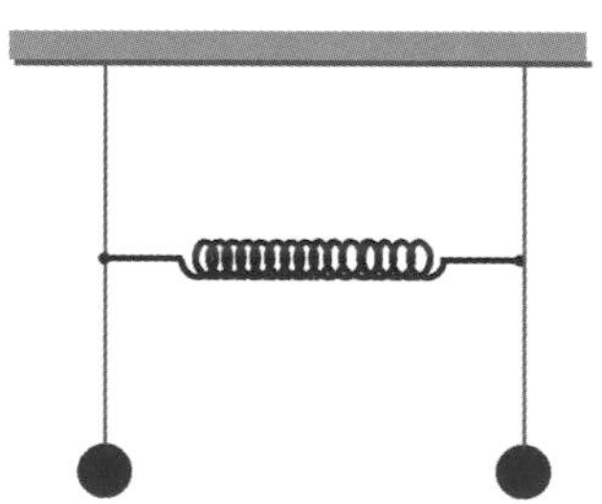

Fig. 3.16 *Exercise 3.2.*

(3.2) Two identical pendulums each of mass m are hanging by a "massless" but rigid rod of length l. A spring of spring constant k is attached between the pendulums at the half-way point of the rods, as shown in Fig. 3.16.

(a) Derive the horizontal component of the equations of motion of the two bobs in the small-angle approximation and work out the normal modes of this system.

(b) Compare your answer here to the modes when the spring is attached at the masses. Are the masses in this system more or less strongly coupled than when the spring is attached at the masses?

(c) What would you expect if the spring was attached at the very top, from where the pendulums are hanging?

(d) What would you get for the normal mode frequencies if the spring were attached at a distance a from the suspension points with $a < l$? Check your answer for $a = \frac{l}{2}$ that you obtained in part (a).

(3.3) **Two mass two spring system.** A block of mass m is connected to a massless spring with spring constant k and hung from a rigid ceiling. Another block of the same mass is hung from the first block by an identical spring. Suppose the blocks are displaced by y_A and y_B from their equilibrium positions at some arbitrary instant t, as shown in Fig. 3.17.

(a) Deduce the equations of motion of y_A and y_B.

(b) Solve the equations of motion to find the normal mode frequencies and amplitude ratios for the modes.

(c) Sketch the configurations of the system for the two normal modes indicating clearly the relative size and phase of the displacements of the two masses in the two modes.

(d) Suppose you let go of the two masses such that at $t = 0$, the displacements and velocities were $y_A(0) = 0$, $y_B(0) = 1$ cm, $v_A(0) = 0$, and $v_B(0) = 0.5$ cm/s. What would be $y_A(t)$ and $y_B(t)$?

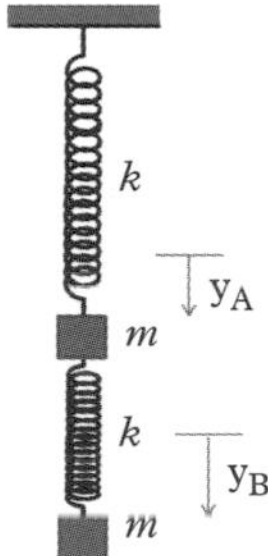

Fig. 3.17 *Exercise 3.3.*

(3.4) **Unsymmetric longitudinal oscillations.** In the text we worked out the longitudinal oscillations of two identical oscillators of same mass m attached to two supports by springs with the same spring constant k and coupled by a spring with spring constant k_c. Now, we wish to make a small modification in this problem. Let the masses of the two oscillators be different, m_A and m_B as shown in Fig. 3.18.

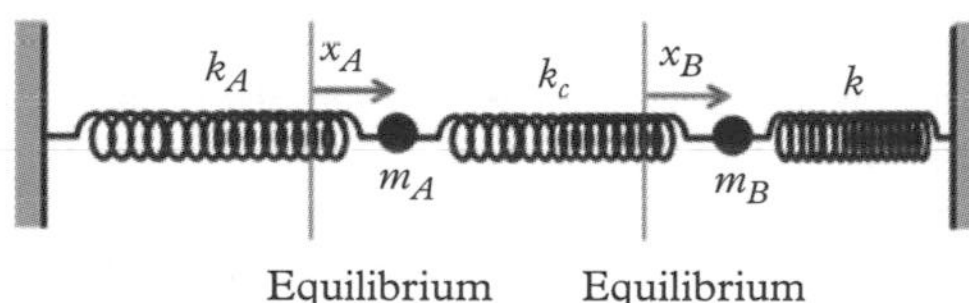

Fig. 3.18 *Exercise 3.4.*

(a) Find the normal modes now. You need to find the normal mode frequencies and the mode amplitude ratios.

(b) Draw modes for the case $m_B = 2m_A$.

(3.5) A block of mass m is hung from the center of a plate of mass M using a massless spring with spring constant k, as shown in Fig. 3.19. The plate is held by four identical springs with spring constant K, above a fixed table. We wish to study the vertical oscillations of m and M about their equilibrium positions.

(a) Find the equations of motion for m and M.

(b) Solve the equations of motion to find the normal mode frequencies and amplitude ratios for the modes for $m = M$ and $k = K$.

(c) Sketch the configurations of the system for the two normal modes, indicating clearly the relative sizes and phases of the displacements of the two masses in the two modes. You may sketch the configurations for the case $M = m$ and $K = k$.

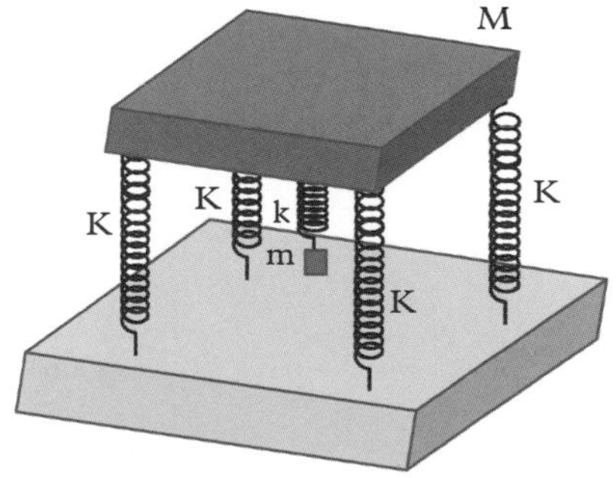

Fig. 3.19 *Exercise 3.5.*

(3.6) A system of two coupled oscillators obeys the following equations of motion:

$$\frac{d^2x_A}{dt^2} = -x_A + \frac{2}{3}x_B$$

$$\frac{d^2x_B}{dt^2} = -x_B + \frac{3}{8}x_A$$

(a) Use the matrix method to determine the eigenvalues and eigenvectors of the system.

(b) What would be $x_A(t)$ and $x_B(t)$ for the initial conditions $x_A(0) = 1$, $v_A(0) = 0$, $x_B(0) = 0$, $v_B(0) = 1$?

(3.7) Derive Eqs. 3.94 and 3.95.

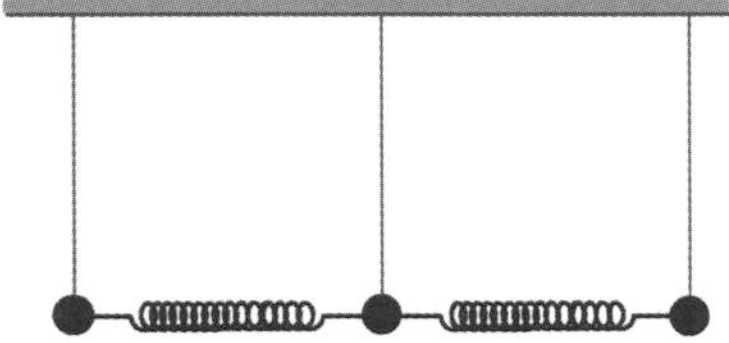

Fig. 3.20 *Exercise 3.8.*

(3.8) **Three coupled pendulums.** Three identical pendulums of mass m and length l are coupled by two identical springs with spring constant k, as shown in Fig. 3.20.

(a) Find the equations of motion for the three masses.

(b) Solve the equations of motion to find the normal mode frequencies and amplitude ratios for the modes. Note: matrix method may be helpful.

(c) Sketch the configurations of the system for the three normal modes indicating clearly the relative size and phase of the displacements of the three masses in the three modes.

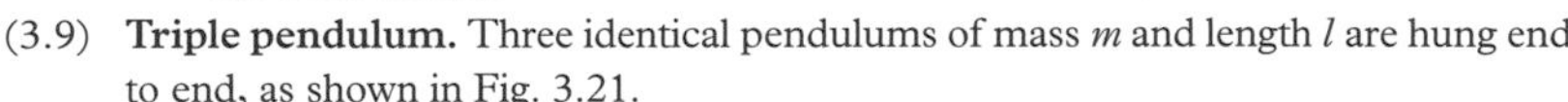

(3.9) **Triple pendulum.** Three identical pendulums of mass m and length l are hung end to end, as shown in Fig. 3.21.

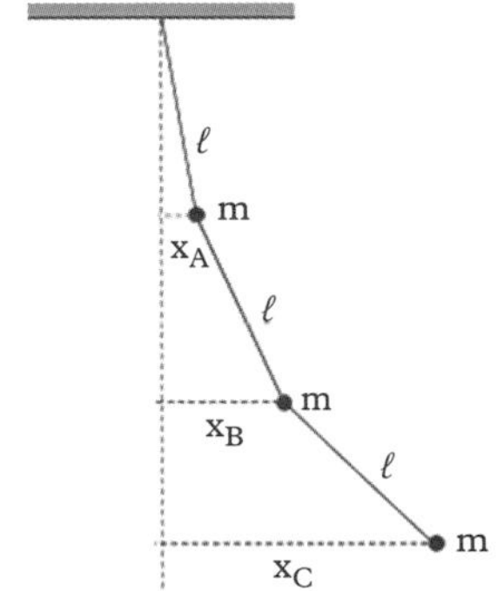

Fig. 3.21 *Exercise 3.9.*

(a) Find the equations of motion for the x-displacements of the three masses from their equilibria.

(b) Solve the equations of motion to find the normal mode frequencies and amplitude ratios for the modes. Note: matrix method may be helpful.

(c) Sketch the configurations of the system for the three normal modes indicating clearly the relative size and phase of the displacements of the three masses in the three modes.

(3.10) Three blocks of the same mass m are connected with four identical massless springs with spring constant k to two fixed supports on the two sides, as shown in Fig. 3.22.

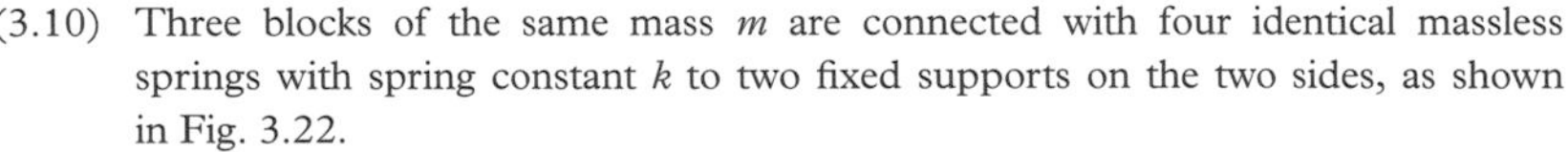

(a) Find the equations of motion for the longitudinal (horizontal) displacements of the three masses from their equilibria.

(b) Solve the equations of motion to find the normal mode frequencies and amplitude ratios for the modes. Note: matrix method may be helpful.

(c) Sketch the configurations of the system for the three normal modes indicating clearly the relative size and phase of the displacements of the three masses in the three modes.

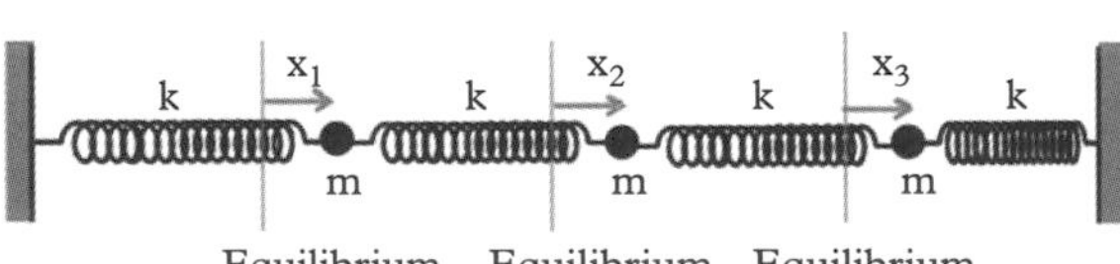

Fig. 3.22 *Exercise 3.10.*

(3.11) **Eigenvalue/Eigenvector Exercises.** Find the eigenvalues and eigenvectors of the following matrices:

(a) $\begin{bmatrix} 1 & 0 \\ 0 & -1 \end{bmatrix}$

(b) $\begin{bmatrix} 0 & 1 \\ 1 & 0 \end{bmatrix}$

(c) $\begin{bmatrix} 1 & 3 \\ 2 & 2 \end{bmatrix}$

(d) $\begin{bmatrix} 1 & -1 & 0 \\ -1 & 2 & -1 \\ 0 & -1 & 1 \end{bmatrix}$

4 Systems with Many Degrees of Freedom

Chapter Goals

In this chapter you will learn how to generalize the normal mode analysis we studied in the last chapter. You will also learn how to derive the classical wave equation and how to solve it. The normal modes of continuous systems will be obtained from solutions of the classical wave equation. We will also study Fourier analysis and apply it to the study of general vibratory motion of continuous bodies.

Physical bodies such as guitar strings are made up of a large number of particles, of the order of 10^{23}. But, so far, we have studied systems with only one or two degrees of freedom. Now, we will study systems that have N degrees of freedom, with N an unspecified number. When N is very large we can treat it as an infinity. Just as two bodies, whose vibrations are coupled, have two normal modes when they move in one dimension, N coupled bodies moving in one dimension have N modes. Similarly to the systems of two degrees of freedom, the general motion of N degrees of freedom can also be studied systematically in terms of the superposition of the normal modes.

In this chapter we will study a prototypical continuous system—the taut string. A taut string can be thought of as consisting of a large number of oscillators with the distance between nearest oscillators being small compared to the distances over which the string oscillates. Thus, by modeling a string with N oscillators and taking N to ∞ limit we can model the string as a continuous system. We will focus our attention on the transverse vibrations of a taut string since the transverse vibrations are easier to display and visualize. The transverse vibrations of a taut string have obvious applications in the music of stringed instruments since the modes of vibration correspond to pure tone notes of the instrument. Analysis of the vibration of strings is also applicable to vibrations of beams and other linear structures, and is indispensable for understanding the stability of buildings and bridges.

4.1 Transverse Oscillations of Beads on a String

As a prelude to discussion of oscillations of a continuous system, such as a guitar string, later in the chapter, we discuss here the normal modes of a system of N coupled oscillators. Let N beads of mass m each be placed on a "massless" string at equal intervals, denoted by l, and the two ends tied to fixed supports, as shown in Fig. 4.1. We will also assume that the tension T in the string is the same throughout and that the tension is sufficiently large that we can ignore the force of gravity on the masses. We study a small displacement of the

A First Course in Vibrations and Waves. First Edition. Mohammad Samiullah.

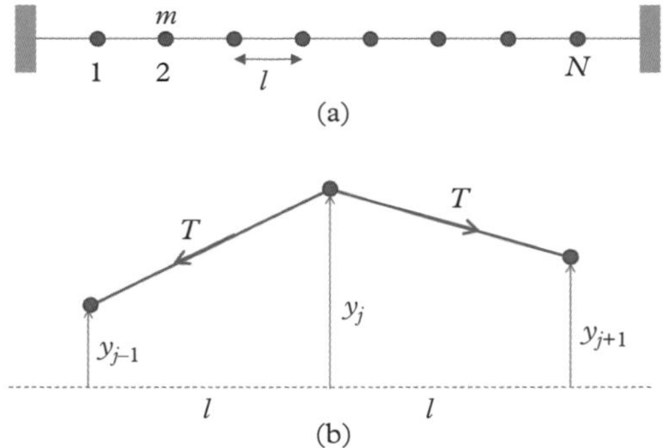

Fig. 4.1 *(a) N beads on a string. (b) Forces on the j-th mass from the tensions of the springs on the two sides.*

masses from their equilibrium positions. Let y_j be the displacement of the jth bead from its equilibrium which is at $y_j = 0$, with $j = 1, 2, 3, \ldots, N$.

We also need to make sure that the shape of the string is consistent with the fixed ends. The $y_j(t)$'s tell us the y-coordinates at some positions of the string at a particular time. Using this notation, we see that y_0 and y_{N+1} would represent the displacements of the points of the string at the supports, and since the strings are not allowed to move at these points we must have the following:

$$y_0 = 0, \tag{4.1}$$

$$y_{N+1} = 0. \tag{4.2}$$

We say that these equations represents **boundary conditions** on the variable y treated as a function of index j. The y-component of the Newton's equation of motion of the beads will give the accelerations of the y-displacements of the beads. To deduce the y-component of the Newton's equations of motion of the beads we need the y-component of the net force on the beads. For the sake of concreteness let us focus on an arbitrary bead labeled the jth bead, as shown in Fig. 4.1. The y-component of forces on this bead will be

$$F_{jy} = -T\left(\frac{y_j - y_{j-1}}{l}\right) + T\left(\frac{y_{j+1} - y_j}{l}\right), \quad (j = 1, 2, \ldots, N)$$

$$= -\left(\frac{2T}{l}\right) y_j + \left(\frac{T}{l}\right)\left(y_{j-1} + y_{j+1}\right). \tag{4.3}$$

Therefore, the equations of motion for the N variables, $y_1, y_2, \ldots, y_N$, are

$$\frac{d^2 y_j}{dt^2} = -2\omega_0^2 y_j + \omega_0^2\left(y_{j-1} + y_{j+1}\right), \; (j = 1, 2, 3, \ldots, N), \tag{4.4}$$

where we have introduced the following symbol for convenience:

$$\omega_0 = \sqrt{\frac{T}{ml}}. \tag{4.5}$$

Therefore, our problem is to use $N + 2$ equations given in Eqs. 4.1, 4.2, and 4.4 and solve for y_j for $j = 0, 1, 2, \ldots, N + 1$ with y_0 and y_{N+1} already known from Eqs. 4.1 and 4.2.

4.1.1 Normal Modes

Since the equations for y_j are coupled, we first seek the normal modes of the system. In any one normal mode, every y_j will oscillate with the same frequency and phase constant, and will have fixed amplitude ratios among different y_j's, except for the possible presence of minus signs in these amplitudes. Therefore, as before, we require that in any normal mode we would have

$$y_j(t) = C_j \cos(\omega t - \phi), \quad (j = 0, 1, 2, 3, \ldots, N, N+1). \tag{4.6}$$

Actually, keeping ϕ in the solution is not necessary when looking for normal modes. With this assumption in Eqs. 4.4, 4.1, and 4.2, we find the following for the coefficients:

$$C_0 = 0, \tag{4.7}$$

$$C_{j-1} + C_{j+1} = \left(-\frac{\omega^2}{\omega_0^2} + 2\right) C_j, \quad (j = 1, 2, 3, \ldots N), \tag{4.8}$$

$$C_{N+1} = 0. \tag{4.9}$$

Equation 4.8 says that the sum of two terms around C_j in the sequence $\{C_1, C_2, \ldots, C_N\}$ is proportional to C_j. This is the characteristic of the sequence of sines or cosines, i.e. $\{\sin(\alpha), \sin(2\alpha), \ldots, \sin(N\alpha)\}$, or, $\{\cos(\alpha), \cos(2\alpha), \ldots, \cos(N\alpha)\}$ for arbitrary α,

$$\sin[(j-1)\alpha] + \sin[(j+1)\alpha] = (2\cos\alpha)\sin(j\alpha), \tag{4.10}$$

$$\cos[(j-1)\alpha] + \cos[(j+1)\alpha] = (2\cos\alpha)\cos(j\alpha). \tag{4.11}$$

Since we also desire $C_0 = 0$, the sine would be the right choice. Therefore, we make the following claim that the solution for C_j should be in the form,

$$C_j = \sin(j\alpha), \quad (j = 0, 1, 2, 3, \ldots, N+1), \tag{4.12}$$

where α is yet to be determined. This form satisfies Eq. 4.7 automatically. To satisfy Eq. 4.8, we need the following condition to hold:

$$\sin[(j-1)\alpha] + \sin[(j+1)\alpha] = \left(-\frac{\omega^2}{\omega_0^2} + 2\right)\sin[j\alpha],$$

$$\text{or,} \quad 2\cos\alpha = \left(-\frac{\omega^2}{\omega_0^2} + 2\right). \tag{4.13}$$

This condition relates the arbitrary α to the unknown ω of the mode. The boundary condition at $j = N+1$ helps us in finding the allowed values of α and hence, the mode frequencies. According to Eq. 4.9 we have for $j = N+1$,

$$\sin[(N+1)\alpha] = 0. \tag{4.14}$$

That is, α must have one of the following values:

$$(N+1)\alpha = n\pi \quad (n = 1, 2, 3, \ldots). \tag{4.15}$$

Note that $n = 0$ is not allowed, since by Eq. 4.12, all C_j would become zero. The negative n will give negative α, which will just multiply all C_j by -1 and therefore would not affect the relative signs of different C_j. Since the overall sign is not physical, we omit $n < 0$ also. This explains $n = 1, 2, 3, \ldots$ in Eq. 4.15. Let us place an index n on α to denote these distinct values,

$$\alpha_n = \frac{n\pi}{N+1}, \quad (n = 1, 2, 3, \ldots). \tag{4.16}$$

These different α_n correspond to the allowed normal mode frequencies according to Eq. 4.13. Here, we identify the normal mode frequencies by placing a mode index on ω also and writing them as ω_n,

$$\left(-\frac{\omega_n^2}{\omega_0^2} + 2\right) = 2\cos(\alpha_n). \tag{4.17}$$

Therefore, the normal frequencies of the system of N beads are

$$\omega_n^2 = 2\omega_0^2\,(1 - \cos\alpha_n), \tag{4.18}$$

which we can rewrite as follows using the double angle formula:

$$\omega_n^2 = 4\omega_0^2 \sin^2(\alpha_n/2). \tag{4.19}$$

Now, we use the values of α_n given in Eq. 4.16, to obtain the normal mode frequencies as follows:

$$\omega_n^2 = 4\omega_0^2 \sin^2\left[\frac{n\pi}{2(N+1)}\right] \implies \boxed{\omega_n = 2\omega_0 \sin\left[\frac{n\pi}{2(N+1)}\right].} \tag{4.20}$$

The mode amplitudes for each mode are determined by using the α_n for the corresponding mode in Eq. 4.12. Therefore, we need to attach a mode index to each C_j, e.g. $C_{n,j}$ for the amplitude of bead number j when moving in mode n. From Eqs. 4.12 and 4.16 we get

$$C_{n,j} = \sin(j\alpha_n) = \sin\left(\frac{jn\pi}{N+1}\right). \tag{4.21}$$

Note that the mode counter n is allowed to take any positive integer $(1, 2, 3, \ldots)$. Not all values of n correspond to a mode. For instance, when $n = (N+1), 2(N+1), 3(N+1), \ldots$, all amplitudes are zero. Therefore, these values of n do not correspond to any normal mode. Furthermore, the following calculation shows that the mode amplitudes for $n - N$ and $n + (N+2)$ are the same, such that all amplitude ratios are identical in the two cases. Furthermore, $C_{(n-N),j}$ and $C_{(N-n),j}$ are of opposite sign and so refer to the same mode.

$$\begin{aligned} C_{n+(N+2),j} &= \sin\left[\frac{j(n+N+2)\pi}{N+1}\right] = \sin\left[\frac{j(n+2(N+1)-N)\pi}{N+1}\right] \\ &= \sin\left[\frac{j(n-N)\pi}{N+1}\right] = C_{(n-N),j} = -C_{(N-n),j}. \end{aligned}$$

Table 4.1 *Normal modes of transverse oscillations of N oscillators.*

Mode frequency	Mode amplitudes
$\omega_1 = 2\omega_0 \sin\left[\frac{\pi}{2(N+1)}\right]$	$y_1(t) = \sin\left(\frac{\pi}{N+1}\right)$ $y_2(t) = \sin\left(\frac{2\pi}{N+1}\right)$ ... $y_N(t) = \sin\left(\frac{N\pi}{N+1}\right)$
$\omega_2 = 2\omega_0 \sin\left[\frac{2\pi}{2(N+1)}\right]$	$y_1(t) = \sin\left(\frac{2\pi}{N+1}\right)$ $y_2(t) = \sin\left(\frac{4\pi}{N+1}\right)$ ... $y_N(t) = \sin\left(\frac{2N\pi}{N+1}\right)$
...	...
$\omega_N = 2\omega_0 \sin\left[\frac{N\pi}{2(N+1)}\right]$	$y_1(t) = \sin\left(\frac{N\pi}{N+1}\right)$ $y_2(t) = \sin\left(\frac{2N\pi}{N+1}\right)$ ... $y_N(t) = \sin\left(\frac{N^2\pi}{N+1}\right)$

If all the signs of all y_j are flipped at the same time, we do not get a new mode, but just the same mode at another time in the oscillation. Therefore, the overall sign of C_j, which is independent of j, does not matter. Hence, the mode $n + (N + 2)$ is same as the modes $n - N$ and $N - n$, both in terms of the frequency and in terms of the mode amplitude ratios. Therefore, there are only N modes for the y-motion of the beads with indices n from 1 to N. For convenience, in Table 4.1 we list the mode frequencies and the corresponding oscillations of the beads.

It is instructive to display the normal modes of some particular values of N. We already know the modes for $N = 2$ from the last chapter. In Fig. 4.2, we display the modes for $N = 2$,

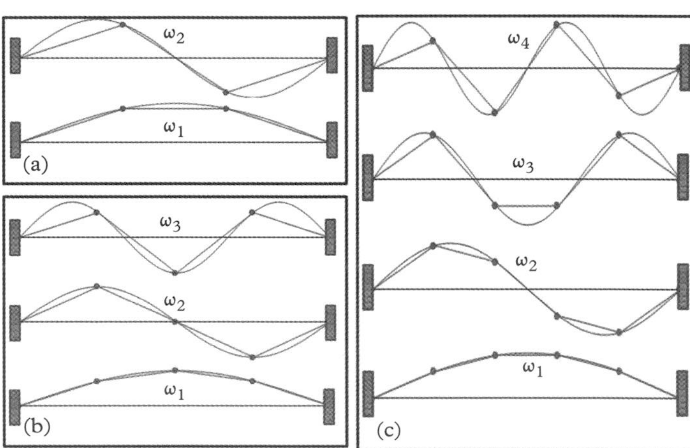

Fig. 4.2 *Transverse modes of (a) $N = 2$, (b) $N = 3$, and (c) $N = 4$. In each case ω_1 is the lowest frequency mode. The points at the nodes do not oscillate; other points of the string oscillate up and down with the frequency of the normal mode.*

$N = 3$, and $N = 4$. You can see that the lowest mode does not have any nodes in between the two ends, and each successively higher frequency mode has one additional node in its profile. The profiles in Fig. 4.2 show the maximum displacements of each bead from the equilibrium. In each mode, the beads execute vertical oscillations of the amplitudes shown in the figure. Only beads that have non-zero amplitude in a mode profile have harmonic motion. For instance, if a bead is at one of the nodes in a profile, then that bead remains at rest. To display the harmonic motion of the beads, one often includes the other extreme of the motion for each bead, as shown for a string in Fig. 4.7 later in the chapter.

4.1.2 General Solution

Using normal mode frequencies and mode ratios for the amplitudes of vibrations of the N beads we can immediately write the general solution for the movement of each bead,

$$y_1 = A_1 \cos(\omega_1 t - \phi_1) + A_2 \cos(\omega_2 t - \phi_2) + \cdots + A_N \cos(\omega_N t - \phi_N), \tag{4.22}$$

$$y_2 = B_{21}^{(1)} A_1 \cos(\omega_1 t - \phi_1) + B_{22}^{(2)} A_2 \cos(\omega_2 t - \phi_2) + \cdots + B_{2N}^{(N)} A_N \cos(\omega_N t - \phi_N), \tag{4.23}$$

$$\vdots$$

$$y_N = B_{N1}^{(1)} A_1 \cos(\omega_1 t - \phi_1) + B_{N2}^{(2)} A_2 \cos(\omega_2 t - \phi_2) + \cdots + B_{NN}^{(N)} A_N \cos(\omega_N t - \phi_N), \tag{4.24}$$

where $B_{ij}^{(n)}$ are the amplitude ratios in mode n,

$$B_{ij}^{(n)} = \frac{C_{n,j}}{C_{n,i}} = \frac{\sin\left(\dfrac{jn\pi}{N+1}\right)}{\sin\left(\dfrac{n\pi}{N+1}\right)}. \tag{4.25}$$

As usual, the $2N$ constants $\{A_j, \phi_j\}$ for $j = 1, 2, \ldots, N$ are determined from the $2N$ initial conditions $\{x_i(0), \dot{x}_i(0)\}$, for $i = 1, 2, \ldots, N$, where the over dot on x represents the derivative d/dt.

Example 4.1 Normal Modes of a Discrete System

Figure 4.3 shows four masses coupled with five springs. For simplicity, we take all masses equal to m and all spring constants equal to k. What are the normal modes for horizontal motion?

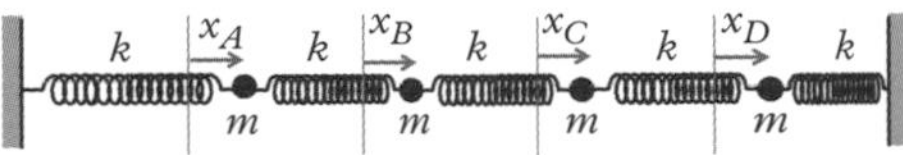

Fig. 4.3 *Example 4.1.*

Solution

Let us introduce the frequency parameter

$$\omega_0 = \sqrt{k/m}.$$

continued

Example 4.1 *continued*

Using dot notation for time derivatives, the equations of motion of the four masses will be

$$\ddot{x_A} = \omega_0^2\,[-2x_A + x_B]$$

$$\ddot{x_B} = \omega_0^2\,[x_A - 2x_B + x_C]$$

$$\ddot{x_C} = \omega_0^2\,[x_B - 2x_C + x_D]$$

$$\ddot{x_D} = \omega_0^2\,[x_C - 2x_D]$$

Now, for normal mode discovery, we express each of the displacements as oscillations at same frequency,

$$x_A = A\cos(\omega t),\;\; x_B = B\cos(\omega t),\;\; x_C = C\cos(\omega t),\;\; x_D = D\cos(\omega t).$$

Putting these into equations of motion results in algebraic equations for A, B, C, D, and ω. For ease of writing, let

$$\lambda = \omega^2/\omega_0^2.$$

We get the following in matrix notation:

$$\begin{bmatrix} 2 & -1 & 0 & 0 \\ -1 & 2 & -1 & 0 \\ 0 & -1 & 2 & -1 \\ 0 & 0 & -1 & 2 \end{bmatrix}\begin{bmatrix} A \\ B \\ C \\ D \end{bmatrix} = \lambda \begin{bmatrix} A \\ B \\ C \\ D \end{bmatrix}.$$

The eigenvalues, organized from lowest to highest are

$$\lambda_1 = \frac{1}{2}(3-\sqrt{5}),\;\; \lambda_2 = \frac{1}{2}(5-\sqrt{5}), \tag{4.26}$$

$$\lambda_3 = \frac{1}{2}(3+\sqrt{5}),\;\; \lambda_4 = \frac{1}{2}(5+\sqrt{5}). \tag{4.27}$$

This corresponds to frequencies

$$\omega_1 = \omega_0\sqrt{\frac{1}{2}(3-\sqrt{5})},\;\; \omega_2 = \omega_0\sqrt{\frac{1}{2}(5-\sqrt{5})}, \tag{4.28}$$

$$\omega_3 = \omega_0\sqrt{\frac{1}{2}(3+\sqrt{5})},\;\; \omega_4 = \omega_0\sqrt{\frac{1}{2}(5+\sqrt{5})}. \tag{4.29}$$

The eigenvectors are

$$\begin{bmatrix} A \\ B \\ C \\ D \end{bmatrix}_1 = \begin{bmatrix} 1 \\ 2+\frac{1}{2}(-3+\sqrt{5}) \\ 2+\frac{1}{2}(-3+\sqrt{5}) \\ 1 \end{bmatrix},\;\; \begin{bmatrix} A \\ B \\ C \\ D \end{bmatrix}_2 = \begin{bmatrix} 1 \\ -2+\frac{1}{2}(5-\sqrt{5}) \\ 2+\frac{1}{2}(-5+\sqrt{5}) \\ 1 \end{bmatrix},$$

$$\begin{bmatrix} A \\ B \\ C \\ D \end{bmatrix}_3 = \begin{bmatrix} 1 \\ 2+\frac{1}{2}(-3-\sqrt{5}) \\ 2+\frac{1}{2}(-3-\sqrt{5}) \\ 1 \end{bmatrix},\;\; \begin{bmatrix} A \\ B \\ C \\ D \end{bmatrix}_4 = \begin{bmatrix} 1 \\ 2-\frac{1}{2}(5+\sqrt{5}) \\ -2+\frac{1}{2}(5+\sqrt{5}) \\ -1 \end{bmatrix}$$

4.2 The Normal Modes in the Continuum Limit

From the results of N beads on a taut string, we can deduce the oscillations of a taut string of total mass M and total length L by modeling the latter as a discrete system. The trick is to distribute the mass of the string into "beads" placed on a massless string. Let us divide the mass M of the string into N parts each of mass $m = M/N$. We place these masses at equal intervals of distance $l = L/(N+1)$, as shown in Fig. 4.4. Now, this system is the N beads on the string system,

$$Nm = M, \tag{4.30}$$

$$(N+1)l = L. \tag{4.31}$$

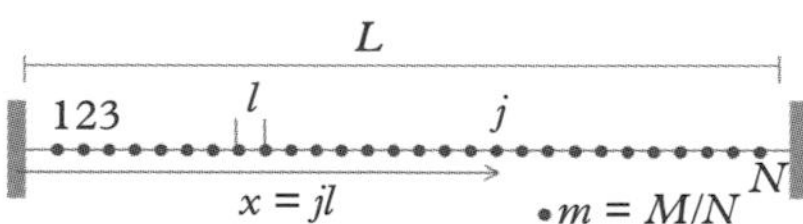

Fig. 4.4 *A discrete model of a continuous string. The string is divided into N parts which are distributed uniformly. We assume that the masses are coupled by linear restoring forces for transverse displacements. In the text, $j = 0$ and $j = N + 1$ are used for the points fixed at the supports.*

As far as vibrations of finite bodies are concerned note that there are two length scales in the system—the elementary distance l between successive masses and the length λ over which significant deformations of the body occur while oscillating in a particular mode. In regular solids $l \sim 10^{-9}$ m and vibration modes of interest have $\lambda \sim 10^{-3}$ m to 10 m. This makes $\lambda \gg l$. If λ of a mode is much larger than the inter-particle separation, i.e. if $\lambda \gg l$, then the mode profile will not change much over many particles in the system, and the physical body will be essentially continuous.

To obtain the modes of a continuous string we will need to decrease the distance l between the beads indefinitely. With the length L of the string fixed this would mean the number of beads N would increase indefinitely. As we increase N, while keeping the total length L and the total mass M fixed at the given values for the string as a whole, mass m of each "bead" becomes smaller and the distance l between them decreases. The distance l between the masses becomes infinitesimal in the limit $N \Rightarrow \infty$. Thus, in this limit the x-coordinate of masses m on the string can be assumed to be a continuous variable.

Now, multiplying the numerator and the denominator of the argument of the sine function for the modes in the N-bead problem given in Eq. 4.21 by l, we can write the formula for the amplitudes of the displacements of the elements of the string as functions of x,

$$C_{n,j} \xrightarrow{x=jl} C_n(x) = A_n \sin\left[\frac{n\pi x}{L}\right]. \tag{4.32}$$

This shows that the space-dependence of the normal modes are a periodic function of x if the domain of x is extended over the entire x-axis. The repeat distance for the nth mode, called the **wavelength** of the mode, will be designated by λ_n. For a string of length L fixed at both ends, the wavelength of the nth mode would be

$$\lambda_n = \frac{2L}{n}. \tag{4.33}$$

A quantity related to the inverse of the wavelength, called the **wave number** and denoted by letter k, is often used when discussing waves. The wave number k_n for the nth mode is related to the wavelength λ_n by

$$k_n = \frac{2\pi}{\lambda_n} = \frac{\pi n}{L}. \tag{4.34}$$

To find the normal mode frequencies we take the continuum limit of Eq. 4.20 by taking $N \to \infty$ with $Nl = L$, the length of the string.

$$\omega_n^2 = 4\omega_0^2 \sin^2\left[\frac{n\pi}{2(N+1)}\right] \longrightarrow 4\omega_0^2 \frac{n^2\pi^2}{4(N+1)^2}$$

$$= 4\left(\frac{T}{\mu l^2}\right)\frac{n^2\pi^2}{4N^2} = \left(\frac{T}{\mu}\right)\left(\frac{n\pi}{L}\right)^2. \tag{4.35}$$

We can express this result as a relation between the mode frequency and the mode wave number,

$$\omega_n^2 = \left(\frac{T}{\mu}\right) k_n^2. \tag{4.36}$$

The relation between ω and k is called the **dispersion relation** of the system. Here the dispersion relation in the continuum limit (Eq. 4.36) is a linear relation between ω and k, while the relation in the discrete system of finite N as given in Eq. 4.20 is nonlinear.

To get a feel for the oscillation modes, in Fig. 4.5 we plot some of the lower modes for large N. The boundary condition places nodes at the ends. There is no node between the ends in the lowest mode. The lowest mode is symmetric about the middle. The second mode has one node, the third mode two nodes, etc. If we *imagine* extending the modes outside of $0 \leq x \leq L$ to $\infty \leq x \leq \infty$, you will find that the lowest mode amplitude function $C_1(x)$ is periodic in space with the **spatial period** or wavelength $\lambda_1 = 2L$ the next mode amplitude function $C_2(x)$ will have a spatial period of $\lambda_2 = 2L/2$, and so on.

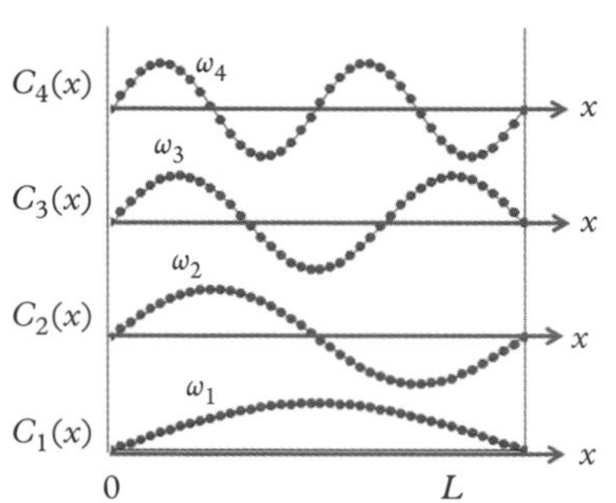

Fig. 4.5 *Four lowest frequency modes of vibration of a string in $N = 40$ parts. We see that the masses start to form almost continuous string and their positions fall on the sine function given in Eq. 4.32.*

4.3 Vibrations of a Taut String—Continuum Model

In the last section, we arrived at the modes of lower frequency vibration modes of a continuous body starting from its discrete structure. However, to study continuous bodies, it is not necessary to start from a discrete structure. In this section, we will present a derivation of the same vibration modes by treating physical bodies in a continuum way from the start.

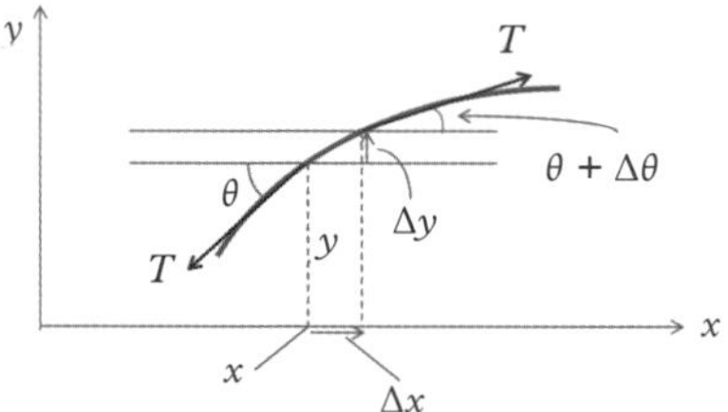

Fig. 4.6 *The tension forces on the two sides of an element of the taut string. The transverse displacements in y are exaggerated to make the figure clearer; in the text, we are interested in y small compared to x. Note the slope $\Delta y/\Delta x = \tan\theta$.*

4.3.1 Derivation of Wave Equation

Consider a taut string of total mass M and length L whose ends are fixed to immovable supports so that there is a tension T throughout. To find the equation of motion of the string, we focus our attention on an infinitesimal element Δl of the string at an arbitrary place along the string. The forces on the two sides of the mass element of the string at an arbitrary instant are shown in Fig. 4.6.

The x- and y-components of forces on the element are

$$F_x = T\cos(\theta + \Delta\theta) - T\cos(\theta), \tag{4.37}$$

$$F_y = T\sin(\theta + \Delta\theta) - T\sin(\theta). \tag{4.38}$$

We can simplify these expressions for the force components in the case of a small transverse displacement. To implement the small transverse displacement condition we require Δy to be much smaller than the length Δl of the element,

$$\Delta y \ll \Delta l, \tag{4.39}$$

which translates into the following approximations for the sine and cosine of $\Delta\theta$ when we keep terms up to one power of $\Delta\theta$ in the expansion of the trigonometric functions:

$$\sin(\Delta\theta) \approx \Delta\theta, \tag{4.40}$$

$$\cos(\Delta\theta) \approx 1. \tag{4.41}$$

Since no point of the string moves significantly away from the horizontal axis, we have the following additional simplifications due to the smallness of the angle θ, not just the change in the angle $\Delta\theta$:

$$\sin(\theta) \approx \theta, \tag{4.42}$$

$$\cos(\theta) \approx 1, \tag{4.43}$$

$$(\theta\Delta\theta) < \min(\theta, \Delta\theta). \tag{4.44}$$

These approximations can be applied to Eqs. 4.37 and 4.38 after expanding the trigonometric expressions using the sum of angles formulas,

$$\cos(\theta + \Delta\theta) = \cos(\theta)\cos(\Delta\theta) - \sin(\theta)\sin(\Delta\theta), \tag{4.45}$$

$$\sin(\theta + \Delta\theta) = \sin(\theta)\cos(\Delta\theta) + \cos(\theta)\sin(\Delta\theta), \tag{4.46}$$

to arrive at the following expressions for the x- and y-components of the force on the element of the string:

$$F_x \approx 0, \tag{4.47}$$

$$F_y \approx T\Delta\theta. \tag{4.48}$$

Therefore, the y-component of the equation of motion for the element under consideration is

$$\Delta m \frac{\partial^2 y}{\partial t^2} \approx T\Delta\theta, \tag{4.49}$$

where Δm is the mass of the element, which can be written in terms of mass per unit length μ,

$$\Delta m = \mu\Delta l \approx \mu\Delta x. \tag{4.50}$$

Note that, here we use partial derivatives since the displacement y depends on two independent variables x and t, and therefore, y should be and is a bivariate function: $y(x, t)$. Making Δx infinitesimal, we find the following equation of motion for the y-displacement of the element located at x, or more specifically located between x and $x + dx$:

$$\frac{\partial^2 y}{\partial t^2} = \left(\frac{T}{\mu}\right)\frac{\partial\theta}{\partial x}. \tag{4.51}$$

The partial derivative of θ with respect to x can be written in terms of y by noting that the slope of y as a function of x is equal to $\tan\theta$, as shown in Fig. 4.6.

$$\tan(\theta) = \frac{\partial y}{\partial x}. \tag{4.52}$$

Taking derivatives with respect to x on both sides we obtain

$$\sec^2(\theta)\,\frac{\partial \theta}{\partial x} = \frac{\partial^2 y}{\partial x^2}. \tag{4.53}$$

Now, we use the small displacement approximation to set $\sec^2(\theta) = 1/\cos^2(\theta) = 1$. This gives us

$$\frac{\partial \theta}{\partial x} = \frac{\partial^2 y}{\partial x^2}. \tag{4.54}$$

Inserting this in Eq.4.51, we find

$$\frac{\partial^2 y}{\partial t^2} = \left(\frac{T}{\mu}\right)\frac{\partial^2 y}{\partial x^2}. \tag{4.55}$$

This equation is called the **classical wave equation** or simply the wave equation. The wave equation shows up in many areas of physics and it will be helpful to you if you study this equation in some detail. The quantity (T/μ) in this equation is particular to the taut string. In other situations we will have other properties in place of (T/μ). Dimensionally, (T/μ) has dimensions of speed squared. It is common to denote T/μ by v^2,

$$v^2 = \frac{T}{\mu}. \tag{4.56}$$

With this substitution we arrive at the more common form for the wave equation,

$$\frac{\partial^2 y}{\partial x^2} = \frac{1}{v^2}\frac{\partial^2 y}{\partial t^2}. \tag{4.57}$$

This equation governs the dynamics of the transverse vibrations of the taut string. At the moment v is just a parameter in the equation with dimensions of speed. Later on we will find that v is equal to the speed of traveling waves on the string. The wave equation (4.57) is the statement of Newton's law for the motion of the mass elements of the string which are coupled to their neighbors. Just as the coupled systems had special motions called the normal modes, the wave equation also has special solutions which we identify as the modes of the system. In this chapter we will solve the wave equation for three boundary conditions and deduce the normal modes corresponding to the system with these boundary conditions:

(1) Both ends fixed
(2) One end fixed, the other end free
(3) Both ends free

4.3.2 Modes of a String Fixed at Both Ends

In this section we will find oscillatory solutions of the wave equation, Eq. 4.57, for a string of length L that is tied to fixed posts at both ends as in Fig. 4.4. That is, we are interested in the following boundary conditions on the solution:

$$y(0, t) = 0, \tag{4.58}$$

$$y(L, t) = 0. \tag{4.59}$$

To solve Eq. 4.57, we will use a technique called the **separation of variables**. In this technique we first find a particular type of solution called a **product solution,**

$$y(x, t) = f(x)g(t). \tag{4.60}$$

The actual solution of the wave equation that satisfies the boundary conditions and the initial conditions may not be of this type. Actually, we will find that there are many solutions of the product type that satisfy most of the conditions of the required solution. We will label them by 1, 2, 3, ..., etc.

$$y(x, t) = f_1(x)g_1(t),\ f_2(x)g_2(t),\ f_3(x)g_3(t), \ldots. \tag{4.61}$$

To obtain the solution that will satisfy all conditions, we take a superposition of these product solutions, and look for the coefficients $\{A_i\}$ so that the resulting $y(x, t)$ satisfies all of the remaining conditions,

$$y(x, t) = \sum_i A_i f_i(x)g_i(t). \tag{4.62}$$

Let us work out the details now for the case of the string tied at both ends. We start by using the product form of the solution given in Eq. 4.60 in the wave equation Eq. 4.57,

$$g(t)\frac{d^2f}{dx^2} = \frac{1}{v^2}f(x)\frac{d^2g}{dt^2}. \tag{4.63}$$

Note that the partial derivatives have now turned into ordinary derivatives.

Now, dividing both sides by fg we obtain an equation where one side depends only on x and the other side only on t,

$$\frac{1}{f}\frac{d^2f}{dx^2} = \frac{1}{v^2}\frac{1}{g}\frac{d^2g}{dt^2}. \tag{4.64}$$

Since the two sides are functions of different independent variables, and hence independent of each other, they can be equal to each other only if they are each equal to some constant, which we will denote by the Greek letter α for now,

$$\frac{1}{f}\frac{d^2f}{dx^2} = \alpha, \tag{4.65}$$

$$\frac{1}{v^2}\frac{1}{g}\frac{d^2g}{dt^2} = \alpha. \tag{4.66}$$

Rearranging these equations we get the following:

$$\frac{d^2 f}{dx^2} = \alpha f, \tag{4.67}$$

$$\frac{d^2 g}{dt^2} = \alpha v^2 g. \tag{4.68}$$

The second of these equations would be an equation for a harmonic oscillator if the constant on the right side αv^2 were a negative quantity. Since we seek oscillatory solutions, we will demand that this be the case and replace α by $-k^2$. We do not need to think in terms of α any more, we will instead think in terms of k, which will turn out to be the wave number,

$$\frac{d^2 f}{dx^2} = -k^2 f, \tag{4.69}$$

$$\frac{d^2 g}{dt^2} = -k^2 v^2 g. \tag{4.70}$$

The general solutions of these equations are easy to write down,

$$f(x) = A\cos(kx) + B\sin(kx), \tag{4.71}$$

$$g(t) = C\cos(kvt) + D\sin(kvt). \tag{4.72}$$

The constants C and D in $g(t)$ will be related to the initial conditions and the constants A and B in $f(x)$ will be related to the boundary conditions. For instance, for the boundary condition $y(0,t) = 0$ at $x = 0$, we must have $f(0) = 0$,

$$0 = f(0) = A\cos(0) + B\sin(0) = A. \implies A = 0.$$

Therefore, $f(x)$ simplifies by the boundary condition at $x = 0$,

$$f(x) = B\sin(kx). \tag{4.73}$$

Now, we attempt to satisfy the condition at $x = L$, which is $y(L,t) = 0$ giving $f(L) = 0$,

$$0 = f(L) = B\sin(kL) \implies \sin(kL) = 0. \tag{4.74}$$

This places restrictions on the possible values for the constant k,

$$kL = \pm\pi, \pm 2\pi, \pm 3\pi, \ldots. \tag{4.75}$$

Note that we have left out $k = 0$ since if $k = 0$ we will have $f(x) = 0$. This will correspond to the string not oscillating at all. We can write these solutions compactly as

$$k_n = \frac{n\pi}{L}, \quad n = \pm 1, \pm 2, \ldots. \tag{4.76}$$

The solution for the time part, Eq. 4.72, shows that the angular frequency would also be restricted to particular values since k is restricted to be one of the k_n's. Writing ω for the angular frequency for $g(t)$ we have

$$\omega = kv \implies \omega_n = k_n v, \quad n = \pm 1, \pm 2, \ldots. \tag{4.77}$$

This gives a negative ω_n value for negative values of k_n. However, we are interested in only the positive frequency values. Therefore, we will further restrict the possible values for k and ω to

$$\begin{aligned} k_n &= n\tfrac{\pi}{L}, \quad n = 1, 2, \ldots, \\ \omega_n &= k_n v. \end{aligned} \tag{4.78}$$

The process above has identified an infinite number of solutions of the wave equation that satisfy the two boundary conditions at $x = 0$ and $x = L$. We can label the constant coefficients and the solution y with n also. We can also absorb the overall factor B into the other constants C and D to obtain the following possible solutions:

$$y_n(x, t) = \sin(k_n x)\,[C_n \cos(k_n vt) + D_n \sin(k_n vt)]\,, \quad n = 1, 2, \ldots. \tag{4.79}$$

These solutions are also normal modes of the string, since when the string is oscillating according to one of these solutions every element of the string, i.e. the string at every point x, oscillates at the same frequency $\omega_n = k_n v$. Thus, the product solution here gives us the normal mode frequencies. We can combine the sines and cosines in the time part and write them as one cosine with an amplitude and a phase constant for each n,

$$y(x, t) = [A_n \sin(k_n x)]\, \cos(k_n vt - \phi_n). \tag{4.80}$$

In this form the solution looks like the solutions of the other discrete systems we have studied before. The factor $[A_n \sin(k_n x)]$ is the amplitude of the element at x. Therefore, in this mode, the transverse displacement amplitude ratios of elements at $x = x_1$ and $x = x_2$ are given by

$$\frac{\text{Amplitude at } x_1}{\text{Amplitude at } x_2} = \frac{\sin(k_n x_1)}{\sin(k_n x_2)} \text{ if } \sin(k_n x_2) \neq 0. \tag{4.81}$$

Shapes of Modes

It is informative to plot some of the lowest frequency modes over a cycle to get a feel for the oscillations of the string. In Fig. 4.7 we plot the oscillations of the lowest frequency mode, $n = 1$, of a 1-m-long taut string. The lowest frequency mode is symmetric about the middle position. Suppose, we start the string such that the points of the string are coincident with the mode shape and let go from rest. Figure 4.7 shows the position of the string at successive intervals of $(1/8)^{th}$ periods. The string moves more slowly in the first $(1/8)^{th}$ period, when it is furthest from the equilibrium, than in the second and third $(1/8)^{th}$ periods when it passes through the equilibrium position. This behavior is similar to the behavior of a mass attached to a spring. The particles near the middle of the string cover large distance in the

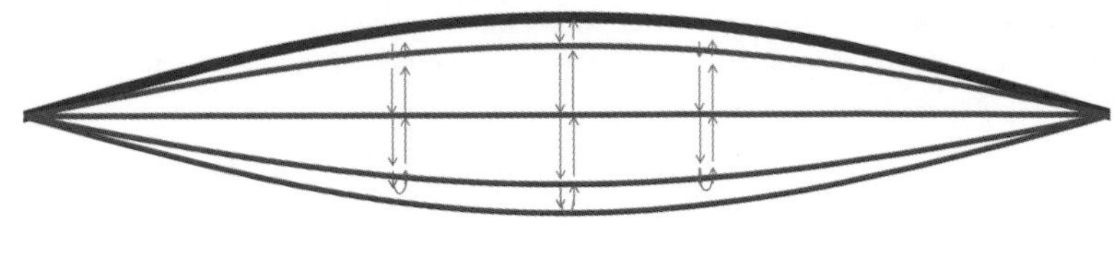

Fig. 4.7 *The fundamental mode of a string fixed at both ends. The string is plucked in the shape shown as a thick black line and let go from rest. The string's location at successive 1/8th period intervals is shown in the figure. The string vibrates such that the ratios of the vertical displacements at different horizontal locations remain independent of time.*

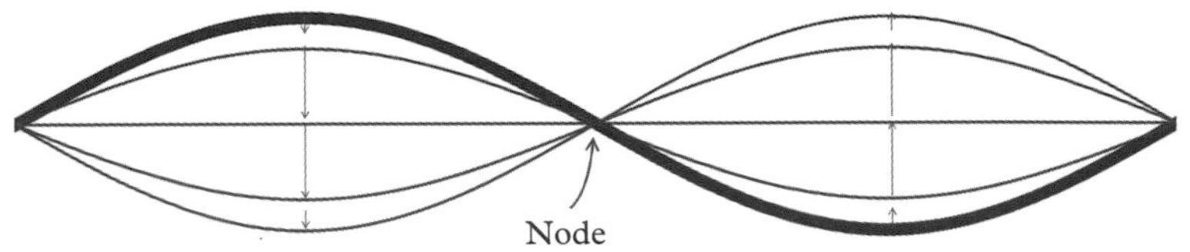

Fig. 4.8 *The second lowest frequency mode, $n = 2$, of a string fixed at both ends. The taut string is pulled in the shape shown as a thick black line and let go from rest. The string's location at successive 1/8th period intervals is shown in the figure. The point labeled "Node" is stationary and does not move.*

same time than particles near the fixed ends, while all particles of the string have the same period of oscillation.

The first overtone, i.e. mode $n = 2$, is shown in Fig. 4.8. The first overtone mode is antisymmetric about the middle position. When the left half of the string moves up, the right half moves down, and vice versa, with every particle oscillating with frequency ω_2. The point in the middle remains at rest and is a node of the mode.

General Solution

The most general solution of the given problem of the oscillations of the taut string tied at two ends can be constructed from a superposition of the normal modes, as we have seen in the study of the coupled systems with two degrees of freedom,

$$y(x,t) = \sum_{n=1}^{\infty} \sin(k_n x)\,[C_n \cos(k_n vt) + D_n \sin(k_n vt)]. \tag{4.82}$$

The constants C_n and D_n are determined from the initial conditions $y(x,0)$ and $\dot{y}(x,0)$ where the overdot stands for the time derivative, as before.

Example 4.2 Starting Vibration in a Normal Mode

Suppose the string were let go from rest while it had a shape given by the function $y(x,0) = \sin(\pi x/L)$. What would be the subsequent motion?

Solution

A quick answer: Since we started the string in the shape of mode 1 at rest and since there are no nonlinear forces among the mass elements of the string, the string will continue to oscillate in mode 1 only. This is the same conclusion that we had in the two-variable coupled systems. Therefore, we expect

$$y(x,t) = \sin(\pi x/L)cos(\pi vt/L),$$

where $v = \sqrt{T/\mu}$.

Longer answer: Now, let us see how this conclusion can also be found by satisfying initial conditions on the general solution. Let us first list the two initial conditions on $y(x, t)$ that we must satisfy in this case,

$$y(x, 0) = \sum_{n=1}^{\infty} \sin(k_n x) C_n = \sin(\pi x/L),$$

$$\dot{y}(x, 0) = \sum_{n=1}^{\infty} \sin(k_n x) k_n v D_n = 0.$$

From the second equation we get

$$D_n = 0, \quad n = 1, 2, 3, \ldots,$$

while the first equation gives the following:

$$C_1 = 1,$$

$$C_n = 0, \quad n \neq 1.$$

Therefore, the oscillations of the string for the given initial conditions in this example will be

$$y(x, t) = \sin\left(\frac{\pi x}{L}\right) \cos\left(\frac{\pi v t}{L}\right),$$

which is same as we found by appealing to the rule: in linear systems, once in a mode, always in that mode.

Example 4.3 Starting Vibration in a Superposition of Normal Modes

Suppose the string were let go from rest while it had a shape given by the function $y(x, 0) = 2\sin(\pi x/L) + 3\sin(2\pi x/L)$. What would be the subsequent motion?

Solution

A quick answer: Since we started the string in the shape of superposition of mode 1 and mode 2 at rest and since there are no nonlinear forces among the mass elements of the string, the string will continue to oscillate in the same superposition. Each mode in the superposition will oscillate at its own frequency Therefore, we expect

$$y(x, t) = 2\sin(\pi x/L) cos(\pi v t/L) + 3\sin(2\pi x/L) cos(2\pi v t/L),$$

where $v = \sqrt{T/\mu}$.
Longer answer: We again start by first listing the initial conditions on displacement and velocity,

$$y(x, 0) = \sum_{n=1}^{\infty} \sin(k_n x) C_n = 2\sin(\pi x/L) + 3\sin(2\pi x/L),$$

$$\dot{y}(x, 0) = \sum_{n=1}^{\infty} \sin(k_n x) k_n v D_n = 0.$$

continued

Example 4.3 *continued*

Equating the coefficients of the same sines on the two sides, we get the constants to be

$$D_n = 0, \quad n = 1, 2, 3, \dots,$$
$$C_1 = 2, \quad C_2 = 3$$
$$C_n = 0, \quad n = 3, 4, \dots.$$

Therefore,

$$y(x,t) = 2\sin\left(\frac{\pi x}{L}\right)\cos\left(\frac{\pi vt}{L}\right) + 3\sin\left(\frac{2\pi x}{L}\right)\cos\left(\frac{2\pi vt}{L}\right).$$

4.4 Transverse Oscillations of a String Free at One End

The vibration problem of a string fixed at only one end differs in boundary condition from the case of both ends fixed. Let us simplify the problem so that the string remains in the xy-plane. Note that we still wish to maintain the string with tension T. One way to physically realize this problem is to attach the string to a light (massless) ring and let the ring glide frictionlessly on a rod while keeping the string taut. This arrangement will ensure that the tension is maintained at some fixed value T, as shown in Fig. 4.9. Since the tension in the string also acts on the massless ring and since there is no friction along the rod, the y-component of the tension at the ring will be zero. This physical requirement on T_y gives the boundary condition at the free end at $x = L$. Let θ be the angle the string makes with the horizontal direction at $x = L$, then we get

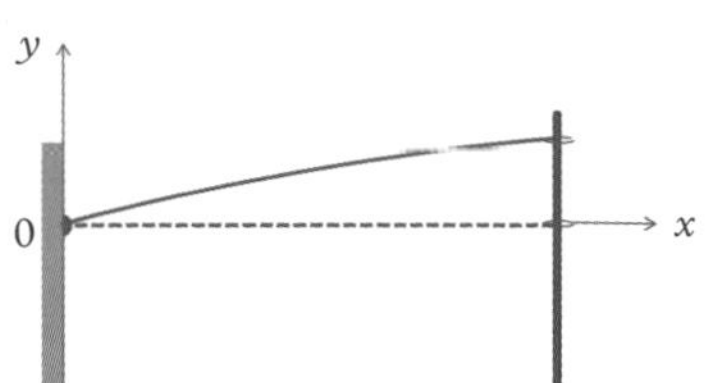

Fig. 4.9 *A physical realization of a string with one free end and one fixed end. At the free end, the string is tied to a light ring which can glide frictionlessly on a rod while maintaining the tension T in the string.*

$$y\text{-component:} \quad -T\sin\theta = 0 \quad (\text{at } x = L). \tag{4.83}$$

Since $T \neq 0$, we must have

$$\sin\theta\,|_{x=L} = 0. \tag{4.84}$$

We use small-angle approximation to replace the sine function with the tangent function so that we can relate the condition on θ to the slope of y,

$$\tan\theta\,|_{x=L} = 0. \tag{4.85}$$

This gives the following condition on the profile $y(x)$ at the free boundary at $x = L$:

$$\left(\frac{\partial y}{\partial x}\right)_{x=L} = 0. \tag{4.86}$$

Thus, our mathematical problem is: Find the solution of the wave equation

$$\frac{\partial^2 y}{\partial x^2} = \frac{1}{v^2}\frac{\partial^2 y}{\partial t^2}, \tag{4.87}$$

with the following boundary conditions:

$$y(0, t) = 0, \tag{4.88}$$

$$\left(\frac{\partial y}{\partial x}\right)_{x=L} = 0. \tag{4.89}$$

You can follow the same procedure as outlined previously when we solved the problem where the string was fixed at both ends. This calculation is left as a fruitful exercise for the student. Here, we quote the answer only,

$$y_n(x, t) = A_n \sin(k_n x) \cos(\omega_n t - \phi_n), \tag{4.90}$$

$$k_n = (2n - 1)\frac{\pi}{2L} \quad (n = 1, 2, \ldots), \tag{4.91}$$

$$\omega_n = v k_n. \tag{4.92}$$

The general solution is written as a superposition of all of the normal modes,

$$y(x, t) = \sum_{n=0}^{\infty} A_n \sin(k_n x) \cos(\omega_n t - \phi_n). \tag{4.93}$$

We use the initial conditions to determine A_n and ϕ_n, as illustrated in the Fourier series technique later in this chapter. For instance, if at $t = 0$ the string has $y(x, 0) = p(x)$ and $\dot{y}(x, 0) = q(x)$ then

$$\sum_{n=0}^{\infty} A_n \sin(k_n x) \cos(\phi_n) = p(x), \tag{4.94}$$

$$\sum_{n=0}^{\infty} A_n \omega_n \sin(k_n x) \sin(\phi_n) = q(x). \tag{4.95}$$

Fourier techniques can be used to determine $A_n \cos(\phi_n)$ from the first equation and $A_n \omega_n \sin(\phi_n)$ from the second, from which we can readily deduce A_n and ϕ_n.

4.4.1 The Modes of a String with Both Ends Free

When both ends of the taut string are free to move in the perpendicular direction to the string while still under tension T, then the boundary conditions will be

$$\left(\frac{\partial y}{\partial x}\right)_{x=0} = 0, \tag{4.96}$$

$$\left(\frac{\partial y}{\partial x}\right)_{x=L} = 0. \tag{4.97}$$

The normal modes for these boundary conditions will be cosine functions in the position variable,

$$y_n(x, t) = A_n \cos(k_n x) \cos(\omega_n t - \phi_n), \tag{4.98}$$

$$k_n = \frac{n\pi}{L} \quad (n = 1, 2, \cdots), \tag{4.99}$$

$$\omega_n = v k_n. \tag{4.100}$$

Figure 4.10 shows some of the low frequency modes for the three boundary conditions we have discussed above. Note that when both ends are free, the lowest mode will have a node between the ends, while there is no node in the lowest mode of the other two boundary conditions.

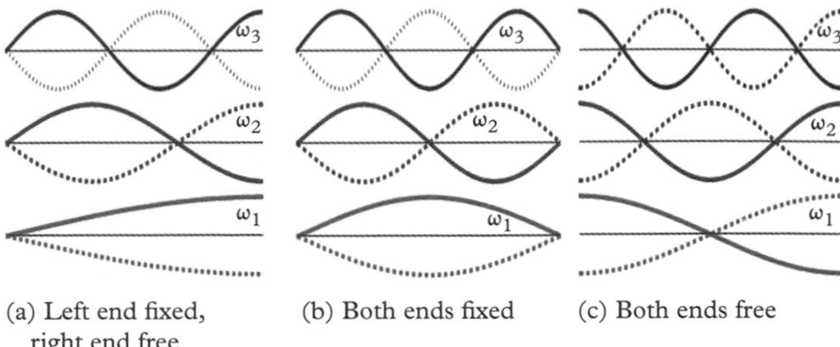

Fig. 4.10 *The three lowest frequency transverse modes of oscillations corresponding to three boundary conditions, one side clamped the other free, both sides clamped, and both sides free. The frequencies of the three modes are given in the text.*

4.5 Longitudinal Oscillations

4.5.1 Stress and Strain

Most physical bodies can be deformed by applying an external force. When a body is compressed or extended, the binding forces between atoms and molecules of the body change to counter the external applied force. In elastic materials, the original shape of the body is recovered when the external force is removed. Therefore, elastic materials are like springs and they can be successfully modeled by assuming that an elastic body is made up of coupled oscillators connected by spring-like forces. Internal forces develop in an elastic material as a response to an applied force that is proportional to the resulting deformation.

Suppose a rod is pulled along its length. Then we find that an internal force F in the rod develops as a response to the applied force. According to Hooke's law (the same law that explains the spring force), the magnitude of the internal force will be proportional to the percentage change in length if the change is not too great,

$$F \propto \frac{\Delta L}{L}, \tag{4.101}$$

where ΔL is the change in length of the rod of length L. We also find that a greater force is needed for a thicker rod, i.e. F is also proportional to the cross-sectional area A,

$$F \propto A. \tag{4.102}$$

Therefore, the internal force F in an elastic rod that has been extended is given as

$$F = YA\frac{\Delta L}{L}, \tag{4.103}$$

where the proportionality constant Y is called the **Young's modulus**. Young's modulus is a characteristic of the material of the rod. Equation 4.103 shows that, instead of change in length, a more relevant quantity is the relative change in length, and, instead of force on the material, the physically more relevant quantity is the force per unit area. The relative change in length, called **strain**, gives us the relevant measure of deformation of the material, and the force per unit area, called **stress** gives a measure of relevant internal forces,

$$\text{strain} = \frac{\Delta L}{L}, \tag{4.104}$$

$$\text{stress} = \frac{F}{A}. \tag{4.105}$$

Therefore, in terms of stress and strain, Eq. 4.103 is also written as

$$\text{stress} = Y \times \text{strain}. \tag{4.106}$$

The stress must be the same throughout a physical body in a static equilibrium. But, when the body is not in static equilibrium the stress inside the body will be position and time dependent, just as was the case with the mass/spring system where force changed with time and displacement. The body is said to be away from equilibrium when the stress varies in the body causing acceleration in the elements of the body. We will see below that particles in a body oscillate if they are displaced from their equilibria due to an imbalance of the stress in the body. If the vibrations are parallel to the long axis of the body, we call them **longitudinal vibrations**, and if they are perpendicular to the length, then we call them **transverse vibrations**, as we have seen above.

4.5.2 Longitudinal Vibrations in a Rod

Suppose a rod is clamped at one end to a rigid support and then pulled at the other end and released from rest. We will find that different parts of the rod will vibrate along the axis of the rod in a complicated motion reminiscent of the arbitrary motion of coupled oscillators, unless the starting configuration was in a normal longitudinal mode. If the rod were pulled and care was taken to excite one of the normal modes only, then the rod would vibrate in that normal mode only, as was the case with the transverse oscillations. The longitudinal vibrations in a rod turn out to be similar to transverse vibrations of a string: they obey a similar wave equation. In this subsection we will derive the relevant equations of motion and present some solutions.

Consider a rod of length L, cross-sectional area A, and Young's modulus Y. We take the x-axis as the axis of the rod with origin at the fixed end. In order to describe the strain at some point x, it is necessary to know how an infinitesimal length Δx, between x and $x + \Delta x$, has changed. That is, we would like to distinguish between the displacements of particles at x and $x + \Delta x$ since we need the change in Δx itself. Therefore, we introduce another symbol $\psi(x, t)$ for the displacement of the particle at x at time t. The velocity and acceleration of the particle at x will be

$$v_x = \frac{\partial \psi}{\partial t}, \tag{4.107}$$

$$a_x = \frac{\partial^2 \psi}{\partial t^2}. \tag{4.108}$$

We wish to write the force on the particle at x so that we can use Newton's second law of motion to obtain the equation of motion. Consider an element of the rod that has a length Δx at equilibrium, as shown in Fig. 4.11. In the non-equilibrium situation, the rod will be deformed such that the particle which was at point x in equilibrium will be at $(x + \psi)$, and

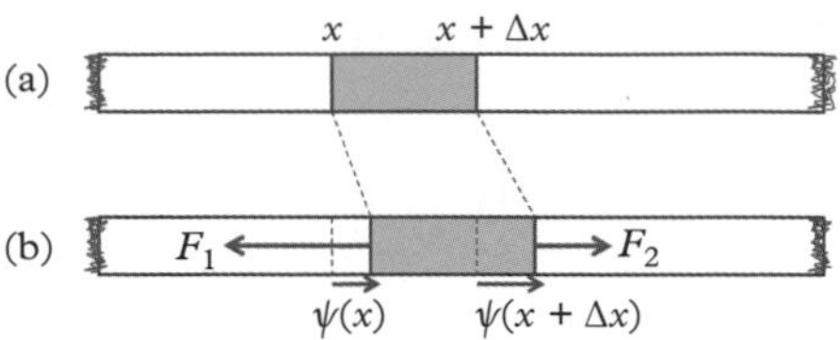

Fig. 4.11 *The longitudinal oscillations of a rod. (a) Two nearby points on the rod at equilibrium. (b) When the rod is away from equilibrium, the particles that used to be at point x and x+ Δx are now displaced by $\psi(x)$ and $\psi(x + \Delta x)$ respectively. The difference between the forces F_1 and F_2 provides the restoring force on the material, shown shaded here.*

the particle that was at $(x + \Delta x)$ will be at $[(x + \psi + \Delta x) + \Delta\psi]$. Therefore, the strain of the element Δx at the instant will be

$$\text{strain} = \frac{\Delta\psi}{\Delta x}. \tag{4.109}$$

The formula for the strain at x, to be denoted by $\epsilon(x)$, can be obtained by taking the $\Delta x \to 0$ limit of this equation,

$$\epsilon(x) = \frac{\partial\psi}{\partial x}. \tag{4.110}$$

The stresses on the two ends of the infinitesimal element Δx, denoted as $\sigma(x)$ and $\sigma(x + \Delta x)$, are

$$\sigma(x) = Y\,\epsilon(x), \tag{4.111}$$

$$\sigma(x + \Delta x) = Y\,\epsilon(x + \Delta x). \tag{4.112}$$

Therefore, the x-component F_x of the net force on the masses in the element Δx is given as

$$F_x = F_2 - F_1 = \sigma(x + \Delta x)A - \sigma(x)A. \tag{4.113}$$

Expanding $\sigma(x + \Delta x)$ about x and keeping only two terms in the expansion, we obtain the following:

$$\sigma(x + \Delta x) \approx \sigma(x) + \frac{\partial\sigma}{\partial x}\,\Delta x. \tag{4.114}$$

Therefore, the x-component of the net force will be

$$F_x = \frac{\partial\sigma}{\partial x}\,\Delta x\,A. \tag{4.115}$$

Now, using $\sigma = Y\,\epsilon$ from Eq.106 and ϵ from Eq. 4.110 in this equation we get

$$F_x = Y\,A\,\Delta x\,\frac{\partial^2\psi}{\partial x^2}. \tag{4.116}$$

We will equate this to ma_x where the mass m in the element is

$$m = \rho A \Delta x, \tag{4.117}$$

where ρ is the density and the average acceleration of these masses is given in Eq. 4.108. Therefore, the equation of motion is given as

$$\frac{\partial^2 \psi}{\partial x^2} = \frac{\rho}{Y} \frac{\partial^2 \psi}{\partial t^2}. \tag{4.118}$$

In deriving Eq. 4.118, Δx cancels out in an intermediate step. We recognize this as a wave equation for the displacement ψ with speed, v, given as

$$v = \sqrt{Y/\rho}. \tag{4.119}$$

Note that here, the speed has a formal resemblance to the speed for the transverse waves on a taut string, which was given by $v = \sqrt{T/\mu}$, where T was the tension and μ the mass per unit length. In both cases, the speed is related to the ratio of an elastic property (T or Y) and an inertial property (μ or ρ) of the medium. Let us write the wave equation in the familiar form for future reference,

$$\frac{\partial^2 \psi}{\partial x^2} = \frac{1}{v^2} \frac{\partial^2 \psi}{\partial t^2}. \tag{4.120}$$

Equation 4.120 along with the boundary condition of ψ form the mathematical problem whose solutions describe the longitudinal vibrations of the rod.

Boundary Conditions 1: Both Ends Fixed

We first look at the boundary condition where both ends of the rod are kept fixed, i.e. the length of the rod is fixed. In this case, the vibration will extend the rod in some parts and contract in other parts. Since both ends are fixed, the boundary conditions are

$$\psi(0, t) = 0, \tag{4.121}$$

$$\psi(L, t) = 0. \tag{4.122}$$

Normal Modes Corresponding to Boundary Conditions 1

The oscillatory solutions of wave equation 4.120 with boundary conditions 4.121 and 4.122 are the normal modes of the longitudinal vibrations of the rod with both ends fixed. We have solved a similar mathematical problem before, and we just give the solution here,

$$\psi_n(x, t) = A_n \sin(k_n x) \cos(\omega_n t - \phi_n), \tag{4.123}$$

$$k_n = \frac{n\pi}{L} \quad (n = 1, 2, \ldots), \tag{4.124}$$

$$\omega_n = v k_n. \tag{4.125}$$

According to Eqs. 4.123 and 4.124, the lowest mode makes one half of a sine period over the length of the rod, the second mode makes a full period, the third mode, three times a half-period, etc. Some of the low frequency modes are shown in Fig. 4.10.

Boundary Conditions 2: One End Fixed and One End Free

At the fixed end, it is obvious that the displacement will be zero,

$$\psi(0,t) = 0. \tag{4.126}$$

At the free end the stress must vanish since there is nothing to pull against. Therefore, from Eqs. 4.111 and 4.110 we find the second boundary condition to be

$$\left.\frac{\partial \psi}{\partial x}\right|_{x=L} = 0. \tag{4.127}$$

Normal Modes Corresponding to Boundary Conditions 2

The oscillatory solutions of wave equation 4.120 with boundary conditions 4.126 and 4.127 are the normal modes of the longitudinal vibrations of a rod with one end fixed and the other end free. We have worked out a similar mathematical problem before, and we just give the solution here,

$$\psi_n(x,t) = A_n \sin(k_n x)\cos(\omega_n t - \phi_n), \tag{4.128}$$

$$k_n = (2n-1)\frac{\pi}{2L} \quad (n = 1, 2, \ldots), \tag{4.129}$$

$$\omega_n = v k_n. \tag{4.130}$$

According to Eqs. 4.128 and 4.129, the lowest mode makes one quarter of a sine period over the length of the rod, the second mode makes three quarters, the third mode, five quarters. Some of the low frequency modes are shown in Fig. 4.10.

Boundary Conditions 3: Both Ends Free

What happens when both of the ends are free? In this case, the force on the rod at the ends must be zero, which requires the space-derivative of the displacement to vanish at the ends,

$$\left.\frac{\partial \psi}{\partial x}\right|_{x=0} = 0, \tag{4.131}$$

$$\left.\frac{\partial \psi}{\partial x}\right|_{x=L} = 0. \tag{4.132}$$

Normal Modes Corresponding to Boundary Conditions 3

The oscillatory solutions of wave equation 4.120 with boundary conditions 4.131 and 4.132 are the normal modes of longitudinal vibrations of the rod with both ends free. It is easy to show that the solution of the wave equation with these boundary conditions is

$$\psi_n(x,t) = A_n \cos(k_n x)\cos(\omega_n t - \phi_n), \tag{4.133}$$

$$k_n = \frac{n\pi}{L} \quad (n = 1, 2, \ldots), \tag{4.134}$$

$$\omega_n = v k_n. \tag{4.135}$$

According to Eqs. 4.133 and 4.134, the lowest mode makes one half of a cosine period over the length of the rod, the second mode makes a full period, the third mode, three half-periods, etc. Some of the low frequency modes are shown in Fig. 4.10.

Example 4.4 Longitudinal Modes of an Aluminum Rod

Find the frequencies of three lowest normal modes of longitudinal vibrations of a 1 m aluminum rod which is (a) clamped at one end and (b) clamped at both ends. Use $\rho = 2.7 \times 10^3$ kg/m^3 and $Y = 70$ GPa.

Solution

(a) The frequencies are given by

$$\begin{aligned} f_n &= \omega_n/2\pi \\ &= (2n-1)\frac{\sqrt{Y/\rho}}{4L}. \end{aligned}$$

Therefore, the lowest frequency is

$$f_1 = \frac{\sqrt{Y/\rho}}{4L} = 1273 \text{ Hz}.$$

The next two normal modes are

$$f_2 = 3f_1 = 3819 \text{ Hz},$$
$$f_3 = 5f_1 = 6365 \text{ Hz}.$$

(b) When both ends are fixed, the frequencies are different:

$$f_1 = \frac{\sqrt{Y/\rho}}{2L} = 637 \text{ Hz},$$
$$f_2 = 2f_1 = 1273 \text{ Hz},$$
$$f_3 = 3f_1 = 1910 \text{ Hz}.$$

4.6 Vibrations of an Air Column

The heart of a wind instrument is usually a vibrating air column. For instance, when a player blows across an open hole of a flute, this sets the air column in the flute in vibration. The player controls the effective length of the air column by opening and closing holes in the instrument. In the case of sound, vibrations of air particles occur along the line of travel of the sound wave. Therefore, we wish to study the longitudinal vibration modes of a column of air, i.e. the modes along the length of the tube.

The reaction of the particles of air to compression or rarefaction will be similar to the reaction of a solid, as studied above. Therefore, the vibrations will be described by the same wave equation with the speed v now related to the elastic and inertial properties of air. For air of density ρ [kg/m^3] and bulk modulus B [N/m^2] (where we had Y [N/m^2] before) the speed v can be shown to be

$$v = \sqrt{B/\rho}, \tag{4.136}$$

where the bulk modulus is defined by Hooke's law for relative volume change under a pressure p,

$$B = -V\frac{dp}{dV}. \tag{4.137}$$

Since the equation of motion of longitudinal vibrations of particles of air is the same as the wave equation worked out above for a rod, the modes of an air column will be the same as the modes of the rod. Just as the longitudinal modes of the vibrating rod depend on the length of the rod, the modes of vibrating air will depend on the length L of the column. In the two cases of particular interest for musical instruments, namely, one end open and both ends open, these will be

One end open and one end closed:

$$k_n = (2n-1)\frac{\pi}{2L} \quad (n = 1, 2, \ldots),$$

$$\lambda_n = \frac{2\pi}{k_n} = \frac{4L}{2n-1}.$$

Both ends open:

$$k_n = \frac{n\pi}{L} \quad (n = 1, 2, \ldots),$$

$$\lambda_n = \frac{2\pi}{k_n} = \frac{2L}{n},$$

with frequencies of modes

$$\omega_n = vk_n,$$

with $v = \sqrt{B/\rho}$. Figure 4.12 displays the lowest three modes for the two types of boundary conditions with the wavelengths of each mode displayed alongside. Successive modes are smaller in wavelength. In the case of one end fixed and the other end open, the length of the tube is an odd multiple of the wavelength for various modes, $L = n\lambda/4$, n odd, and in the case of both ends open, the length of the tube is an integer multiple of half a wavelength for the modes, $L = n\lambda/2$, n integer.

We can express the bulk modulus B in terms of the pressure and temperature if we assume that air acts as an ideal gas. Since the velocities of the air particles are much greater than the speed of the wave, there is no transfer of internal energy in heat, and wave propagation is an adiabatic process. In an adiabatic process for an ideal gas, the pressure and volume are related as

$$pV^{\gamma} = \text{constant}, \tag{4.138}$$

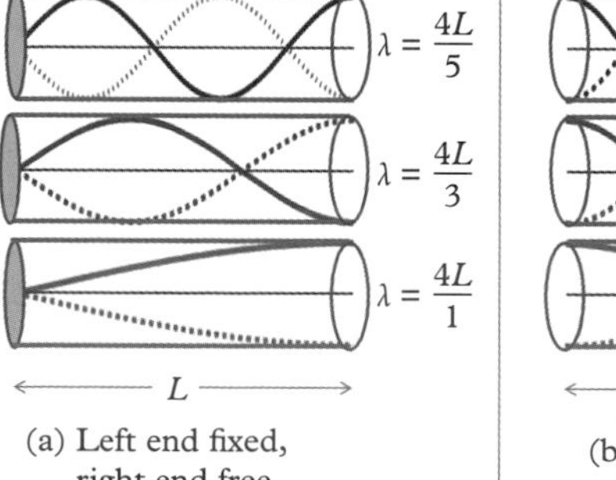

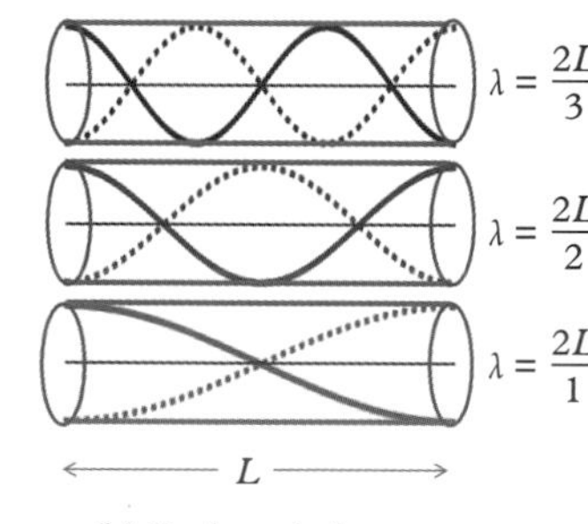

Fig. 4.12 *The three lowest frequency modes of air columns for two boundary conditions: (a) one end open and the other end closed, and (b) both ends open. The solid line shows the vibration mode at one instant of time when the displacements are largest and the dashed line shows the other instant of the largest displacement from equilibrium.*

where $\gamma = C_p/C_v$, the ratio of specific heats at constant pressure (C_p) and constant volume (C_v). Therefore,

$$V^\gamma \, dp + \gamma p V^{\gamma-1} = 0. \tag{4.139}$$

Hence, the bulk modulus is given as follows:

$$B = -V \frac{dp}{dV} = \gamma \, p. \tag{4.140}$$

This gives the speed of the sound as

$$v = \sqrt{\frac{\gamma p}{\rho}}. \tag{4.141}$$

Since air is composed of mostly nitrogen and oxygen, you may use γ for a diatomic molecule, which is $\frac{5}{3}$. Same speed applies for other boundary conditions, closed on both ends and open at both ends.

Example 4.5 Modes of an Air Column

Find the frequencies of three lowest allowed modes in a tube of length 1 m which is open at one end and closed at the other. Use standard pressure 1.0×10^5 Pa, the density of air 1.0 kg/m^3, and $\gamma = 1.44$ for air.

Solution

From the data, the speed of the wave is

$$v = \sqrt{\frac{\gamma \, p}{\rho}} = \sqrt{\frac{1.44 \times 1.0 \times 10^5}{1.0}} = 379.5 \text{ m/s}.$$

We need three lowest frequencies, which means we need three smallest wave numbers. That means we need three largest wavelengths. For the given boundary conditions, the wavelengths are $\lambda_n = 4L/(2n-1)$, which gives the three largest wavelengths as

$$\lambda_1 = 4 \text{ m}; \; \lambda_2 = 1.33 \text{ m}; \; \lambda_3 = 0.8 \text{ m}.$$

Or, the wave numbers as

$$k_1 = \frac{2\pi}{4} \text{ m}^{-1}; \; k_2 = \frac{2\pi}{1.33} \text{ m}^{-1}; \; k_3 = \frac{2\pi}{0.8} \text{ m}^{-1}.$$

Multiplying the k's with the speed, we obtain

$$\omega_1 = 379.5 \times \frac{2\pi}{4} \text{ s}^{-1}; \; \omega_2 = 379.5 \times \frac{2\pi}{1.33} \text{ s}^{-1}; \; \omega_3 = (379.5 \times \frac{2\pi}{0.8} \text{ s}^{-1}.$$

The frequencies are

$$f_1 = \frac{379.5}{4} = 95 \text{ Hz}; \; f_2 = \frac{379.5}{1.33} = 285.3 \text{ Hz}; \; f_3 = \frac{379.5}{0.8} = 474.3 \text{ Hz}.$$

4.7 Vibrations of Two- and Three-Dimensional Systems

The vibrations of two- and three-dimensional systems such as drums and solid bodies are more complex than for the one-dimensionsl systems we have studied so far since they possess additional degrees of freedom. Let us denote the displacement from equilibrium by the symbol ψ. While in one dimension with the system along the x-axis in equilibrium, we had the displacement variable a function of x and t only, in two dimensions with the system in the xy-plane in equilibrium, we will have the displacement ψ a function of x, y, and t, i.e. $\psi(x, y, t)$. In three dimensions we will have $\psi(x, y, z, t)$. A similar derivation as that conducted for the oscillations of a one-dimensional system leads to the wave equations in two and three dimensions,

$$\text{Two dimensions: } \frac{\partial^2\psi}{\partial x^2} + \frac{\partial^2\psi}{\partial y^2} = \frac{1}{v^2}\frac{\partial^2\psi}{\partial t^2}, \tag{4.142}$$

$$\text{Three dimensions: } \frac{\partial^2\psi}{\partial x^2} + \frac{\partial^2\psi}{\partial y^2} + \frac{\partial^2\psi}{\partial z^2} = \frac{1}{v^2}\frac{\partial^2\psi}{\partial t^2}, \tag{4.143}$$

where v is related to the surface tension S and surface density σ [mass/area] in the case of the two-dimensional system and to the bulk modulus B and volume density ρ [mass/volume] in the case of the three-dimensional system,

$$\text{Two dimensions: } v^2 = \frac{S}{\sigma}, \tag{4.144}$$

$$\text{Three dimensions: } v^2 = \frac{B}{\rho}. \tag{4.145}$$

In the following two subsections we will work out two examples of two-dimensional systems, one with a rectangular geometry and the other with a circular geometry. We will learn how to handle the boundary condition in these two elementary types of geometries.

4.7.1 Transverse Oscillations of a Rectangular Plate

Consider a rectangular plate that is fixed at the edges but otherwise allowed to vibrate everywhere else. What will be the normal modes?

As in the one-dimensional case we can obtain this information by solving the wave equation Eq. 4.142 for the $\psi(x, y, t) = 0$ at the edges as the boundary condition. To be concrete, consider a $a \times b$ plate placed in the xy-plane between $0 \leq x \leq a$ and $0 \leq y \leq a$. Then, we need to solve Eq. 4.142 with the following boundary conditions:

$$\psi(x = 0, y, t) = 0, \tag{4.146}$$

$$\psi(x = a, y, t) = 0, \tag{4.147}$$

$$\psi(x, y = 0, t) = 0, \tag{4.148}$$

$$\psi(x, y = b, t) = 0. \tag{4.149}$$

To find the modes we employ the product solution for ψ in the following form:

$$\psi(x, y, t) = f(x)g(y)h(t). \tag{4.150}$$

Let us insert the product solution in the two-dimensional wave equation, Eq. 4.142, perform the derivatives, and then divide every term by *fgh*. We will get

$$\frac{1}{f}\frac{d^2f}{dx^2}+\frac{1}{g}\frac{d^2g}{dy^2}=\frac{1}{v^2}\frac{1}{h}\frac{d^2h}{dt^2}. \tag{4.151}$$

The right side of this equation is only a function of t and the left side that of (x, y). The only way the two sides will be equal to each other is if they are both equal to the same constant. Since, we are looking for oscillating solutions, the constant has to be a negative number. We will choose $-k^2$ for the constant,

$$\frac{1}{f}\frac{d^2f}{dx^2}+\frac{1}{g}\frac{d^2g}{dy^2}=-k^2, \tag{4.152}$$

$$\frac{d^2h}{dt^2}=-k^2v^2h(t), \tag{4.153}$$

where k is introduced here as in the one-dimensional case. The general solution for $h(t)$ will be

$$h(t)=A\cos(kvt)+B\sin(kvt). \tag{4.154}$$

We need to separate the x and y parts in Eq. 4.152. Sending the y-dependent quantities to the right side of the equation gives an equation in which the left side depends only on x and the right side depends only on y. Therefore, we can equate each side to a constant. Looking ahead, we choose the constant to be a negative number since that would give the required oscillating behavior of the modes. We will denote this constant by $-p^2$,

$$\frac{d^2f}{dx^2}=-p^2 f(x), \tag{4.155}$$

$$\frac{d^2g}{dy^2}=-\left(k^2-p^2\right)g(y). \tag{4.156}$$

The solution of Eq. 4.155 satisfying the boundary conditions 4.146 and 4.147 is

$$f_n(x)=C\sin(p_n x),\ \ p_n=\frac{n\pi}{a},\ \ n=1,2,3,\ldots, \tag{4.157}$$

keeping only the positive solutions for p. The solution of Eq. 4.156 satisfying the boundary conditions 4.148 and 4.149 is

$$g_m(y)=C\sin(q_m y),\ \ q_m=\left[\sqrt{k^2-p^2}\right]_m=\frac{m\pi}{b},\ \ m=1,2,3,\ldots, \tag{4.158}$$

keeping only the positive solutions for q. We note that each mode will be specified by two indices, one for the x-part and the other for the y-part. Therefore, we label k by two indices m and n as

$$k_{n,m}^2=p_n^2+q_m^2=\pi^2\left[\frac{n^2}{a^2}+\frac{m^2}{b^2}\right]. \tag{4.159}$$

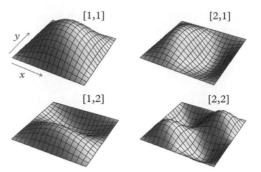

Fig. 4.13 *The four lowest modes of vibrations of a plate. The modes are labeled with two mode indices, one for each side of the plate.*

Putting the product solution together, we find that the normal modes of the plate will be given by the following frequencies and mode amplitudes:

$$\omega_{n,m} = v\, k_{n,m}, \quad k_{n,m} = \pi \sqrt{\frac{n^2}{a^2} + \frac{m^2}{b^2}}, \quad n, m = 1, 2, 3, \ldots, \tag{4.160}$$

$$\psi_{n,m} = \sin\left(\frac{n\,\pi}{a}\, x\right) \sin\left(\frac{m\,\pi}{b}\, y\right) \left[A_{n,m}\, \cos(k_{n,m} vt) + B_{n,m}\, \sin(k_{n,m} vt)\right]. \tag{4.161}$$

A general motion of the plate will be given by superposing these solutions as

$$\psi(x, y, t) = \sum_{n=1}^{\infty} \sum_{m=1}^{\infty} \sin\left(\frac{n\,\pi}{a}\, x\right) \sin\left(\frac{m\,\pi}{b}\, y\right) \left[A_{n,m}\, \cos(kvt) + B_{n,m}\, \sin(kvt)\right], \tag{4.162}$$

where the constants $A_{n,m}$ and $B_{n,m}$ are fixed by the initial conditions on the plate. The mathematical method of Fourier analysis is used to extract the constants from initial conditions of the starting shape and speed at each point of the plate. We will study the Fourier analysis a little later in this chapter where we will do a problem in which we determine the coefficients in the general solution. The profiles of four low modes, $[n, m]$ = [1,1],[2,1],[1,2], [2,2], are displayed in Fig. 4.13.

4.7.2 Free Vibrations of a Drum

Suppose an elastic membrane is stretched and fixed to a circular tube of radius a. What will be the normal modes of the vibrations of the membrane? The boundary condition in this situation can be expressed more readily in a polar coordinate with the origin at the center of the membrane. Let the polar coordinates be denoted by (r, θ). The transverse displacement of an element of the membrane at the point (r, θ) at time t will be given by $\psi(r, \theta, t)$. The boundary condition will be

$$\psi(a, \theta, t) = 0, \quad \text{(since fixed boundary at } r = a.\text{)} \tag{4.163}$$

The wave equation for ψ will still be the same equation as Eq. 4.142 but now written in the polar coordinates,

$$\frac{1}{r}\frac{\partial}{\partial x^2}\left[r \frac{\partial \psi}{\partial r}\right] + \frac{1}{r^2}\frac{\partial^2 \psi}{\partial \theta^2} = \frac{1}{v^2}\frac{\partial^2 \psi}{\partial t^2}. \tag{4.164}$$

So, the task before us is to solve this partial differential equation, making sure that the solution satisfies the boundary condition in Eq. 4.163. Since we will only work out the normal modes, we do not need to worry about the initial conditions here.

Again, we implement the separation of variables technique by seeking the solution in the form of a product of separate independent variables. Let us set $\psi(r, \theta, t) = R(r)\Theta(\theta)T(t)$ in Eq. 4.164. The separated equations can be written as three ordinary differential equations,

$$\frac{d^2T}{dt^2} = -k^2v^2\, T, \tag{4.165}$$

$$\frac{d^2\Theta}{d\theta^2} = -\mu^2\, \Theta, \tag{4.166}$$

$$\frac{d^2R}{dr^2} + \frac{1}{r}\frac{dR}{dr} + \left(k^2 - \frac{\mu^2}{r^2}\right) R = 0, \tag{4.167}$$

where constants k and μ will be determined below. We have seen how to solve equations of the type Eq. 4.165 and 4.166—they have the same form as that of simple harmonic oscillators. Therefore, we can write their general solution as

$$T(t) = C_1\, \cos(kvt) + C_2\, \sin(kvt), \tag{4.168}$$

$$\Theta(\theta) = C_3\, \cos(\mu\theta) + C_4\, \sin(\mu\theta). \tag{4.169}$$

Now, we note that every function of θ must be periodic in θ, since $\theta = 0$ and $\theta = 2\pi$ are identical. This means that μ must be an integer. Since negative μ does not give new modes, we will keep μ in the semi-positive integer range,

$$\mu = 0, 1, 2, \ldots. \tag{4.170}$$

Now, the radial equation, Eq. 4.167, has more information about the constant μ,

$$\frac{d^2R}{dr^2} + \frac{1}{r}\frac{dR}{dr} + \left(k^2 - \frac{\mu^2}{r^2}\right) R = 0, \quad (\mu = 0, 1, 2, \ldots). \tag{4.171}$$

This equation is called the Bessel equation. For each value of μ, there are two independent solutions of the Bessel equation: the Bessel functions $\mathcal{J}_\mu$ and the Neumann function N_μ. They are written as functions of kr rather than a function of r. The general solution will be a superposition of these independent solutions,

$$R(r) = C_5\, \mathcal{J}_\mu(kr) + C_6\, N_\mu(kr). \tag{4.172}$$

The Bessel functions $\mathcal{J}_\mu$ are well behaved at $r = 0$, but Neumann functions blow up at $r = 0$. In our geometry we expect regular behavior at the center of the drum. Therefore, $C_6 = 0$ here. This means that the radial part contains only Bessel functions,

$$R(r) = C_5\, \mathcal{J}_\mu(kr). \tag{4.173}$$

The boundary condition at $r = a$ given in Eq. 4.163 yields the following requirement:

$$\mathcal{J}_\mu(ka) = 0. \tag{4.174}$$

Table 4.2 *Zeros of $\mathcal{J}_{\mu n}(\rho)$. The values in the table are the values of ρ for which $\mathcal{J}_{\mu n}(\rho) = 0$. In the text $\rho = ka$.*

	$n = 1$	$n = 2$	$n = 3$	$n = 4$
$\mu = 0$	2.40483	5.52008	8.65373	11.7915
$\mu = 1$	3.83171	7.01559	10.1735	13.3237
$\mu = 2$	5.13562	8.41724	11.6198	14.796
$\mu = 3$	6.38016	9.76102	13.0152	16.2235
$\mu = 4$	7.58834	11.0647	14.3725	17.616

This says that k are no longer arbitrary but are obtained from the zeros of the Bessel function of the argument ka. For each μ, there is an infinite number of zeros of $\mathcal{J}_\mu$: whenever ka equals one of these zeros, $\mathcal{J}_\mu$ will be zero. Table 4.2 shows the first four zeros of the first five Bessel functions. Let us denote the zeros of the Bessel function $\mathcal{J}_\mu$ by $\alpha_{\mu n}$ with $n = 1, 2, 3, \ldots$.

Attaching the two indices to the k also we write

$$k_{\mu n} = \frac{\alpha_{\mu n}}{a}. \tag{4.175}$$

Now, we can write the product solution which would be also labeled with the subscript μn,

$$\psi_{\mu n} = \mathcal{J}_\mu(k_{\mu n} r) \left[A_{\mu n} \cos(\mu\theta) + B_{\mu n} \sin(\mu\theta)\right] \left[C_{\mu n} \cos(\omega_{\mu n} t) + D_{\mu n} \sin(\omega_{\mu n} t)\right], \tag{4.176}$$

where we have renamed constants. These are the normal modes labeled by two indices and $\omega_{\mu n}$ are the normal mode frequencies. Some of the lower frequency modes are given in Table 4.3.

Note that for $\mu \neq 0$, there are two modes for each frequency, one with $\cos(\mu\theta)$ and the other with $\sin(\mu\theta)$: this is called **degeneracy**. In Fig. 4.14, density plots of some modes of a drum are shown.

Table 4.3 *Some normal modes of a circular drum.*

μ	n	$\psi_{\mu n}$	$\omega_{\mu n}$
0	1	$\psi_{01} = \mathcal{J}_0(k_{01} r)\,[C_{01} \cos(\omega_{01} t) + D_{01} \sin(\omega_{01} t)]$	$2.404/av$
0	2	$\psi_{02} = \mathcal{J}_1(k_{02} r)\,[C_{02} \cos(\omega_{02} t) + D_{02} \sin(\omega_{02} t)]$	$5.520/av$
1	1	$\psi_{11} = \mathcal{J}_1(k_{11} r)\cos(\theta)\,[C_{11} \cos(\omega_{11} t) + D_{11} \sin(\omega_{11} t)]$	$3.832/av$
1	1	$\psi'_{11} = \mathcal{J}_1(k_{11} r)\sin(\theta)\left[C'_{11} \cos(\omega_{11} t) + D'_{11} \sin(\omega_{11} t)\right]$	$3.832/av$
2	1	$\psi_{21} = \mathcal{J}_2(k_{21} r)\cos(2\theta)\,[C_{21} \cos(\omega_{21} t) + D_{21} \sin(\omega_{21} t)]$	$5.135/av$
2	1	$\psi'_{21} = \mathcal{J}_2(k_{21} r)\sin(2\theta)\left[C'_{21} \cos(\omega_{21} t) + D'_{21} \sin(\omega_{21} t)\right]$	$5.135/av$

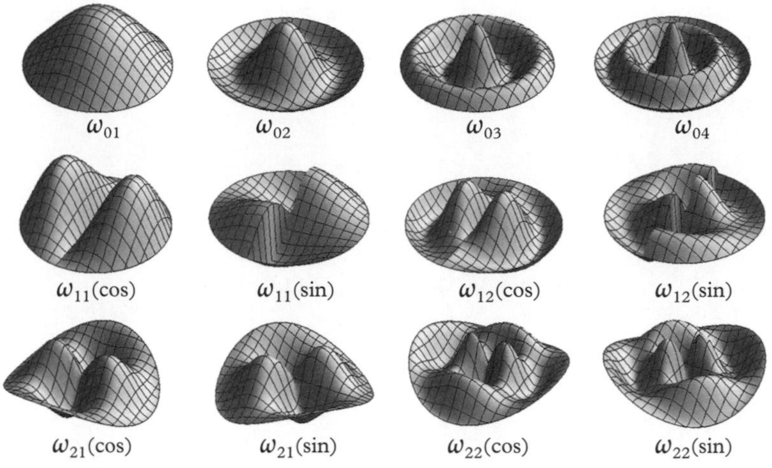

ω_{01} ω_{02} ω_{03} ω_{04}

$\omega_{11}(\cos)$ $\omega_{11}(\sin)$ $\omega_{12}(\cos)$ $\omega_{12}(\sin)$

$\omega_{21}(\cos)$ $\omega_{21}(\sin)$ $\omega_{22}(\cos)$ $\omega_{22}(\sin)$

Fig. 4.14 *Low frequency normal modes of a circular drum.*

4.8 Fourier Analysis

Recall that normal modes are independent ways a system can vibrate. An arbitrary motion, periodic or nonperiodic, can be constructed from the normal modes of the system. In this section, we will demonstrate how to construct an arbitrary motion of an oscillating string from a superposition of its normal modes.

4.8.1 Fourier Series

A Fourier series is a series expansion of a periodic function $f(x)$ of period L in terms of sines and cosines of periods $\{L, L/2, L/3, \ldots\}$.

$$f(x) = \frac{A_0}{2} + \sum_{n=1}^{\infty} A_n \cos\left(\frac{2\pi nx}{L}\right) + \sum_{n=1}^{\infty} B_n \sin\left(\frac{2\pi nx}{L}\right), \tag{4.177}$$

where the Fourier coefficients A_n and B_n are determined by

$$A_0 = \frac{2}{L}\int_0^L f(x)\,dx, \tag{4.178}$$

$$A_n = \frac{2}{L}\int_0^L f(x)\cos\left(\frac{2\pi nx}{L}\right)dx, \quad (n = 1, 2, 3, \ldots), \tag{4.179}$$

$$B_n = \frac{2}{L}\int_0^L f(x)\sin\left(\frac{2\pi nx}{L}\right)dx, \quad (n = 1, 2, 3, \ldots). \tag{4.180}$$

Note that A_0 comes with a factor of $1/2$ so that the same integral Eq. 4.179 can be used for all n. To apply a Fourier series to a non-periodic function, say $g(x)$, defined over an interval $[0, L]$, we construct another periodic function $h(x)$ which is periodic with period L and agrees with $g(x)$ in the interval $[0, L]$.

Example 4.6 Fourier Series of Square Wave

Find the Fourier series of the wave shown in Fig. 4.15.

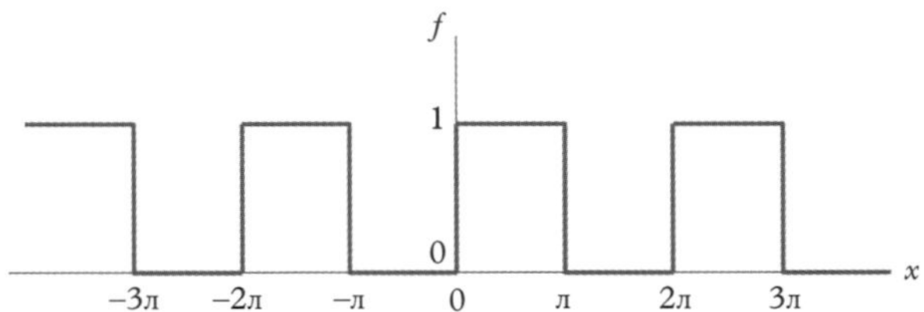

Fig. 4.15 *Square wave.*

Solution

First we write the analytic expression for the given function in one period. We choose to work with the period $0 \le x < 2\pi$,

$$f(x) = \begin{cases} 1 & 0 \le x < \pi \\ 0 & \pi \le x < 2\pi \end{cases}. \tag{4.181}$$

Now, using Eqs. 4.179 and 4.180 for $L = 2\pi$, we determine the Fourier coefficients,

$$A_0 = \frac{2}{2\pi}\int_0^{2\pi} f(x)dx = \frac{1}{2}, \tag{4.182}$$

$$A_n = \frac{2}{2\pi}\int_0^{2\pi} f(x)\ \cos(nx)dx = 0,\ (n = 1, 2, 3, \ldots) \tag{4.183}$$

$$B_n = \frac{2}{2\pi}\int_0^{2\pi} f(x)\ \sin(nx)dx = \begin{cases} \dfrac{2}{n\pi} & n \text{ odd} \\ 0 & n \text{ even} \end{cases}. \tag{4.184}$$

Therefore, the step-wise periodic function given in Fig. 4.15 has the following Fourier series representation:

$$f(x) = \frac{1}{2} + \frac{2}{\pi}\sum_{n=1,3,5,\ldots}^{\infty} \frac{\sin(nx)}{n},\quad (-\infty < x < \infty). \tag{4.185}$$

How closely does this series resemble the original function? Mathematically, to be an exact representation of the original function, you need all the terms in the sum. However, in practice, just a few terms are enough, as shown in Fig. 4.16, where we plot up to four terms and 25 terms. It is seen that even a few terms of the Fourier series give a fairly good representation of the original function.

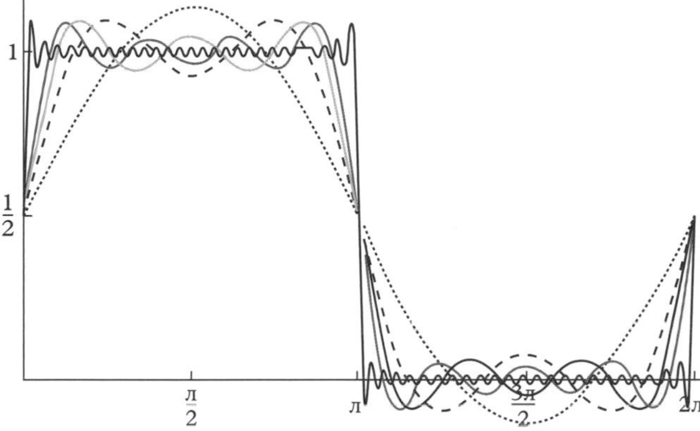

Fig. 4.16 *Partial Fourier series representations of a square wave with 1, 2, 3, 4, and 25 terms. Aside from ringings at the sharp edges, the 25-term approximation is very close to the original function.*

4.8.2 Fourier Analysis in Terms of Normal Modes

We have seen above that modes of strings of length tied at both ends are

$$\psi_n(x) = \sin(k_n x), \quad k_n = \frac{n\pi}{L}, \quad n = 1, 2, 3, \ldots. \tag{4.186}$$

Notice that these mode functions are terms of the Fourier sine series for period $2L$. Any periodic function of period $2L$ that is an odd function of x with zeros at $x = 0$ and $x = L$ will have a Fourier series with only the sine terms of modes in Eq. 4.186. That is,

$$\text{Given } f(x + 2L) = f(x),\ f(0) = 0,\ f(L) = 0, \text{and } f(-x) = -f(x),$$

$$\text{the Fourier series will be } f(x) = \sum_{n=1}^{\infty} B_n \sin\left(\frac{n\pi x}{L}\right). \tag{4.187}$$

Any shape of the string will be given by some continuous function $f(x)$ in $0 \leq x \leq L$. In order to make use of the Fourier series representation of that function as in Eq. 4.187 we will extend the function to $-L \leq x \leq L$ such that the extended function is an odd function, and, then copy the function over the entire range $-\infty < x < \infty$ to construct a periodic function which agrees with the desired function in the physically relevant domain. I will now illustrate the expansion of an arbitrary shape into the normal modes of the system with the following example.

Example 4.7 Fourier Series in Normal Modes

A string tied at both ends is pulled in the shape shown in Fig. 4.17. Express the mode shape $y(x)$ in terms of the normal modes of the system.

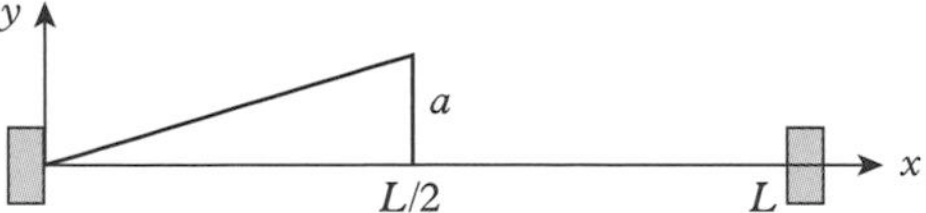

Fig. 4.17 *Triangular wave.*

Solution

First we will write down the analytic form of the given shape. The function $y(x)$ is a piece-wise continuous function given by

$$y(x) = \begin{cases} \dfrac{2a}{L}x, & 0 \le x < \dfrac{L}{2}, \\ 0, & \dfrac{L}{2} \le x \le L. \end{cases} \tag{4.188}$$

As written here, this function is not periodic in x. We need to replace this function by a periodic function of period $2L$ so that only sine terms will arise in the Fourier expansion. This can be obtained by creating an odd function of period $2L$ that agrees with $y(x)$ in the domain $0 \le x \le L$.

$$f(x) = \begin{cases} 0, & -L \le x < -\dfrac{L}{2}, \\ \dfrac{2a}{L}x, & -\dfrac{L}{2} \le x < \dfrac{L}{2}, \\ 0, & \dfrac{L}{2} \le x \le L, \end{cases} \tag{4.189}$$

with copies appended ad infinitum such that

$$f(x + 2L) = f(x).$$

A small part of this function is shown in Fig. 4.18. Now, since $f(x)$ is periodic and odd we will have only the sine series terms non-zero, which will all be the normal mode functions of this system. Using Eq. 4.189 in Fourier series with period $2L$ we obtain

$$B_n = \frac{2}{2L}\int_{-L}^{L} f(x)\sin\left(\frac{n\pi x}{L}\right)dx = \frac{2}{L}\int_{0}^{L} f(x)\sin\left(\frac{n\pi x}{L}\right)dx,$$

where I have made use of the fact that the product of two odd functions is an even function and converted the limits in the range of 0 to L. Performing this integral we get

$$B_n = -\frac{4a}{L^2}\left[\frac{L}{n\pi}x\cos\left(\frac{n\pi x}{L}\right) - \left(\frac{L}{n\pi}\right)^2\sin\left(\frac{n\pi x}{L}\right)\right]_0^{L/2}$$

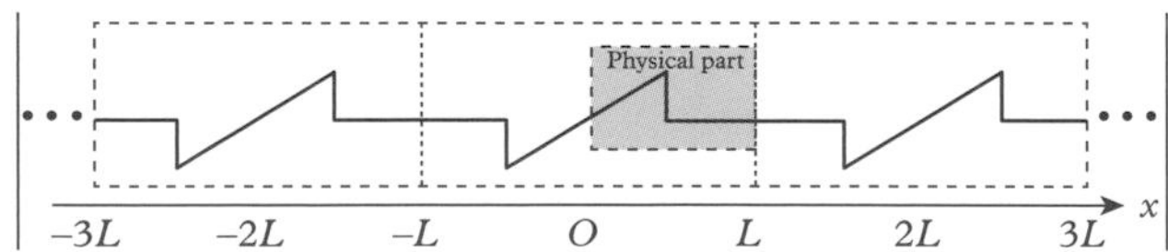

Fig. 4.18 *Triangular wave.*

Note that $\cos(n\pi) = (-1)^n$ and $\sin(n\pi/2) = \pm 1$ for n odd. Therefore, some of the B_n's are:

$$B_1 = \frac{4a}{\pi^2}, \quad B_2 = \frac{a}{\pi}, \quad B_3 = -\frac{4a}{9\pi^2}, \quad B_4 = -\frac{a}{2\pi}, \text{ etc.}$$

The expansion of the given shape $y(x)$ in the modes of the string will be

$$y(x) = \left\{ \begin{array}{ll} \frac{2a}{L}x, & 0 \le x < \frac{L}{2}, \\ 0, & \frac{L}{2} \le x \le L. \end{array} \right\} = \frac{4a}{\pi^2}\sin\left(\frac{\pi x}{L}\right) + \frac{a}{\pi}\sin\left(\frac{2\pi x}{L}\right) - \frac{4a}{9\pi^2}\sin\left(\frac{3\pi x}{L}\right) + \cdots$$

4.8.3 Dynamics of Taut String Using Modes

Recall that each mode of a vibrating system oscillates at its own frequency. That is, if you have a system in a mode ψ_n of frequency ω_n, then it will evolve as

$$\psi(t) = \psi_n \cos(\omega_n t - \phi_n).$$

We saw above that any shape of a string can be written as a sum of modes. Thus, the shape of a string at time $t = 0$ may be

$$y(x, 0) = \sum_n B_n \psi_n(x),$$

where B_n are the coefficients, which could be A_n or B_n in the Fourier series, although it was only the B_n in the case of strings tied at $x = 0$ and $x = L$ as we have seen. Now, from this we can immediately state the shape of the string at an arbitrary time by multiplying the time evolving part based on the frequency of each mode,

$$y(x, t) = \sum_n A_n \psi_n(x) \cos(\omega_n t - \phi_n), \tag{4.190}$$

where $\psi_n(x) = \sin(n\pi x/L)$. The A_n and ϕ_n are obtained from the $y(x, 0)$ and dy/dt at $t = 0$. Now, we will work out a calculation that will show how to use the initial conditions $y(x, 0)$ and $(dy/dt)_{x,0}$ to determine A_n and ϕ_n for all n. Suppose the initial conditions are given to be

$$y(x, 0) = f(x), \tag{4.191}$$

$$(dy/dt)_{x,0} = g(x). \tag{4.192}$$

To implement the condition on y we just set $t = 0$ in Eq. 4.190. To implement the condition on $\dot{y}$ we take one time derivative on Eq. 4.190 and then set $t = 0$. We get

$$\sum_{n=1}^{\infty} A_n \sin\left(\frac{n\pi x}{L}\right) \cos(\phi_n) = f(x), \tag{4.193}$$

$$\sum_{n=1}^{\infty} \omega_n A_n \sin\left(\frac{n\pi x}{L}\right) \sin(\phi_n) = g(x). \tag{4.194}$$

Let us make the following substitutions to simplify writing:

$$C_n = A_n \cos(\phi_n), \quad D_n = \omega_n A_n \sin(\phi_n). \tag{4.195}$$

Eqs. 4.193 and 4.194 become

$$\sum_{n=1}^{\infty} C_n \sin\left(\frac{n\pi x}{L}\right) = f(x), \tag{4.196}$$

$$\sum_{n=1}^{\infty} D_n \sin\left(\frac{n\pi x}{L}\right) = g(x). \tag{4.197}$$

How do we determine C_n and D_n? The following identity is crucial for determining these coefficients:

$$\frac{2}{L}\int_0^L \sin\left(\frac{m\pi x}{L}\right)\sin\left(\frac{n\pi x}{L}\right) dx = \begin{cases} 1 & m = n \\ 0 & m \neq n. \end{cases} \tag{4.198}$$

To extract the mth coefficient, C_m, we multiply both sides of Eq. 4.196 by $\sin(m\pi x/L)$, and then integrate both sides. Note that we use the subscript m because n is the dummy index for the sum on the right side of Eq. 4.196,

$$C_m = \frac{2}{L}\int_0^L f(x) \sin\left(\frac{m\pi x}{L}\right) dx. \tag{4.199}$$

Similarly, we can work with Eq. 4.197 to obtain D_m,

$$D_m = \frac{2}{L}\int_0^L g(x) \sin\left(\frac{m\pi x}{L}\right) dx. \tag{4.200}$$

From C_m and D_m we obtain A_m and ϕ_m as follows:

$$A_m = \sqrt{C_m^2 + (D_m/\omega_m)^2}, \tag{4.201}$$

$$\tan\phi_m = \frac{D_m}{\omega_m C_m}. \tag{4.202}$$

The information for C_m and D_m comes from the way the string was set into vibration at $t = 0$. With these A_m and ϕ_m the solution of $y(x, t)$ gives the motion of the string at all subsequent instants.

EXERCISES

(4.1) **Longitudinal oscillations of springs and beads.** In the text we studied the transverse modes of beads on a stretched massless string. In this problem you will study the longitudinal modes of beads attached to identical massless springs. Suppose N beads of equal mass m are attached to $N + 1$ identical springs with spring constant k_s and length l, as shown in Fig. 4.19. Let the distance between the fixed supports be L.

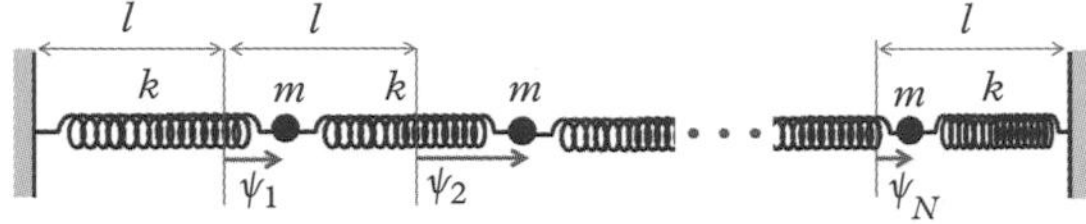

Fig. 4.19 *Exercise 4.1.*

(a) Set up Newton's equation of motion for the displacement of an arbitrary bead.
(b) Solve the equations of motion of all beads to deduce the normal mode frequencies and the normal mode amplitude ratios.
(c) Check your answer for $N = 2$ with the answer for this part studied in the last chapter.
(d) Sketch the configuration for the four lowest frequency modes for the $N = 5$ system.

(4.2) **Normal modes of a massive spring.** A spring of mass M, length L, and spring constant k is tied at both ends to fixed supports, as shown in Fig. 4.20. We wish to find the normal modes of the spring by mapping this system to the N bead and springs system in Exercise 4.1 by distributing the total mass of the system on the beads and leaving the springs massless and then taking the $N \to \infty$ limit.

In the $N \to \infty$ limit, $l = \frac{L}{N+1} \to 0$ and $m = \frac{M}{N} \to 0$, but $L = (N + 1)l$ and $M = Nm$ remain fixed. By taking appropriate limits find the normal mode frequencies and normal mode amplitudes for the massive spring. You can express the normal modes in terms of the x-coordinate of the jth particle of the spring at $x = jl$. In the continuum limit, you should get $\psi_n(x)$ from the nth mode of the jth particle.

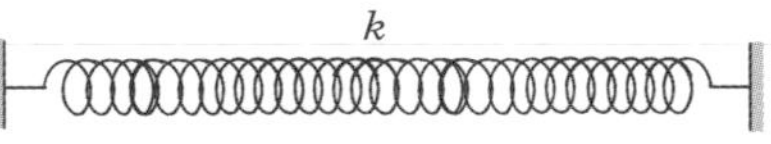

Fig. 4.20 *Exercise 4.2.*

(4.3) **Mass correction of spring.** The angular frequency of oscillation of a mass M attached to a spring with spring constant k_s is usually given to be $\omega = \sqrt{k_s/M}$. This formula assumes that the spring is massless. In physics lab you find that if the mass of the spring is m, then the frequency is modified to $\omega \approx \sqrt{k_s/(M + \frac{1}{3}m)}$. This mass correction is only approximate. In this problem you will be guided to arrive at the correct expression for the frequency.

Consider a spring of unstretched length L and mass m. Let us attach a block of mass M and place the block on a horizontal frictionless surface, as shown in Fig. 4.21. With the origin at the fixed support and the positive x-axis pointed towards the block, the displacement of any point of the spring will be denoted by the function $\psi(x, t)$. A small element of the spring which used to be between x and

$x + dx$ at equilibrium will now be between $x + \psi(x, t)$ and $x + dx + \psi(x + dx, t)$. That is, the element of length dx will be stretched by $d\psi = \psi(x + dx, t) - \psi(x, t)$ as the spring of size dx is stretched to size $d\psi$. This will cause a tension in the spring that will be a function of position and time, denoted by $T(x, t)$.

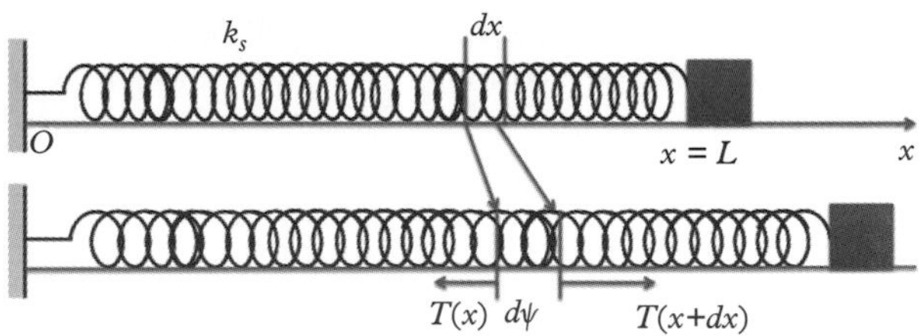

Fig. 4.21 *Exercise 4.3.*

(a) Show that the tension at x due to additional stretch $d\psi$ will be

$$T(x, t) = k_s L \frac{\partial \psi}{\partial x}.$$

(b) Analyze forces on the element dx and show that the equation of motion of the element can be simplified to yield the following equation for $\psi(x, t)$:

$$\frac{\partial^2 \psi}{\partial t^2} = \frac{k_s L^2}{m} \frac{\partial^2 \psi}{\partial x^2}. \tag{4.203}$$

(c) We need to solve Eq. 4.203 subject to conditions at the boundaries at $x = 0$ and $x = L$. What is the boundary condition on $\psi(x, t)$ at $x = 0$?

(d) To obtain the boundary condition at $x = L$ we look at the equation of motion of the block subject to the force by the spring at that point, which is just the tension evaluated at $x = L$, namely, $T(L, t)$. Show that the equation of motion of the displacement of the block will be

$$M \frac{\partial^2 \psi(L, t)}{\partial t^2} = -k_s L \left(\frac{\partial \psi}{\partial x} \right)_{x=L}. \tag{4.204}$$

(e) For normal modes, we try a solution of the form

$$\psi(x, t) = [A \cos(kx) + B \sin(kx)] \cos(\omega t),$$

where k is a wave number, not a spring constant.

Using this show that (i) $A = 0$, and (ii) ω must satisfy

$$\tan\left(\frac{\sqrt{\epsilon}\omega}{\omega_0} \right) = \frac{\omega_0 \sqrt{\epsilon}}{\omega}, \tag{4.205}$$

where $\omega_0 = \sqrt{k_s/M}$ and $\epsilon = m/M$.

(f) To obtain the frequency of oscillation in the case of a spring with negligible mass, you can take $\epsilon \to 0$ limit. Show that in this limit $\omega = \omega_0$, the expected result.

(g) Solve Eq. 4.205 numerically for $\omega_0 = 1\ \text{sec}^{-1}$ and (i) $\epsilon = 0.1$, (ii) $\epsilon = 1$, and (iii) $\epsilon = 10$ and compare your answers with what you get from the $\frac{1}{3}m$ mass correction formula.

(h) To find an approximate solution of Eq. 4.205 we make use of the following approximate expansion of $1/\tan(x)$:

$$\frac{1}{\tan(x)} \approx \frac{1}{x} - \frac{1}{3}x.$$

Show that using this approximation leads to the solution

$$\omega = \sqrt{\frac{k_s}{M + \frac{1}{3}m}}.$$

(4.4) A spring of mass M, length L, and spring constant k_s is fixed to a rigid support at one end. The spring is then placed in a tube so that it can move only in one direction. The tube is placed horizontally on a table with one end of the spring fixed and the other end free to oscillate, as shown in Fig. 4.22.

The modes of the oscillations of the spring can be obtained by drawing pictures of modes similar to the case for the transverse motion of a string with one end fixed and the other end free. Let $\psi(x, t)$ be the displacement of the element of the spring at point x at time t.

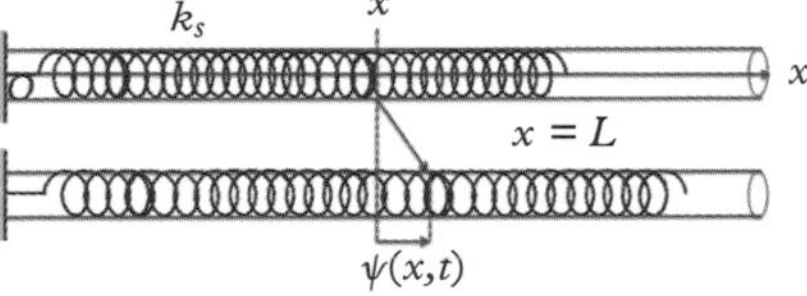

Fig. 4.22 *Exercise 4.4.*

(a) Draw figures showing the variation of $\psi(x)$ at an arbitrary instant along the length of the spring for three lowest frequency modes. These variations should be shown as a plot of $\psi(x)$ versus x.

(b) From your figures deduce the wave numbers corresponding to the three modes.

(c) Write a general formula for the wave number k_n for mode n.

(d) Based on your k_n and the boundary conditions, what are the mode shape functions $\psi_n(x)$.

(e) The displacement $\psi(x, t)$ obeys the following classical wave equation as derived in the last problem:

$$\frac{\partial^2 \psi}{\partial t^2} = \frac{k_s L^2}{m} \frac{\partial^2 \psi}{\partial x^2}.$$

What are the mode frequencies?

(4.5) Consider three identical beads, each of mass m, separated by equal distances a on a massless string of length $4a$. The string is taut with tension T and the ends tied to a fixed support. A hammer hits the middle bead and gives it an initial speed of v_0 while its initial displacement is zero. The initial displacements and velocities of the other beads are zero. Find the subsequent motion of the beads.

(4.6) **Coupled pendulums.** Consider N identical pendulums of mass m and length l each, coupled by identical springs with spring constant k, as shown in Fig. 4.23. When the pendulum strings are in the vertical position, the bobs are separated by a distance a, which is equal to the unstretched length of the springs.

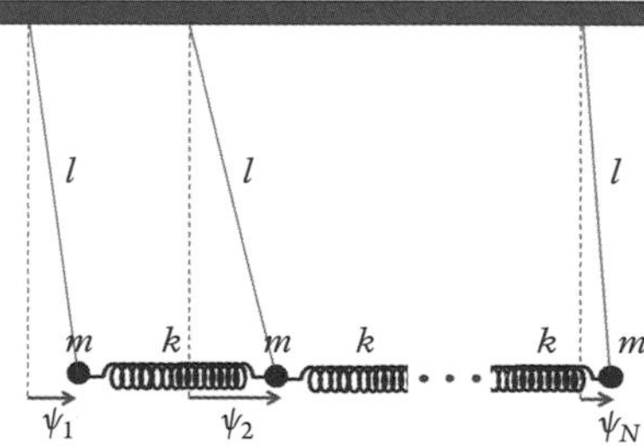

Fig. 4.23 *Exercise 4.6.*

(a) Set up Newton's equation of motion for the small horizontal displacement ψ_j of an arbitrary bead labeled j which is not at one of the two ends.

(b) Suppose you ignore the motion of the beads at the ends, and assume the solution of the bead is in the form

$$\psi_j(t) = \cos(\omega t)\,[A\cos(Kja) + B\sin(Kja)],$$

where ω is the frequency and K the wave number. Find the relation between ω and K, i.e. the dispersion relation of the system.

(c) What is the frequency of the lowest frequency mode and how do the bobs oscillate when in this mode?

(4.7) Find the Fourier series for the following functions given in the domain $(0 \le x \le L)$ and periodically repeated with period L, i.e. for each function $f(x+L) = f(x)$ in the entire domain $-\infty < x < \infty$:

(a) $f(x) = A\,\sin(2\pi x/L),\ 0 \le x \le L;\ f(x+L) = f(x)$
(b) $f(x) = A\,\sin(\pi x/L),\ 0 \le x \le L;\ f(x+L) = f(x)$
(c) $f(x) = x,\ 0 \le x \le L;\ f(x+L) = f(x)$
(d) $f(x) = x(L-x),\ 0 \le x \le L;\ f(x+L) = f(x)$
(e) $f(x) = A\,\sin(2\pi x/L)$ when $(0 \le x \le L/2)$ and $f(x) = 0$ when $(L/2 \le x \le L)$; $f(x+L) = f(x)$.

(4.8) (a) Prove that the Fourier series of an even function will contain only the cosine terms and the constant term.

(b) Prove that the Fourier series of an odd function will contain only the sine terms.

(4.9) Consider a taut string of mass density $\mu = 0.1$ kg/m and tension $T = 10$ N with both ends attached to fixed supports at $x = 0$ and $x = 1$ m. The string is bent around so that its shape can be given by the transverse displacement $\psi(x, 0)$ at $t = 0$, given by the function

$$\psi(x, 0) = 2 \times 10^{-2}\ \text{cm}\ \sin(7\pi x)\,.$$

The string is released at rest from this configuration. Find the function $\psi(x, t)$ that would describe the subsequent transverse vibratory motion of the string.

(4.10) Consider a taut string of mass density $\mu = 0.1$ kg/m and tension $T = 40$ N, with both ends attached to fixed supports at $x = 0$ and $x = 1$ m. The string is bent around so that the transverse displacement $\psi(x, 0)$ at $t = 0$ is given by the function

$$\psi(x, 0) = 2 \times 10^{-2}\ \text{cm}\ \sin(4\pi x) + 3 \times 10^{-2}\ \text{cm}\ \sin(6\pi x)\,.$$

The string is released at rest from this configuration. Find the function $\psi(x, t)$ that would describe the subsequent motion of the string.

(4.11) **Plucked string 1.** Consider a taut string of mass density μ and tension T with both ends attached to fixed supports. The string is pulled in the middle so that it makes the shape shown in Fig. 4.24 and then let go at rest from this configuration at $t = 0$.

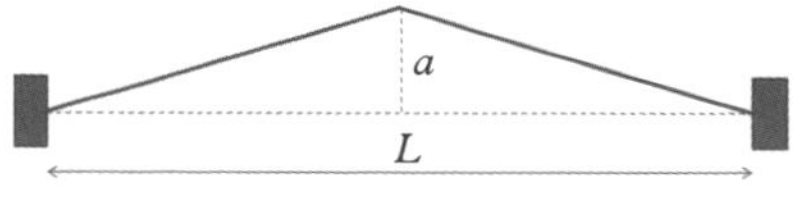

Fig. 4.24 *Exercise 4.11.*

(a) Choose a coordinate system with the origin at the left support, the positive x-axis horizontally towards the other support, and the positive y-axis pointed vertically up. What will be the function $\psi(x)$ that describes the string shape at $t = 0$?

(b) Express $\psi(x)$ found in (a) as a linear superposition of the normal modes of the string fixed at the two ends.

(c) Suppose $L = 10$ and $a = 1$. Compare the plot of $\psi(x)$ from (a) to the plot from (b) with only the (i) one normal mode of the lowest wave number in the expansion, (ii) five normal modes of the lowest wave numbers in the expansion, and (iii) ten normal modes of the lowest wave numbers in the expansion.

(d) What will be the configuration of the string at an arbitrary instant t?

(4.12) **Plucked string 2.** Consider a taut string of mass density μ and tension T with both ends attached to fixed supports. The string is pulled in the middle so that it makes the shape shown in Fig. 4.25 and then let go at rest from this configuration at $t = 0$.

(a) Choose a coordinate system with the origin at the left support, the positive x-axis horizontally towards the other support, and the positive y-axis pointed vertically up. What will be the function $y(x)$ that describes the string shape at $t = 0$?

(b) Express $y(x)$ found in (a) as a linear superposition of the normal modes of the string fixed at the two ends.

(c) Suppose $L = 10$ and $a = 1$. Compare the plot of $y(x)$ from (a) to the plot from (b) with only the (i) one normal mode of the lowest wave number in the expansion, (ii) five normal modes of the lowest wave numbers in the expansion, and (iii) ten normal modes of the lowest wave numbers in the expansion.

(d) What will be the configuration of the string at an arbitrary instant t?

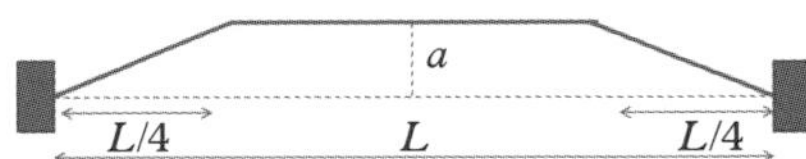

Fig. 4.25 *Exercise 4.12.*

(4.13) **Hammered string.** Consider a taut string of mass density μ and tension T with both ends attached to fixed supports. The string is struck in the middle so that the particle at the mid-point gets a velocity v_0 upward while the rest of the string is still at rest, as shown in Fig. 4.26. You can assume that the velocity at $t = 0$ is a Dirac delta function,

$$v(x, 0) = v_0 \delta\left(\frac{x}{L} - \frac{1}{2}\right) = v_0 L \delta\left(x - \frac{L}{2}\right).$$

(a) Find the subsequent motion of the string.

(b) How often will the string reappear horizontal in its motion?

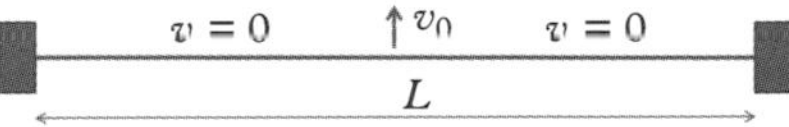

Fig. 4.26 *Exercise 4.13.*

(4.14) **Kundt's tube** A Kundt's apparatus is shown in Fig. 4.27. The apparatus consists of a glass tube closed at one end and open at the other end. Cork dust or a similarly light powder is spread in the tube as a visual aid to observe the vibrations of the air in the tube. A metal rod with a piston is then inserted from the open end. The middle of the rod is clamped. Now, when the rod is scored with a piece of cloth or leather, the rod is set into the fundamental vibration mode of the rod. The vibration of the rod sets the piston into vibration which excites the air particles inside the tube. If the frequency of vibration of the rod is equal to the frequency of one of the vibration modes of the air column then a stationary wave is set up in the air tube. As a result the dust particles in the tube agitate and get thrown away from the antinodes and settle in heaps near the nodes (Fig. 4.27).

By observing the pattern of the dust, you can determine the mode of the air column that was excited. To find the proper condition, it is necessary to move the piston in the tube, thus changing the length of the air column, and hence the wavelength of the modes. The vibration of the rod occurs at very high frequency which tends to excite higher modes of the air column. The distance between successive heaps of dust is equal to half the wavelength. It is usually difficult to measure the distance between successive heaps; instead, we measure the distance between several heaps and obtain the distance between successive heaps, equating it to half the wavelength of the sound wave excited in air.

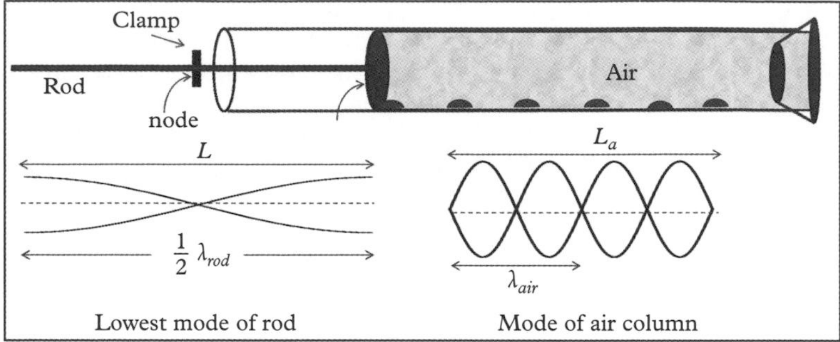

Fig. 4.27 *Exercise 4.14. Kundt's Apparatus.*

(a) Suppose, that in one experiment, you measure the distance between 11 heaps in the tube to be 40 cm. Let the speed of sound in air be 340 m/s, and the length of the rod be 0.8 m. Find the speed of the sound wave in the metal rod.

(b) Deduce a formula for the speed of the sound wave in a metal rod of length L if the wavelength of sound excited in the tube is λ_0 from the vibration of the rod in the fundamental mode. Use the symbol v_s for the speed of sound in air.

(4.15) **Simple flute.** In woodwind instruments the normal mode of the air in the instrument is excited by the player. Consider a simple flute with the openings at A near the mouthpiece and D at the other end. Thus the flute has openings at both ends and will have pressure nodes at the two ends. The simple flute for this problem also has holes at B and C, as shown in Fig. 4.28.

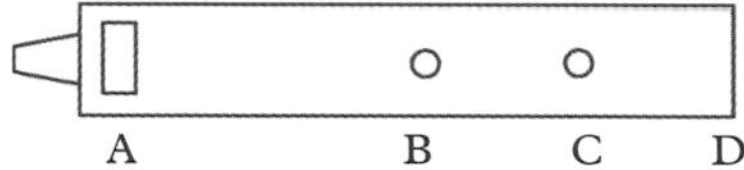

Fig. 4.28 *Exercise 4.15.*

Note that when the holes are open, there will be pressure nodes (zero pressure difference from the ambient pressure) at those locations. Thus, if you have both B and C closed, then the sound wave generated will have zero pressure only at A and D. When B is open and C is closed, you will generate pressure nodes at A, B, and D, etc. Let AB = BD and BC = CD. What frequencies will be generated if AD = 40 cm and the speed of sounds 345 m/s in the following situations: (i) both B and C are closed, (ii) C is closed but B is open, and (iii) both B and C are open.

Driven Oscillations

5

Chapter Goals

In this chapter you will learn to quantify the response of an oscillator to a harmonic driving force. You will also study the condition under which an oscillator can be driven into resonance and the roles of natural frequency and damping in the phenomenon of resonance. The response of systems with two and more degrees of freedom will also be described.

So far in this book we have studied free oscillations of ideal and damped oscillators. By free oscillations we mean the oscillation due to a restoring force alone. The free oscillation of an ideal oscillator is special in the sense that once the oscillator has been set in motion, it will oscillate for ever. The motion of a damped oscillator, on the other hand, is more realistic—after the oscillator is set in motion, the amplitude of oscillations decreases over time, which brings the oscillator to rest.

In this chapter we will study ideal and damped oscillators when they are subject to a time-dependent force. The oscillator is then said to be a **forced or driven oscillator**. For the driving force we will use a **harmonically oscillating force**, i.e. a force that is either a cosine or a sine function of time. It turns out that a force that is not a harmonic function of time can always be expressed as a sum (or an integral) of harmonic functions by using a Fourier series or Fourier integral. Since our system is linear, the response from each harmonic component of a non-harmonic force can just be summed over to get the net response of the system. Therefore, our use of a harmonic force does not limit our study as we can extend our analysis to any arbitrary force by Fourier analysis. We will present the Fourier analysis in the next chapter when we study traveling waves.

We will first study the response of a one-dimensional harmonic oscillator to a harmonic force in some detail. We find that if an oscillator is driven for a long time it reaches a steady state and oscillates at the frequency of the driving force. We will study the steady state response at various frequencies of the driving force, which will show the appearance of resonance phenomena in the driven system. We will then apply the same techniques to the response of systems with two degrees of freedom and study how the two normal modes respond as we drive the system at a particular frequency. Finally, we will study the response of systems with several degrees of freedom, at frequencies that are near the normal modes, below the lowest frequency mode, and above the largest frequency mode.

A First Course in Vibrations and Waves. First Edition. Mohammad Samiullah.

5.1 Damped Driven One-dimensional Harmonic Oscillator

Consider a block of mass m attached to a spring with spring constant k and subject to a viscous damping force that is linear in velocity with magnitude given by bv, where b is the damping constant and v the speed of the oscillator. We have studied the free oscillations of this type of oscillator before. Depending on the relative values of $b/2m$ and $\sqrt{k/m}$, the oscillator will oscillate or not,

$$b/2m < \sqrt{k/m} \quad \text{Under-damped oscillator}$$

$$b/2m = \sqrt{k/m} \quad \text{Critically damped case}$$

$$b/2m > \sqrt{k/m} \quad \text{Over-damped case}$$

Only the under-damped oscillator oscillates, the other two cases damp out too quickly. Now we will study the response of the under-damped oscillator when it is subject to a harmonically varying force of frequency ω. To be concrete, we will assume that the oscillator moves along the x-axis and all forces are along the x-axis as well. Let the driving force have the largest value F_0 and be pointed towards the positive x-axis at $t = 0$. Then, the mathematical form of the x-component of the sinusoidal force will be given by

$$F_x(\text{driving}) = F_0 \cos(\omega t), \tag{5.1}$$

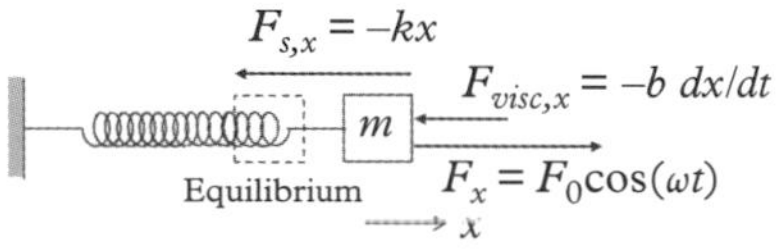

$F_{s,x} = -kx$

$F_{visc,x} = -b\, dx/dt$

m

Equilibrium

$F_x = F_0 \cos(\omega t)$

x

Fig. 5.1 *The forces on a damped driven oscillator.*

The three forces on the damped driven oscillator are shown schematically in Fig. 5.1. A simple physical realization of a damped driven oscillator is shown in Fig. 5.2.

Then the x-component of Newton's second law of motion gives the following equation of motion for the x-coordinate of the block, which is taken from the equilibrium point of the motion of the block:

$$m\frac{d^2x}{dt^2} = -kx - b\frac{dx}{dt} + F_0 \cos(\omega t). \tag{5.2}$$

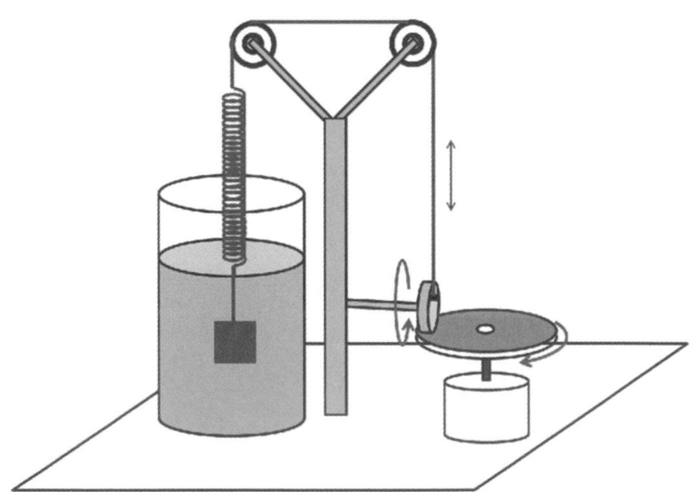

Fig. 5.2 *A physical realization of the damped driven oscillator, called the "Texas Tower" has been developed by J.G. King at the Education Research Center, Massachusetts Institute of Technology. A variable motor drives the up and down motion of the block in the fluid sinusoidally. The fluid provides the damping force and the spring the restoring force. Adapted from A.P. French, Vibrations and Waves, MIT, 1971.*

Now we divide both sides of the equation by m and define the following composite parameters, as we have done previously when discussing the free damped oscillator:

$$\omega_0 = \sqrt{k/m}, \text{ the natural frequency,} \tag{5.3}$$

$$\Gamma = b/m, \text{ the damping constant.} \tag{5.4}$$

Using the new parameters the equation of motion can be rewritten as follows,

$$\frac{d^2x}{dt^2} + \Gamma\frac{dx}{dt} + \omega_0^2 x = \frac{F_0}{m}\cos(\omega t). \tag{5.5}$$

The motion of the oscillator is completely described by the solutions of this equation. The quantities ω_0 and Γ characterize the oscillator. We are interested in finding $x(t)$ that obeys this equation and satisfies initial conditions $x(0)$ and $(dx/dt)_{t=0}$.

The differential equation, Eq. 5.5, is considerably more difficult to solve than those for undamped and damped oscillators. A function of time that satisfies Eq. 5.5 is called a **particular solution,** $x_p(t)$,

$$\frac{d^2x_p}{dt^2} + \Gamma\frac{dx_p}{dt} + \omega_0^2 x_p = \frac{F_0}{m}\cos(\omega t). \tag{5.6}$$

Since, there is no x on the right side of Eq. 5.5, we can add another function $x_c(t)$ to $x_p(t)$, and the sum will also satisfy Eq. 5.5 as long as x_c satisfies

$$\frac{d^2x_c}{dt^2} + \Gamma\frac{dx_c}{dt} + \omega_0^2 x_c = 0. \tag{5.7}$$

The function $x_c(t)$ is called the **complementary solution**. Thus, **the complete (or the full) solution** of Eq. 5.5 consists of two parts, $x_p(t)$ and $x_c(t)$,

$$\text{General solution: } x(t) = x_p(t) + x_c(t), \tag{5.8}$$

where $x_p(t)$ and $x_c(t)$ satisfy Eqs. 5.6 and 5.7 respectively. From our discussions on the damped oscillator, we recognize that x_c obeys the equation for a damped oscillator without the driving force. We know that x_c will damp out with time with a time constant $\frac{2}{\Gamma}$. We will find below that x_p does not damp out with time. Therefore, for large enough time, the relative contribution of x_c to the full solution $x(t)$ will become negligible compared to x_p. Therefore, the particular solution x_p will dominate at large times compared to $\frac{2}{\Gamma}$. We say that for large times the system reaches a **steady state**, and the displacement x is given by x_p alone. To emphasize this aspect of the complete solution, the part x_p is also called the **steady state solution**. To keep this point evident we will often denote this solution by x_s rather than x_p,

$$\text{For } t \gg \frac{2}{\Gamma}, \quad x(t) \approx x_p(t) \equiv x_s(t), \quad \text{(the steady state).} \tag{5.9}$$

We will find below that, in the steady state, the oscillator executes a simple harmonic motion at the driving frequency ω. The amplitude of the oscillations depends on the strength F_0 of the applied force and the driving frequency. The phase of the displacement of the oscillator has a definite relation to the phase of the driving force. The phase lag of the displacement with respect to the driving force depends on the driving frequency.

5.2 Steady State Solution

5.2.1 Amplitude and Phase Constant in Steady State

Since the driving force in Eq. 5.5 is sinusoidal in time, we demand that $x(t)$ in Eq. 5.5 be a sum of a cosine and a sine term of the same frequency, or alternatively a cosine of the same frequency ω with a phase constant,

$$x(t) \xrightarrow{t >> \frac{2}{\Gamma}} x_s(t) = A\cos(\omega t - \delta), \tag{5.10}$$

where A is the amplitude of the steady oscillations and δ is the phase constant, which is written as a **phase lag** with respect to the phase of the driving force. Note that we will use the symbol δ for the phase constant in this chapter, while we have used the symbol ϕ in previous chapters. We will do that to emphasize the phase lag notation in the present context. Inserting the assumed solution into Eq. 5.5, and then separating terms multiplying $\cos(\omega t)$ from those multiplying $\sin(\omega t)$ we find that

$$\left[\left(\omega_0^2 - \omega^2\right) A\cos\delta + \Gamma\omega A\sin\delta - \frac{F_0}{m}\right]\cos(\omega t) + \left[\left(\omega_0^2 - \omega^2\right) A\sin\delta - \Gamma\omega A\cos\delta\right]\sin(\omega t) = 0. \tag{5.11}$$

Since $\sin(\omega t)$ and $\cos(\omega t)$ are linearly independent functions, the coefficients multiplying them must be zero independently for their sum to be zero,

$$\left(\omega_0^2 - \omega^2\right) A\cos\delta + \Gamma\omega A\sin\delta - \frac{F_0}{m} = 0, \tag{5.12}$$

$$\left(\omega_0^2 - \omega^2\right) A\sin\delta - \Gamma\omega A\cos\delta = 0. \tag{5.13}$$

From the second equation, we obtain δ, which we use in the first equation to get A. The details of the calculations are left for the student to complete. The result of these calculations are

$$A = \frac{F_0/m}{\sqrt{\left(\omega_0^2 - \omega^2\right)^2 + (\Gamma\omega)^2}}, \tag{5.14}$$

$$\tan\delta = \frac{\Gamma\omega}{\omega_0^2 - \omega^2}. \tag{5.15}$$

Often, we are interested in these expressions in terms of the quality factor Q of the underdamped oscillator defined by

$$Q = \frac{\omega_0}{\Gamma}.$$

Replacing Γ in Eqs. 5.14 and 5.15 we get

$$A = \frac{F_0/m}{\sqrt{\left(\omega_0^2 - \omega^2\right)^2 + (\omega_0\omega/Q)^2}}, \tag{5.16}$$

$$\tan\delta = \frac{\omega_0\omega}{Q\left(\omega_0^2 - \omega^2\right)}. \tag{5.17}$$

5.2.2 Complex Exponential Method for Steady State Solution

We obtained the steady state solution $x_s = A\cos(\omega t - \delta)$ above by imposing physical expectation on the solution and we worked with all real functions. In this subsection, we will show that the steady state solution can be obtained much more readily when we work in the complex exponential method, since the algebra is much easier. In this method we replace the physical displacement x in the differential equation by a complex displacement z, and also replace the cosine function of time on the right side of Eq. 5.5 by a complex exponential,

$$\frac{d^2z}{dt^2} + \Gamma\frac{dz}{dt} + \omega_0^2 z = \frac{F_0}{m}e^{-i\omega t}. \tag{5.18}$$

Note that in this equation only z and i are complex, all other quantities, namely, t, Γ, ω_0, m, and F_0 are real. Now, we assume a complex form for z,

$$z = Ce^{-i\omega t}, \tag{5.19}$$

where C is a complex amplitude to be determined from the calculations here. After putting this z into Eq. 5.18 we can solve for C,

$$C = \frac{F_0/m}{\omega_0^2 - \omega^2 - i\Gamma\omega}. \tag{5.20}$$

To obtain the amplitude and phase of C we perform elementary complex algebra on the right side of this equation and express the right side in the following form:

$$C = |C|e^{i\theta}. \tag{5.21}$$

Details of the complex algebra are left as an exercise for the student. After these calculations you will find that

$$|C| = \frac{F_0/m}{\sqrt{\left(\omega_0^2 - \omega^2\right)^2 + \Gamma^2\omega^2}}, \tag{5.22}$$

$$\tan\theta = \frac{\Gamma\omega}{\omega_0^2 - \omega^2}. \tag{5.23}$$

Using Eq. 5.21 in Eq. 5.19 we obtain x from the real part of z,

$$x = Re[z] = |C|\cos(\omega t - \theta). \tag{5.24}$$

This is identical to the solution given above in Eq. 5.14 if we identify $A = |C|$ and $\delta = \theta$.

5.2.3 Absorptive and Elastic Amplitudes

In the foregoing discussion of the steady state of an oscillator we have described the displacement in terms of its amplitude A and the phase lag δ. As you know from your studies of the free oscillator we can equally well describe the oscillations in terms of the superposition of a cosine and a sine term.

Since the driving force $F_0 \cos(\omega t)$ is a pure cosine, the superposition of cosine and sine for $x(t)$ will correspond to the parts of x that are in phase and 90° out of phase, respectively, with respect to the driving force. The part that is in phase with the driving force is called the **elastic amplitude** and the part that is 90° out of phase is called the **absorptive amplitude**. Let A_{el} denote the elastic amplitude and A_{ab} the absorptive amplitude. Then, the displacement of the oscillator can be written as

$$x(t) = A_{el}\ \cos(\omega t) + A_{ab}\ \sin(\omega t). \tag{5.25}$$

To obtain the expressions for A_{el} and A_{ab}, we equate $x(t)$ in Eq. 5.25 to the steady state solution $A\ \cos(\omega t - \delta)$, and solve for A_{el} and A_{ab},

$$A_{el} = \frac{F_0}{m}\left[\frac{\omega_0^2 - \omega^2}{\left(\omega_0^2 - \omega^2\right)^2 + \Gamma^2\omega^2}\right], \tag{5.26}$$

$$A_{ab} = \frac{F_0}{m}\left[\frac{\Gamma\omega}{\left(\omega_0^2 - \omega^2\right)^2 + \Gamma^2\omega^2}\right]. \tag{5.27}$$

From the two representations of x given in polar form in Eq. 5.10 and rectangular form in Eq. 5.25 we can immediately deduce the relation between the elastic and absorptive amplitudes and the amplitude A and the phase constant δ as follows:

$$A = \sqrt{A_{el}^2 + A_{ab}^2},\ \ \tan\delta = \frac{A_{ab}}{A_{el}}. \tag{5.28}$$

5.2.4 Power of the Driving Force

In order to drive a steady motion of the oscillator the driving force must do work to provide the energy that is being dissipated due to viscous forces. The instantaneous power $P(t)$ or the rate of the energy input by the driving force at any instant t is equal to the scalar product of the driving force and the velocity vectors. For the purposes of calculation of power we find it useful to write the velocity of the oscillator in terms of the elastic and absorptive amplitudes rather than the amplitude and phase constant,

$$P(t) = \vec{F}\cdot\vec{v} = F_x v_x = F_0\cos(\omega t)\frac{dx}{dt}$$

$$= \omega F_0\left[-\frac{1}{2}A_{el}\sin(2\omega t) + A_{ab}\cos^2(\omega t)\right]. \tag{5.29}$$

In the steady state, the quantity of importance is the power expended by the force over a full cycle—this is the energy that the driving force must supply per cycle to maintain the steady motion of the oscillator. The time-averaged power, to be denoted by placing angle brackets, $\langle\cdot\rangle$, around P, is obtained by integrating Eq. 5.29 over one cycle and dividing the result by the time period. Integration over the term multiplying A_{el} gives zero, and integration over the square of the cosine gives $\frac{1}{2}$,

$$\langle\sin(2\omega t)\rangle = \frac{\omega}{2\pi}\int_0^{2\pi/\omega}\sin(2\omega t)\ dt = 0, \tag{5.30}$$

$$\langle\cos^2(\omega t)\rangle = \frac{\omega}{2\pi}\int_0^{2\pi/\omega}\cos^2(\omega t)\ dt = \frac{1}{2}. \tag{5.31}$$

Therefore, we find the following for the time-averaged power:

$$\langle P \rangle = \frac{1}{2} F_0 \omega A_{ab}. \tag{5.32}$$

This shows that the power input to the oscillator is proportional to the absorption amplitude and not to the elastic amplitude—the power goes into the displacement that is 90° out of phase with the driving force.

Average Power Dissipated by the Damping Force

From the conservation of energy it is apparent that the average power $\langle P \rangle$ input by the driving force found in Eq. 5.32 must equal the time-averaged rate of energy dissipated by the damping force, which is given by $F_{visc} = -b\, dx/dt$. Let us calculate the latter and verify this expectation. The instantaneous power of the viscous force will be

$$P_{\text{visc}}(t) = -b \left(\frac{dx}{dt} \right)^2 . \tag{5.33}$$

The negative sign here refers to the fact that this power is the power leaving the oscillator, while the power of the driving force was the power input to the oscillator. For the steady state we use the solution in terms of the elastic and absorptive amplitudes to obtain the following:

$$\langle P_{\text{visc}} \rangle = -\Gamma m \left\langle \left(\frac{dx}{dt} \right)^2 \right\rangle = -\frac{1}{2} \Gamma m \omega^2 \left(A_{el}^2 + A_{ab}^2 \right) . \tag{5.34}$$

This expression does not look like the expression in Eq. 5.32. The average power dissipated by the damping force appears to involve both the absorptive and elastic amplitudes, while the power by the driving force had only the absorptive amplitude. However, by using the explicit expressions for A_{el} and A_{ab} given in Eqs. 5.26 and 5.27 you can show that the two expressions are actually the same.

Proof: From Eqs. 5.26 and 5.27 we find that A_{el} and A_{ab} are related by

$$\frac{A_{el}}{\omega_0^2 - \omega^2} = \frac{A_{ab}}{\Gamma \omega}. \tag{5.35}$$

Using this in Eq. 5.34 we can write $\langle P_{\text{visc}} \rangle$ in terms of A_{ab} alone,

$$\begin{aligned} \langle P_{\text{visc}} \rangle &= -\frac{1}{2} \Gamma m \omega^2 \left(A_{el}^2 + A_{ab}^2 \right) \\ &= -\frac{1}{2} \Gamma m \omega^2 \left[\left(\frac{\omega_0^2 - \omega^2}{2 \frac{1}{2} \Gamma \omega} \right)^2 \right] A_{ab}^2 . \end{aligned}$$

Replacing one factor of A_{ab} by the expression in Eq. 5.27 we can show that this expression is identical to that found for $\langle P \rangle$ of the driving force,

$$\langle P_{\text{visc}} \rangle = -\frac{1}{2} F_0 \omega A_{ab} = -\langle P \rangle . \tag{5.36}$$

5.2.5 Resonance Curve of Power

Equation 5.32 shows that the time-averaged power input depends on the frequency ω of the driving force. To see this dependence more explicitly let us write the full expression for the time-averaged power by replacing A_{ab} by its expression given in Eq. 5.27,

$$\langle P \rangle = P_{\max} \left[\frac{(\Gamma\omega)^2}{\left(\omega_0^2 - \omega^2\right)^2 + (\Gamma\omega)^2} \right], \tag{5.37}$$

where

$$P_{\max} = \frac{F_0^2}{2\Gamma m}. \tag{5.38}$$

A plot of $\langle P \rangle$ versus ω is shown in Fig. 5.3. In the figure we see that the power input in the steady state has a maximum at natural frequency ω_0. We say that the oscillator is **at resonance** when we drive the oscillator at this frequency. We can demonstrate this fact mathematically by setting the first derivative of $\langle P \rangle$ with respect to ω to zero to find the maximum,

$$\frac{d\langle P \rangle}{d\omega} = 0. \tag{5.39}$$

Solving the complicated expression yields $\omega = \omega_0$ when $\langle P \rangle$ is maximum. We denote this resonance frequency by ω_R. Thus,

$$\omega_R = \omega_0. \tag{5.40}$$

The maximum power at the resonance is obtained by setting $\omega = \omega_R$ in Eq. 5.37, which shows immediately that $P_{\max}$, given in Eq. 5.37, is the power input at the resonance.

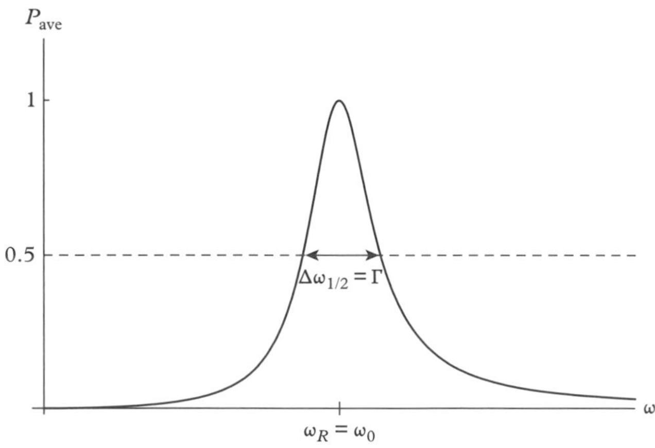

Fig. 5.3 *The power curve of a driven oscillator. The resonance of the power occurs at* $\omega = \omega_0$ *and the width at half of the maximum,* $\Delta\omega_{1/2}$*, which is also denoted by* $(\Delta\omega)_{FWHM}$*, for a lightly damped oscillator is determined by the damping constant* Γ*.*

Resonance Peak Width

The resonance curve is characterized by the height and width of the peak. It is easy to understand the importance of the peak height, which gives us the maximum power the force can deliver. A common measure of width is the width at the half-height of the peak, called **full width at half maximum (FWHM)**. It is hard to appreciate the importance of the peak width; but it is quite important as we see below. The peak width gives us information about the rate at which energy is dissipated. The resonance peak is sharper when the damping of an oscillator is less; that is, the resonance peak is sharper for a better oscillator.

We will now calculate the FWHM of $\langle P\rangle$ to show that the width is actually equal to the damping parameter Γ. We seek the values of ω for which $\langle P\rangle$ is half of P_{max}. This can be obtained by solving the following equation for ω, obtained by setting $\langle P\rangle = \frac{1}{2}P_{\text{max}}$:

$$\frac{(\Gamma\omega)^2}{\left(\omega_0^2 - \omega^2\right)^2 + (\Gamma\omega)^2} = \frac{1}{2}. \tag{5.41}$$

Simplifying this equation, we obtain two quadratic equations for ω,

$$\omega^2 - \omega_0^2 = \Gamma\omega, \quad \omega^2 - \omega_0^2 = -\Gamma\omega. \tag{5.42}$$

Each of these quadratic equations in ω gives one positive and one negative solution. Discarding the negative solutions we find

$$\omega = \sqrt{\omega_0^2 + \frac{1}{4}\Gamma^2} \pm \frac{1}{2}\Gamma. \tag{5.43}$$

These two positive values of ω correspond to two points on the resonance curve where power is half of the maximum. Therefore, the difference of the two solutions gives us the FWHM, $(\Delta\omega)_{FWHM}$, also denoted by $\Delta\omega_{1/2}$,

$$(\Delta\omega)_{\text{FWHM}} = \Gamma. \tag{5.44}$$

In terms of the quality factor Q of the oscillator, which is $Q = \omega_0/\Gamma$,

$$(\Delta\omega)_{\text{FWHM}} = \omega_0/Q. \tag{5.45}$$

The oscillator with a higher Q will have sharper resonance peak.

Resonance Width and Decay Time Constant

Recall that the time constant τ_E for the decay of the energy of a damped oscillator is equal to the inverse of Γ,

$$\tau_E = \frac{1}{\Gamma}. \tag{5.46}$$

Therefore, we can write the result of the width of the resonance peak, Eq. 5.66, in terms of the time constant as

$$(\Delta\omega)_{\text{FWHM}} \times \tau_E = 1. \tag{5.47}$$

This equation says that the product of the frequency width of the resonance peak of the driven system and the time constant of the free system is equal to 1. Although this result was derived for a one-dimensional harmonic oscillator, the result turns out to be applicable to many systems. This result also provides an important way for experiments to gain access to the properties of a system. For instance, to gain a direct reading of τ_E, one would need to observe the decay of energy with time, which is difficult to do experimentally. Equation 5.47 says that one can get the information about τ_E from the frequency width of the plot of the average power versus the frequency, which is a much easier experiment to perform.

5.2.6 Variation of the Elastic and Absorptive Amplitudes with Frequency

Let us rewrite the expressions for A_{el} and A_{ab} from Eqs. 5.26 and 5.27:

$$A_{el} = \frac{F_0}{m}\left[\frac{\omega_0^2-\omega^2}{\left(\omega_0^2-\omega^2\right)^2+\Gamma^2\omega^2}\right], \tag{5.48}$$

$$A_{ab} = \frac{F_0}{m}\left[\frac{\Gamma\omega}{\left(\omega_0^2-\omega^2\right)^2+\Gamma^2\omega^2}\right]. \tag{5.49}$$

At the resonance of the power the frequency is $\omega = \omega_0$. Setting $\omega = \omega_0$ in these expressions we find that at the resonance of the power the elastic amplitude is zero. That is, in the steady state at the resonance of the power, the elastic amplitude does not take part in the motion of the oscillator. Does that mean that the elastic amplitude is unimportant? Far from it. Let us look at the ratio of the two amplitudes to gain an understanding of their relative importance at various frequencies:

$$\frac{A_{el}}{A_{ab}} = \frac{\omega_0^2-\omega^2}{\Gamma\omega}. \tag{5.50}$$

Since $A_{ab} > 0$ always, the elastic amplitude A_{el} would have to be positive below the resonance, i.e. for $\omega < \omega_0$, and negative above the resonance, $\omega > \omega_0$. In Fig. 5.4 we show plots of A_{el} and A_{ab} with various important points on the plots. It is clear that while the absorptive amplitude is maximum at the resonance frequency, it decreases away from the resonance. Therefore, away from the resonance, it is the elastic amplitude that dominates the motion of the oscillator.

5.2.7 Variation of Amplitude and Phase Constant with Frequency

Note that in the steady state the oscillator's displacement varies as $x = A\cos(\omega t - \delta)$, while the time-dependent driving force goes as $F = F_0\cos(\omega t)$. A rotating vector representation of these sinusoidal functions shows their phase relation pictorially, as in Fig. 5.5.

In spite of the fact that in the steady state, the driving force and the oscillator displacement both change sinusoidally with the same frequency ω, they are not necessarily in phase with each other since the phase difference δ is usually not zero. If $\delta = 0$, then the driving force and the displacement are in phase, i.e. the block moves in tandem with the force.

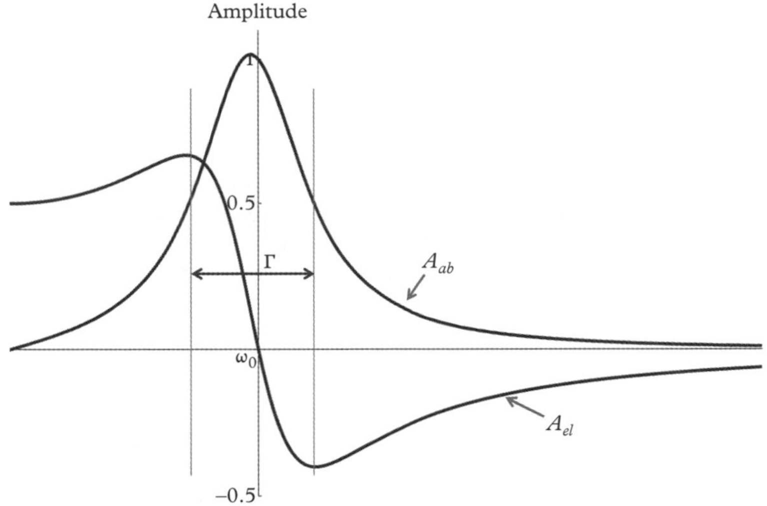

Fig. 5.4 *The variation of elastic (A_{el}) and absorptive (A_{ab}) amplitudes with the frequency (ω) of the driving force. The amplitudes have been scaled so that the value of the absorptive amplitude at $\omega = \omega_0$ is equal to 1.*

However, if $\delta \neq 0$, the block's motion would not be in sync with the driving force and there would be a phase lag δ of the displacement relative to the driving force.

Low and High Frequency Behaviors

Phase lag From Eq. 5.15, we see that if the driving force has a low frequency, i.e. when $\omega < \omega_0$, the phase lag is positive since $\delta > 0$, and for high driving frequencies when $\omega > \omega_0$, the phase lag is negative. For really low driving frequency the phase lag would be very small and can be approximated to zero,

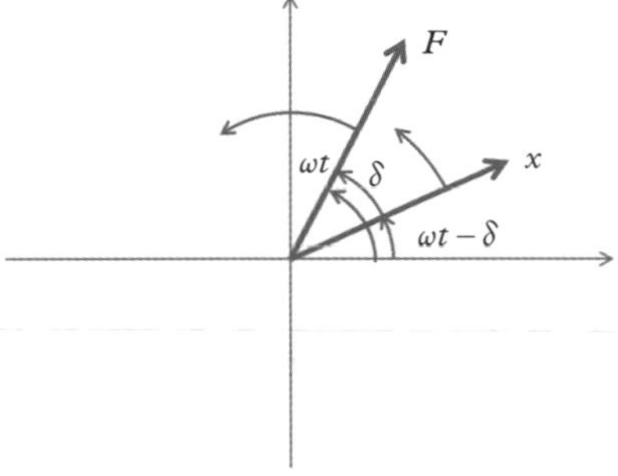

Fig. 5.5 *Rotating vector representations of the harmonic driving force $F = F_0 \cos(\omega t)$ and the displacement $x = A\cos(\omega t - \delta)$ in the steady state. If you think of the harmonic functions as the real part of complex exponentials, then $F = F_0 e^{i\omega t}$ and $x = Ae^{i(\omega t - \delta)}$. In that case the horizontal axis is the real part and the vertical axis the imaginary part.*

Very low driving frequencies:

$$\delta = \lim_{\omega/\omega_0 << 1} \tan^{-1}\left(\frac{\Gamma\omega}{\omega_0^2 - \omega^2}\right) = 0. \tag{5.51}$$

Therefore, at low frequencies the oscillator would move with the driving force. At very high frequencies, there is a phase lag of π radians as the corresponding limit of Eq. 5.15 shows:

Very high driving frequencies:

$$\delta = \lim_{\omega/\omega_0 >> 1} \tan^{-1}\left(\frac{\Gamma\omega}{\omega_0^2 - \omega^2}\right) = \pi. \tag{5.52}$$

Therefore, the oscillator is completely out of step with the driving force at high frequencies.

Amplitude Similarly, the amplitude A of steady state oscillations varies with the driving frequency according to Eq. 5.14. At very low driving frequency, the amplitude is nearly independent of the driving frequency, and at high frequency the amplitude goes to zero:

Very low driving frequencies:

$$A = \lim_{\omega/\omega_0 << 1}\left[\frac{F_0/m}{\sqrt{\left(\omega_0^2 - \omega^2\right)^2 + (\Gamma\omega)^2}}\right] = \frac{D}{\omega_0^2}. \tag{5.53}$$

Very high driving frequencies:

$$A = \lim_{\omega/\omega_0 >> 1} \left[\frac{F_0/m}{\sqrt{\left(\omega_0^2 - \omega^2\right)^2 + (\Gamma\omega)^2}} \right] = 0. \tag{5.54}$$

Note that Eq. 5.14 for the amplitude shows that when the driving frequency is close to the natural frequency, the denominator becomes small and hence the amplitude becomes large. We will show below that there is an optimal driving frequency at which the amplitude is the largest. This optimal frequency is called the resonance frequency of the amplitude, or simply the resonance frequency, which occurs at a different frequency than the resonance of the power, as we will now derive.

Resonance of the amplitude The steady state amplitude of the oscillator varies with the strength of the force F_0, as you would expect, but also on the frequency ω of the harmonic driving force, as shown in Fig. 5.6. When you vary the frequency ω of the driving force, you find that the amplitude A of the oscillations of the block also varies. That is, the amplitude A is a function of the driving frequency, ω. To indicate the functional relation we sometimes write the amplitude as $A(\omega)$. The amplitude A takes its maximum value at the **resonance frequency of the amplitude,** which we will denote by $\hat{\omega}_R$. Since $A(\omega)$ has its maximum at $\omega = \hat{\omega}_R$, to find $\hat{\omega}_R$ we will set the first derivative of A with respect to ω to zero,

$$\left. \frac{dA}{d\omega} \right|_{\omega = \hat{\omega}_R} = 0. \tag{5.55}$$

Rather than do the calculation indicated in this equation, we note that the extrema of A and $1/A^2$ will occur for the same ω since $A \neq 0$ for finite positive ω,

$$\frac{d}{d\omega}\frac{1}{A^2} = -2\frac{1}{A^3}\frac{dA}{d\omega} = 0. \tag{5.56}$$

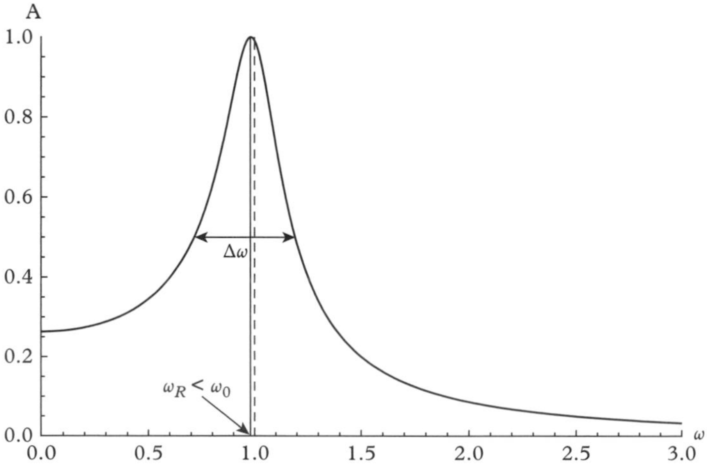

Fig. 5.6 *The amplitude of steady state oscillations as a function of driving frequency. Here the normalized amplitude A/A_R is plotted along the vertical axis and the driving frequency in units of ω_0 is plotted along the horizontal axis for an oscillator of damping constant $\Gamma = 0.2\ \omega_0$. Note that the resonance of the amplitude, occurs at a frequency $\hat{\omega}_R$ that is less than the resonance frequency of the average power, which occurs at $\omega_R = \omega_0$. The width at half-height $\Delta\omega$ characterizes the sharpness of the peak.*

Thus, we calculate the extrema of the quantity inside the radical in the denominator of A given in Eq. 5.14 to find the resonance frequency of A. This provides a much simpler calculation for the same result,

$$\left.\frac{d[(\omega_0^2-\omega^2)^2+(\Gamma\omega)^2]}{d\omega}\right|_{\omega=\hat{\omega}_R}=0. \tag{5.57}$$

We can solve this equation for $\hat{\omega}_R$,

$$\hat{\omega}_R=\sqrt{\omega_0^2-\frac{1}{2}\Gamma^2}. \tag{5.58}$$

The frequency $\omega=\hat{\omega}_R$ is not only an extremum of $A(\omega)$ but also its maximum, as you can easily verify from the second derivative of $A(\omega)$. Therefore, the oscillator will vibrate with the largest amplitude if it is driven at this frequency. This phenomenon is called the **resonance of the amplitude**. Equation 5.58 shows that, normally, the resonance frequency $\hat{\omega}_R$ is below the natural frequency ω_0 of the oscillator at which the average power was found to resonate. However, for a lightly damped oscillator ($\omega_0 \gg \frac{1}{2}\Gamma$) the resonance frequency $\hat{\omega}_R$ will be very near its natural frequency, $\omega_0 \equiv \sqrt{k/m}$,

$$\text{Lightly damped oscillator: } \hat{\omega}_R \approx \omega_0 = \omega_R. \tag{5.59}$$

Therefore, in a lightly damped oscillator, an experimental determination of the resonance frequency $\hat{\omega}_R$ gives a good indicator of the natural frequency of the system.

Phase lag and amplitude at resonance You saw above that the phase lag of the displacement with respect to the driving force is zero if driven at low frequencies compared to the natural frequency ω_0 of the oscillator. We also found that the phase lag is π if driven at higher frequencies. What happens if driven at the resonance frequency? We can find the answer to this question by putting $\omega=\hat{\omega}_R$ directly into Eq. 5.15 and taking the limit of a lightly damped oscillator,

$$\delta_{\text{resonance}}=\tan^{-1}\left(\frac{\Gamma\hat{\omega}_R}{\omega_0^2-\hat{\omega}_R^2}\right)\sim\tan^{-1}\left(\frac{\omega_0}{\Gamma}\right)\rightarrow\frac{\pi}{2}\ \text{as}\ \frac{\omega_0}{\Gamma}\rightarrow\infty. \tag{5.60}$$

Therefore, at resonance, the positions of a lightly damped oscillator and the harmonic driving force are 90° out of phase with each other. This says that at resonance, while the driving force varies as $\cos(\hat{\omega}_R t)$, the position of the mass varies as $\sin(\hat{\omega}_R t)$,

$$x_s|_{\omega=\hat{\omega}_R}=A(\hat{\omega}_R)\cos\left(\hat{\omega}_R t-\frac{\pi}{2}\right)=A_R\sin\left(\hat{\omega}_R t\right), \tag{5.61}$$

where A_R is the amplitude at resonance frequency given by

$$A_R=A(\hat{\omega}_R)=\frac{F_0}{m\Gamma}\frac{1}{\sqrt{\omega_0^2-\frac{1}{4}\Gamma^2}}. \tag{5.62}$$

Amplitude resonance peak width The variation of amplitude with the frequency of the driving force is seen in many areas of physics. As we have discussed above, the amplitude is the largest if driven at its resonance frequency $\hat{\omega}_R$. Figure 5.6 shows a plot of the amplitude as a function of the driving frequency. The peak in the amplitude is characterized by the height of the peak and the width of the peak, as we have seen.

Experimentally, the resonance peak is sharper for a better oscillator. We will now calculate the FWHM of the $A(\omega)$ function to show that the width is actually proportional to the damping parameter Γ. Recall that the amplitude in the steady state is given by

$$A(\omega) = \frac{F_0}{m}\frac{1}{\sqrt{\left(\omega_0^2-\omega^2\right)^2+(\Gamma\omega)^2}}, \tag{5.63}$$

whose peak occurs at $\omega = \hat{\omega}_R$ with amplitude at resonance, A_R, given in Eq. 5.62. Here, we seek the values of ω for which the amplitude is half of A_R. This can be obtained by solving the following equation for ω, obtained by setting $A(\omega) = A_R/2$:

$$\frac{F_0}{m}\frac{1}{\sqrt{\left(\omega_0^2-\omega^2\right)^2+(\Gamma\omega)^2}} = \frac{A_R}{2}. \tag{5.64}$$

The solutions are

$$\omega = \pm\sqrt{\omega_0^2-\Gamma^2 \pm 2\sqrt{3}\frac{1}{2}\Gamma\sqrt{\omega_0^2-\frac{1}{4}\Gamma^2}}. \tag{5.65}$$

We keep only the positive root for ω, and then the difference of the two solutions, one with the plus sign inside the radical and the other with the minus sign, gives us the FWHM, $\Delta\omega$,

$$\Delta\omega = \sqrt{\omega_0^2-\Gamma^2+2\sqrt{3}\frac{1}{2}\Gamma\sqrt{\omega_0^2-\frac{1}{4}\Gamma^2}}-\sqrt{\omega_0^2-\Gamma^2-2\sqrt{3}\frac{1}{2}\Gamma\sqrt{\omega_0^2-\frac{1}{4}\Gamma^2}}. \tag{5.66}$$

The expression for FWHM can be simplified for a lightly damped oscillator to obtain the following expression:

$$\Delta\omega \approx \sqrt{3}\Gamma. \tag{5.67}$$

Role of damping in amplitude resonance We saw above that the width of the resonance peak is directly proportional to the damping constant. Now, we study the effect of damping in a little more detail. We have introduced the quality factor $Q = \omega_0/\Gamma$ to characterize the effect of damping on harmonic oscillations. Therefore, let us rewrite our formulas for amplitude and phase lag in terms of Q, by substituting $\Gamma = \omega_0/Q$ in the corresponding formulas, Eqs. 5.14 and 5.15. After some algebra, which is left as an exercise for the student, we find that

$$A = \frac{F_0}{m}\frac{1}{\omega_0^2}\frac{1}{\sqrt{\left(1-\Omega^2\right)^2+\Omega^2/Q^2}}, \tag{5.68}$$

$$\delta = \tan^{-1}\left[\frac{\Omega/Q}{1-\Omega^2}\right], \tag{5.69}$$

where for brevity we have introduced Ω for ω/ω_0. The quantity Ω is dimensionless and is equal to the frequency of the driving force in units of the natural frequency ω_0 of the oscillator. The resonance frequency $\hat{\omega}_R$ in units of ω_0, to be denoted by $\hat{\Omega}_R$, takes the following form:

$$\Omega_R \equiv \frac{\hat{\omega}_R}{\omega_0} = \sqrt{1 - \frac{1}{2Q^2}}. \tag{5.70}$$

We plot amplitude from Eq. 5.68 and resonance frequency from Eq. 5.70 in Figs. 5.7 and 5.8, and the phase lag from Eqn. 5.69 in Fig. 5.9. The resonance peaks of the amplitude versus frequency show that the resonance becomes taller and sharper with lowering of damping, that is, with increasing Q factor. Unlike the resonance of the average power input, the resonance peak of the resonance of the amplitude A does not coincide with the natural frequency ω_0 as seen in Fig. 5.8. The resonance frequency is less than ω_0 unless Q is very large. For a good oscillator with low damping we can usually take the resonance frequency to be reasonably close to the natural frequency.

Figure 5.9 shows the variation of phase lag for different Q factors. The phase lag goes from zero to π for low to high frequencies, with a transition at the resonance frequency. The transition at resonance frequency becomes sharper for higher Q oscillators. For an undamped oscillator, the transition is a step function at $\Omega = \Omega_R$.

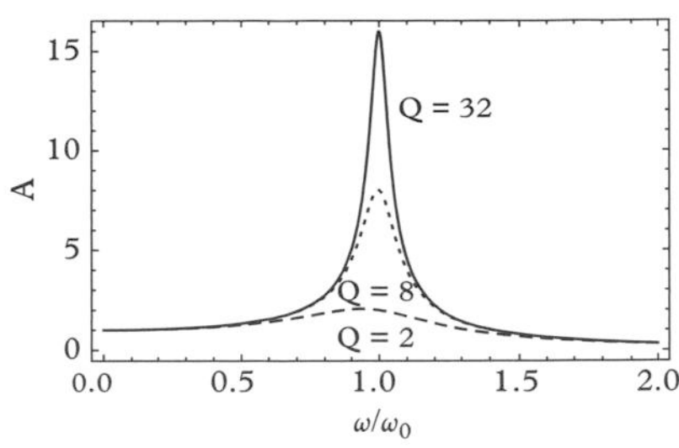

Fig. 5.7 *The resonance for different Q. Plot of amplitude A as a function of driving frequency for $Q = 2$, 4, and 32 for $F_0/m\omega_0^2 = 1$.*

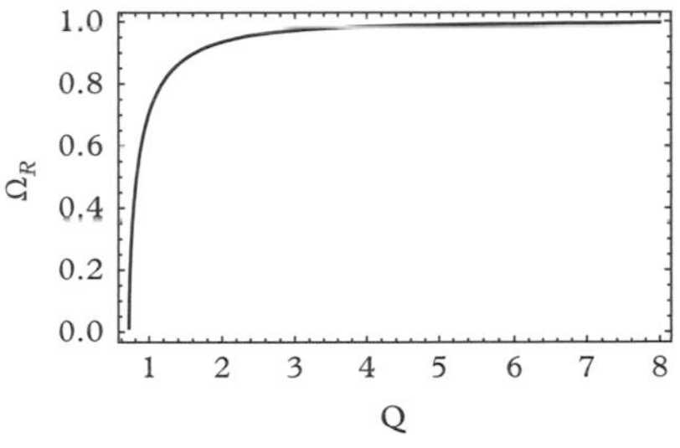

Fig. 5.8 *The resonance frequency $\Omega_R = \hat{\omega}_R/\omega_0$ as a function of Q shows that resonance frequency tends to the natural frequency ω_0 as quality Q of the oscillator rises.*

5.3 Transient Solution

5.3.1 General Solution

So far our focus in this chapter has been on the steady state of a harmonically driven oscillator. An under-damped oscillator reaches steady state after several time constants ($\tau \sim 1/\Gamma$) of the free motion have elapsed. When the system is not in a steady state we need the complete solution. The complete solution of the problem will satisfy the equation of motion and take into account the initial conditions $x(0)$ and $v(0)$ as well. We have already discussed that the complete solution consists of two parts, the particular solution x_p, which is the same as the steady state solution, and the complimentary solution x_c which is the solution in the absence of the driving force,

$$x(t) = x_c(t) + x_p(t). \tag{5.71}$$

We have already solved the differential equation (for the under-damped oscillator) satisfied by the complimentary solution, and we can write x_c directly as

$$x_c(t) = e^{-\frac{1}{2}\Gamma t}\left[C_1 \cos(\omega_1 t) + C_2 \sin(\omega_1 t)\right], \tag{5.72}$$

where C_1 and C_2 are constants to be determined from the initial conditions on the displacement and velocity, and ω_1 is the frequency of oscillation of the free oscillator,

$$\omega_1 = \sqrt{\omega_0^2 - \frac{1}{4}\Gamma^2}. \tag{5.73}$$

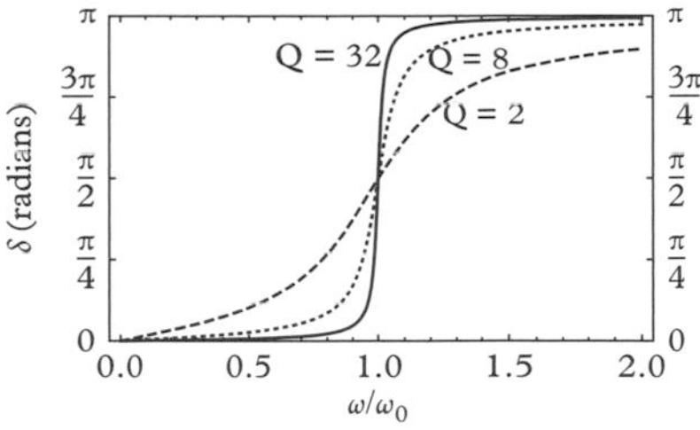

Fig. 5.9 *The phase lag as a function of driving frequency for different Q.*

Adding the steady state solution to x_c we obtain the general solution as

$$x(t) = e^{-\frac{1}{2}\Gamma t}\left[C_1 \cos(\omega_1 t) + C_2 \sin(\omega_1 t)\right] + A_{el} \cos(\omega t) + A_{ab} \sin(\omega t). \tag{5.74}$$

In this solution the only quantities not yet fixed are the constants C_1 and C_2. As stated above these are determined from $x(0)$ and $(dx/dt)_{t=0}$. As you know, the solution can also be written in other ways such as in terms of amplitudes and phase constants,

$$x(t) = e^{-\frac{1}{2}\Gamma t}\left[C_1 \cos(\omega_1 t) + C_2 \sin(\omega_1 t)\right] + A_{el}\cos(\omega t) + A_{ab}\sin(\omega t), \tag{5.75}$$

$$x(t) = C\, e^{-\frac{1}{2}\Gamma t}\cos(\omega_1 t - \phi) + A_{el}\cos(\omega t) + A_{ab}\sin(\omega t), \tag{5.76}$$

$$x(t) = C\, e^{-\frac{1}{2}\Gamma t}\cos(\omega_1 t - \phi) + A\cos(\omega t - \delta), \tag{5.77}$$

$$x(t) = e^{-\frac{1}{2}\Gamma t}\left[C_1 \cos(\omega_1 t) + C_2 \sin(\omega_1 t)\right] + A\cos(\omega t - \delta). \tag{5.78}$$

Using Initial Conditions

The constants C_1 and C_2 in the transient solution given in Eq. 5.74 are fixed by given position and velocity at some time, usually, the initial time. Let x_0 and v_0 be the position and velocity of the oscillator at $t = 0$. First, we find the expression for velocity by taking the derivative of $x(t)$ with respect to time,

$$\begin{aligned} v(t) = &-\frac{1}{2}\Gamma e^{-\frac{1}{2}\Gamma t}\left[C_1 \cos(\omega_1 t) + C_2 \sin(\omega_1 t)\right] \\ &+ \omega_1 e^{-\frac{1}{2}\Gamma t}\left[-C_1 \sin(\omega_1 t) + C_2 \cos(\omega_1 t)\right] \\ &+ \omega\left[-A_{el}\sin(\omega t) + A_{ab}\cos(\omega t)\right]. \end{aligned} \tag{5.79}$$

Setting $t = 0$ in the expressions for $x(t)$ and $v(t)$, we find the following equations for C_1 and C_2:

$$x_0 = C_1 + A_{el}, \quad v_0 = -\frac{1}{2}\Gamma C_1 + \omega_1 C_2 + \omega A_{ab}. \tag{5.80}$$

We solve these equations for C_1 and C_2 to find

$$C_1 = x_0 - A_{el}, \quad C_2 = \frac{1}{\omega_1}\left[v_0 + \frac{1}{2}\Gamma\,(x_0 - A_{el}) - \omega A_{ab}\right]. \tag{5.81}$$

An Example Calculation of Complete Solution

The complete solution of an under-damped driven oscillator is a sum of two oscillating parts: x_c oscillates at the frequency ω_1 and x_p at the driving frequency ω. Just as the superposition of two vibrations leads to beats, we expect to see beats here also. We will see this by performing a numerical solution of the complete problem.

For the numerical problem we will choose the following values:

$$\begin{aligned} &m = 1 \text{ kg},\ \omega_0 = 1 \text{ rad/sec},\ \Gamma = 0.02 \text{ rad/sec}, \\ &\omega = 0.9 \text{ rad/sec},\ F_0 = 1 \text{ N},\ x_0 = 1 \text{ m},\ v_0 = 0. \end{aligned} \tag{5.82}$$

Figure 5.10 shows the component solutions, x_c and x_p, and the complete solution x. The complementary solution x_c decays with time, while the particular solution x_p remains steady right from the start, as expected. The sum of the complementary and particular solutions is complicated and non-periodic for small times compared to the time constant $\tau = 2/\Gamma$. With increasing time, the particular solution begins to dominate, and by four or five time constants, the steady solution dominates. Thereafter the oscillator oscillates at the frequency of the driving force.

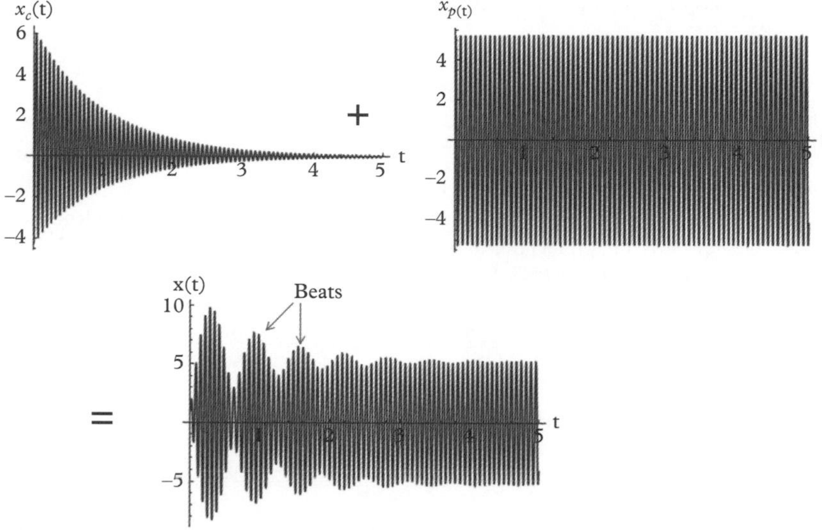

Fig. 5.10 *Complete solution of a damped driven oscillator. The following parameter values were used: $m = 1$ kg, $\omega_0 = 1$ rad/s, $\Gamma = 0.02$ rad/s, $\omega = 0.9$ rad/s, $F_0 = 1$, $x_0 = 1$, $v_0 = 0$. The time axis is in units of $2/\Gamma$, i.e. $\Delta t = 1$ on the axis represents 100 s. The complimentary solution x_c damps out with time while the particular solution x_p remains steady with time. The oscillator reaches steady state in less than five time constants.*

5.4 Resonance in Coupled Systems

5.4.1 Normal Modes and Harmonic Driving Force

A coupled system can be driven by placing an external force on any part of the system. For instance in the system in Fig. 5.11 you can apply a force on m_A or m_B or both. To be concrete, we will examine the case in which a harmonic force acts on the mass m_A, as shown in the figure. Furthermore, for simplicity, we assume the masses to be equal $m_A = m_B = m$ and the damping constants to be equal also, $b_1 = b_2 = b$. In this case, the equations of motion of the two oscillators are

$$\frac{d^2x_A}{dt^2} = -\left(\omega_0^2 + \omega_c^2\right) x_A - \Gamma\frac{dx_A}{dt} + \omega_c^2 x_B + \frac{F_0}{m}\cos(\omega t), \tag{5.83}$$

$$\frac{d^2x_B}{dt^2} = -\left(\omega_0^2 + \omega_c^2\right) x_B - \Gamma\frac{dx_B}{dt} + \omega_c^2 x_A, \tag{5.84}$$

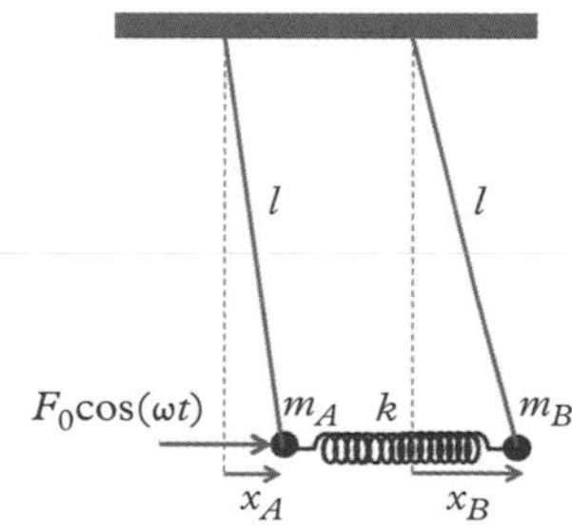

Fig. 5.11 *Two coupled pendulums driven by a harmonically varying time-dependent force on one of them. Even though the driving force acts on one of the oscillators, it drives the motion of both oscillators due to the coupling between the two.*

where

$$\omega_0^2 = \frac{g}{l}, \quad \omega_c = \frac{k}{m}, \quad \Gamma = \frac{b}{m}. \tag{5.85}$$

We wish to study only the steady state of this system. Recall that the steady state will take hold if we wait long enough compared to the time constant of the damping, i.e. in the present case, when $t \gg 1/\Gamma$. To make progress in a coupled system it is often better to go over to the normal modes, therefore, let us first write these equations for the normal modes of the system. You may recall that the normal coordinates q_1 and q_2 of this system are related to the displacements x_A and x_B by the following relations:

$$q_1 = x_A + x_B \text{ and } q_2 = x_A - x_B. \tag{5.86}$$

The equations of motion for q_1 and q_2 are obtained by adding and subtracting Eqs. 5.83 and 5.84,

$$\frac{d^2 q_1}{dt^2} = -\omega_0^2 q_1 - \Gamma \frac{dq_1}{dt} + \frac{F_0}{m} \cos(\omega t), \tag{5.87}$$

$$\frac{d^2 q_2}{dt^2} = -\left(\omega_0^2 + 2\omega_c^2\right) q_2 - \Gamma \frac{dq_2}{dt} + \frac{F_0}{m} \cos(\omega t). \tag{5.88}$$

Note that the normal coordinates have separated the system into two independent oscillators. We now have the problem of two independent forced oscillators. We know how to solve these equations. The steady state solutions of these uncoupled equations are

$$q_1 = \frac{F_0}{m} \frac{1}{\sqrt{\left(\omega_1^2 - \omega^2\right)^2 + \Gamma^2 \omega^2}} \cos(\omega t - \delta_1), \tag{5.89}$$

$$q_2 = \frac{F_0}{m} \frac{1}{\sqrt{\left(\omega_2^2 - \omega^2\right)^2 + \Gamma^2 \omega^2}} \cos(\omega t - \delta_2), \tag{5.90}$$

where $\omega_1 = \omega_0$ and $\omega_2 = \sqrt{\omega_0^2 + 2\omega_c^2}$, the normal mode frequencies of the free system, and the phase lags are given by

$$\tan \delta_1 = \frac{\Gamma \omega}{\omega_1^2 - \omega^2}, \tag{5.91}$$

$$\tan \delta_2 = \frac{\Gamma \omega}{\omega_2^2 - \omega^2}. \tag{5.92}$$

From the normal coordinates q_1 and q_2 we can extract the displacements x_A and x_B of the two oscillators by

$$x_A = \frac{q_1 + q_2}{2}, \quad x_B = \frac{q_1 - q_2}{2}. \tag{5.93}$$

For simplicity let us ignore damping and consider the motion of the two masses when $\Gamma = 0$. In this case, the phase constants of the normal modes become

$$\delta_1 = \delta_2 = 0,$$

and the normal modes take the following simpler expressions:

$$q_1 = \frac{F_0}{m} \frac{1}{\left|\omega_1^2 - \omega^2\right|} \cos(\omega t), \quad q_2 = \frac{F_0}{m} \frac{1}{\left|\omega_2^2 - \omega^2\right|} \cos(\omega t).$$

The displacements of the two pendulums will be

$$x_A = \left(\frac{1}{\left|\omega_1^2 - \omega^2\right|} + \frac{1}{\left|\omega_2^2 - \omega^2\right|} \right) \frac{F_0}{m} \cos(\omega t), \tag{5.94}$$

$$x_B = \left(\frac{1}{\left|\omega_1^2 - \omega^2\right|} - \frac{1}{\left|\omega_2^2 - \omega^2\right|} \right) \frac{F_0}{m} \cos(\omega t). \tag{5.95}$$

This shows explicitly that just driving A makes B oscillate also. However, the way B responds is quite different from A, especially when we look at x_A and x_B at different frequencies of the driver. To see this, let us write the ratio $x_B : x_A$. The expression takes different forms in three domains $\omega \leq \omega_1$, $\omega_1 < \omega < \omega_2$, and $\omega \geq \omega_2$:

$$\frac{x_B}{x_A} = \begin{cases} (\omega_2^2 - \omega_1^2)/(\omega_1^2 + \omega_2^2 - 2\omega^2) & \omega \leq \omega_1 \\ (\omega_1^2 + \omega_2^2 - 2\omega^2)/(\omega_2^2 - \omega_1^2) & \omega_1 < \omega < \omega_2 \\ -(\omega_2^2 - \omega_1^2)/(2\omega^2 - \omega_1^2 - \omega_2^2) & \omega \geq \omega_2 \end{cases}$$

The amplitude of pendulum B is equal to the amplitude of A at the normal mode frequencies, but for other frequencies $|x_B| < |x_A|$. If A is driven at large frequencies, B has a very small amplitude compared to A as though B cannot keep up with the oscillations of A in terms of amplitudes although both oscillate at the same frequency. We say that the high frequency oscillations at the "input" A are not transmitted to the "output" B.

5.4.2 Power and Normal Modes

From the amplitudes of the normal coordinates it is clear that, as the frequency of the driving force is varied, the system will resonate when the frequency gets close to either of the two normal mode frequencies. When the frequency of the driving force is close to ω_1 the system resonates in mode 1 and we say that mode 1 of the system is excited. Similarly, when the frequency of the driving force is close to ω_2, mode 2 will be excited and the system will resonate in mode 2. The average power of the driving force can be computed by time-averaging the rate at which the force applied on m_A does work on m_A,

$$\langle P \rangle = \frac{\omega}{2\pi} \int_0^{2\pi/\omega} F_0 \cos(\omega t)\, v_A dt, \tag{5.96}$$

where

$$v_A = \frac{dx_A}{dt} = \frac{1}{2}\left(\frac{dq_1}{dt} + \frac{dq_2}{dt}\right).$$

After a lengthy calculation one can show that the average power has two terms, each corresponding to a mode, clearly showing that power is delivered to the oscillators by way of mode 1 when the frequency ω of the driver is near ω_1 and through mode 2 when ω is near ω_2,

$$\langle P \rangle = \frac{F_0^2}{4m\Gamma}\left[\frac{\Gamma^2\omega^2}{\left(\omega_1^2 - \omega^2\right)^2 + \Gamma^2\omega^2} + \frac{\Gamma^2\omega^2}{\left(\omega_2^2 - \omega^2\right)^2 + \Gamma^2\omega^2}\right]. \tag{5.97}$$

Therefore, the average power will have resonance peaks in this system at the normal mode frequencies $\omega = \omega_1$ and $\omega = \omega_2$, as illustrated in Fig. 5.12.

Infrared spectroscopy uses the resonance of the power to find the normal mode frequencies in molecules. For instance, a water molecule has three vibrational modes at $f_1 = 47.855$ THz, $f_2 = 109.73$ THz, and $f_3 = 112.70$ THz. These modes are given in the inverse of the wavelength of the light corresponding to these frequencies. In the experimental units of cm^{-1}, obtained by dividing the frequencies by the speed of light 3×10^{10} cm/s, these modes are located at the wave numbers 1595.15 cm^{-1}, 3657.64 cm^{-1},

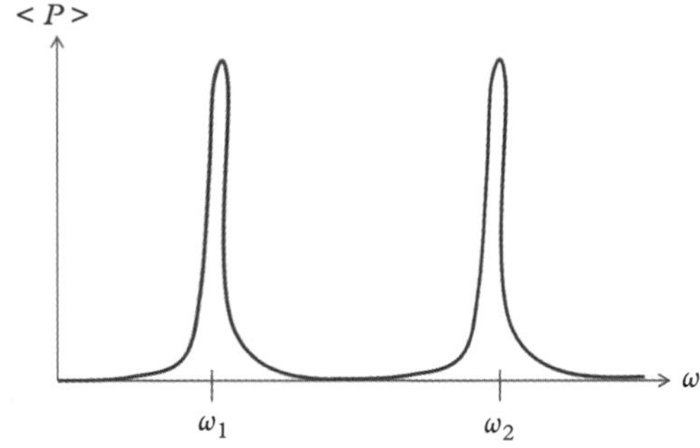

Fig. 5.12 *The power of the driving force peaks at the two normal mode frequencies of the system.*

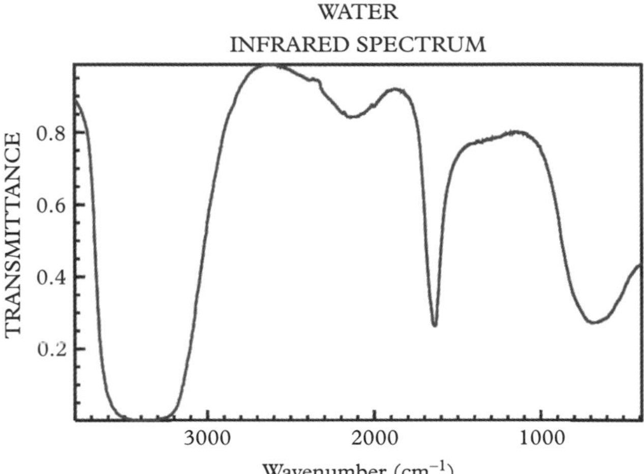

NIST Chemistry WebBook <http://webbook.nist.gov/chemistry>

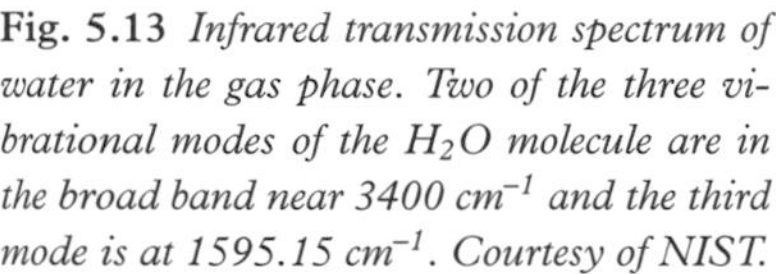

Fig. 5.13 *Infrared transmission spectrum of water in the gas phase. Two of the three vibrational modes of the H_2O molecule are in the broad band near 3400 cm^{-1} and the third mode is at 1595.15 cm^{-1}. Courtesy of NIST.*

and 3756.57 cm^{-1} respectively. Figure 5.13 shows the transmission spectrum in which we see that the resonances at 3657.64 cm^{-1} and 3756.57 cm^{-1} are unresolved and the peak at 1595.15 cm^{-1} is clearly visible.

Example 5.1 Forcing One Mass in a Coupled System

Consider a coupled pendulum system with identical pendulums A and B of mass m and length l coupled by a spring with spring constant k. Suppose bob A is moved harmonically so that its horizontal displacement from the equilibrium is given by

$$x_A = A\cos(\omega t).$$

Find the steady state displacement of bob B.

Solution

The equations of motion of the two bobs are now

$$x_A = A\cos(\omega t),$$
$$\frac{d^2x_B}{dt^2} = -(\omega_0^2 + \omega_c^2)x_B + \omega_0^2 x_A,$$

where $\omega_0 = \sqrt{g/l}$, and $\omega_c = \sqrt{k/m}$. Using the given expression for x_A in the equation of motion of x_B and rearranging the equation we find that x_B is a harmonically driven oscillator,

$$\frac{d^2x_B}{dt^2} + (\omega_0^2 + \omega_c^2)x_B = \omega_0^2 A\cos(\omega t).$$

The steady state solution of this equation will be

$$x_B = B\cos(\omega t - \delta),$$

with

$$B = \frac{\omega_0^2 A}{\left|\omega_0^2 + \omega_c^2 - \omega^2\right|},$$
$$\delta = 0.$$

We get $\delta = 0$ since we have ignored damping. When $\omega \gg \sqrt{\omega_0^2 + \omega_c^2}$, the second bob hardly moves since the amplitude of B will then be

$$B \to 0, \text{ as } \omega \to \infty.$$

Another interesting observation is that for lightly coupled oscillators, i.e. when $\omega_c \ll \omega_0$, we can have B oscillating with a much higher amplitude than A if we move A at natural frequency,

$$B \to \frac{\omega_0^2}{\omega_c^2} A >> A, \text{ when } \omega \to \omega_0.$$

5.5 Driving a Coupled System with Many Degrees of Freedom

5.5.1 Upper and Lower Cutoffs

The spatial domain of modes of a closed continuous system are restricted within the space occupied by the medium. For instance, in the case of a taut string the modes are confined to stay between the two ends of the string. The spatial confinement of the medium gives rise to discreteness in the mode wave numbers and mode frequencies. In the case of the taut string of length L, the wave numbers are multiples of a lowest wave number given by

$$k_{\min} = \frac{\pi}{L}. \tag{5.98}$$

This lowest wave number corresponds to the mode of the smallest frequency given by

$$\omega_{\text{min}} = v\frac{\pi}{L}. \tag{5.99}$$

The lowest frequency is also called the **lowest cutoff frequency** since no wave of lower frequency can propagate in the medium. If the string is vibrated at one end with a frequency less than ω_{min}, the displacement in the string is not given by a sine or cosine but by exponential functions of space. Suppose all moving parts oscillate with the same frequency ω and some phase constant, say zero, then the displacement at point x will take the following form for $\omega < \omega_{\text{min}}$:

$$\psi(x, t) = (A\, e^{\kappa x} + B\, e^{-\kappa x})\, \cos(\omega t), \quad (\omega < \omega_{\text{min}}) \tag{5.100}$$

where κ depends on the boundary condition. The range of frequency $\omega < \omega_{\text{min}}$ is called the **lower reactive range**. There is also an upper reactive range since there is also an upper cutoff. The upper cutoff occurs because the largest wave number of a mode would correspond to a node between every successive particle of the system. Since there is a smallest distance l between the elementary masses that make up the string, there would be a largest $k_{\text{max}} \sim 1/l$. In the case of strings, we obtain a more exact answer for k_{max} as

$$k_{\text{max}} = \frac{\pi}{l}, \tag{5.101}$$

which corresponds to the mode with a node between every pair of successive masses, as shown in Fig. 5.14. An ideal continuous system that has masses continuously with zero distance between them will not have an upper cutoff since all real numbers will be allowed for k. The **highest frequency cutoff** is given as

$$\omega_{\text{max}} = v\frac{\pi}{l}. \tag{5.102}$$

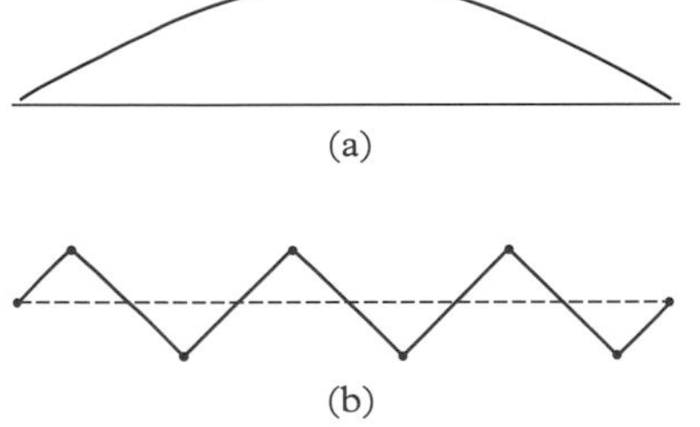

Fig. 5.14 *(a) The lowest frequency cutoff mode. The lowest cutoff wave number is $k_{min} = \pi/L$, where L is the length of the string. (b) The highest frequency cutoff mode. The wave number for the largest frequency cutoff is $k_{max} = \pi/l$, where l is the smallest distance between constituent particles.*

What happens if we drive the system at a frequency ω that is larger than ω_{max} of the system? The displacement of the system again propagates, not in an oscillatory way, but rather as exponentials in space with amplitudes changing sign between successive particles, as in the highest frequency modes. The range of frequency $\omega > \omega_{\text{max}}$ is called the **upper reactive range**.

5.5.2 Solving Multi-particle Systems

Problems With the Driving Force on One Particle Given

The driven oscillations of a system with many degrees of freedom can be understood in a similar way to systems with two degrees of freedom. Let there be N displacement variables $x_1, x_2, \ldots, x_N$ of a system of particles with the same mass m that are coupled so that only nearest neighbors interact, as in the coupled pendulums or mass/spring systems we have studied. Let object 1 be subject to a harmonic force $F_0 \cos(\omega t)$. Then, we will find the following equations of motion:

$$m\frac{d^2x_1}{dt^2} = F_1(x_1, x_2) + F_0 \cos(\omega t),$$

$$m\frac{d^2x_2}{dt^2} = F_2(x_1, x_2, x_3),$$

$$\ldots$$

$$m\frac{d^2x_{N-1}}{dt^2} = F_N(x_{N-2}, x_{N-1}, x_N),$$

$$m\frac{d^2x_N}{dt^2} = F_N(x_{N-1}, x_N),$$

where F_i is the net elastic force on the ith mass. The normal coordinates of the system without the driving force are $q_1, q_2, \ldots, q_N$. Then, just as we found for the coupled pendulum problem in the last section, we will find that the normal coordinates act as independent driven oscillators corresponding to the normal mode frequencies,

$$\frac{d^2q_1}{dt^2} = -\omega_1^2 q_1 + f_1 \cos(\omega t),$$

$$\frac{d^2q_2}{dt^2} = -\omega_2^2 q_1 + f_2 \cos(\omega t),$$

$$\ldots$$

$$\frac{d^2q_N}{dt^2} = -\omega_N^2 q_1 + f_N \cos(\omega t),$$

where $f_1, f_2, \ldots, f_N$ are constant amplitudes with the unit of acceleration.

After we solve these equations, we can deduce the oscillations of the individual masses. We find that the resulting motion depends on the frequency of the driving force as compared to the mode frequencies. We can divide the motion into three regimes.

(1) $\omega < \omega_{\text{min}}$: When particle 1 is driven with a frequency in this regime, the amplitudes of the displacements of other particles are exponential functions of the distance from particle 1. That is, the oscillations are either decaying or building up exponentially in space when driven in this regime. This regime is called the **lower reactive range**.

(2) $\omega_{\text{min}} < \omega < \omega_{\text{max}}$: When particle 1 is driven with a frequency in this regime, the amplitudes of the displacements of other particles are sine or cosine functions of the distance from particle 1 and from them. That is, the oscillations are sinusoidal in space when driven in this regime. This regime is called the **dispersive range**.

(3) $\omega > \omega_{\text{max}}$: When particle 1 is driven with a frequency in this regime, the amplitudes of the displacements of other particles are exponential functions of the distance from particle 1 and from them. That is, the oscillations are either decaying or building up exponentially in space when driven in this regime. This regime is called the **upper reactive range**.

Problems with the Displacement on One Particle Given

In a previous example we saw that in a two-particle coupled system if you drove the system such that the displacement of one of the particles was maintained in a harmonic motion, then it led to the driven oscillation of the other particle. When you have N coupled particles in which neighbors are coupled, and you maintain, say, the displacement of particle 1 in a harmonic motion, then, you will have a system of $N - 1$ equations to solve,

$$x_1 = A\cos(\omega t),$$

$$m\frac{d^2x_2}{dt^2} = F_2(x_1, x_2, x_3),$$

$$m\frac{d^2x_3}{dt^2} = F_2(x_2, x_3, x_4),$$

$$\ldots$$

$$m\frac{d^2x_{N-1}}{dt^2} = F_N(x_{N-2}, x_{N-1}, x_N),$$

$$m\frac{d^2x_N}{dt^2} = F_N(x_{N-1}, x_N).$$

When variable x_1 is put into the equation for x_2, then the x_1 part of the latter acts as a harmonic driving force. Thus, this case is similar to that of the last subsection with $N - 1$ coupled oscillators in which the first particle in the sequence is driven harmonically. Suppose the coupling of x_2 is by a coupling constant c. Separating out the x_1 part of the force on x_2 and treating the x_1 part as the driving force we obtain

$$m\frac{d^2x_2}{dt^2} = g_2(x_2, x_3) + cA\cos(\omega t),$$

$$m\frac{d^2x_3}{dt^2} = F_2(x_2, x_3, x_4),$$

$$\ldots$$

$$m\frac{d^2x_{N-1}}{dt^2} = F_N(x_{N-2}, x_{N-1}, x_N),$$

$$m\frac{d^2x_N}{dt^2} = F_N(x_{N-1}, x_N).$$

This set of equations can be solved by the method of normal coordinates for $N - 1$ coupled oscillators.

Example 5.2 Driving N Coupled Oscillators

Consider N masses, each of mass m, coupled by identical springs with spring constant k. One end of the chain is fixed to a support and the other end oscillates in a harmonic motion. Let the chain be along the x-axis with one end at the origin and the other end at $x = (N + 1)a$ and masses at $x_j = ja$. The end at $x = 0$ is shaken back and forth in a harmonic manner so that $x(t) = A\cos(\omega t)$ for that end. Determine the motion of other masses in the steady state.

Solution

Let us start with the equations of motion of each mass in the system. For brevity we will write $k = m\omega_0^2$,

$$\frac{d^2x_1}{dt^2} = -2\omega_0^2 x_1 + \omega_0^2 x_2 + \omega_0^2 A\cos(\omega t),$$

$$\frac{d^2x_2}{dt^2} = -2\omega_0^2 x_2 + \omega_0^2 x_3 + \omega_0^2 x_1,$$

$$\ldots$$

$$\frac{d^2x_j}{dt^2} = -2\omega_0^2 x_j + \omega_0^2 x_{j+1} + \omega_0^2 x_{j-1}$$

$$\ldots$$

$$m\frac{d^2x_{N-1}}{dt^2} = -2\omega_0^2 x_{N-1} + \omega_0^2 x_N + \omega_0^2 x_{N-2},$$

$$m\frac{d^2x_N}{dt^2} = -2\omega_0^2 x_N + \omega_0^2 x_{N-1}.$$

We have studied normal modes of this system in an earlier chapter. There, to make progress, we looked at the equation of motion of the jth particle. Let us do the same here:

$$\frac{d^2x_j}{dt^2} = -2\omega_0^2 x_j + \omega_0^2 x_{j+1} + \omega_0^2 x_{j-1}. \tag{5.103}$$

The time behavior of each oscillator will be the same for all oscillators in the steady state. Since we have ignored damping in the problem, the phase lag will be zero and we write the following for the displacement of each oscillator:

$$x_j = A_j\cos(\omega t). \tag{5.104}$$

The coefficients A_j may or may not be a periodic function of the particle number j depending upon whether we are in the dispersive or reactive range, as we will discuss below. Putting the time behavior of Eq. 5.104 into Eq. 5.103 and rearranging terms we get the following:

$$A_{j+1} + A_{j-1} = \left[\frac{2\omega_0^2 - \omega^2}{\omega_0^2}\right] A_j. \tag{5.105}$$

Thus, A_j are functions of the index j, or equivalently the position ja of the mass m_j, such that adding two neighboring terms gives a result that is proportional to the present term. When we studied normal modes of an N-particle system we were looking for oscillatory solutions and claimed that A_j would be a sine or cosine of ja. As a matter of fact, the A_j could also be an exponential function of ja if the driving frequency is in the reactive region, i.e. outside of $\omega_{\min}$ and $\omega_{\max}$ of the normal modes of the system. The $\omega_{\min}$ and $\omega_{\max}$ can be determined by looking at the $\lambda_{\min}$ and $\lambda_{\max}$ of the modes. The largest wavelength mode has a wavelength equal to twice the length of the string, which here is $(N+1)a$,

$$\lambda_{\max} = 2(N+1)a. \tag{5.106}$$

Therefore, the corresponding wave number will be a minimum wave number,

$$k_{\min} = \frac{2\pi}{\lambda_{\max}} = \frac{\pi}{(N+1)a}. \tag{5.107}$$

The minimum frequency mode has the frequency

$$\omega_{\min} = v\frac{\pi}{(N+1)a}, \tag{5.108}$$

continued

Example 5.2 *continued*

where v is the speed of waves on the string, which for a taut string will be

$$v = \sqrt{T/\mu}, \tag{5.109}$$

where T is the tension and μ is the mass per unit length. The smallest wavelength happens when the subsequent masses oscillate in an opposite phase. This gives

$$\lambda_{\text{min}} = 2a. \tag{5.110}$$

Therefore, the corresponding wave number will be a maximum wave number,

$$k_{\text{max}} = \frac{2\pi}{\lambda_{\text{min}}} = \frac{\pi}{a}. \tag{5.111}$$

The minimum frequency mode has the frequency

$$\omega_{\text{min}} = v\frac{\pi}{a}. \tag{5.112}$$

In the dispersive range of driving frequencies between ω_{min} and ω_{max}, the amplitudes A_j are a sine or cosine function of ja, since in the dispersive range, not only will each mass oscillate about its equilibrium position, but different masses on the chain would make a wave pattern along the chain. This is the dispersive solution we have mentioned above. In the lower reactive range, $\omega < \omega_{\text{min}}$, all masses are in phase as in the lowest frequency mode, but their amplitudes are not related by sines and cosines but by exponentials. In the upper reactive range, $\omega > \omega_{\text{max}}$, neighboring masses are completely out of phase as in the highest frequency mode, but their amplitudes are not related by sines and cosines but by exponentials. Thus, we have the following forms of the amplitudes in the three regimes:

$$\text{Dispersive,} \quad \omega_{\text{min}} \leq \omega \leq \omega_{\text{max}}: \qquad A_j = B\sin(Kja) + C\cos(Kja), \tag{5.113}$$

$$\text{Lower reactive,} \quad \omega < \omega_{\text{min}}: \qquad A_j = De^{\alpha ja} + Ee^{-\alpha ja}, \tag{5.114}$$

$$\text{Upper reactive,} \quad \omega > \omega_{\text{max}}: \qquad A_j = (-1)^j\left(Fe^{\beta ja} + Ge^{-\beta ja}\right), \tag{5.115}$$

where k, α, and β are constants with units of inverse length. They are called wave numbers, even though only k is truly associated with a wave. The reactive solution is also called an **exponential or evanescent wave**. By setting $A_{N+1} = 0$ since the other end at $x = (N+1)a$ is fixed, we get the coefficients of the sines and cosines in these expressions,

$$\text{Dispersive}: \qquad B = -\frac{\cos(K[N+1]a)}{\sin(K[N+1]a)}C, \tag{5.116}$$

$$\text{Lower reactive}: \qquad E = -De^{(2\alpha[N+1]a)}, \tag{5.117}$$

$$\text{Upper reactive}: \qquad E = -De^{(2\alpha[N+1]a)}. \tag{5.118}$$

5.5.3 Driving Continuous Systems

Vibrations in continuous systems, such as a beam or air, have many applications. Just as a discrete system has normal modes, a continuous body also has normal modes. For instance, the normal modes of transverse vibrations of a string of length L and mass per unit length μ tied at both ends so that the tension in the string is T are

$$\omega_n = n\frac{\pi}{L}, \quad \psi_n(x,t) = A\sin(k_n x)\cos(\omega_n t), \tag{5.119}$$

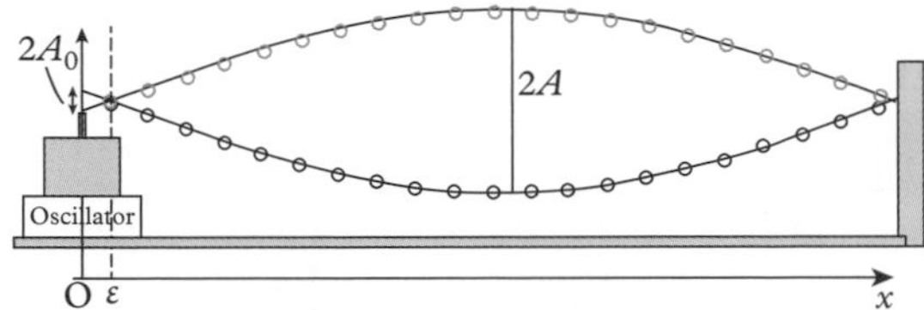

Fig. 5.15 *Generation of standing wave by driving vibrations by an oscillator near the node of a normal mode.*

where

$$k_n = \omega_n/v, \quad v = \sqrt{T/\mu}.$$

To drive this system, let us attach one end of the string to the plunger of a vibrator and hold the other end fixed so that the string is taut and has a tension T, as shown in Fig. 5.15. The harmonic driving force of the vibrator at $x = 0$ will cause the element of the string at $x = 0$ to oscillate harmonically with the same frequency as the plunger. Let ω be the frequency of the vibrator and $\psi(x, t)$ denote the transverse displacement at time t of a particle of the string which is located at $x = x$. Then, we have the following boundary condition at $x = 0$:

$$\psi(x = 0, t) = A_0 \cos(\omega t), \tag{5.120}$$

where A_0 is the amplitude of the vibrations of the plunger of the vibrator. You already know from the last chapter that the equation of motion of the vibrating string is the classical wave equation,

$$\frac{\partial^2 \psi}{\partial t^2} - v^2 \frac{\partial^2 \psi}{\partial x^2}, \tag{5.121}$$

with $v = \sqrt{T/\mu}$. The boundary condition in Eq. 5.120 serves as a driving force on the system, and just as was the case with discrete coupled mass systems, the particles of the string will be driven into resonance if the driving frequency is equal to one of the normal mode frequencies,

$$\text{Resonance when } \omega = \omega_n. \tag{5.122}$$

If the frequency is below the lowest cutoff, which here is ω_1, the string will swing up and down with very small amplitude, which is the lower reactive range,

$$\text{Lower reactive range: } \omega < \omega_1 = v\frac{\pi}{L}. \tag{5.123}$$

If the string is π m long and the speed of the wave in the string 1000 m/s, then the lower cutoff will be

$$\omega_{\text{min}} = 1000 \text{ m/s} \times \frac{\pi}{\pi \text{ m}} = 1000 \text{ s}^{-1}. \tag{5.124}$$

The upper reactive range occurs at very high frequency, since the distance between successive masses in a continuous system will be the inter-atomic distance between neighboring

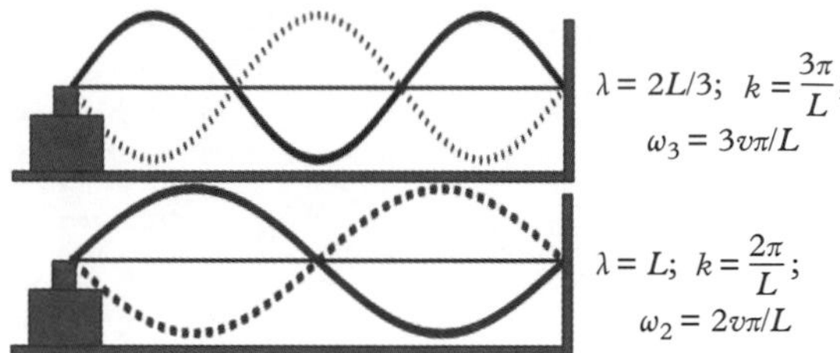

Fig. 5.16 *Mode of frequency* $\omega_2 = 2\frac{\pi v}{L}$ *(lower figure) is excited when the vibration occurs at this frequency. Similarly, for the excitation of a mode of frequency* $\omega_3 = 3\frac{\pi v}{L}$ *(upper figure).*

atoms. For instance, if the interatomic distance is π Å and the speed of the wave in the string is 1000 m/s, then the upper cutoff will be

$$\omega_{\max} = 1000 \text{ m/s} \times \frac{\pi}{\pi \text{ Å}} = 10^{13} \text{ s}^{-1}. \tag{5.125}$$

The dispersive range for this system will be ω between 1000 s^{-1} and 10^{13} s^{-1}. All of its normal modes are in this range. When the string is driven with a frequency in the dispersive range we get a sinusoidal solution in both space and time. In the steady state the solution takes the form of a standing wave even when the frequency is not at any of the normal mode frequencies. The amplitude of motion of a particle at x at time t would be given by

$$\psi(x,t) = [A\sin(kx) + B\cos(kx)]\cos(\omega t), \tag{5.126}$$

where ω is the frequency of the driving force and $k = \omega/v$. Figure 5.15 shows the resonance of the lowest mode and Fig. 5.16 illustrates modes for ω_2 and ω_3 .

You should note that the $\lambda_{\max}$ in Fig. 5.15 is only approximately $2L$, since the mode function $\psi(x,t)$ cannot have a node at the location of the vibrator because the vibrator will clearly be moving the part it is attached to—the node will be a little displaced from the vibrator, as shown in the figure. Let ϵ be the displacement of the node from the vibrator, then the standing wave on the string will be

$$\psi(x,t) = A\ \sin[k(x-\epsilon)]\cos(\omega\ t), \tag{5.127}$$

where $k = \omega/v$, and v is the speed of the wave on the string. When we try to satisfy the boundary condition, Eq. 5.120, at $x = 0$ we find

$$|-A\ \sin(k\epsilon)| = A_0, \tag{5.128}$$

Since ϵ is small, A will be large compared to A_0. At resonance, the string vibrates with much larger amplitudes than that of the driver.

5.6 Electrical Resonance—RLC circuit

An electric circuit containing a capacitance and an inductance oscillates like a simple harmonic oscillator, as we have seen previously. The resistance in the circuit dissipates energy away from the circuit and damps the oscillations in current and voltage. If we include an alternating current or voltage source, then the RLC circuit shows resonance similar to the damped driven harmonic oscillator.

5.6.1 Single Variable Driven Circuit

Consider a series RLC circuit (see Fig. 5.17) with an EMF source V that is a sinusoidal function of time of frequency ω and amplitude V_0,

$$V(t) = V_0 \cos(\omega t). \tag{5.129}$$

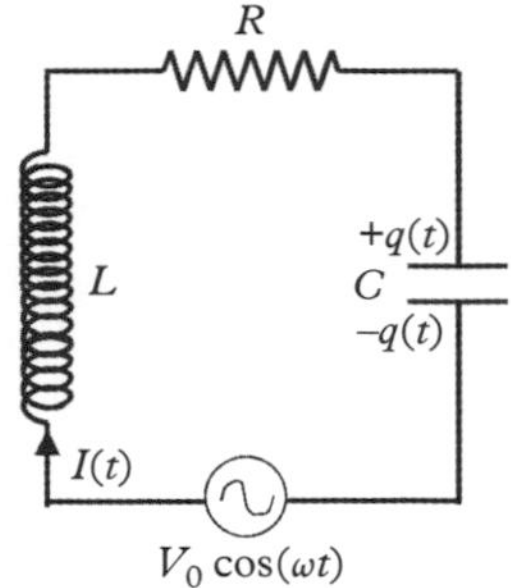

Fig. 5.17 *An RLC circuit driven by a sinusoidal source.*

The angular frequency ω of the source EMF is the driving frequency. We can study the resonance of several different quantities in the circuit: charge q on the capacitor, the current I through the inductor, the voltage drop across the resistor, the power delivered to the circuit, etc. Let us illustrate the resonance of the charge on the capacitor, which is analogous to the displacement of the mass in the mass/spring system. The voltage loop equation for the driven circuit yields the following "equation of motion" for the charge $q(t)$ on the capacitor plates:

$$L\frac{d^2q}{dt^2} + R\frac{dq}{dt} + \frac{q}{C} = V_0 \cos(\omega t). \tag{5.130}$$

This equation for the charge on the capacitor is analogous to the equation for the displacement of the harmonic oscillator. We can deduce the analogy between the mechanical and electrical systems shown in Table 2.2. By analogy, we introduce the following constants:

$$\omega_0 = \frac{1}{\sqrt{LC}}, \quad \Gamma = \frac{R}{L}. \tag{5.131}$$

To study the resonance in the steady state charge q, we assume that we have an underdamped RLC circuit by requiring the following condition:

$$\Gamma < 2\omega_0 \Rightarrow R < 2\sqrt{\frac{L}{C}}. \tag{5.132}$$

The quality factor (Q) of the circuit is given by

$$Q = \frac{\omega_0}{\Gamma} = \frac{1}{R}\sqrt{\frac{L}{C}}. \tag{5.133}$$

Do not confuse this Q with the charge on the capacitor. Using the analogy with the mechanical system, the steady state solution of the driven RLC circuit can be written as

$$q = q_0 \cos(\omega t - \delta), \tag{5.134}$$

where

$$q_0 = \frac{V_0/L}{\sqrt{\left(\omega_0^2 - \omega^2\right)^2 + (\Gamma\omega)^2}}, \tag{5.135}$$

$$\tan\delta = \frac{\Gamma\omega}{\omega_0^2 - \omega^2}. \tag{5.136}$$

The charge on any one of the plates of the capacitor oscillates between $+q_0$ and $-q_0$ with frequency ω. The current driven in the circuit by the source is

$$I(t) = \frac{dq}{dt} = -\frac{\omega V_0}{L\sqrt{\left(\omega_0^2-\omega^2\right)^2 + (\Gamma\omega)^2}}\sin(\omega t - \delta). \tag{5.137}$$

The ratio of the amplitude of the current through the source and the voltage of the source is called the **impedance** of the circuit, which is usually denoted by the letter Z,

$$Z = \frac{\text{Amplitude of voltage of source}}{\text{Amplitude of current of source}} = \frac{1}{\omega}L\sqrt{\left(\omega_0^2-\omega^2\right)^2 + (\Gamma\omega)^2}. \tag{5.138}$$

Note that the impedance defined in this way does not capture the information in the phases of the current and voltage, which are different. To take this into account a complex impedance $\tilde{Z}$ is defined by the ratio of the complex representation of the current through the source and the voltage of the source,

$$\tilde{Z} = \frac{\text{Complex voltage of source}}{\text{Complex current of source}}. \tag{5.139}$$

To get the complex representation we replace $\cos(\theta)$ by $e^{i\theta}$. The sine and the overall sign are put together and converted into the cosine as

$$-\sin\theta = \cos(\theta + \pi/2). \tag{5.140}$$

Therefore, we get the following for the complex impedance.

$$\tilde{Z} = Z\, e^{i\phi}, \tag{5.141}$$

with $\phi = -\delta + \pi/2$ and Z is the real impedance defined above. The instantaneous power delivered by the source is

$$P(t) = V(t)I(t) = -\frac{V_0^2}{Z}\cos(\omega t)\sin(\omega t - \delta). \tag{5.142}$$

The average power over one cycle can be obtained from this by integrating over one cycle. This gives

$$\langle P \rangle = \frac{V_0^2}{2Z}\sin\delta = \frac{V_0^2}{2R}\left[\frac{\Gamma^2\omega^2}{\left(\omega_0^2-\omega^2\right)^2 + \Gamma^2\omega^2}\right]. \tag{5.143}$$

By setting the derivative of this with ω to zero we find that the resonance frequency ω_R of power in the circuit occurs at the natural frequency of the circuit,

$$\omega_R = \omega_0 = \frac{1}{\sqrt{LC}}. \quad \text{(Resonance of average power } \langle P \rangle\text{)} \tag{5.144}$$

The maximum power is delivered to the circuit at this frequency. Putting $\omega = \omega_0$ in Eq. 5.143 gives the maximum power to be

$$P_{\max} = \frac{V_0^2}{2R}. \tag{5.145}$$

The student is encouraged to determine the resonance frequencies of the charge on the capacitor plate by looking at the way the amplitude of oscillations of q varies with frequency, and the resonance of the current in the circuit in a similar way. You will find some surprises here,

$$\text{Resonance frequency of } q: \ \omega_q = \frac{1}{L}\sqrt{LC - 2R^2}, \tag{5.146}$$

$$\text{Resonance frequency of } I: \ \omega_I = \omega_0. \tag{5.147}$$

5.6.2 Electrical Filters and Driven Coupled Circuits

In the driven two-pendulum system we found that if we drive pendulum A by a harmonic force, then the response of pendulum B depends on the frequency compared to the normal mode frequencies. If the driving force is higher than the larger of the two normal modes then the amplitude of B is quite suppressed. At low frequencies the amplitude of B keeps up with that of A.

That is, in the coupled pendulum higher frequency effort by the driving force is "filtered" by the pendulums and only low frequencies are effective in driving the second pendulum. The same phenomenon is possible in two coupled circuits in which one is driven by an AC source, as shown in Fig. 5.18. For simplicity we will ignore any resistance in the circuit.

The loop equations give us the following:

$$V_s - \frac{q_A}{C_A} - L_A\frac{dI_A}{dt} - \frac{q_C}{C_C} = 0, \tag{5.148}$$

$$+\frac{q_C}{C_C} - L_B\frac{dI_B}{dt} - \frac{q_B}{C_B} = 0. \tag{5.149}$$

The current conservation at the nodes gives

$$I_A = I_B + I_C.$$

With time $t = 0$ chosen such that charge on the capacitor is zero at that instant, we get

$$q_A = q_B + q_C.$$

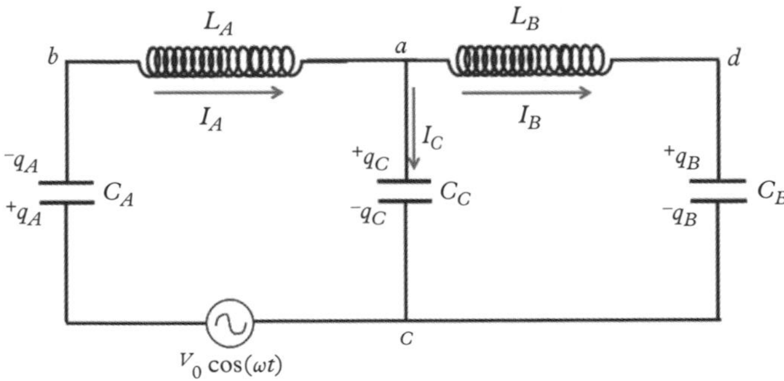

Fig. 5.18 *Two coupled RLC circuits driven by a sinusoidal source.*

We can eliminate q_C from the loop equations, and use dq/dt for the two currents to obtain the following equations:

$$L_A \frac{d^2 q_A}{dt^2} + \left(\frac{1}{C_A} + \frac{1}{C_C}\right) q_A - \frac{1}{C_C} q_B = V_0 \cos(\omega t), \tag{5.150}$$

$$L_B \frac{d^2 q_B}{dt^2} + \left(\frac{1}{C_B} + \frac{1}{C_C}\right) q_B - \frac{1}{C_C} q_A = 0. \tag{5.151}$$

Now, let us consider the following simpler case:

$$L_A = L_B \equiv L, \;\; C_A = C_B \equiv C, \tag{5.152}$$

and introduce the following parameters:

$$\omega_0^2 = \frac{1}{LC}, \;\; \omega_c^2 = \frac{1}{LC_C}. \tag{5.153}$$

The equations of motion are now identical to the equations of motion of two coupled pendulums:

$$\frac{d^2 q_A}{dt^2} = -\left(\omega_0^2 + \omega_c^2\right) q_A + \omega_c^2 q_B + \frac{V_0}{L} \cos(\omega t), \tag{5.154}$$

$$\frac{d^2 q_B}{dt^2} = -\left(\omega_0^2 + \omega_c^2\right) q_B + \omega_c^2 q_A. \tag{5.155}$$

These coupled equations can be decoupled when written in terms of the normal modes. Let us denote the normal coordinates here by ψ_1 and ψ_2, since we are using q's for charges on the capacitors,

$$\psi_1 = q_A + q_B, \;\; \psi_2 = q_A - q_B. \tag{5.156}$$

The equations of motion now separate into independent equations for ψ_1 and ψ_2,

$$\frac{d^2 \psi_1}{dt^2} = -\omega_0^2 \psi_1 + \frac{V_0}{L} \cos(\omega t), \tag{5.157}$$

$$\frac{d^2 \psi_2}{dt^2} = -\left(\omega_0^2 + 2\omega_c^2\right) \psi_2 + \frac{V_0}{L} \cos(\omega t). \tag{5.158}$$

This shows that in the electrical system the resonances of q_A and q_B are mathematically identical to the resonances of x_A and x_B, treated in Section 5.4. When you vary the frequency of the AC generator, there will be resonance around the normal mode frequencies $\omega_1 = \omega_0$ and $\omega_2 = \sqrt{\omega_0^2 + 2\omega_c^2}$. The power delivered to the circuit by the source will similarly have resonances at these normal mode frequencies.

The steady state solutions of the normal modes are

$$\psi_1 = \frac{V_0/L}{\left|\omega_1^2 - \omega^2\right|} \cos(\omega t), \tag{5.159}$$

$$\psi_2 = \frac{V_0/L}{\left|\omega_2^2 - \omega^2\right|} \cos(\omega t). \tag{5.160}$$

From these we obtain the charges on the capacitors in the two circuits,

$$q_A = \frac{V_0}{L}\left[\frac{1}{|\omega_1^2-\omega^2|} + \frac{1}{|\omega_2^2-\omega^2|}\right]\cos(\omega t), \tag{5.161}$$

$$q_B = \frac{V_0}{L}\left[\frac{1}{|\omega_1^2-\omega^2|} - \frac{1}{|\omega_2^2-\omega^2|}\right]\cos(\omega t). \tag{5.162}$$

The ratio of the amplitudes in each of the three frequency domains is

$$\frac{q_B}{q_A} = \begin{cases} (\omega_2^2-\omega_1^2)/(\omega_2^2+\omega_1^2-2\omega^2), & \omega < \omega_1, \\ (\omega_2^2+\omega_1^2-2\omega^2)/(\omega_2^2-\omega_1^2), & \omega_1 < \omega < \omega_2, \\ -(\omega_2^2-\omega_1^2)/(\omega_2^2+\omega_1^2-2\omega^2), & \omega > \omega_2. \end{cases} \tag{5.163}$$

For large ω we find that the response of the second circuit drops as the square of the driving frequency,

$$\left|\frac{q_B}{q_A}\right| \sim \frac{1}{\omega^2},$$

while at low frequency, say at $\omega = 0$, we have

$$\left|\frac{q_B}{q_A}\right| \sim 1.$$

That is, the coupled oscillator in Fig. 5.18 lets through a low frequency signal and blocks the high frequency signals. This type of filter is a **low-pass filter**.

5.6.3 Driven LC Network

A coaxial cable can be considered to be a series of inductors in series and capacitors in parallel. We will consider driving a coaxial cable whose other end is simply shorted. Let there be N segments from $x = 0$ to $x = l$, the length of the cable so that $Na = l$. We short the $x = l$ end and drive at the $x = 0$ end using an AC source such that the voltage between the two conductors of the cable has the following values at the two ends:

$$V(x=0,t) = V_0\cos(\omega t), \tag{5.164}$$

$$V(x=l,t) = 0. \tag{5.165}$$

Let C and L be the capacitance and inductance per unit length of the coaxial cable. For convenience of calculation, we will model the cable by placing inductance La and capacitance Ca at intervals of $\Delta x = a$ along the length of the cable, as shown in Fig. 5.19.

The voltage at any point in the cable is the voltage across the internal and external wires of the cable at that point. From the figure, this voltage will be that across the capacitor at that point. To find the voltage at any point in the cable we will first derive the equation of motion of the voltage as a function of node j along the cable and time t. For this purpose we only need to look at two neighboring segments, as in Fig. 5.20.

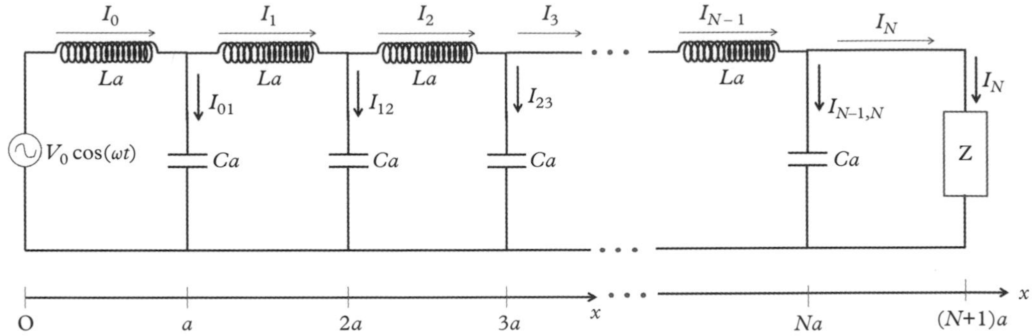

Fig. 5.19 *An LC network shorted at one end is driven by an AC source at the other end.*

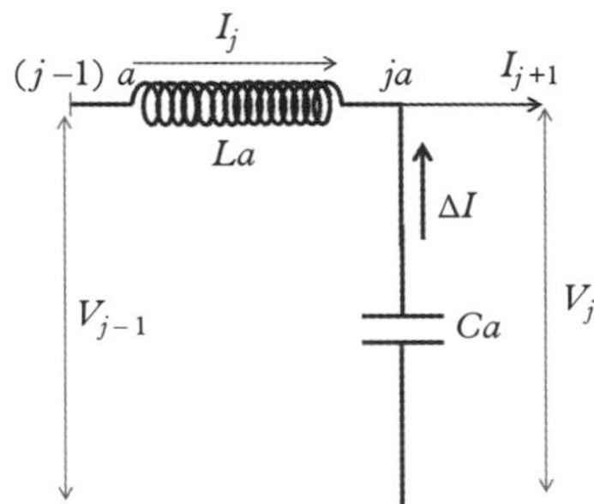

Fig. 5.20 *An arbitrary segment of the LC network.*

The voltage across the inductor gives

$$(La)\frac{\partial I_j}{\partial t} = V_j - V_{j-1}, \tag{5.166}$$

and the conservation of charge at the node *ja* gives

$$(Ca)\frac{\partial V_j}{\partial t} = I_{j+1} - I_j. \tag{5.167}$$

Now, we can take a time derivative of this equation and use the other equation to obtain

$$\frac{\partial^2 V_j}{\partial t^2} = \frac{1}{LC}\left(\frac{V_{j+1} - 2V_j + V_{j-1}}{a^2}\right). \tag{5.168}$$

When we take the $a \to 0$ limit and write the coordinate of the point j which is ja as x, we will obtain the equation for voltage as a function of (x, t),

$$\frac{\partial^2 V(x,t)}{\partial t^2} = \frac{1}{LC}\frac{\partial^2 V(x,t)}{\partial x^2}. \tag{5.169}$$

This equation is called the **cable equation**. The cable equation is identical in form to the classical wave equation. The quantity $1/\sqrt{LC}$ is a characteristic frequency per unit length, which we will denote by ω_0,

$$\omega_0 = \frac{1}{\sqrt{LC}}. \quad \text{(characteristic frequency)} \tag{5.170}$$

Then, the cable equation takes the form

$$\frac{\partial^2 V(x,t)}{\partial t^2} = \omega_0^2\frac{\partial^2 V(x,t)}{\partial x^2}. \tag{5.171}$$

We need to solve this equation for the boundary conditions,

$$V(x=0,t) = V_0\cos(\omega t), \quad V(x=l,t) = 0.$$

Since we are looking for a steady state solution, voltage across all parts of the cable will vary sinusoidally, and since there is no damping the phase constant will be zero. Therefore, we try the product form

$$V(x,t) = f(x)\cos(\omega t). \tag{5.172}$$

When we substitute this form of solution in Eq. 5.171 we get the following differential equation for $f(x)$:

$$\frac{d^2f(x)}{dx^2} = -\frac{\omega^2}{\omega_0^2} f(x), \tag{5.173}$$

with the boundary values of f,

$$f(0) = V_0, \ f(l) = 0.$$

The general solution of this equation is

$$f(x) = A\cos(\omega x/\omega_0) + B\sin(\omega x/\omega_0). \tag{5.174}$$

Fitting the boundary values gives A and B to be

$$A = V_0, \ B = -\frac{\cos(\omega l/\omega_0)}{\sin(\omega l/\omega_0)} V_0. \tag{5.175}$$

Therefore, the voltage at an arbitrary point in the cable will be

$$V(x,t) = \frac{V_0}{\sin(\omega l/\omega_0)} \sin\left[\frac{\omega}{\omega_0}(l-x)\right]\cos(\omega t). \tag{5.176}$$

This is an expression for a standing wave. When we study a traveling wave in the next chapter we will see that this solution says that there are two voltage waves riding on the cable at the same time, one moving towards the positive x-axis and the other moving towards the negative x-axis, such that their superposition creates a steady vibration given by the profile from the x-part of this solution. If the cable were left unconnected at the far end, the boundary condition there would be given in terms of the current being zero at that end. In that case, it is better to write the equation of motion of the cable for current at an arbitrary point. The a student is encouraged to show that the equation for the current is the same as the cable equation for the voltage.

EXERCISES

(5.1) In the text we have solved the driven damped one-dimensional harmonic oscillator with a driving force given by $F_0\cos(\omega t)$. Suppose the driving force were given instead by $F_0\sin(\omega t)$. Find the steady state solution for this form of the driving force in the equation of motion

$$m\frac{d^2x}{dt^2} = -kx - b\frac{dx}{dt} + F_0\sin(\omega t).$$

(5.2) The driving force on a damped one-dimensional harmonic oscillator is given by $F_c\cos(\omega t) + F_s\sin(\omega t)$. Find the steady state solution for this form of the driving force. Hint: Use the superposition principle for linear systems.

(5.3) A damped one-dimensional harmonic oscillator with $m = 5$ kg, $\Gamma = 1$ sec^{-1}, and $\omega_0 = 20$ sec^{-1} is driven by a harmonic driving force 5 N $\cos(\omega t)$ with $\omega = 10$ sec^{-1}.

(a) Set up the differential equation of motion for the oscillator.

(b) Solve the equation and determine the amplitude for the steady motion of the oscillator.

(c) Determine the phase lag of the oscillator.

(d) Determine the average energy per unit time the driving force puts into the oscillator.

(5.4) A sinusoidal driving torque acts on a pendulum of mass m and length l, whose expression about the point of suspension is given by

$$\tau(t) = F_0 l \cos(\omega t).$$

The air drag on the pendulum applies the following torque about the point of suspension:

$$\tau_d = bl^2 \frac{d\theta}{dt},$$

where b is a constant.

(a) Set up the equation of motion of the pendulum using the angle variable θ in the small angle approximation.

(b) Find the elastic and absorptive amplitudes.

(c) Find the time-averaged power delivered by the driving force.

(5.5) **Seismograph.** A simple seismograph is made by hanging a block from a spring attached to a rigid support that is fixed to the Earth, as shown in Fig. 5.21. In an earthquake, when the surface of the Earth moves, the support moves with it.

Let m be the mass of the block and k the spring constant of the spring. Let Q be the quality of the oscillator. Suppose in an earthquake the Earth moves up and down harmonically, with frequency ω, so that the vertical displacement y_E of the surface of the Earth with respect to its position at equilibrium moves as $y_E = A_E \cos(\omega t)$.

(a) Set up an equation of motion of the displacement of the block from the equilibrium in an inertial frame.

(b) Find the steady state amplitude of the oscillation of the block during an earthquake.

(c) How would you use this solution to determine the frequency ω of the earthquake?

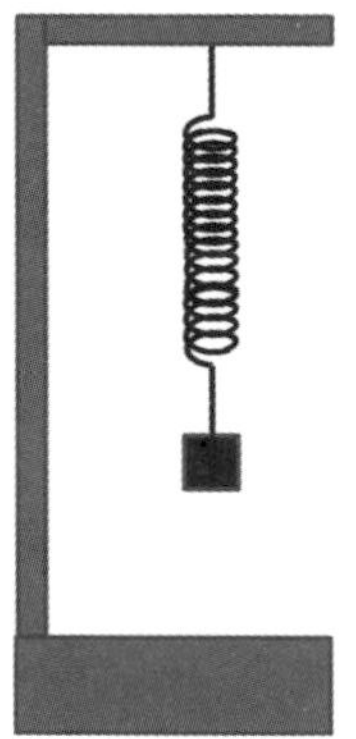

Fig. 5.21 *Exercise 5.5.*

(5.6) Two blocks of mass m_1 and m_2 are connected by a spring with spring constant k and placed on a frictionless surface, as shown in Fig. 5.22. A rigid massless rod of length l is connected to m_1. Suppose the block and rod are placed along the x-axis with the origin at the other end of the rod. The end of the rod away from m_2 is moved back and forth harmonically at frequency ω so that the position of mass m_1 is given by $x_1(t) = \eta_0 \sin(\omega t)$.

(a) Ignoring damping, what is the equation of motion of m_2?

(b) Find the steady state solution of the displacement of m_2 from the equilibrium.

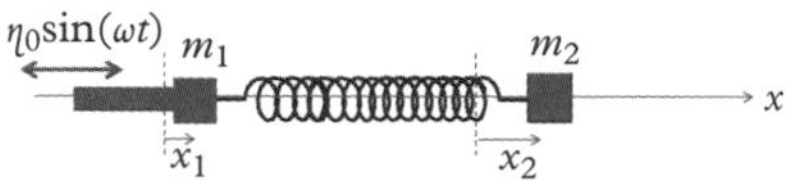

Fig. 5.22 *Exercise 5.6.*

(5.7) Two identical springs with spring constant k connect three identical blocks of mass m, as shown in Fig. 5.23. Block m_1 is attached to a rigid rod and vibrated sinusoidally at a constant frequency, so that its displacement from the equilibrium is given by $x_1 = \eta_0 \cos(\omega t)$. Ignore any damping.

(a) Find the particular solution of the equation of motion of the blocks m_2 and m_3.

(b) Describe what would happen if the driving frequency were near one of the normal mode frequencies based on the steady state solution.

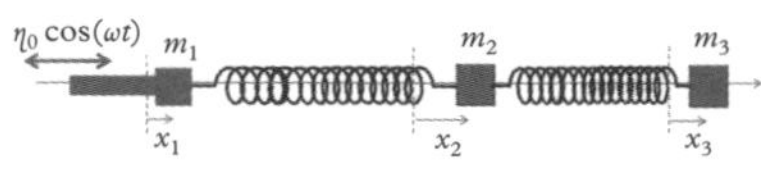

Fig. 5.23 *Exercise 5.7.*

(5.8) The support of a pendulum of mass m and length l is moved harmonically, as shown in Fig. 5.24. What is the particular solution of the equation of motion of the pendulum for small angles? Ignore any damping.

(5.9) The support of a double pendulum with equal masses m and equal length l, is moved harmonically, as shown in Fig. 5.25. Find the particular solutions of the equations of motion of the two pendulums for small angles. Ignore damping.

(5.10) A resonance power curve is fitted to the following function of frequency ω:

$$P(\omega) = 20\ \text{W}\,\frac{\Gamma^2\omega^2}{(\omega^2 - 25)^2 + \Gamma^2\omega^2},$$

where ω is in units of sec^{-1}, and $\Gamma = 1\ \text{sec}^{-1}$.

(a) Plot $P(\omega)$.
(b) Find the frequency at which power is a maximum.
(c) Find the full width at half maximum (FWHM).
(d) Find the time constant of the oscillator.

(5.11) A steel wire of length 100 cm is tied between two posts. Striking the wire in the middle with a hammer sets the wire vibrating in many modes at the same time.

(a) What is the wave number of the lowest frequency mode of which that the wire is vibrating?
(b) If the tension in the wire is 200 N and the density of the wire is 100 g/m, what is the frequency of the lowest mode?

(5.12) A cylindrical tube of glass is filled with water such that there is a column of air of length 70 cm above the water level. A vibrator is placed at the opening which vibrates in a broad range of continuous frequencies between 300 Hz and 3000 Hz, with more or less equal amplitude for each frequency. Assume the speed of sound to be 340 m/s.

(a) Would the lowest mode of the sound in the air column be excited by the vibrator? Why or why not?
(b) What frequencies of the sound will be excited? List them.

(5.13) A damped harmonic oscillator of mass $m = 1$ kg, spring constant $k = 1000$ N/m, and damping constant $b = 20$ kg/s, is driven by a sinusoidal force $F(t) = 10\ \text{N}\cos(200\pi t)$, where t is in seconds. At $t = 0$, the oscillator has a displacement of $x(0) = 0$ and velocity $v(0) = 0$.

(a) Find the complete solution $x(t)$.
(b) Identify the steady state solution.
(c) What is the criterion on time after which we can use the steady state solution for $x(t)$? Identify the physical reason for this criterion.

(5.14) Two sinusoidal driving forces of same magnitude but different frequencies act on a damped one-dimensional harmonic oscillator of mass m, spring constant k, and damping constant b,

$$F(t) = F_0\cos(\omega_1 t) + F_0\cos(\omega_2 t).$$

(a) Find the steady state solution. Hint: Use the superposition principle for linear systems.
(b) Find the instantaneous power delivered by this force.
(c) Find the average power delivered.

(5.15) Two sinusoidal driving forces of different magnitude and frequency act on a damped one-dimensional harmonic oscillator of mass m, spring constant k, and damping constant b,

$$F(t) = F_1\cos(\omega_1 t) + F_2\cos(\omega_2 t).$$

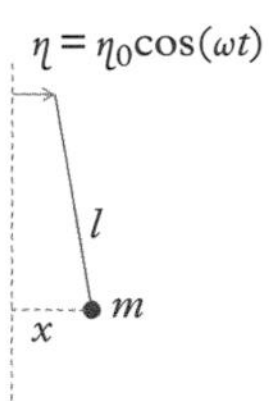

Fig. 5.24 *Exercise 5.8.*

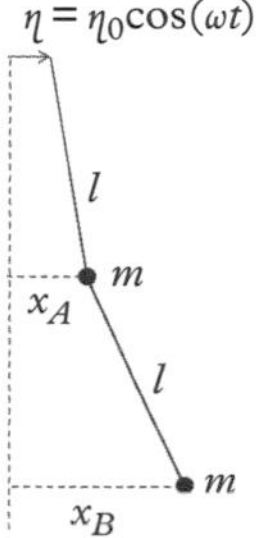

Fig. 5.25 *Exercise 5.9.*

(a) Find the steady state solution.

(b) What will be the steady state solution if the driving force were the following infinite sum?

$$F(t) = \sum_{i=1}^{\infty} F_i \cos(\omega_i t).$$

(5.16) A periodic but not sinusoidal driving force acts on a damped one-dimensional harmonic oscillator of mass m, spring constant k, and damping constant b. The force has the square wave shape in time with the analytic form given by

$$F(t) = \begin{cases} 0 & 0 \le t \le T/2 \\ 1 & T/2 \le t \le T \end{cases},$$

with

$$F(t+T) = F(t), \quad \text{for} \ -\infty < t < \infty.$$

(a) Find the Fourier series representation of $F(t)$.

(b) Find the steady state solution $x(t)$ of the displacement of the oscillator.

(5.17) Three identical pendulums are coupled by springs, as shown in Fig. 5.26. A metal rod is glued to the pendulum on the left which is then vibrated sinusoidally so that its horizontal displacement from the equilibrium is $\psi_1(t) = \eta_0 \cos(\omega t)$. Find the steady state solutions for $\psi_2(t)$ and $\psi_3(t)$ in the small angle approximation and without any damping.

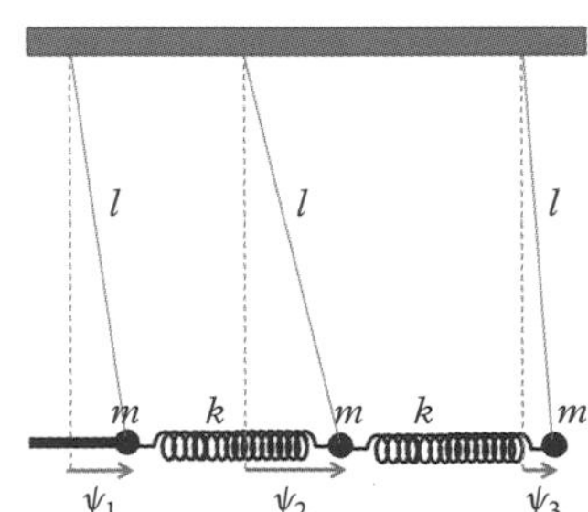

Fig. 5.26 *Exercise 5.17.*

(5.18) The power delivered by the AC source in a series RLC circuit has resonance at 20 kHz and a peak width 10 Hz.

(a) What is the Q of the circuit?

(b) What will happen to the resonance frequency and peak width if resistance R is doubled?

(c) What will happen to the resonance frequency and peak width if capacitance C is doubled?

(d) What will happen to the resonance frequency and peak width if inductance L is doubled?

(e) What should be the peak voltage if we need to provide a power of 30 W at the resonance frequency to the resistor that has a value of 10 Ohm?

(5.19) Find the steady state current in each branch of the circuit shown in Fig. 5.27.

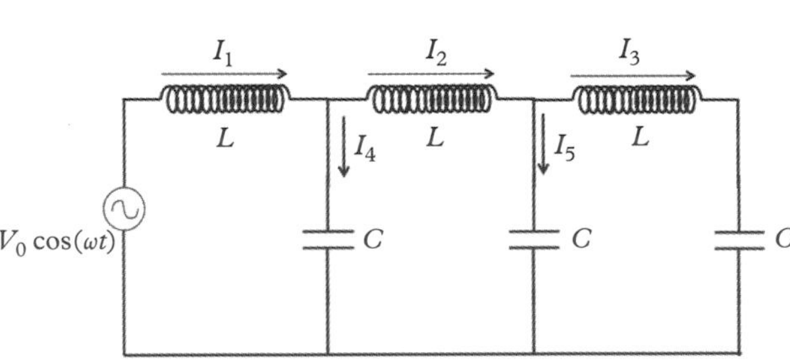

Fig. 5.27 *Exercise 5.19.*

Part III

Waves

Traveling Waves in One Dimension

6

Chapter Goals

In this chapter you will study traveling waves in a one-dimensional medium that continues to infinity. You will learn basic characteristics of a harmonic traveling wave, energy in the wave, and impedance of a medium. Superposition of harmonic waves will show that waves can beat against each other and move as a group. You will also learn how to study wave pulses by a Fourier integral method. A Fourier integral method will be used to learn about the frequency content of a pulse. The phenomenon of dispersion will also be discussed.

To most people, waves conjure up images of water waves striking a lake shore or an oceanfront. When you throw a stone into water you can clearly see the circular water waves traveling outward from the point of impact (Fig. 6.1). The energy in the stone is first given to the water molecules at the point of impact of the stone, which vibrate about their equilibrium position, much like masses attached to springs. Water molecules act similarly to the coupled oscillators we have discussed before, and the motion of one molecule is transferred to neighboring molecules through intermolecular coupling. In this way, the energy, initially at the point of impact, travels to distant places without any actual transport of any of the water molecules.

In addition to these familiar water waves, there are many types of waves in nature, e.g. sound waves, seismic waves, waves on a string, electromagnetic waves, etc. just to a name a few of those commonly encountered. In spite of this tremendous variety, most waves share some common characteristics: waves originate with a vibrating material, and transport

Fig. 6.1 *Picture of water waves.*

energy and momentum from one place to another by disturbing the medium at successive places locally. Thus, there is a close connection between waves traveling in a medium and the oscillations of the medium.

Traveling wave is the name given to a wave that travels in open space from some point or region in a medium where it is generated. The property characterizing the disturbance depends on the nature of the wave. For instance, in the case of a mechanical wave or sound wave, the disturbance refers to the displacement of particles of the medium from equilibrium, and in the case of an electromagnetic wave, the disturbance is the change in electric and magnetic fields in space. The disturbance varies both in space and time, and the function describing the disturbance is called the **wave function**.

Waves moving in two or three dimensions in space, such as light and sound waves, have more features than waves that can only move along one dimension, such as a wave on a string. For instance, waves in two and three dimensions can spread out from the source such that the energy in the wave dilutes out over larger and larger space, which results in decreased amplitude as the distance from the source increases. The wave in one dimension does not have any other degree of freedom into which the energy can spread, if we neglect any dissipation from viscosity and damping forces. Thus, in the absence of damping forces in the medium, a wave moving in one dimension will tend to maintain its amplitude.

In this chapter we will study waves in one dimension only. This choice is made to keep the mathematics simple and to discuss the fundamental characteristics of waves in more detail. In the next chapter we will address particular aspects of waves in two and three dimensions that are unique to those cases. Even though traveling waves in two or three dimensions would spread out in space, it is nevertheless helpful to study idealized waves in two and three dimensions, called straight waves and plane waves respectively, whose amplitudes do not drop with distance from the source. These idealized waves model behavior of many real waves whose amplitudes do not vary much over the distance of interest and could be treated as if they were one-dimensional waves.

6.1 Harmonic Traveling Waves

If the driving force at the source of a wave oscillates harmonically, the generated wave will also be harmonic. A harmonic traveling wave has a particularly simple wave function which can always be written as a sine or cosine function of space and time, which are very useful for mathematical analysis. Thus, harmonic waves are characterized by periodicities in space and time, as reflected in their representation in terms of periodic functions of sines and cosines.

The study of harmonic waves also provides a necessary foundation for waves that are not harmonic, since Fourier analysis can be used to cast the wave function of arbitrary waves, such as pulses, in terms of superposition of harmonic waves. Harmonic waves are also called **sinusoidal waves**. **Plane waves** in three-dimensional space and straight waves in two-dimensional space are also harmonic waves.

Supposing we wish to study a harmonic wave riding on a taut string. Such a wave can be generated in the string if you swing one end of the string continuously up and down at a definite frequency. The wave generated will move down the string and after you have swung the string a number of cycles a stable wave will be set up that will be moving away from the point of its generation.

To visualize a wave we need to draw multiple images of the shape of the string at different times so that the motion of the wave can be understood from the images. Figure 6.2 shows snapshots of a harmonic wave at three different times, traveling to the right. Particles of the

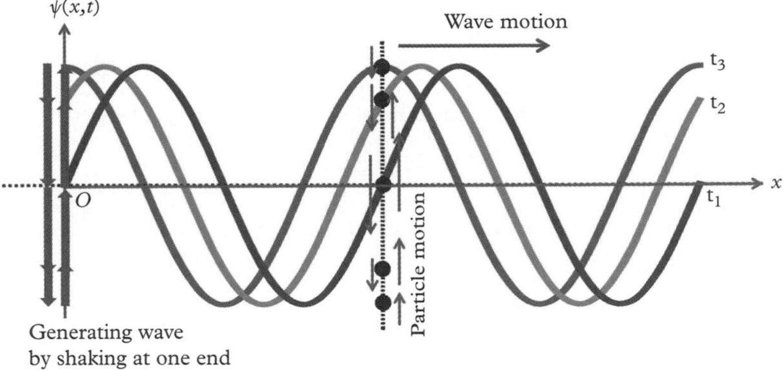

Fig. 6.2 *A sinusoidal traveling wave on a taut string for three different instants ($t_3 > t_2 > t_1$). The particles of the string move harmonically up and down as the wave passes to the right. The wave function for the wave moving towards the positive x-axis has the form $\psi(x,t) = A\cos(kx + \omega t)$.*

string execute a simple harmonic motion as shown for one particle in the figure. The frequency of vibration of the particles of the string is the same as the frequency of the wave.

Shaking the string at $x = 0$ will result in the displacement $\psi(0,t)$ of the particle at $x = 0$ to be

$$\psi(0,t) = A\cos(\omega t), \tag{6.1}$$

where ω is the frequency of the harmonic shaking with the assumption that at $t = 0$, the wave function has the value of the amplitude A,

$$\psi(0,0) = A. \tag{6.2}$$

There will be a phase constant in the argument of the cosine if $\psi(0,0)$ is different from the amplitude A of the vibration. A harmonic wave travels at a constant velocity in a given medium, called the phase velocity, which we will denote by v_p. When the wave moves through the string with speed v_p, the wave will arrive at the point where $x = x$ at time $t = x/v_p$. Therefore, the motion of the particle at x at time t will be the same as the motion of the particle at $x = 0$ at an *earlier time* t_{ret}, also called the **retarded time**,

$$t_{\text{ret}} = t - \frac{x}{v_p}. \tag{6.3}$$

Thus, the displacement $\psi(x,t)$ of the particle at x will be equal to the displacement $\psi(0,t)$ for $t = t_{\text{ret}}$,

$$\psi(x,t) = \psi(0,t_{\text{ret}}) = A\cos(\omega t_{\text{ret}}) = A\cos\left(\omega t - \frac{\omega}{v_p}x\right). \tag{6.4}$$

The result can equivalently be written as

$$\psi(x,t) = A\cos\left(\frac{\omega}{v_p}x - \omega t\right), \tag{6.5}$$

by using the identity $\cos(-\theta) = \cos(\theta)$. In this book we will use this latter form since that is more common. Note the amplitude of vibration at $x = x$ will be same as the vibration at $x = 0$ since in one dimension the wave cannot spread out. The wave function $\psi(x, t)$ tells us that at a particular point on the string the vibrations are oscillatory with frequency ω. Recall the relations of angular frequency ω to time period T and regular frequency f,

$$T = 2\pi/\omega = 1/f. \tag{6.6}$$

The wave function in Eq. 6.5 also tells us that at a *particular time*, the shape of the string is periodic in space with a repeat distance λ, called the **wavelength**, given by

$$\lambda = 2\pi \frac{v_p}{\omega}. \tag{6.7}$$

Recall that the wave number k is related to the wavelength by $k = 2\pi/\lambda$,

$$k = \frac{2\pi}{\lambda} = \frac{\omega}{v_p}. \tag{6.8}$$

The wave function in Eq. 6.5 is more commonly written in terms of angular frequency and wave number,

$$\psi(x, t) = A\cos(kx - \omega t). \tag{6.9}$$

The quantity in the argument of the cosine is called the phase or phase function of the wave and is often denoted as φ,

$$\varphi(x, t) = kx - \omega t. \tag{6.10}$$

The term phase velocity comes from examining this function. We ask: what is the speed with which a point of same phase (function) on the wave moves in space? To find this we just set the differential of the phase to zero and solve for dx/dt,

$$d\varphi = kdx - \omega dt = 0 \implies \frac{dx}{dt} = \frac{\omega}{k}, \tag{6.11}$$

which is the phase velocity v_p. If the string is from $x = 0$ to $x = -\infty$ and vibrated at $x = 0$, as in Fig. 6.3, the generated wave will move towards the negative x-axis. In that case the wave function will be given by

$$\psi(x, t) = A\cos(kx + \omega t). \tag{6.12}$$

Thus, if you vibrate a taut string harmonically at $x = 0$, that extends from $x = -\infty$ to $x = +\infty$, you will generate two waves, one will move towards $x = +\infty$ and the other towards $x = -\infty$, which are also called the right-moving and the left-moving waves,

$$\psi(x, t) = \begin{cases} A\cos(kx + \omega t) & (-\infty < x \leq 0), \\ A\cos(kx - \omega t) & (0 \leq x < \infty). \end{cases} \tag{6.13}$$

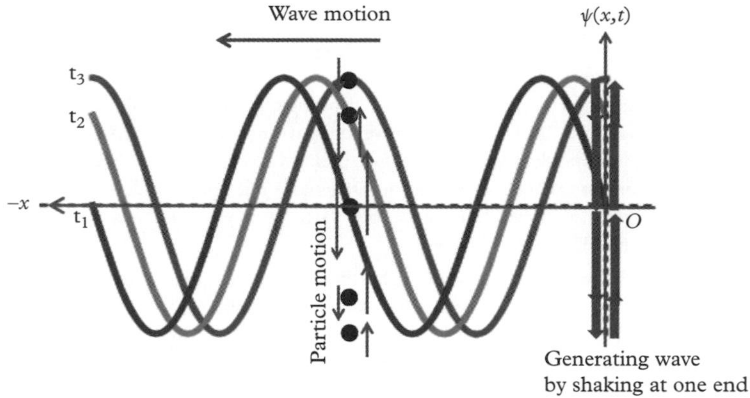

Fig. 6.3 *A sinusoidal traveling wave is generated at x = 0 and moves on the string towards the negative x-axis. The particles of the string move harmonically up and down as the wave passes to the left. The wave function for the wave moving towards the positive x-axis has the form* $\psi(x, t) = A\cos(kx + \omega t)$.

6.2 Standing Waves

In previous chapters when studying the free vibrations of finite-sized bodies we found some special motions called modes or normal modes. We also found that a body can be driven to oscillate at the driving frequency. In a mode or in a steady state of the driven body, different parts of the body move in a synchronized way—at the time one point of the body is at extreme displacement, all other parts of the body are also at extreme displacement and when displacement is zero at one point, the displacements of all other parts are also zero. These motions are also called **standing waves**. In the case of a vibrating string of length L placed between $x = 0$ and $x = L$, the following general expression describes the displacements of different parts of the string in a standing wave:

$$\psi(x, t) = A\sin(kx)\cos(\omega t), \quad 0 \leq x \leq L, \tag{6.14}$$

where I have set the phase constant to be zero, which is possible if the velocity everywhere is taken to be zero at $t = 0$. In the case of normal modes, k would be one of the wave numbers k_n of the normal mode and ω would be the corresponding normal mode frequencies ω_n. In the case of a driven finite string, ω is the driving frequency and k the wave number such that ω/k corresponds to the wave speed. For a wave on a taut string we had

$$\frac{\omega}{k} = \sqrt{T/\mu},$$

where T was the tension and μ the mass per unit length.

6.2.1 Similarities and Differences Between Standing Waves and Traveling Waves

The wave function in Eq. 6.14 for the standing wave shares two characteristics with a traveling wave on a string given in Eq. 6.9:

(i) The particle at any x executes harmonic motion.

(ii) The shape of the string at any time is periodic in space.

However, the standing wave in Eq. 6.14 differs from the traveling wave in the following important ways.

(1) In the standing wave, the amplitudes of particle motions at different x are different but they are the same in the traveling wave in the one-dimensional case. For instance, in a standing wave the particle at $x = a$ has a harmonic motion with amplitude $A\sin(ka)$, while the particle at $x = b$ has a harmonic motion with amplitude $A\sin(kb)$. But, in a traveling wave, both particles at $x = a$ and $x = b$ oscillate with the same amplitude A.

(2) In a standing wave, the phase of the harmonic motion of two particles at two different places on the string are either the same or they differ by π, which corresponds to the different sign in Eq. 6.14 at different places. But, in a traveling wave, the phase difference between the harmonic motions of the two particles is proportional to the distance between the two particles. For two particles at $x = a$ and $x = b$ we have

$$\text{Standing: } \Delta\varphi = 0 \text{ or } \pi, \quad \text{Traveling: } \Delta\varphi = |k(a-b)|.$$

(3) Wave functions for standing waves are a product of space and time parts while the wave functions for harmonic traveling waves have space and time parts together, so that the phase of the wave moves at the phase speed,

$$\text{Standing: } \psi(x,t) = A\sin(kx)\cos(\omega t), \tag{6.15}$$

$$\text{Traveling: } \psi(x,t) = A\cos(kx-\omega t). \tag{6.16}$$

(4) The standing waves are set up in a finite closed system while traveling waves are defined for open systems. In an open system the traveling wave can continue for ever, starting from a source somewhere.

Trigonometric identities can be employed to write a standing wave as a sum of appropriate traveling waves and vice versa. For instance, we can use the following trigonometric identity to replace the product of sine and cosine in Eq. 6.15 by pure sines:

$$2\sin A\cos B = \sin(A+B) + \sin(A-B). \tag{6.17}$$

This gives the following alternative expression for the standing wave:

$$\psi(x,t) = A\sin(kx)\cos(\omega t) = \frac{A}{2}\sin(kx+\omega t) + \frac{A}{2}\sin(kx-\omega t),$$

which is a superposition of two traveling waves, one moving towards the negative x-axis and the other towards the positive x-axis. In a similar way, we can express the traveling wave in Eq. 6.16 as a sum of two standing waves by expanding the cosine in that equation,

$$\psi(x,t) = A\cos(kx-\omega t) = A\cos(kx)\cos(\omega t) + A\sin(kx)\sin(\omega t).$$

6.3 Dispersion and Group Velocity

The dynamics of waves are given by the wave equation for that type of wave. For instance, the wave on a taut string obeys the classical wave equation

$$\frac{\partial^2 \psi}{\partial t^2} = v^2 \frac{\partial^2 \psi}{\partial x^2}, \tag{6.18}$$

where ψ is the displacement and v the speed of the wave, and the transverse waves in a metal bar obey the Euler–Bernoulli beam equation,

$$\frac{\partial^2 \psi}{\partial t^2} = -b^2 \frac{\partial^4 \psi}{\partial x^4}, \tag{6.19}$$

where $b = YI/\rho A$ with Y the Young's modulus, I the area moment, ρ the material density, and A the area of cross-section. For each type of wave, oscillating motion of a particular frequency has a definite dependence on the wave number, known as the **dispersion relation**. You can obtain the dispersion relation of a wave by plugging either the traveling or standing wave solution into the appropriate wave equation. Thus, the dispersion relation for the classical wave equation Eq. 6.18, will be

$$\omega^2 = v^2 k^2, \tag{6.20}$$

and that of the transverse wave on the rod, Eq. 6.19, will be

$$\omega^2 = b^2 k^4. \tag{6.21}$$

Dispersion relations such as Eqs. 6.20 and 6.21 encapsulate the wave equation in a symbolic form—the derivatives of time in the wave equation give factors of ω and derivatives of space in the wave equation give factors of k. In this way, the dynamics of the wave of a single frequency and the corresponding wave number is captured by the dispersion relation. The ratio of ω to k is called the phase velocity v_p,

$$v_p = \frac{\omega}{k}. \tag{6.22}$$

A plot of the dispersion relation often gives a visual sense of the nonlinearity of the dispersion relation. An example of plots of $\omega = k$ and $\omega = k^2$ is shown in Fig. 6.4.

The dispersion relation of waves on a string has a linear relation between k and ω. Therefore, every wave, regardless of the frequency (or equivalently, wave number or wavelength) will travel with the same speed. We saw above that a wave that is a superposition of two frequencies has the shape of beats in space. Such a wave on a string will travel with the shape intact. If you make a pulse on a string by giving it a snap, the pulse will move with its shape unchanged. We say that the string as a medium is a **non-dispersive medium**.

The dispersion relation for the transverse wave on a beam, as given in Eq. 6.21, does not have a linear relation between frequency and wave number. Therefore, the phase velocity of the transverse wave on a beam depends on the frequency,

$$v_p = \frac{\omega}{k} = bk = \sqrt{b\omega}. \tag{6.23}$$

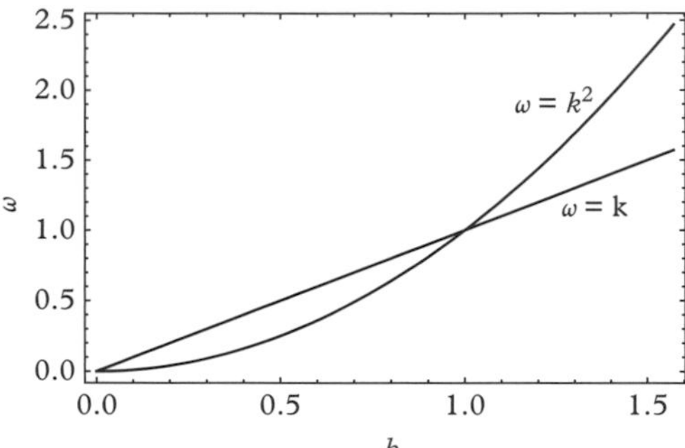

Fig. 6.4 *Example dispersion relations:* $\omega = kv$ *for transverse oscillations on a string and* $\omega = bk^2$ *for transverse waves on a beam.*

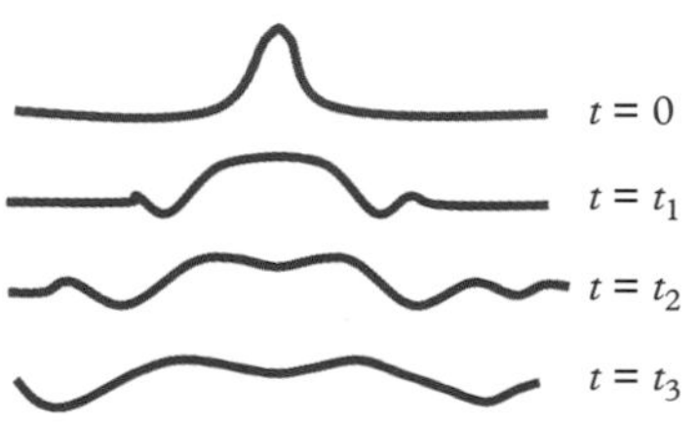

Fig. 6.5 *Dispersion of a transverse wave on a beam.*

That is, harmonic waves of different frequencies will travel with different speeds. Therefore, if you have a wave that is a superposition of two or more frequencies, then, the shape of the wave will change over time, as illustrated in Fig. 6.5 for a pulse. Fourier analysis tells us that any shape can be thought of as a superposition of harmonic functions. Therefore, we can think of the pulse wave at $t = 0$ in Fig. 6.5 to be made up of waves of different frequencies, and since the dispersion relation is nonlinear, different frequencies move at different speeds. Therefore, the wave will not keep its shape as it travels through the medium.

The dependence of phase speed on frequency is called **dispersion**. Since each frequency moves at its phase speed, the meaning of wave speed becomes difficult to assign to the complete wave if that happens. When you study the speed with which the energy in the wave moves in space, you find that that speed is given by the slope of the dispersion curve. This speed is called the group speed or **group velocity**,

$$v_g = \frac{d\omega}{dk}. \tag{6.24}$$

The dispersion plays an important role in many waves. For instance, dispersion is the key to understanding the origin of a rainbow from white light. The bending of light waves is caused by the refractive index n of the medium, which is related to the speed v of light in the medium compared to the speed c of light in a vacuum,

$$n = \frac{c}{v}. \tag{6.25}$$

Transparent media such as glass are dispersive, since the speed of light in the medium depends upon the wavelength as given by the Cauchy equation,

$$n(\lambda) = A + \frac{B}{\lambda^2}, \tag{6.26}$$

where A and B are constants. For instance, for borosilicate glass,

$$n(\lambda) = 1.5046 + \frac{0.00420\ \mu\text{m}^2}{\lambda^2}. \tag{6.27}$$

Because of different n for different λ the bending of light depends on the color or wavelength of light, which is responsible for separation of colors when white light is sent through a prism.

Example 6.1 Group Velocity of Light

Find the group velocity of a light wave in a medium for which the refractive index is given as

$$n(\lambda) = A + \frac{B}{\lambda^2}.$$

Solution

Let us first deduce the function $\omega(k)$ from the given phase velocity,

$$v_p = \frac{c}{n} = \frac{c}{A + \frac{B}{\lambda^2}}.$$

Now, we have

$$v_p = \frac{\omega}{k}, \quad \lambda = \frac{2\pi}{k}.$$

Therefore, the dispersion relation for the light wave is

$$\omega = \frac{ck}{A + \frac{Bk^2}{4\pi^2}}.$$

The group velocity is obtained by taking the derivative,

$$v_g = \frac{d\omega}{dh} = \frac{c}{A + \frac{Bk^2}{4\pi^2}} - \frac{ck}{\left(A + \frac{Bk^2}{4\pi^2}\right)^2} \frac{Bk}{2\pi^2}.$$

Example 6.2 Water Waves in Finite Depth

Water waves in a tank or pond of depth h have the following dispersion relation:

$$\omega = \sqrt{g\,k\tanh(kh)}, \tag{6.28}$$

where g is the acceleration due to gravity. Are the waves dispersive for long waves when $\lambda \gg h$?

Solution

Let us write k in the equation as $2\pi/\lambda$ and write the square of the given dispersion relation,

$$\omega^2 = \frac{2\pi g}{\lambda}\tanh\left(\frac{2\pi h}{\lambda}\right). \tag{6.29}$$

In the limit $\lambda \gg h$, the argument of the function tanh() becomes small. The $\tanh(x)$ has the following McLaurin series:

$$\tanh(x) = x - \frac{1}{3}x^3 + \frac{2}{15}x^5 + \cdots \tag{6.30}$$

Keeping the leading term in Eq. 6.29 gives

$$\omega^2 = \frac{2\pi g}{\lambda} \times \frac{2\pi h}{\lambda} = ghk^2. \tag{6.31}$$

Therefore,

$$\omega = \sqrt{gh}k. \tag{6.32}$$

Since ω is directly proportional to k, the phase velocity ω/k will be independent of k. Hence, the long water waves will be non-dispersive with phase velocity $v_p = \sqrt{gh}$.

Example 6.3 Water Waves in Deep Water

The dispersion relation for the water waves in deep water differ from the dispersion relation given above. For deep water the water waves obey the following approximate dispersion relation:

$$\omega^2 = gk + \frac{S}{\rho}k^3, \tag{6.33}$$

where g is the acceleration due to gravity, S the surface tension of water, and ρ the density of water. Find the phase and group velocities of waves in deep water.

Solution

The phase velocity will be ω/k and group velocities will be $d\omega/dk$. Carrying out the calculations we see that these waves are dispersive,

$$v_p = \sqrt{\frac{g}{k} + \frac{S}{\rho}k}, \tag{6.34}$$

$$v_g = \frac{g + 3\frac{S}{\rho}k^2}{2\omega} \neq v_p. \tag{6.35}$$

6.4 Energy Transport by Traveling Wave

6.4.1 Energy in a Wave

When a mechanical wave such as sound moves through a medium it sets particles of the medium in motion. Thus, the kinetic and potential energies of a medium that has a wave traveling through it would be different from the same medium with no wave. To be concrete, consider a transverse wave $\psi(x, t)$ on a taut string (the medium) traveling in the direction of the positive x-axis. We wish to calculate the energy of the string due to the presence of the wave.

The wave function $\psi(x, t)$ is the y-displacement of an infinitesimal element of the string located at position x (more appropriately between x and $x + dx$) at time t. Let the mass of the element be dm and its length when horizontal be dx. The y-component of the velocity of the mass element is simply the time derivative of the wave function,

$$v_y(x, t) = \frac{\partial \psi(x, t)}{\partial t}. \tag{6.36}$$

Since we are looking at only the transverse wave, the elements of the string have only y-velocity, given in Eq. 6.36. Therefore, the kinetic energy dK of the element at x will be

$$dK = \frac{1}{2}(dm)\, v^2 = \frac{1}{2}(\mu dx)\left(\frac{\partial \psi}{\partial t}\right)^2, \tag{6.37}$$

where μ is the mass density, which is the mass per unit length here. When the string is vibrating it will be displaced and stretched such that the length of the string will be different from dx. Let us denote the length of the same element dx at time t by ds. The change in length occurs against the tension T of the string. Therefore, the potential energy dU of the string element will be the work in stretching the string from dx to ds against force T,

$$dU = T(ds - dx). \tag{6.38}$$

Writing ds in terms of displacements dy and dx, we have from Pythagoras's theorem,

$$ds = \sqrt{dx^2 + dy^2} = dx\sqrt{1 + \left(\frac{dy}{dx}\right)^2}. \tag{6.39}$$

Now, the vertical displacement of the element dy is the same as the change in wave function,

$$\frac{dy}{dx} = \frac{\partial\psi}{\partial x}, \tag{6.40}$$

where we have changed the derivative to a partial derivative. This gives the element length when not horizontal to be

$$ds = dx\sqrt{1 + \left(\frac{\partial\psi}{\partial x}\right)^2}. \tag{6.41}$$

Assuming the vertical displacement to be small compared to the element length, i.e.

$$\left|\frac{\partial\psi}{\partial x}\right| \ll 1, \tag{6.42}$$

we can expand the radical in Eq. 6.41 and obtain the following approximate expression:

$$ds - dx \approx \frac{1}{2}\left(\frac{\partial\psi}{\partial x}\right)^2 dx. \tag{6.43}$$

Therefore, we obtain the following expression for the potential energy associated with the element at x:

$$dU = \frac{1}{2}T\left(\frac{\partial\psi}{\partial x}\right)^2 dx. \tag{6.44}$$

Adding the kinetic and potential energies of the string element under consideration, we find that the wave has the following energy dE between x and $x + dx$:

$$dE = dK + dU = \left[\frac{1}{2}\mu\left(\frac{\partial\psi}{\partial t}\right)^2 + \frac{1}{2}T\left(\frac{\partial\psi}{\partial x}\right)^2\right] dx. \tag{6.45}$$

This equation is applicable to all waves on a string. Specifically, let us work out the energy of a sinusoidal wave traveling towards the positive x-axis over a length of the string equal to one wavelength, $\lambda = 2\pi/k$. In this case, the wave function is given by

$$\psi(x, t) = A\cos(kx - \omega t). \tag{6.46}$$

For this sinusoidal wave function we get the following:

$$\frac{\partial \psi}{\partial t} = \omega A \sin(kx - \omega t), \quad \frac{\partial \psi}{\partial x} = -kA \sin(kx - \omega t), \tag{6.47}$$

$$dK = \frac{1}{2}\mu A^2 \omega^2 \sin^2(kx - \omega t)dx, \quad dU = \frac{1}{2} TA^2 k^2 \sin^2(kx - \omega t)dx, \tag{6.48}$$

$$dE = A^2 \mu \omega^2 \sin^2(kx - \omega t)dx, \tag{6.49}$$

where in the last line we have used $v = \omega/k = \sqrt{T/\mu}$. We can find the energy in the wave at $t = 0$ over a space of one wavelength by integrating from $x = 0$ to $x = \lambda = 2\pi/k$. This gives the following result:

$$E_\lambda = \frac{1}{2}\lambda \mu \omega^2 A^2. \tag{6.50}$$

Where does this energy in the wave come from? To address this question, we need to look at the source of generation of the wave. We do this next.

6.4.2 Power of the Wave Generator

Characteristic Impedance

Suppose a wave in the string is generated by attaching one end of the string to a vibrator at $x = 0$, as shown in Fig. 6.6 and the other end of the string held at a far away point so that there is a constant tension T in the string. Because of the tension T in the string, it applies a force equal to T on the vibrator, which in turn must apply a force F on the string equal in magnitude to T and opposite in direction

$$F = T.$$

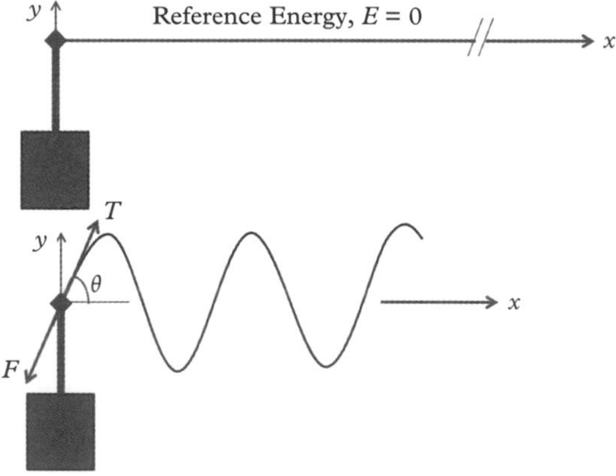

Fig. 6.6 *Generation of a traveling wave by driving vibrations at one end. The work done by the vibrator goes into the energy of the wave. Here T is the tension in the string and F the force the vibrator applies on the string. By Newton's third law, the magnitude of F is equal to the magnitude of T. The work done by the vibrator for the displacement dy of the string at $x = 0$ will be $F_y\, dy$. In the text we denote y of the string at a particular x by $\psi(x, t)$.*

Since the displacement of the element of the string is parallel to the y-axis, we need the y-component of the force F to calculate the work done on the string. Referring to the drawing in Fig. 6.6 we note that

$$F_y = -T\sin\theta \approx -T\tan\theta = -T\left[\frac{\partial\psi}{\partial x}\right]_{x=0}. \tag{6.51}$$

This force is actually proportional to the velocity of the particles of the string at $x = 0$, as we can show by using the wave function of the traveling wave. The wave on the string will be given by the wave function moving to the right on the x-axis with wave number k and frequency ω,

$$\psi(x,t) = A\,\cos(k\,x - \omega\,t). \tag{6.52}$$

To relate F_y to the velocity $\partial\psi/\partial t$ we need to find the relation between $\partial\psi/\partial x$ and $\partial\psi/\partial t$. Taking the derivatives of $\psi(x,t)$ we find that

$$\frac{\partial\psi}{\partial x} = -k\,A\,\sin(k\,x - \omega\,t), \tag{6.53}$$

$$\frac{\partial\psi}{\partial t} = \omega\,A\,\sin(k\,x - \omega\,t). \tag{6.54}$$

Therefore,

$$\frac{\partial\psi}{\partial x} = -\frac{k}{\omega}\frac{\partial\psi}{\partial t} = -\frac{1}{v_p}\frac{\partial\psi}{\partial t}, \tag{6.55}$$

where I have replaced ω/k by v_p, the phase velocity. Now using this relation in Eq. 6.51, and for brevity, dropping the brackets and $x = 0$, we find

$$F_y = \frac{T}{v_p}\frac{\partial\psi}{\partial t}. \tag{6.56}$$

The proportionality constant is called the characteristic impedance Z of the medium,

$$Z = \frac{T}{v_p}, \tag{6.57}$$

and we write Eq. 6.56 as

$$F_y = Z\frac{\partial\psi}{\partial t}. \tag{6.58}$$

This says that the string applies a velocity-dependent force on the vibrator,

$$T_y = -F_y = -Z\frac{\partial\psi}{\partial t}. \tag{6.59}$$

We see that this force is similar to the drag force $-b\dot{x}$ on a harmonic oscillator as we have studied previously with the impedance Z taking the role of b. That is, the medium applies a drag force on the wave generator and takes energy away from the generator. Using $v_p = \sqrt{T/\mu}$ in Eq. 6.57 we can write the impedance for the transverse waves in a string in terms of the tension and the density,

$$Z = \frac{T}{v_p} = \sqrt{T\,\mu} = v_p\mu. \tag{6.60}$$

Power Output of the Wave Generator

Now, we calculate the work done by the force F_y of the wave generator on the medium of the wave, which here is the string. The instantaneous power $P(t)$ associated with this force will be

$$P(t) = F_y v_y = Z\left(\frac{\partial\psi}{\partial t}\right)^2. \tag{6.61}$$

To compare the energy output of the vibrator to the energy E_λ in the wave over a distance equal to the wavelength we have calculated above, let us calculate the work done over one time period of the wave, from $t = 0$ to $t = 2\pi/\omega$ at $x = 0$,

$$W = \int_0^\tau P(t)\,dt = \pi Z A^2\omega. \quad (\text{period, } \tau = 2\pi/\omega) \tag{6.62}$$

Using $Z = \sqrt{T\mu} = v_p\mu$, $v_p = \omega/k$, and $k = 2\pi/\lambda$ this simplifies to

$$W = \frac{1}{2}\lambda\mu\omega^2 A^2, \tag{6.63}$$

which is identical to E_λ given in Eq. 6.50.

That is, the work done by the wave generator in one full cycle of the vibration is equal to the energy of the wave in a space of distance equal to one wavelength

$$W = \frac{1}{2}\lambda\mu\omega^2 A^2 = E_\lambda. \tag{6.64}$$

Power Flow at an Arbitrary Point

The calculation of the power of the wave generator at $x = 0$ can actually be carried out at any point on the string. Let us examine the situation at $x = x_0$. We will call the string to the left of $x = x_0$ the system L (for left) and the string to the right as system R (for right), as shown in Fig. 6.7.

A comparison of Fig. 6.7 to Fig. 6.6 shows that the force exerted by system L on system R has the same relation to the tension in the string as was the case at $x = 0$, where the system L consisted of only the wave generator. Therefore the y-component of the tension force F_y on the left of the point $x = x_0$ will have the same relation to the force T on the right side of $x = x_0$, as we had found at $x = 0$ in Fig. 6.6,

$$F_y = -T\,\frac{\partial\psi(x,t)}{\partial x}, \tag{6.65}$$

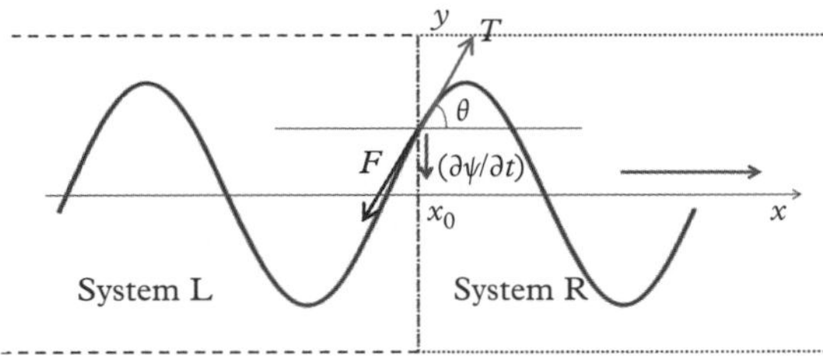

Fig. 6.7 *The applied force and resulting velocity at an arbitrary point of a taut string which carries a wave. The string to the left of the point $x = x_0$ is labeled system L and that to the right is labeled system R. The oscillations on the left of the point $x = x_0$ provide the driving force for generation of the wave on the right side of the point. The tension on the right is labeled T and that on the left F. The force T is applied by R on L and the force F is applied by L on R. The two forces are equal in magnitude but opposite in direction.*

where the derivative of ψ is to be evaluated at $x = x_0$. The velocity of the string element at $x = x_0$,

$$v_y = \frac{\partial \psi(x,t)}{\partial t}. \tag{6.66}$$

From the expression for the traveling wave $\psi(x,t) = A\cos(kx - \omega t)$, we get the following relation:

$$\frac{\partial \psi(x,t)}{\partial x} = \frac{\omega}{k}\frac{\partial \psi(x,t)}{\partial t} = \frac{1}{v_p}\frac{\partial \psi(x,t)}{\partial t}. \tag{6.67}$$

Now writing the y-component of the force by system L on system R we get

$$F_y = \frac{T}{v_p}\frac{\partial \psi(x,t)}{\partial t} = Z\frac{\partial \psi(x,t)}{\partial t}, \tag{6.68}$$

where Z is the characteristic impedance, which was found earlier to be equal to T/v_p or $\sqrt{T\mu}$ for transverse waves on a string. Using this expression for the force by L on R we find that the power input by L into R at the point $x = x_0$ will be

$$P(t)\,|_{x=x_0} = F_y\, v_y = Z\left[\frac{\partial \psi(x,t)}{\partial t}\right]^2_{x=x_0}. \tag{6.69}$$

The time-averaged power for a harmonic traveling wave we have here is easily calculated by performing the following integral:

$$\langle P\rangle = Z\,\frac{1}{\tau}\int_0^{\tau}\left[\frac{\partial \psi(x,t)}{\partial t}\right]^2_{x=x_0} dt, \tag{6.70}$$

where $\tau = 2\pi/\omega$, the time period. The integral is left as an exercise. The result will be independent of x_0,

$$\langle P\rangle = \frac{1}{2}\,Z\,\omega^2 A^2. \tag{6.71}$$

Example 6.4 Longitudinal Wave on a Beaded Slinky

Perhaps you have played with longitudinal waves that you can send down an extended slinky by compressing some of the rings and then letting go of it. We can understand the wave by considering a discrete model of the slinky. Suppose the mass of the slinky is distributed along it as beads of mass m connected by springs with spring constant k and length a, as shown in Fig. 6.8, where the slinky is along the x-axis and ψ_j is the displacement of the jth mass from its equilibrium position.

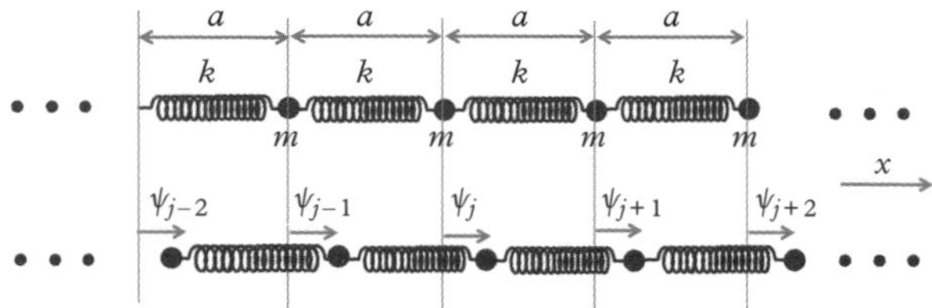

Fig. 6.8 *A longitudinal wave in a beaded slinky.*

Then, the equation motion of the jth mass will be

$$\frac{d^2\psi_j}{dt^2} = \frac{k}{m}\left(\psi_{j+1} - 2\psi_j + \psi_{j-1}\right). \tag{6.72}$$

Divide and multiply the right side by a^2 and take the limit of a tending to zero to obtain

$$\frac{\partial^2\psi}{\partial t^2} = \frac{ka}{m/a}\frac{\partial^2\psi}{\partial x^2}. \tag{6.73}$$

Introducing mass per unit length $\mu = m/a$ as the density, we find that

$$\frac{\partial^2\psi}{\partial t^2} = \frac{ka}{\mu}\frac{\partial^2\psi}{\partial x^2}. \tag{6.74}$$

This is the classical wave equation, which becomes identical to the equation of motion for the transverse motion of a string if we make the following substitution:

$$ka \leftrightarrow T \text{ (the tension)}. \tag{6.75}$$

Therefore, we can take the results from previous sections in this chapter and translate to the present case. Let us denote the wave vector by κ (read: kappa), since we are already using k for the spring constant. The wave function for the longitudinal wave moving toward the positive x-axis will be

$$\psi(x, t) = A\cos(\kappa x - \omega t), \tag{6.76}$$

and here the dispersion relation is linear, as you can verify by plugging in the harmonic solution in the equation of motion,

$$\omega = \left(\frac{ka}{\mu}\right)^{1/2}\kappa. \tag{6.77}$$

Therefore, longitudinal waves will be non-dispersive with the phase velocity,

$$v_p = \sqrt{\frac{ka}{\mu}}. \tag{6.78}$$

The characteristic impedance will be

$$Z = \sqrt{ka\mu}. \tag{6.79}$$

The instantaneous power flow at any point will be

$$P(t)\,|_{x=x_0} = Z\left[\frac{\partial\psi(x,t)}{\partial t}\right]^2_{x=x_0}. \tag{6.80}$$

The average power obtained by integrating over one cycle for the wave given in Eq. 6.76 will be

$$\langle P\rangle = \frac{1}{2}\,Z\,\omega^2 A^2. \tag{6.81}$$

6.5 Traveling Wave in a Transmission Line

Electric transmission lines are used in communication for sending signals, often as radio waves. A transmission line is a pair of conductors that has a uniform cross-section throughout its length. Co-axial cable with two wires, one at the center and the other around it, is a good example of a transmission line. When the two wires are driven by a time-dependent voltage or current source, the current and voltage waves flow in the transmission line.

A transmission line can lose power due to resistance in the cable and also due to radiation escaping the cable. In this section we will study an ideal lossless transmission line which can be modeled as an LC network. In the last chapter we studied a driven LC network that was finite in size. Here, we will look at the same system but now we suppose that the system is infinite in size so that there is only one wave traveling and there is no return wave. A small segment of the cable at an arbitrary time is shown in Fig. 6.9.

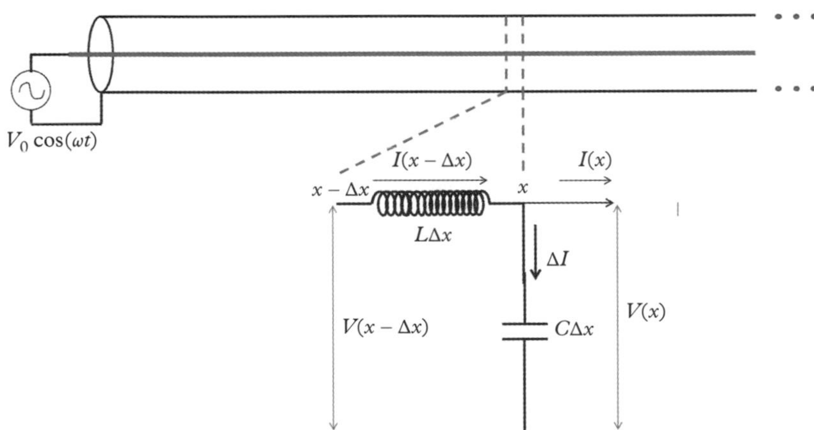

Fig. 6.9 *An arbitrary segment of a lossless transmission line as an LC network.*

As before, let L and C be the inductance and capacitance *per unit length*. You can show that the current and voltage at any point along the transmission line obey

$$L\frac{\partial I(x,t)}{\partial t} = -\frac{\partial V(x,t)}{\partial x}, \tag{6.82}$$

$$C\frac{\partial V(x,t)}{\partial t} = -\frac{\partial I(x,t)}{\partial x}. \tag{6.83}$$

Combining these shows that both I and V along the transmission line obey the same classical wave equation,

$$\frac{\partial^2 I(x,t)}{\partial t^2} = \frac{1}{LC}\frac{\partial^2 I(x,t)}{\partial x^2}, \tag{6.84}$$

$$\frac{\partial^2 V(x,t)}{\partial t^2} = \frac{1}{LC}\frac{\partial^2 V(x,t)}{\partial x^2}. \tag{6.85}$$

Let us look at the current wave in more detail. The dispersion relation of the current wave, obtained by plugging in a harmonic wave solution, $I(x,t) = I_0\cos(kx - \omega t)$,

$$\omega = \frac{1}{\sqrt{LC}}k, \tag{6.86}$$

shows that the current wave is non-dispersive with phase and group wave speeds given by

$$v_p = v_g = \frac{1}{\sqrt{LC}}. \tag{6.87}$$

A similar conclusion can be reached for the voltage wave using $V(x,t) = V_0\cos(kx - \omega t)$. From either Eq. 6.82 or 6.83, the current and voltage amplitudes at any point in the cable are related by

$$V_0 = \frac{k}{\omega C}I_0 \longrightarrow V_0 = \sqrt{\frac{L}{C}}\,I_0. \tag{6.88}$$

Therefore the impedance of the transmission line Z is

$$Z = \sqrt{\frac{L}{C}}. \tag{6.89}$$

The power flow in the transmission line at any place x will be

$$P(x,t) = I(t)V(t) = I_0V_0\cos^2(kx - \omega t). \tag{6.90}$$

The average power obtained by integrating over one cycle will be

$$\langle P\rangle = \frac{1}{2}I_0V_0 = \frac{1}{2}ZI_0^2 = \frac{1}{2}\frac{V_0^2}{Z}. \tag{6.91}$$

Example 6.5 Co-axial cable

A co-axial cable, also called a co-ax, with an inner conductor of diameter a and outer conductor of diameter b has a dielectric with dielectric constant ϵ_r between the conductors, as in Fig. 6.10. The capacitance and inductance per unit length is given by

$$C = \frac{2\pi\epsilon_0\epsilon_r}{\ln(b/a)}, \quad L = \frac{\mu_0}{2\pi}\ln(b/a). \tag{6.92}$$

Here ϵ_0 and μ_0 are the electric permittivity and magnetic susceptibility of a vacuum with the following values:

$$\epsilon_0 \approx 8.85 \times 10^{-12}\ \mathrm{C^2/N.m^2}, \quad \mu_0 = 4\pi \times 10^{-7}\ \mathrm{T.m/A}.$$

Find the characteristic impedance of the cable and the speed of the current wave in the cable if the dielectric constant of the packing material is 3 and $a = 1$ mm and $b = 4$ mm.

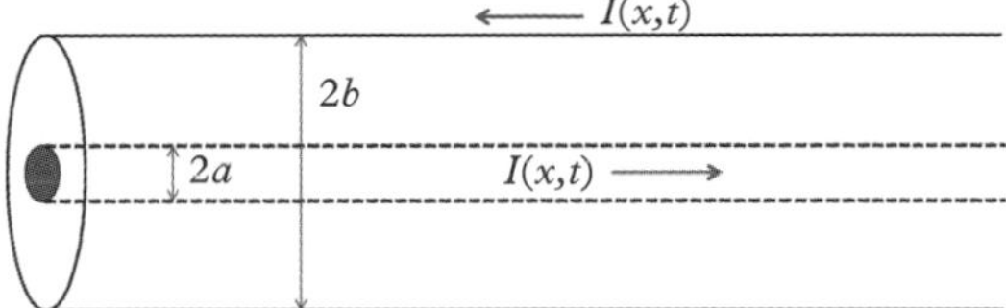

Fig. 6.10 *A co-axial cable as a transmission line. One end of the cable is connected to a power source and the other end is connected to a load.*

Solution

The characteristic impedance Z is given by $\sqrt{L/C}$. Therefore,

$$Z = \sqrt{\frac{\mu_0}{2\pi}\ln(b/a)}\sqrt{\frac{\ln(b/a)}{2\pi\epsilon_0\epsilon_r}} = \frac{\ln(b/a)}{2\pi}\sqrt{\frac{\mu_0}{\epsilon_0\epsilon_r}}.$$

The speed of the current wave in the cable will be

$$v = \frac{1}{\sqrt{LC}} = \frac{1}{\sqrt{\mu_0\epsilon_0\epsilon_r}}.$$

Putting in numerical values, we get $Z = 50\ \Omega$ and $v = 1.73 \times 10^8$ m/s $= 0.58c$, where c is the speed of light in a vacuum.

Example 6.6 Two-wire Transmission Line

Two parallel wires carrying current in opposite directions from a source to a load act as a power transmission line. Let the radius of the two wires each be a and the distance between them d, as in Fig. 6.11. Let the ϵ_r be the dielectric constant of the material, which is usually air in the power transmission lines in which the cables are located. The capacitance and inductance per unit length for this system are given by

$$C = \frac{\pi\epsilon_0\epsilon_r}{\ln\left[\frac{d}{2a} + \sqrt{\left(\frac{d}{2a}\right)^2 - 1}\right]}, \quad L = \frac{\mu_0}{\pi}\ln\left[\frac{d}{2a} + \sqrt{\left(\frac{d}{2a}\right)^2 - 1}\right]. \tag{6.93}$$

continued

Example 6.6 *continued*

Here ϵ_0 and μ_0 are the electric permittivity and magnetic susceptibility of a vacuum with the following values:

$$\epsilon_0 \approx 8.85 \times 10^{-12}\ \text{C}^2/\text{N.m}^2, \quad \mu_0 = 4\pi \times 10^{-7}\ \text{T.m/A}.$$

Find the characteristic impedance of the transmission line and the speed of the current wave in the wires.

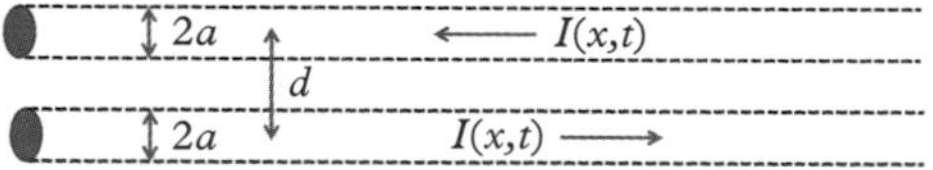

Fig. 6.11 *Two parallel wires as a transmission line. At one end the wires are connected to a power source and at the other, to a load.*

Solution

The characteristic impedance Z is given by $\sqrt{L/C}$. Therefore,

$$Z = \frac{\ln\left[\frac{d}{2a} + \sqrt{\left(\frac{d}{2a}\right)^2 - 1}\right]}{\pi}\sqrt{\frac{\mu_0}{\epsilon_0\epsilon_r}}.$$

The speed of the current wave in the cable will be

$$v = \frac{1}{\sqrt{LC}} = \frac{1}{\sqrt{\mu_0\epsilon_0\epsilon_r}}.$$

6.6 Superposition of Harmonic Waves

So far we have discussed harmonic traveling waves almost exclusively. Harmonic waves have only one frequency ω and one wave number k, and we have written the following analytic form for a harmonic wave traveling towards the positive x-axis:

$$\psi(x,t) = A\cos(kx - \omega t). \tag{6.94}$$

The speeds of these waves are given by their phase velocity ω/k. What will happen if a wave contains many frequencies? We have seen that when you superimpose oscillations of two frequencies ω_1 and ω_2, then the net result is a new vibration that beats in time. Now, in harmonic waves, the oscillations occur in both time and space. Therefore, we will observe beats in time as well as in space if we overlap two harmonic waves of different frequencies.

6.6.1 Beats in Waves of Two Different Frequencies

To illustrate the consequences of the superposition of two waves of different frequencies we will consider the one-dimensional example of the transverse wave on a string. Although we will work with a one-dimensional example, our conclusions will be applicable to two and three dimensions as well. To be concrete, let a taut string be placed along the x-axis from $x = 0$ to $x = \infty$, with a vibrator attached to the string at $x = 0$ with the other end tied to some post very far away, which we have chosen to call $x = \infty$. Unlike our examples of harmonic waves on this system, studied previously, suppose the vibrator now vibrates at two frequencies ω_1 and ω_2 such that the y-displacement of the string at $x = 0$ is a superposition of two vibrations of the same amplitude and same phase,

$$\psi(x = 0, t) = A\cos(\omega_1 t) + A\cos(\omega_2 t). \tag{6.95}$$

I have chosen the amplitudes and phases of the two frequencies to be the same for simplicity and have taken the phase constant to be zero also. Now, we ask: what kind of wave will travel down the string? The wave equation governing the dynamics of waves on a taut string is the classical wave equation, which is a linear homogeneous equation in the wave function ψ. Therefore, we expect that the solution will be the sum of two parts—one wave generated from vibration $A\cos(\omega_1 t)$ and the other from $A\cos(\omega_2 t)$. This is an important point to understand: the system acts as if there are two vibrators each transmitting waves at its own frequency and the net wave is just the sum of the two. Since we have already solved the one-frequency problem, we can write the answer immediately,

$$\psi(x, t) = A\cos(k_1 x - \omega_1 t) + A\cos(k_2 x - \omega_2 t), \tag{6.96}$$

where k_1 and k_2 are the wave numbers for the two waves. Now, a trigonometric identity turns this sum into a product which has a slower varying amplitude multiplying a faster oscillating function,

$$\psi(x, t) = [2A\cos(k_{\text{mod}} x - \omega_{\text{mod}} t)]\cos(k_{\text{av}} x - \omega_{\text{av}} t), \tag{6.97}$$

where

$$k_{\text{mod}} = \frac{k_1 - k_2}{2}, \quad \omega_{\text{mod}} = \frac{\omega_1 - \omega_2}{2}, \tag{6.98}$$

$$k_{\text{av}} = \frac{k_1 + k_2}{2}, \quad \omega_{\text{av}} = \frac{\omega_1 + \omega_2}{2}. \tag{6.99}$$

Note that the quantity in brackets in Eq. 6.97 acts as a time-dependent amplitude to the more rapidly oscillating function. Let us replace the quantity in the brackets [.] by one symbol $A_{\text{mod}}(x, t)$ to indicate this aspect of the result,

$$\psi(x, t) = A_{\text{mod}}(x, t)\cos(k_{\text{av}} x - \omega_{\text{av}} t), \tag{6.100}$$

with

$$A_{\text{mod}}(x, t) = 2A\cos(k_{\text{mod}} x - \omega_{\text{mod}} t). \tag{6.101}$$

This is similar to the beat phenomena when superposing two vibrations. In the case of a wave, the beat phenomenon occurs both in time and in space. Figure 6.12 shows the beating in space as the amplitude modulated wave travels to the right in the figure. As is

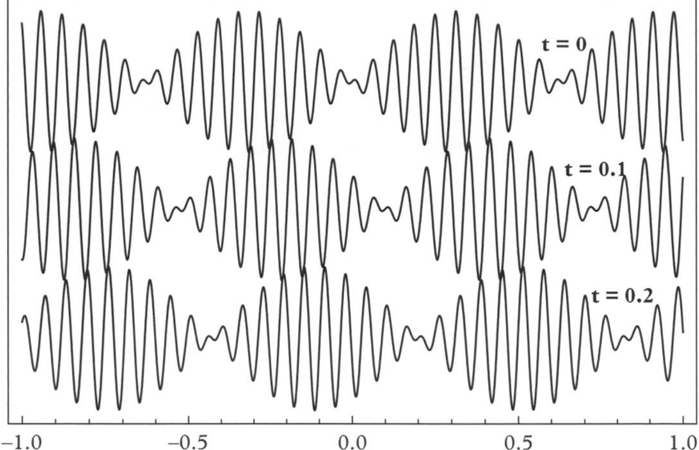

Fig. 6.12 *Movement of the wave resulting from the beating of waves of two nearby frequencies. The plot was made with $\omega_{av} = 100$, $\Delta\omega = 5$, and $v = 1$.*

clear from the figure, the resulting wave in the string here is not absolutely harmonic but close to a harmonic wave with frequency ω_{av} and wave number k_{av}.

If $A_{mod}(x,t)$ did not depend on x and t we would have a simple harmonic wave of frequency ω_{av} and wave number k_{av}. In actuality, the amplitude $A_{mod}(x,t)$ does vary with time. But does it vary by much over a time of one period of the average frequency wave? It turns out that $A_{mod}(x,t)$ does not vary much over a period $2\pi/\omega_{av}$ or distance $2\pi/k_{av}$, and we may treat $A_{mod}(x,t)$ as almost constant. In this sense, we say that the harmonic wave $\cos(k_{av}x - \omega_{av}t)$ is **amplitude modulated**.

The modulated amplitude in the case of the superposition of two waves, $A_{mod}(x,t)$, is itself a harmonic wave. The velocity of the $A_{mod}(x,t)$ wave is not the same as the velocity of the more rapidly varying wave $\cos(k_{av}x - \omega_{av}t)$. The velocity of $A_{mod}(x,t)$ wave is called the **modulation velocity** v_{mod}, while the velocity of $\cos(k_{av}x - \omega_{av}t)$ of a faster varying wave is called the phase velocity,

$$v_{mod} = \frac{\omega_{mod}}{k_{mod}} = \frac{\omega_1 - \omega_2}{k_1 - k_2}. \tag{6.102}$$

In most applications ω_1 and ω_2 differ only slightly so that their difference is a small fraction of their average. Let

$$\omega_1 = \omega_2 + \Delta\omega, \quad k_1 = k_2 + \Delta k. \tag{6.103}$$

Then, we will have

$$v_{mod} = \frac{\Delta\omega}{\Delta k}. \tag{6.104}$$

Note that ω is a function of k through the dispersion relation, $\omega(k)$, which for the taut string of tension T and mass per unit length μ is

$$\omega = k\sqrt{T/\mu}. \tag{6.105}$$

In other media and/or with a different wave equation we will have a different relation between ω and k. Generally, we will write ω as $\omega(k)$ to indicate the relation between ω and k of a wave given by the dispersion relation of the wave. Taking the infinitesimal limit of Eq. 6.104 we obtain the modulation velocity for superposition of two waves whose frequencies and wave numbers are infinitesimally close. We call the result the **group velocity** v_g,

$$v_g = \frac{d\omega}{dk}. \tag{6.106}$$

In electrical signals when two frequencies are superimposed to produce amplitude-modulated signals, the frequency ω_{av} is called the **carrier frequency** and the frequency ω_{mod} the **signal frequency**. Note that without the modulation the single-frequency wave does not carry any information from one place to another, since each cycle of the harmonic wave looks the same as any other and the wave goes on endlessly. By modulating the wave, which can be done by changing either the amplitude or the frequency or the phase constant, we introduce variability in the wave which can be used to code information in the wave. We saw in this section that the modulation travels at a different velocity, namely the group velocity, than the phase velocity. Since information is coded in the modulations, the information will travel at the group velocity. This is why, even though the phase velocity may exceed the speed of light in vacuum c, the group velocity must strictly be less than c so as not to violate the tenets of the special theory of relativity.

6.6.2 Superposition of *N* Harmonic Waves

Suppose the wave generator at $x = 0$ vibrates at several frequencies, $\omega_{\text{min}} = \omega_1, \omega_2, \ldots, \omega_{\text{max}} = \omega_N$, with equal amplitudes and phases producing the following vibration of the string at $x = 0$:

$$\psi(0,t) = A\cos(\omega_1 t) + A\cos(\omega_2 t) + \cdots + A\cos(\omega_N t). \tag{6.107}$$

The traveling wave will also be a sum of the contributions from each harmonic component, giving us the following wave downstream at x at time t:

$$\psi(x,t) = A\cos(k_1 x - \omega_1 t) + A\cos(k_2 x - \omega_2 t) + \cdots + A\cos(k_N x - \omega_N t), \tag{6.108}$$

where $k_i = \omega_i / v_p$ with v_p the phase velocity. Can we sum the cosines in Eq. 6.107 or 6.108? Rather than attempt the sum for general values of the frequencies, we look at a case where the difference between successive frequencies is the same,

$$\text{Case:} \quad \omega_{i+1} - \omega_i = \epsilon. \tag{6.109}$$

Since there are $N-1$ intervals between ω_{min} and ω_{max}, we have the following relation for the range of frequencies $\Delta\omega$ in the system:

$$\Delta\omega \equiv \omega_N - \omega_1 = (N-1)\epsilon. \tag{6.110}$$

Let us work with Eq. 6.107 since the summation on wave functions in Eq. 6.108 can be obtained in a similar way. In the case given in Eq. 6.109, we will have the following equation to work with:

$$\psi(0,t) = A\cos\omega_1 t + A\cos(\omega_1+\epsilon)t + A\cos(\omega_1+2\epsilon)t + \cdots + A\cos[\omega_1+(N-1)\epsilon]t. \tag{6.111}$$

Finite N

Let us first find the sum when the number of terms in Eq. 6.111 is finite. This summation is more easily done using complex algebra, where we replace $\cos(\theta)$ by $e^{i\theta}$ and take the real part of the final result. The complexified $\psi(0,t) \equiv \psi(t)$ would be

$$\psi(t) = A\,e^{i\omega_1 t} + A\,e^{i\omega_1 t + i\epsilon t} + A\,e^{i\omega_1 t + 2i\epsilon t} + \cdots + A\,e^{i\omega_1 t + i(N-1)\epsilon t}. \tag{6.112}$$

In this expression we find that $Ae^{i\omega_1 t}$ is a common factor to all terms,

$$\psi(t) = A\,e^{i\omega_1 t}\left[1 + e^{i\epsilon t} + e^{2i\epsilon t} + \cdots + Ae^{i(N-1)\epsilon t}\right]. \tag{6.113}$$

Writing a for $e^{i\epsilon t}$ we notice that the quantities in the bracket $[\cdots]$ form a geometric series of a,

$$\psi(t) = A\,e^{i\omega_1 t}\left[1 + a + a^2 + \cdots + a^{(N-1)}\right]. \tag{6.114}$$

Therefore, we get

$$\psi(t) = A\,e^{i\omega_1 t}\frac{a^N - 1}{a - 1} = Ae^{i\omega_1 t}\left[\frac{e^{iN\epsilon t} - 1}{e^{i\epsilon t} - 1}\right]. \tag{6.115}$$

We need the real part of ψ. If the quantity in Eq. 6.115 can be expressed in the polar form of a complex number, $r\,e^{i\theta}$, we can easily figure out the real part. The following trick of factoring out 1/2 of the exponential parts in both the numerator and denominator helps here:

$$\frac{e^{iN\epsilon t} - 1}{e^{i\epsilon t} - 1} = \frac{e^{i(N/2)\epsilon t}}{e^{i(1/2)\epsilon t}}\left[\frac{e^{i(N/2)\epsilon t} - e^{-i(N/2)\epsilon t}}{e^{i(1/2)\epsilon t} - e^{-i(1/2)\epsilon t}}\right]. \tag{6.116}$$

Now, the quantity in the bracket $[\cdots]$ is a real number when we use the Euler formula. Therefore,

$$\frac{e^{iN\epsilon t} - 1}{e^{i\epsilon t} - 1} = e^{i\frac{N-1}{2}\epsilon t}\left[\frac{\sin\left(\frac{N}{2}\epsilon t\right)}{\sin\left(\frac{1}{2}\epsilon t\right)}\right]. \tag{6.117}$$

Putting this in Eq. 6.115 we find the following for ψ:

$$\psi(t) = A\,e^{i\omega_1 t}e^{i\frac{N-1}{2}\epsilon t}\left[\frac{\sin\left(\frac{N}{2}\epsilon t\right)}{\sin\left(\frac{1}{2}\epsilon t\right)}\right]. \tag{6.118}$$

Using Eq. 6.110 we can write this as

$$\psi(t) = A\left[\frac{\sin\left(\frac{N}{2}\epsilon t\right)}{\sin\left(\frac{1}{2}\epsilon t\right)}\right]e^{i\frac{(\omega_N + \omega_1)}{2}t}. \tag{6.119}$$

The real part of this gives

$$\psi(0,t) = Re[\psi(t)] = \left[A\,\frac{\sin\left(\frac{N}{2}\epsilon t\right)}{\sin\left(\frac{1}{2}\epsilon t\right)}\right]\cos\left(\omega_{\text{av}} t\right), \tag{6.120}$$

where ω_{av} is the average of the minimum and maximum frequencies, $\omega_{av} = \frac{1}{2}(\omega_N + \omega_1)$. The quantity in brackets varies much more slowly with time than does function $\cos(\omega_{av} t)$. Therefore, the net vibration is similar to the two-wave case we worked out before, in the sense that a fast oscillation is amplitude modulated. Let us write the quantity in the bracket as $B(t)$, which stands for the slowly varying amplitude,

$$\psi(0,t) = B(t)\,\cos(\omega_{av} t), \quad \text{with} \quad B(t) = A\,\frac{\sin\left(\frac{N}{2}\epsilon t\right)}{\sin\left(\frac{1}{2}\epsilon t\right)}. \tag{6.121}$$

Very Large N

What if the vibration at $x = 0$ were generated by a very large number of frequencies between two frequencies ω_{min} and ω_{max} with equal amplitude and equal phase? In this case, the number of terms in Eq. 6.111 would be very large. Furthermore, since ϵ is the difference between successive frequencies in the range $\Delta\omega = \omega_{max} - \omega_{min}$, when N becomes large ϵ would become small if the range $\Delta\omega$ were fixed. Therefore, we can obtain the sum for this situation by taking a large N and small ϵ limit of the results in Eqs. 6.121. This would be a brute force solution to the problem. Another elegant solution for this case is possible here. Let us multiply each term in the sum for $\psi(0,t)$ by ϵ and divide out by an overall factor of ϵ. We will also multiply and divide by N so that we can cast our result in terms of NA or $B(0)$. Let us write the result using the summation symbol:

$$\psi(0,t) = \frac{AN}{N\epsilon}\sum_{n=0}^{N-1}\cos\left[(\omega_1 + n\epsilon)t\right]\,\epsilon. \tag{6.122}$$

When we take $N \to \infty$ and $\epsilon \to 0$, the integral is just an integration over frequency from ω_{min} to ω_{max},

$$\psi(0,t) = \frac{B(0)}{\omega_{max} - \omega_{min}}\int_{\omega_{min}}^{\omega_{max}}\cos(\omega t)\,d\omega. \tag{6.123}$$

This equation represents the vibrations as an integral over a cosine function and is an example of a Fourier integral, to be discussed below. The integral can be performed exactly and cast as a product of a slowly varying amplitude and a harmonic function of time,

$$\psi(0,t) = \frac{B(0)}{\Delta\omega}\,\frac{(\sin\omega_{max}t - \sin\omega_{min}t)}{t} = \left[B(0)\,\frac{\sin\left(\frac{\Delta\omega}{2}t\right)}{\frac{\Delta\omega}{2}t}\right]\cos(\omega_{av}t), \tag{6.124}$$

where $\omega_{av} = \frac{1}{2}(\omega_{max} + \omega_{min})$ is the average frequency. The quantity in the bracket is the time-dependent amplitude of the vibration at the average frequency. Unlike the case of two frequencies, where the beating of the waves led to a sinusoidal time dependence of amplitude, here the beating of infinitely many waves has led to an amplitude itself decreasing with time due to the t in the denominator. The ratio of $\sin(x)/x$ is denoted by one symbol $\text{sinc}(x)$,

$$\psi(0,t) = \left[B(0)\,\text{sinc}\left(\frac{\Delta\omega}{2}t\right)\right]\cos(\omega_{av}t). \tag{6.125}$$

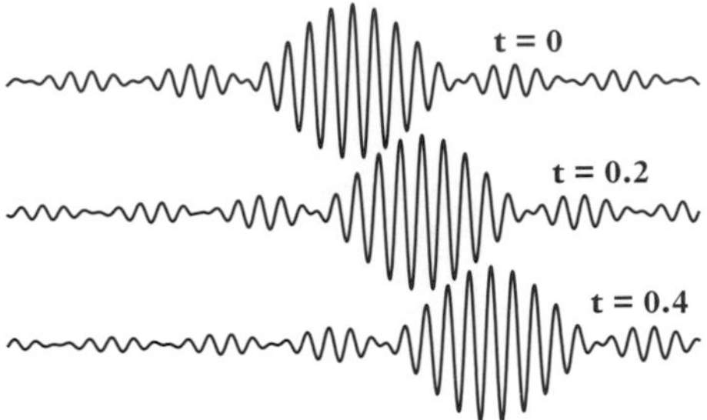

Fig. 6.13 *Movement of the wave resulting from the beating of waves of many frequencies. The plot was made with* $\omega_{av} = 100$, $\Delta\omega = 10$, *and* $v = 1$.

Since the wave equation for this wave has a linear relation between ω and k of the wave, the waves of different frequencies will travel at the same wave velocity, which would mean that the wave packet moves with speed $v = \omega/k$,

$$x = vt.$$

Therefore, the vibration at $x = 0$ will move to x at $t_{ret} = t - x/v$. Therefore, the wave function $\psi(x, t)$ can be obtained from this formula for the vibration at $t = 0$ by substituting $t - x/v$ for t,

$$\psi(x, t) = B(0)\,\text{sinc}\left[\frac{\Delta\omega}{2}\left(t - \frac{x}{v}\right)\right]\cos\left[\omega_{\text{av}}\left(t - \frac{x}{v}\right)\right]. \tag{6.126}$$

A plot of this moving wave, given in Fig. 6.13, shows that the beating of infinitely many *continuous frequencies* leads to a **wave packet**, which is a localized nonperiodic wave. The wave packet serves as a model of a free particle in quantum mechanics.

6.7 Spectrum Analysis of Waves

We found above that superposition of harmonic waves can result in a periodic wave as well as a nonperiodic wave such as a wave packet. Now, if we have an arbitrary wave, could we determine the frequency content of the wave? That is, could we reverse engineer the given wave and figure out the amplitudes, frequencies, and phases of the harmonic waves whose superposition will give rise to the given wave? The answer is provided by the Fourier technique. You have already studied Fourier series in an earlier chapter in connection with periodic vibrations. Fourier series allow one to write a *periodic function* into a series of sines and cosines of frequencies that are integral multiples of the frequency of the periodic function. When the technique is extended to nonperiodic functions, the series turns into integrals, called Fourier integrals. The mathematics of Fourier integrals allows us to study any wave packet. In this section we will review the Fourier series first and then study nonperiodic waves.

6.7.1 Non-harmonic Periodic Waves

Let us once again consider the one-dimensional example of the transverse wave on a string which is placed along the x-axis from $x = 0$ to $x = \infty$, with a vibrator at $x = 0$ which transmits a wave towards the positive x-axis. Now, we suppose that the vibrator oscillates such that the transverse displacement $\psi(x = 0, t)$ at $x = 0$ is an arbitrary periodic function of time t, which is not necessarily sinusoidal,

$$\psi(0, t) = f(t), \quad f(t + T) = f(t). \tag{6.127}$$

Here T is the period. We will find it useful to introduce the fundamental angular frequency ω_1 of the system by

$$\omega_1 = \frac{2\pi}{T}. \tag{6.128}$$

By the **Fourier theorem**, since $f(t)$ is periodic with angular frequency ω_1, it will have the following Fourier series representation:

$$f(t) = A_0 + \sum_{n=1,2,\cdots} A_n \cos(n\omega_1 t) + \sum_{n=1,2,\cdots} B_n \sin(n\omega_1 t), \tag{6.129}$$

where the coefficients are given by

$$A_0 = \frac{1}{T}\int_0^T f(t)\, dt, \tag{6.130}$$

$$A_n = \frac{2}{T}\int_0^T f(t)\cos(n\omega_1 t)\, dt, \tag{6.131}$$

$$B_n = \frac{2}{T}\int_0^T f(t)\sin(n\omega_1 t)\, dt. \tag{6.132}$$

Note that A_0 is a constant background and does not lead to an oscillating wave. The amplitudes A_n and B_n correspond to various frequencies and are a function of the mode counter n. They are called amplitude spectra of the wave. Plots of A_n and B_n versus $\omega_n = n\omega_1$ or n are also referred to as cosine and sine amplitude spectra.

We are often interested in power in various modes. The average power in a vibration given by

$$f(t) = C\cos(\omega t) + S\cos(\omega t),$$

is given by

$$< P >= \frac{1}{2} Z\omega^2 \left(C^2 + S^2\right),$$

where Z is the impedance, which may have frequency dependence. Therefore, the average power in any mode n will be

$$< P_n >= \frac{Z_n}{2} n^2\omega_1^2 \left(A_n^2 + B_n^2\right). \tag{6.133}$$

The average power as a function of mode number or frequency is called the **power spectrum** of the wave. We often plot $< P_n >$ versus n or ω_n to display the relative power contribution of different frequencies. Such a plot is also called the power spectrum of the wave.

By replacing t in Eq. 6.129 by $(t - x/v)$ we get the wave function $\psi(x,t)$ for the wave moving towards the positive x-axis that satisfies the classical wave equation, with v being the speed of the wave,

$$\psi(x,t) = A_0 + \sum_{n=1,2,\cdots} A_n \cos(n\omega_1 t - nk_1 x) - \sum_{n=1,2,\cdots} B_n \sin(n\omega_1 t - nk_1 x), \tag{6.134}$$

with $k_1 = \dfrac{\omega_1}{v}$, where v is the phase velocity.

Example 6.7 Spectrum of a Wave on a String

Find the amplitude spectrum and wave function of the wave traveling down the string if the string is pulsed as given in Fig. 6.14.

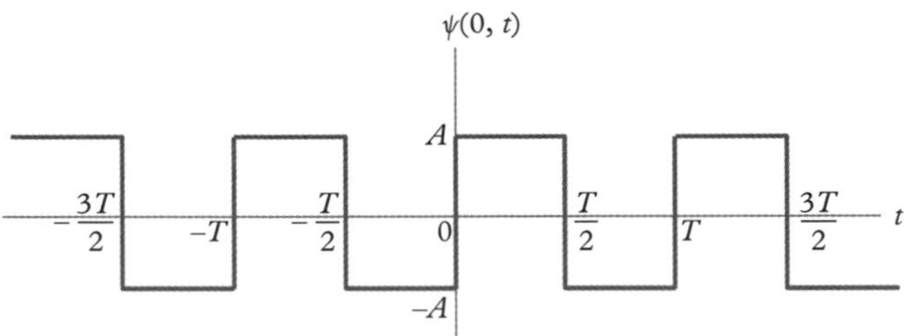

Fig. 6.14 *Periodic square pulses imparted to the string at x = 0.*

Solution

Let us first write the analytic expression corresponding to the vibration at $x = 0$ given in the plot for one period:

$$\psi(0,t) = \begin{cases} -A & -T/2 \leq t < 0 \\ 0 & t = 0 \\ A & 0 < t \leq T/2 \end{cases}. \tag{6.135}$$

Now, we put this into Eqs. 6.130–6.132 and carry, out the integrations quite easily. Since the function in Eq. 6.135 is odd, the cosine coefficients A_n and the constant coefficient A_0 are zero. The sine coefficients are

$$B_n = \frac{2}{T}\left[\int_{-T/2}^{0} (-A)\sin(n\omega_1 t)dt + \int_{0}^{T/2} (-A)\sin(n\omega_1 t)dt\right],$$

where $\omega_1 = 2\pi/T$. This gives the sine amplitude spectrum to be

$$B_n = \frac{2A}{n\pi}\left[1 - \cos(n\pi)\right],\ n = 1, 2, 3, \ldots.$$

Therefore, the sine coefficients of the Fourier series are non-zero only for n an odd integer,

$$B_1 = \frac{4A}{\pi},\ B_2 = 0,\ B_3 = \frac{4A}{3\pi},\ B_4 = 0,\ B_5 = \frac{4A}{5\pi}, \ldots.$$

Therefore, we have the following Fourier series representation of $\psi(0,t)$:

$$\psi(0,t) = \frac{4A}{\pi}\left[\sin(\omega_1 t) + \frac{1}{3}\sin(3\omega_1 t) + \frac{1}{5}\sin(5\omega_1 t) + \cdots\right].$$

Now, we can obtain the wave traveling down the string towards the positive x-axis by replacing $(n\omega_1 t)$ by $(n\omega_1 t - nk_1 x)$ to obtain $\psi(x,t)$,

$$\psi(x,t) = \frac{4A}{\pi}\sum_{n=1,3,\cdots}\frac{\sin(n\omega_1 t - nk_1 x)}{n},$$

where $k_1 = \omega_1/v$ with v the phase velocity.

6.7.2 Nonperiodic Pulses and Fourier Integral Technique

Real waves, such as water waves, a pulse traveling on a string, or the sound of a drum, are not sinusoidal waves. For instance, suppose a taut string is given a snap at one end once and then the end is held fixed in one place, as shown in Fig. 6.15. This results in a bump in the string, which moves towards the other end. Unlike the harmonic wave of a single frequency, the fate of a finite duration pulse depends on whether the medium is dispersive or non-dispersive.

As we have studied previously, the wave speed of a harmonic wave is independent of frequency (or, equivalently, the wavelength) in a non-dispersive medium, but depends on the frequency in a dispersive medium. For instance, the speed of light in a vacuum is the same for all frequencies but in glass it is different for different frequencies. We will see below that a finite-duration pulse can be shown to be a superposition of harmonic waves of infinitely many frequencies. In a dispersive wave each of the component harmonic waves will travel at a different speed and hence they will move to different distances in the same time. Thus, a pulse will be distorted as it moves in a dispersive medium. However, in a non-dispersive medium every component harmonic wave in a pulse will move at the same speed and hence the shape of the pulse will not change, as shown in Fig. 6.15.

Movement of a Pulse in a Non-dispersive Medium

The dynamics of a pulse in a non-dispersive medium is simple since the pulse of the shape will not change with time or distance traveled. Generally, suppose we have a one-dimensional wave with the pulse at time $t = 0$ given by a function $f(x)$. That is, the wave function $\psi(x, t)$ at $t = 0$ is

$$\psi(x, 0) = f(x).$$

Suppose the pulse is moving towards the positive x-axis with speed v. Since the medium is non-dispersive we need only one speed, since every wave of every frequency will move at this speed. In time t, each point of the pulse would have moved a distance

$$\Delta x = vt.$$

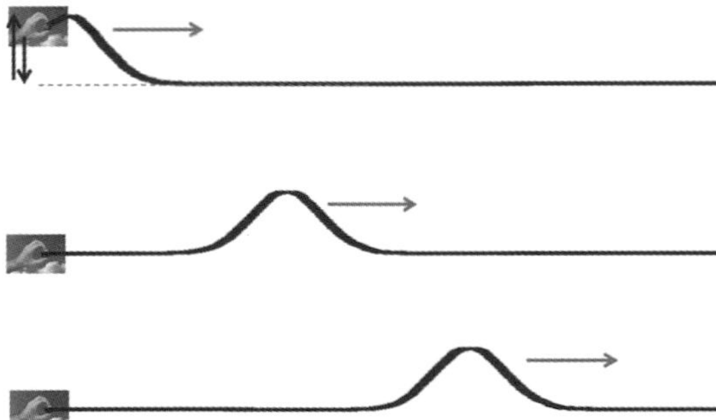

Fig. 6.15 *Generation and movement of a wave pulse on a string.*

Therefore, if the pulse were centered at $x = 0$ at $t = 0$, the wave would be centered at $x = vt$ at time t. Therefore, the wave function at time t will be

$$\psi(x,t) = f(x - vt) \quad \text{(Non-dispersive)}.$$

If the wave were moving towards the negative x-axis, the wave function at time t would be

$$\psi(x,t) = f(x + vt) \quad \text{(Non-dispersive)}.$$

That is,

$$\text{if } \psi(x,0) = f(x), \text{ then } \psi(x,t) = f(x \mp vt). \quad \text{(Non-dispersive)} \tag{6.136}$$

Similarly, suppose a pulse were generated at $x = 0$ by some motion given by

$$\psi(0,t) = g(t).$$

Then, the wave moving towards the positive x-axis would be

$$\psi(x,t) = g\left(t - \frac{x}{v}\right),$$

and the wave moving towards the negative x-axis would be

$$\psi(x,t) = g\left(t + \frac{x}{v}\right).$$

That is,

$$\text{if } \psi(0,t) = g(t), \text{ then } \psi(x,t) = g\left(t \mp \frac{x}{t}\right). \quad \text{(Non-dispersive)} \tag{6.137}$$

Example 6.8 Pulse in a Non-dispersive Medium

A taut string is fixed between two posts that are far away from a pulse generator. Let the string be along the x-axis with the generator driving a pulse of finite duration. For simplicity, let the pulse be of the shape shown in Fig. 6.16, which is generated by pushing the string towards the positive y-axis uniformly for a duration $\Delta t = \tau$ and then pulling the string at $x = 0$ back to $y = 0$ and keeping it there from then on. Two waves are generated, one traveling towards $x = -\infty$ and the other towards $x = \infty$, both at speed v.

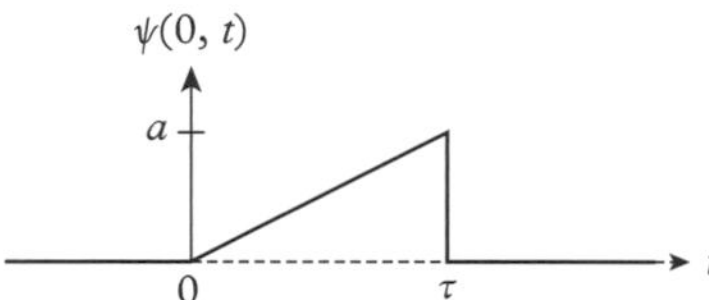

Fig. 6.16 *Displacement at $x = 0$ caused by a pulse generator.*

(1) Draw the shape of the pulse at the following instants $t = \tau$, and an arbitrary instant $t > \tau$.

(2) What is the analytic expression of the displacement at $x = 0$ as a function of time t?

(3) What is the expression for the wave function $\psi(x,t)$?

Solution

(1) Figure 6.17 shows the two waves generated by the wave generator at $x = 0$. The tips of the pulses on both sides are moving away from the origin and at time t, they reach $\pm vt$. The displacement $\psi = a$ is generated at instant $t = \tau$, which moves to a distance $v(t-\tau)$ by the instant t.

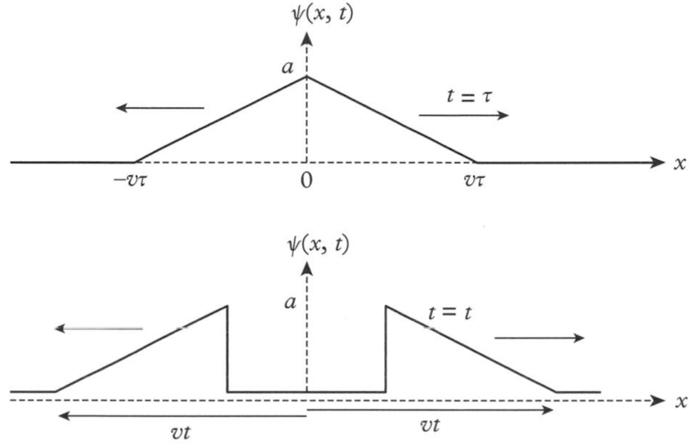

Fig. 6.17 *The waves at $t = \tau$ and an arbitrary time t.*

(2) The analytic expression of the pulse generated at $x = 0$ given in Fig. 6.16 will have three pieces:

$$\psi(0,t) = \begin{cases} 0 & -\infty < t < 0 \\ \dfrac{a}{\tau}t & 0 \le t \le \tau \\ 0 & t > \tau \end{cases} . \tag{6.138}$$

(3) Since we know the pulse at $x = 0$ and the speed of the pulse v, we can replace t by $t - x/v$ to obtain the pulses at the right and at the left,

$$\psi(x > 0, t) = \begin{cases} 0 & -\infty < t - \dfrac{x}{v} < 0 \\ \dfrac{a}{\tau}\left(t - \dfrac{x}{v}\right) & 0 \le \left(t - \dfrac{x}{v}\right) \le \tau \\ 0 & \left(t - \dfrac{x}{v}\right) > \tau \end{cases} ,$$

$$\psi(x < 0, t) = \begin{cases} 0 & -\infty < t + \dfrac{x}{v} < 0 \\ \dfrac{a}{\tau}\left(t + \dfrac{x}{v}\right) & 0 \le \left(t + \dfrac{x}{v}\right) \le \tau \\ 0 & \left(t + \dfrac{x}{v}\right) > \tau \end{cases} .$$

Sometimes, it is helpful to re-write these wave functions with the pieces in x. For the $x > 0$ part this will work out to be

$$\psi(x > 0, t) = \begin{cases} 0 & vt < t < \infty \\ -\dfrac{a}{v\tau}(x - vt) & v(t-\tau) \le x \le vt \\ 0 & 0 < x < v(t-\tau) \end{cases} .$$

continued

Example 6.8 *continued*

To arrive at this, I have made use of the following:

$$-\infty < t - \frac{x}{v} < 0 \implies -t - \infty < -\frac{x}{v} < -t \implies \infty > x > vt,$$

$$0 \leq t - \frac{x}{v} < \tau \implies -t \leq -\frac{x}{v} < -t + \tau \implies vt \geq x \geq v(t - \tau),$$

$$\left(t - \frac{x}{v}\right) > \tau \implies 0 < x < v(t - \tau).$$

The student should work out the $x < 0$ part similarly.

Fourier Integrals and Frequency Content of Waves

The movement of a pulse in a dispersive medium is done by expressing the finite pulse in terms of the frequency content of the pulse, evolving each harmonic component according to its speed, and then recovering the pulse at an arbitrary instant by combining the harmonic components.

To find the harmonic content of a pulse we need Fourier analysis. The Fourier analysis of a periodic pulse is in terms of discrete frequencies. But, the Fourier analysis of a pulse requires continuous frequencies. Thus, the Fourier representation of a nonperiodic pulse will be in terms of integrals over sine and cosine functions rather than as a series of discrete terms of sines and cosines. This representation of a nonperiodic function is called the Fourier integral technique. In this technique, the sine and cosine coefficients are continuous functions of frequency and the representation is in terms of integrals. In this section we will illustrate the method of Fourier integrals by applying the method to a few examples of wave pulses.

According to the **Fourier integral theorem**, the Fourier representation of a piece-wise smooth function $f(t)$ over the domain $-\infty < t < \infty$ is given by the following **Fourier integrals** in terms of frequencies $0 \leq \omega < \infty$:

$$f(t) = \int_0^\infty A(\omega)\cos(\omega t)d\omega + \int_0^\infty B(\omega)\sin(\omega t)d\omega, \tag{6.139}$$

where the cosine and sine coefficients are related to the function $f(t)$ by the **inverse Fourier integrals**,

$$A(\omega) = \frac{1}{\pi}\int_{-\infty}^{\infty} f(t)\cos(\omega t)dt, \tag{6.140}$$

$$B(\omega) = \frac{1}{\pi}\int_{-\infty}^{\infty} f(t)\sin(\omega t)dt. \tag{6.141}$$

The function $A(\omega)$ is called the **cosine transform** of $f(t)$ and function $B(\omega)$ the **sine transform**. The functions $A(\omega)$ and $B(\omega)$ give us the frequency content of the disturbance given by $f(t)$. Fourier trasforms are also defined using complex exponentials as in Eqs. 6.166 and 6.177 below.

The width $\Delta\omega$ of the distribution of $A(\omega)$ and $B(\omega)$ describes a **frequency bandwidth**. We will find in our examples that often these Fourier transforms have a peak at some frequency and are significant within a range of frequencies. In that case the bandwidth is easily obtained. Furthermore, if the pulse is spread over a time interval Δt, then the bandwidth $\Delta\omega$ is inversely proportional to Δt. That is, in general,

$$\Delta\omega = \frac{a}{\Delta t}, \tag{6.142}$$

with a a numerical factor which depends on the details of the pulse.

Frequency Content of a Square Pulse

Suppose a square pulse is generated by a vibrator attached to a taut string at $x = 0$ and the wave is sent towards the positive x-axis. For the sake of getting only the cosine coefficient to be non-zero, we assume the pulse to be symmetric about $t = 0$, as shown in Fig. 6.18. Let the pulse have displacement D from the equilibrium and last for a time Δt,

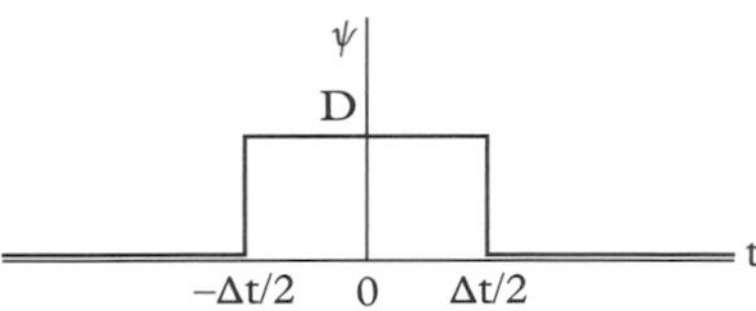

Fig. 6.18 *A square pulse generated during $-\Delta t/2 \le t \le \Delta t/2$ at $x = 0$ is sent towards the positive x-axis.*

$$\psi(0,t) = \begin{cases} 0 & -\infty < t < -\Delta t/2 \\ D & -\Delta t/2 \le t \le \Delta t/2\ . \\ 0 & \Delta t/2 < t < \infty \end{cases} \tag{6.143}$$

Since the function is an even function, we will have $B(\omega) = 0$ and $A(\omega)$ will be,

$$A(\omega) = \frac{1}{\pi}\int_{-\Delta t/2}^{\Delta t/2} D\cos(\omega t)dt = \frac{2D}{\pi}\left[\frac{\sin(\omega\Delta t/2)}{\omega}\right]. \tag{6.144}$$

The quantity within the brackets $[\cdots]$ contains the frequency spectrum of the cosine transform. We can multiply and divide the expression on the right by $\Delta t/2$ so that the quantity within the bracket can be written as a sinc function that we have encountered before,

$$A(\omega) = \frac{D\Delta t}{\pi}\left[\frac{\sin(\omega\Delta t/2)}{\omega\Delta t/2}\right] = \frac{D\Delta t}{\pi}\ \text{sinc}\,(\omega\Delta t/2)\ . \tag{6.145}$$

A plot of $A(\omega)$ is shown in Fig. 6.19. The plot shows that the most important frequencies are in the range $\omega = 0$ to around $\omega = 2\pi/\Delta t$. The frequency bandwidth can be defined from this plot to be

$$\Delta\omega = \frac{2\pi}{\Delta t}. \tag{6.146}$$

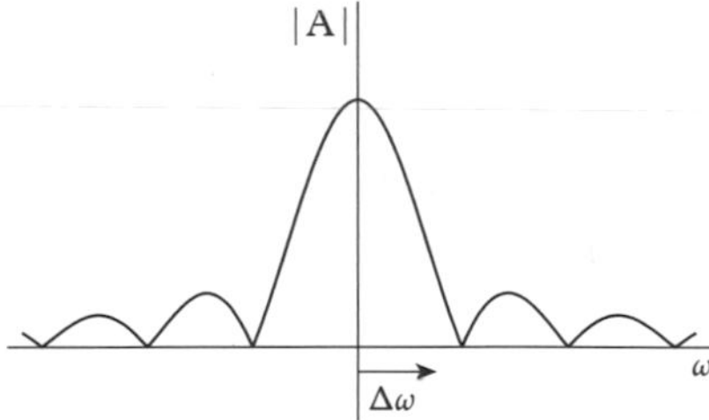

Fig. 6.19 *A plot of $|A(\omega)|$ versus ω shows the frequency bandwidth $\Delta\omega$. Note: $|A| = 0$, when $\omega\Delta t = 2\pi$.*

Using the expression for $A(\omega)$ and with $B(\omega) = 0$ we find that the square pulse spread symmetrically about $t = 0$ has the following Fourier representation:

$$\psi(0,t) = \int_0^\infty \frac{D\Delta t}{\pi}\ \text{sinc}\,(\omega\Delta t/2)\ \cos(\omega t)d\omega. \tag{6.147}$$

If you carry out this integral you will obtain the piece-wise function given in Eq. 6.143. Since the wave on a strings is a non-dispersive wave, each frequency component wave will travel with the same speed. Therefore, the wave generated at $x = 0$ at time $t = 0$ will reach point $x = x$ at time $t_{ret} = t - x/v$, where v is the phase velocity. This implies that the wave function at x will be

$$\psi(x,t) = \int_0^\infty \frac{D\Delta t}{\pi}\ \text{sinc}\,(\omega\Delta t/2)\ \cos[\omega(t - x/v)]d\omega. \tag{6.148}$$

If the wave were non-dispersive, we would still have this expression, except that v on the right side would be the function $v(\omega)$.

Frequency Content of a Wavetrain

Suppose that the vibrator in Fig. 6.6 oscillates for a duration $-\Delta t/2 < t < \Delta t/2$ at a fixed frequency ω_0 and then stops. How would the disturbance so generated travel in the string? Let the wave created on the string be sinusoidal, of frequency ω_0, as given in Fig. 6.20, but lasting for only a limited time. These waves are called **wavetrains**. For the sake of simplicity in calculations, let us assume that the wavetrain is an even function of time t lasting for some duration Δt,

$$\psi(0,t) = \begin{cases} 0 & -\infty < t < -\Delta t/2 \\ \psi_0 \cos(\omega_0 t) & -\Delta t/2 \le t \le \Delta t/2 \; . \\ 0 & \Delta t/2 < t < \infty \end{cases} \qquad (6.149)$$

Will this vibration produce a harmonic vibration of frequency ω_0? This is a tricky question since, if the wavetrain went through a very large number of cycles, the waves from these vibrations would be very similar to a harmonic wave. However, if the wavetrain lasts for only a few cycles, then the wave is not going to be a harmonic wave, even though it might appear harmonic were you to look at the wiggles only. Instead, a finite wavetrain given in Fig. 6.20 consists of many frequencies centered about ω_0, as we will see below when we calculate the Fourier transform.

Since the function $\psi(0,t)$ is an even function (our choice) the sine transform will be zero and the cosine transform will be

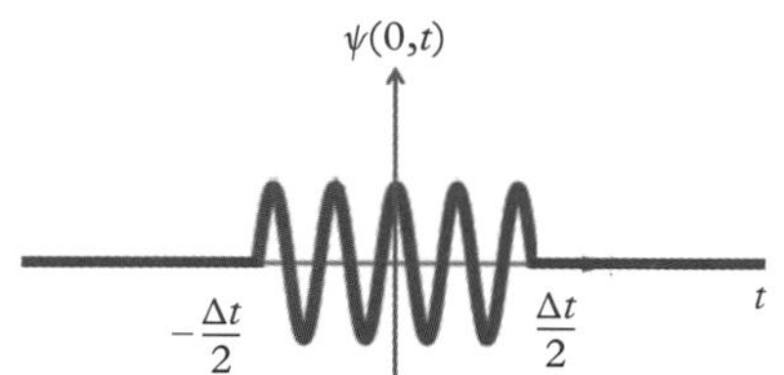

Fig. 6.20 *A wavetrain containing five periods. Such a wavetrain can be generated by oscillating one end of a string for five full cycles.*

$$A(\omega) = \frac{1}{\pi}\int_{-\infty}^{\infty} \psi(0,t)\,\cos(\omega t)\,dt = \frac{\psi_0}{\pi}\int_{-\Delta t/2}^{\Delta t/2} \cos(\omega_0 t)\cos(\omega t)\,dt. \qquad (6.150)$$

We can express the product of cosines as the sum of cosines using the following trigonometric identity:

$$2\cos A \cos B = \cos(A+B) + \cos(A-B). \qquad (6.151)$$

Then, we can integrate the two cosine terms to obtain the following expression for the cosine transform:

$$A(\omega) = \frac{\psi_0 \Delta t}{2\pi}\left[\mathrm{sinc}\{(\omega_0+\omega)\Delta t/2\} + \mathrm{sinc}\{(\omega_0-\omega)\Delta t/2\}\right], \qquad (6.152)$$

where

$$\mathrm{sinc}(x) = \frac{\sin x}{x}.$$

The first term here will be peaked at $\omega = -\omega_0$ and the second at $\omega = \omega_0$. Within a width of about $4\pi/\Delta t$, the two peaks decay away rapidly. Since we are only interested in positive frequencies, the contribution of the first term in the positive frequency part will be negligible. Therefore, we can neglect the first term and we obtain a simpler expression,

$$A(\omega) = \frac{\psi_0 \Delta t}{2\pi}\,\mathrm{sinc}\left[(\omega_0-\omega)\,\Delta t/2\right]. \qquad (6.153)$$

From the zeros of the sinc function we can assign a width to the central peak, which shows that significant contributions are in the approximate range $\omega_0 - 2\pi/\Delta t$ and $\omega_0 + 2\pi/\Delta t$, giving the bandwidth to be

$$\Delta\omega = \frac{4\pi}{\Delta t}. \tag{6.154}$$

Using $B = 0$ and $A(\omega)$ obtained here, we can write the wave at $x = x$ by substituting t in $\psi(0, t)$ by $t - x/v$,

$$\psi(x, t) = \int_0^\infty A(\omega) \cos(\omega[t - x/v]) d\omega. \tag{6.155}$$

Since, the energy transported by a harmonic wave is proportional to the square of the amplitude of the harmonic wave, the power flow at any point will also be proportional to the square of the amplitude, as given by

$$P(x, t) = Z \left(\frac{\partial \psi}{\partial t} \right)^2. \tag{6.156}$$

Averaging over time we find that the average power in the range ω and $\omega + \Delta\omega$ is given by

$$\langle P(\omega) \rangle = \frac{Z}{2} \omega^2 A(\omega)^2. \tag{6.157}$$

Figure 6.21 displays the power spectrum for three different wavetrains of durations 5 periods, 10 periods, and 100 periods. It is clear that the 5-period wavetrain contains significant amounts of other frequencies. It is also clear that longer wavetrains have lesser amounts of other frequencies. The wave with 100 periods is sharply peaked about the frequency of the oscillation that generated the wavetrain.

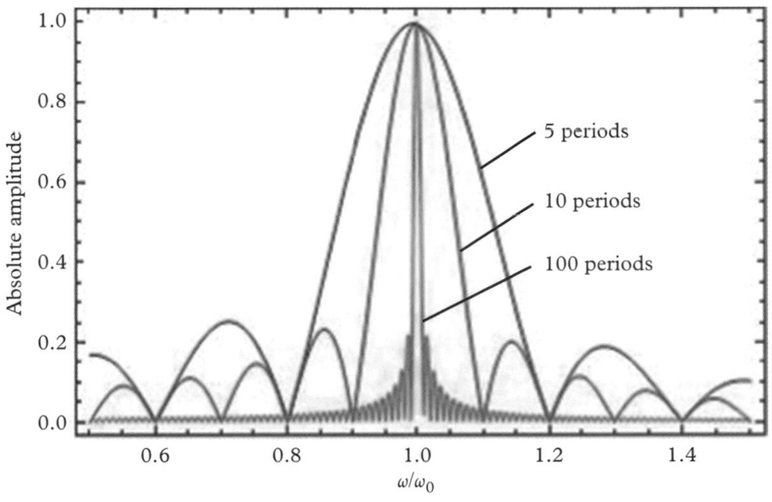

Fig. 6.21 *The frequency content of wavetrains. The intensity of finite sinusoid wavetrains of 5 periods, 10 periods, and 100 periods are plotted against frequency in terms of the frequency of the sine. Duration of the pulse Δt and the width of the frequency spectrum ω are inversely related, $\Delta\omega \sim \frac{1}{\Delta t}$. That is, if the wavetrain lasts longer, then it contains fewer frequencies around the central frequency.*

In the two examples above we have done Fourier analysis in time–frequency (t, ω) variables. Since time and space are linked in a wave motion by the argument $t - x/v$ we can equally well perform the calculations in space–wave number variables (x, k). Similarly to the Fourier transform of $f(t)$ in terms of frequency ω, we have a Fourier transform of a function $f(x)$ in terms of wave number k, since wave number is the spatial frequency, counting the number of wavelengths in a distance 2π [meters] if wavelength is expressed in meters just as ω is the number of cycles in a time 2π [seconds] if the period is expressed in seconds. The Fourier and inverse Fourier pairs of $f(x)$ can be written as

$$f(x) = \int_0^\infty A(k)\cos(kx)dk + \int_0^\infty B(k)\sin(kx)dk, \tag{6.158}$$

where the cosine and sine coefficients are related to the function $f(x)$ by the inverse Fourier integrals:

$$A(k) = \frac{1}{\pi}\int_\infty^\infty f(x)\cos(kx)dt, \tag{6.159}$$

$$B(k) = \frac{1}{\pi}\int_\infty^\infty f(x)\sin(kx)dt. \tag{6.160}$$

Given the wave profile $\psi(x, 0)$ at $t = 0$ we can determine its cosine and sine transforms $A(k)$ and $B(k)$. This gives us the Fourier representation of the wave profile. The traveling wave at time t would be obtained by replacing x by $x - vt$ if the wave is traveling towards the positive x-axis,

$$\psi(x, t) - \int_0^\infty A(k)\cos(k[x - vt])dk + \int_0^\infty B(k)\sin(k[x - vt])dk, \tag{6.161}$$

where the phase velocity $v = \omega/k$ may be a function of k if the wave is non-dispersive.

Frequency Content of a Gaussian Pulse

In the two examples above we have done Fourier analysis in time–frequency (t, ω) variables. Since time and space are linked in a wave motion by the argument $t - x/v$ we can equally well perform the calculations in space–wave number variables (x, k). As an example of an analysis in the (x, k) pair, consider a wave pulse that has a Gaussian shape in the x-coordinate centered at $x = 0$ at $t = 0$, as shown in Fig. 6.22. We will obtain the frequency content of this pulse and its movement towards the positive x-axis.

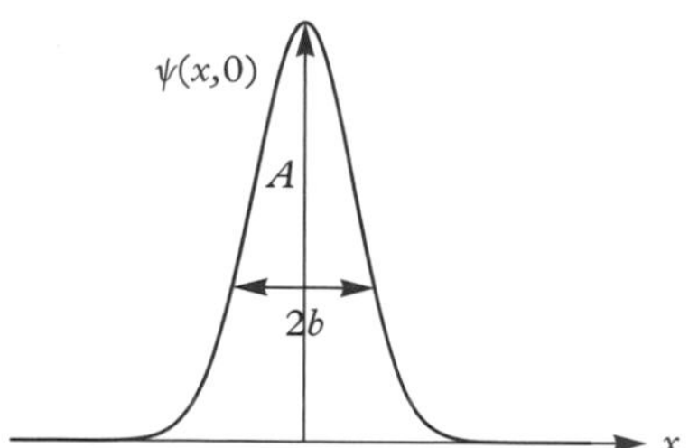

Fig. 6.22 *A Gaussian pulse symmetric about $x = 0$ is given by $\psi(x) = \psi_0\, e^{-x^2/b^2}$, where A is the height at the center and $2b$ is the width of the pulse when $\psi(x)$ is $\frac{1}{e} \times A$.*

Let the peak value of the displacement $\psi(x, 0)$ be ψ_0 and the width of the pulse be b. The wave function at this instant $\psi(x, 0)$ will be given by a Gaussian function of x,

$$\psi(x, 0) = \psi_0\, e^{-x^2/b^2}. \tag{6.162}$$

Similarly to the Fourier transform of $f(t)$ in terms of frequency ω, we have a Fourier transform of a function $f(x)$ in terms of wave number k since the wave number is a spatial frequency, counting the number of wavelengths in a distance 2π [meters] if wavelength is expressed in meters, just as ω is the number of cycles in a time 2π [seconds] if the period is expressed in seconds. The Fourier and inverse Fourier pairs of $f(x)$ can be written as

$$f(x) = \int_0^\infty A(k)\cos(kx)dk + \int_0^\infty B(k)\sin(kx)dk, \tag{6.163}$$

where the cosine and sine coefficients are related to function $f(x)$ by the inverse Fourier integrals,

$$A(k) = \frac{1}{\pi}\int_{\infty}^{\infty} f(x)\cos(kx)dt, \tag{6.164}$$

$$B(k) = \frac{1}{\pi}\int_{\infty}^{\infty} f(x)\sin(kx)dt. \tag{6.165}$$

To understand the evolution and traveling of this pulse in time, we first need to find out how much of each normal mode is contained in the pulse. Since the pulse is given in space, we will conduct the Fourier analysis in space, i.e. for the variable x. This is similar to the wavetrain problem where we performed the Fourier analysis with respect to t.

Rather than perform the sine and cosine transforms we will do the calculations in complex notation. In complex notation we write the Fourier integral pair as $[\psi(x), \Psi(k)]$ as follows:

$$\psi(x) = \frac{1}{\sqrt{2\pi}}\int_{-\infty}^{\infty} \Psi(k)\, e^{ikx}\, dk, \tag{6.166}$$

$$\Psi(k) = \frac{1}{\sqrt{2\pi}}\int_{-\infty}^{\infty} \psi(x)\, e^{-ikx}\, dx, \tag{6.167}$$

where $1/\sqrt{2\pi}$ is included to make the formulas symmetric in the calculations. The Fourier transform of $\psi(x,0)$ given in Eq. 6.162 gives

$$\Psi(k,0) = \frac{1}{\sqrt{2\pi}}\int_{-\infty}^{\infty} \psi(x,0)\, e^{-ikx}\, dx = \frac{bA}{\sqrt{2}}\, e^{-b^2k^2/4}. \tag{6.168}$$

The quantity $\Psi(k,0)$ gives the contribution of the mode with wave number k to the Gaussian pulse. Here, it turns out that $\Psi(k,0)$ is a Gaussian function of k similarly to the way $\psi(x,0)$ is a Gaussian function of x. We say that $\Psi(k,0)$ is a Gaussian in the k-space. The width of $\Psi(k,0)$ in the k-space tells us how many different k are contained in $\psi(x,0)$. From Eq. 6.162 we see that $2b$ is a measure of the width of the pulse in real space, which is the width along the x-axis, also called the x-space,

$$\Delta x = 2b. \tag{6.169}$$

From Eq. 6.168 we deduce that $\frac{1}{b}$ is the width in the k-space,

$$\Delta k = \frac{1}{b}. \tag{6.170}$$

We note that the width in the x-space and width in the k-space are inversely related, similarly to the relation between the width in ω and t found when we studied the wavetrain in time,

$$\Delta k = \frac{2}{\Delta x}, \quad \text{or,} \quad \Delta k \sim \frac{1}{\Delta x}. \tag{6.171}$$

Now, from the known mode content $\Psi(k, 0)$ at time $t = 0$, we can find the mode content at a later time t by evolving each mode in time according to its corresponding frequency,

$$\Psi(k, t) = \Psi(k, 0)\, e^{-i\omega t}, \tag{6.172}$$

where ω depends on k through the dispersion relation of waves on the medium, which for a taut string was found to be

$$\omega = vk, \tag{6.173}$$

where $v = \sqrt{T/\mu}$, and where T is the tension in the string and μ is the mass per unit length. Replacing ω in Eq. 6.172 by vk we obtain the following for the Fourier transform of wave function at time t:

$$\Psi(k, t) = \Psi(k, 0)\, e^{-ivkt}, \tag{6.174}$$

which can be transformed back to find the wave function at an arbitrary time t,

$$\psi(x, t) = \frac{1}{\sqrt{2\pi}} \int_{-\infty}^{\infty} \Psi(k, t)\, e^{ikx}\, dk, \tag{6.175}$$

$$= A\, e^{-\left(\frac{x - vt}{b}\right)^2}, \tag{6.176}$$

which shows that the Gaussian wave travels on the string towards the positive x-axis with velocity v and maintains its shape. The maintenance of shape in the present example is possible because every mode travels at the same speed v, which is a direct consequence of the dispersion relation $\omega^2 = v^2 k^2$.

6.8 Doppler Effect

So far in our discussion, there has been no motion of the detector or the source of the wave with respect to the medium. Hence, in all the foregoing studies, we have assumed that the frequency of the wave detected is the same as that of the source. However, if either the source or the detector moves with respect to the medium, there will be a difference between the frequency detected at the detector and that produced at the source. This effect is called the **Doppler effect**, named after the Austrian physicist Christian Johann Doppler (1803–1853), who in 1842 proposed the effect in his book entitled, *Uber das farbige Lict der Dopperlterne*, which means "the colored light of double stars."

The Doppler effect also provides an explanation for the high pitch of an incoming train whistle and low pitch for a train moving away. To be sure, the incoming train whistle also appears louder since the whistle is blown in the direction of the platform, but here we are not concerned with the loudness, rather the pitch or frequency. This is also the reason why the spectrum of light is shifted to the lower frequency, the so-called red-shift, if a star or a galaxy is moving away.

6.8.1 Non-relativistic Doppler Effect

The formula for the Doppler effect depends on whether any of the speeds involved are relativistic. Let us first consider the case of speeds that are much smaller than the speed of light. The non-relativistic result obtained here will not work for a light wave, but will be quite adequate for most other waves.

To understand the Doppler effect we will first consider two instances, one in which the source is stationary with respect to the medium while the detector moves at a constant velocity, and the other in which the detector is stationary with respect to the medium and the source moves with a constant velocity. Then we will combine the results to figure out what would happen when both the source and the detector move with respect to the medium. To keep it simple the relative motion will be kept along the line joining the source and the detector, which will be taken to coincide with the x-axis, with the positive x-axis pointing from detector to source.

Moving Detector and Stationary Source

Let v_D be the speed of the detector while the source is fixed with respect to the medium, and v be the speed of the wave in the still medium (Fig. 6.23). Note that if the detector is moving away from the source with a speed greater than the speed v of the wave in the medium, then the wave will never catch up. Therefore, we will impose the restriction that if the detector is moving away from the source, its speed be less than the wave velocity,

$$v_D < v \text{ if detector moving away.}$$

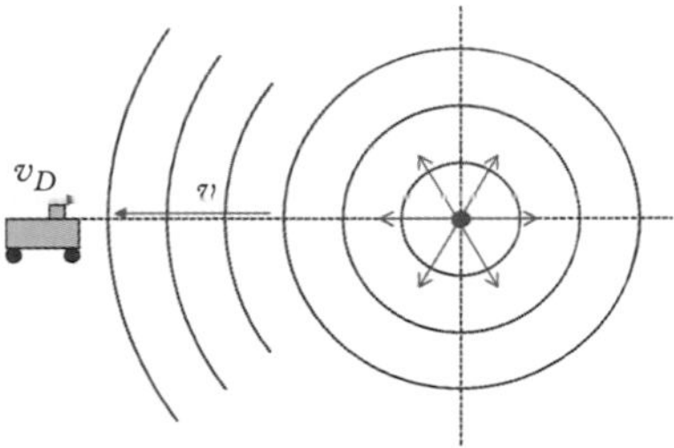

Fig. 6.23 *Doppler effect when the detector moves while the source remains stationary. When the detector is moving towards the stationary source, the detector encounters the wavefronts with greater frequency and therefore it will register a higher frequency for the wave. If the detector were to move away from the source, the wave fronts would reach the detector over longer time intervals and hence the detector would register a smaller frequency.*

We do not need any restriction if the detector is moving towards the source. Let f_0 be the frequency of the wave emitted by the source. That is, there will be a new wave front every $1/f_0$ second. Wavefronts from the source will travel towards the detector at speed v with the distance between the wavefronts equal to $\lambda = v/f_0$, which is the wavelength λ.

For non-relativistic speeds, e.g. v and $v_D \ll c$, the speed of light in a vacuum, the speed of the detector with respect to the moving wavefront will be $v + v_D$ if moving towards the source, and $v - v_D$ if moving away from the source. Hence, the distance λ between the wavefronts will be covered by the detector in a different time than $1/f_0$. Let T_D be the time the detector encounters the successive wavefronts. Then T_D will be

$$T_D = \frac{\lambda}{v \pm v_D} \begin{cases} + \text{ when detector moves towards the source} \\ - \text{ when detector moves away from the source.} \end{cases} \quad (6.177)$$

The frequency of the wave detected by the detector will be the inverse of this time period. Therefore, the frequency detected by the moving detector will be

$$f = \frac{1}{T_D} = \left(\frac{v \pm v_D}{v}\right) f_0 \begin{cases} + \text{ when detector moves towards the source} \\ - \text{ when detector moves away from the source.} \end{cases} \quad (6.178)$$

Thus, when the detector moves towards the stationary source, the frequency at the detector is higher than that produced by the source, and when the detector is moving away from the stationary source, the frequency detected is lower than that produced. This explains why the pitch appears higher if you run towards a stationary horn and lower if you run away from the stationary horn.

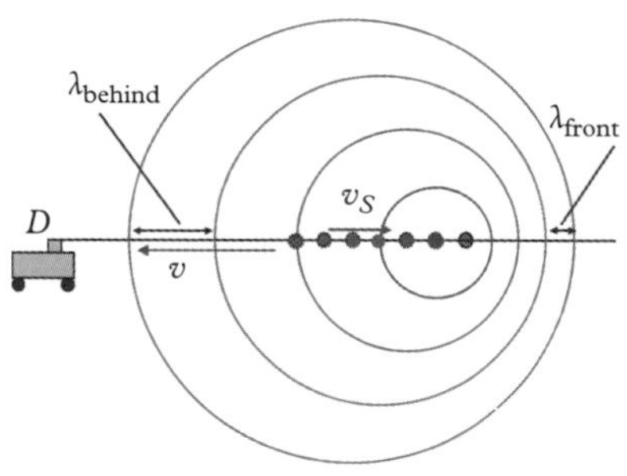

Fig. 6.24 *Doppler effect when the detector is stationary while the source is moving. The source sends out wavefronts spaced equally in time at the rate of $1/f_0$ per wavefront. But, these waves are centered at different places because they are generated at the instantaneous position of the source, which is changing with time. The distance between the wavefronts in front would be less than the distance behind the wavefront. The wavefronts travel at the same speed in the medium regardless of the state of motion of the source. This means that the detector behind the source will register longer time periods or lower frequencies and a detector in front of the source will show higher frequency with respect to the frequency with which the waves were emitted at the source.*

Moving Source and Stationary Detector

Consider a source of a wave that moves with speed v_S with respect to the medium in which the wave travels at speed v while the detector is fixed in place with respect to the medium, as shown in Fig. 6.24.

Even though the source is moving, it still emits waves at the same frequency f_0. Waves move out in expanding spheres centered at the instantaneous location of the source, which is changing in time due to the movement of the source. Hence, to an observer stationary with the medium, the wavefronts will be spaced differently on the two sides of the source. In front of the source, they will appear more closely spaced and behind, more widely separated, because as the source moves, it reduces the distance between a previously emitted wavefront and itself before laying the next wavefront's source point. Therefore, the wavelength will be different in front of the source than behind it,

$$\lambda = \begin{cases} (v - v_S)/f_0 & \text{in front of the source, } v_S < v \\ (v + v_S)/f_0 & \text{behind the source.} \end{cases} \tag{6.179}$$

Since the speed of the wave is v in the medium, the frequency observed at the detector will be v/λ,

$$f = \left(\frac{v}{v \pm v_S}\right) f_0 \begin{cases} + & \text{behind the source} \\ - & \text{in front of the source, } v_S < v. \end{cases} \tag{6.180}$$

Both Source and Detector Moving

It is quite common for both the source and detector to move with respect to the medium. In that case we can imagine a stationary frame of reference in the medium, and then perform a two-step process. First we find the frequency in the imagined stationary frame using the source-moving transformation, and then find the frequency detected by the detector-moving transformation. Let f_0 be the frequency of the source, and v_S, v_D, and v the speeds of the source, the detector, and wave with respect to the stationary medium. Let f_M be the frequency in the imagined frame stationary with respect to the medium, and f be the frequency observed in the detector.

The following two-step process will give the frequency observed by the detector in terms of the frequency produced at the source:

$$f_M = \left(\frac{v}{v \pm v_S}\right) f_0 \begin{cases} + & \text{behind the source} \\ - & \text{in front of the source, } v_S < v, \end{cases} \tag{6.181}$$

$$f = \left(\frac{v \pm v_D}{v}\right) f_M \begin{cases} + & \text{when detector moves towards the source} \\ - & \text{when detector moves away from the source.} \end{cases} \tag{6.182}$$

We can combine these equations to obtain one equation giving us the frequency detected to the frequency produced, as long as we remember the meaning of the different signs,

$$f = \left(\frac{v \pm v_D}{v \pm v_S}\right) f_0. \tag{6.183}$$

6.8.2 Relativistic Doppler Effect

The non-relativistic Doppler effect shows that the effect is different when the source is moving or when the observer is moving. Thus, using the non-relativistic Doppler effect it is possible to tell which one of the two frames, the source frame or the observer frame, is moving even though they may be moving at uniform velocity with respect to the medium. This would not happen if you did not have a medium which takes part in the motion of waves.

In the case of light waves no medium is necessary. When we work out the frequency of light observed by a detector the shift in frequency is the same whether the source is moving or the observer is moving—the effect depends only on the relative motion of the source and the observer. Once again, let f_0 be the frequency of the emitted light in the rest frame of the emitter and f the frequency of the light detected, regardless of the state of motion of the detector. Then, we will find below that the Doppler effect, now called the **relativistic Doppler effect**, give the following relation:

$$f = f_0\sqrt{\frac{1-V/c}{1+V/c}} \quad \begin{cases} \text{Observer receding: } V > 0 \\ \text{Observer approaching: } V < 0. \end{cases} \tag{6.184}$$

To derive this result we will look at a plane wave source traveling along the x-axis. To study a moving source we will place the source at the origin of the S' frame which moves relative to the S frame towards the positive x-axis with speed V as before, as shown in Fig. 6.25.

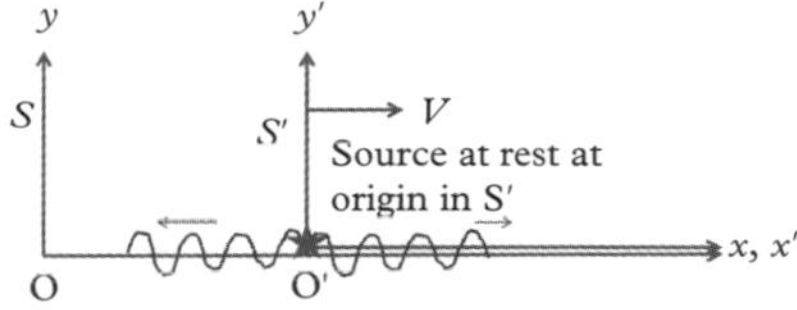

Fig. 6.25 *To calculate the Doppler effect we place the source of the sinusoidal wave at rest at the origin of the S' frame.*

The wave emitted by the source will be traveling towards the negative x-axis to reach the detector at the origin of the S frame, which is to the left of the origin of the source at the origin of the S' frame for $t > 0$. Therefore, the wave at the detector has the following wave function at points on the x'-axis in the S' frame:

$$\psi'(x', t') = \psi_0' \cos\left(\frac{2\pi}{\lambda_0}x' + 2\pi f_0 t'\right). \tag{6.185}$$

The Lorentz transformation gives the following relations among the coordinates of S and S' frames:

$$t' = \gamma\left(t - \frac{V}{c^2}x\right), \quad x' = \gamma(x - Vt), \quad y' = y, \quad z' = z, \tag{6.186}$$

where $\gamma = 1/\sqrt{1 - V^2/c^2}$.

Now, we express x' and t' in Eq. 6.185 in terms of x and t and collect terms in the argument of the cosine as a term multiplying x and another term multiplying t,

$$\psi'(x, t) = \psi_0' \cos\left[2\pi\gamma\left(\frac{1}{\lambda_0} - \frac{Vf_0}{c^2}\right)x + 2\pi\gamma\left(f_0 - \frac{V}{\lambda_0}\right)t\right]. \tag{6.187}$$

In the S frame the wave will be represented by

$$\psi(x, t) = \psi_0 \cos\left(\frac{2\pi}{\lambda}x + 2\pi f t\right). \tag{6.188}$$

We can ignore the amplitudes and look at the phase to determine the frequency of the wave in the S frame in relation to the frequency of the same wave in the S' frame. Equating the factor that multiplies $2\pi t$ in the argument of the cosine function in Eqs. 6.187 and 6.188 gives the desired relation of the frequency detected in the lab (S) frame in terms of the frequency in the rest (S') frame,

$$f = \gamma \left(f_0 - \frac{V}{\lambda_0} \right). \tag{6.189}$$

We can write the right side of this equation in terms of f_0 by noting that $f_0\lambda_0 = c$, where c is the speed of light. Thus,

$$f = f_0 \sqrt{\frac{1 - V/c}{1 + V/c}} \quad \begin{cases} \text{Observer receding: } V > 0 \\ \text{Observer approaching: } V < 0. \end{cases} \tag{6.190}$$

Example 6.9 Doppler Shift in Astronomy

The spectrum of light from radio galaxy 8C1435+635 is shifted in wavelength considerably from what you see when emitting atoms are at rest or moving slowly. For instance, the Lyman α line from the hydrogen atom spectrum is observed at 639.2 nm, instead of 121.6 nm observed in the lab. Since the wavelength is increased, which is same as frequency decreasing, the galaxy must be receding with respect to us. Suppose, the galaxy is receding directly away from us, what will be the speed of recession?

Solution

We use the Doppler formula in the wavelength form,

$$\frac{c}{\lambda} = \frac{c}{\lambda_0} \sqrt{\frac{1 - V/c}{1 + V/c}}.$$

Rearranging we find

$$\frac{1 + V/c}{1 - V/c} = (\lambda/\lambda_0)^2 = (639.2/121.6)^2 = 27.6.$$

Therefore,

$$V = \left(\frac{27.6 - 1}{27.6 + 1} \right) c = 0.93\, c.$$

This says that the radial velocity of the galaxy is 93% of the speed of light. In astronomy, one defines a related quantity called the red-shift or z of recession by

$$z = \frac{\Delta\lambda}{\lambda_0}.$$

This galaxy has a z of

$$z = \frac{638.2 \text{ nm} - 121.6 \text{ nm}}{121.6 \text{ nm}} = 4.25.$$

6.8.3 Doppler Effect and Aberration

In the last subsection we studied the relativistic Doppler effect for the relative motion along the line joining the source and the observer. Now, we suppose the source of the wave is in the xy-plane and is at rest in the S' frame at an angle θ' in the first quadrant from the origin and counterclockwise from the x'-axis as shown in Fig. 6.26.

The propagation vector for the ray from the source will be

$$\vec{k} = \frac{2\pi}{\lambda_0}\left(-\cos\theta'\,\hat{i} - \sin\theta'\,\hat{j}\right), \tag{6.191}$$

where k is the wave number, and $\hat{i}$ and $\hat{j}$ are unit vectors towards the positive x' and y' axes respectively. The wave number k is related to the wavelength λ by $k = 2\pi/\lambda$. Using a general form for the oscillating part of plane waves to be discussed in the next chapter, $\sim \cos(\vec{k}\cdot\vec{r} - 2\pi ft)$, the wave function for this wave will be

$$\psi'(x', y', t') = \psi'_0 \cos\left[\frac{2\pi}{\lambda_0}(-x'\cos\theta' - y'\sin\theta') - 2\pi f_0 t'\right]. \tag{6.192}$$

Now, we express x', y', and t' in this equation in terms of x, y, and t using Lorentz transformations, and collect terms in the argument of the cosine as a term multiplying x, a term multiplying y, and another term multiplying t. We also use $\lambda_0 f_0 = c$ to simplify,

$$\psi'(x,t) = \psi'_0 \cos\left[2\pi\gamma\frac{1}{\lambda_0}\left(-\cos\theta' + \frac{V}{c}\right)x - 2\pi\frac{\sin\theta'}{\lambda_0}y - 2\pi\gamma f_0\left(1 - \frac{V}{c}\cos\theta'\right)t\right]. \tag{6.193}$$

The detector/observer at the origin of the S frame will find this wave coming from the direction θ and assign the following wave function:

$$\psi(x,t) = \psi_0 \cos\left[-\frac{2\pi}{\lambda}(x\cos\theta + y\sin\theta) - 2\pi f t\right]. \tag{6.194}$$

Now, we equate the coefficients of x, y, and t in the argument of the cosine in the last two equations to obtain

$$-\frac{\cos\theta}{\lambda} = \gamma\frac{1}{\lambda_0}\left(-\cos\theta' + \frac{V}{c}\right), \tag{6.195}$$

$$\frac{\sin\theta}{\lambda} = \frac{\sin\theta'}{\lambda_0}, \tag{6.196}$$

$$f = \gamma\left(1 - \frac{V}{c}\cos\theta'\right)f_0. \tag{6.197}$$

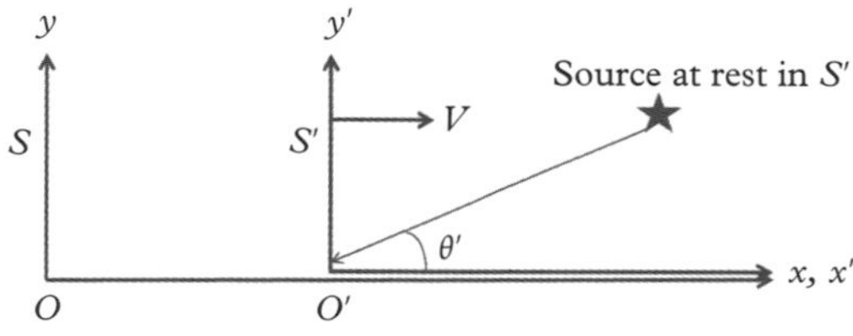

Fig. 6.26 *To calculate the aberration we place the source of the sinusoidal wave at rest in the S' frame in the direction θ'.*

We can eliminate $\cos\theta'$ in Eq. 6.197 in favor of $\cos\theta$ by using Eq. 6.195. This gives the Doppler shift from a source whose direction is θ in the frame of the observer, while the source is moving at velocity in the direction of $\theta = 0$, i.e. the positive x-axis,

$$f = \left[\frac{\sqrt{1 - V^2/c^2}}{1 + \frac{V}{c}\cos\theta}\right] f_0. \tag{6.198}$$

There is a Doppler effect even when the source is moving in the perpendicular direction. To see that, we set $\theta = 90°$ in this equation giving the Doppler shift of an object that is towards the y-axis and moving towards the positive x-axis. This effect is called the **transverse Doppler effect,**

$$f = f_0\sqrt{1 - V^2/c^2}. \tag{6.199}$$

The transverse Doppler effect is much weaker than the longitudinal Doppler effect obtained in Eq. 6.190. Let us expand Eqs. 6.190 and 6.199 in powers of V/c. Let us designate the longitudinal Doppler by $f_\parallel$ and the transverse Doppler by $f_\perp$,

$$f_\parallel = f_0\left[\left(1 - \frac{V}{c}\right)\frac{1}{\sqrt{1 - V^2/c^2}}\right] = f_0\left(1 - V/c + V^2/2c^2 + \cdots\right), \tag{6.200}$$

$$f_\perp = f_0\sqrt{1 - V^2/c^2} = f_0\left(1 - V^2/2c^2 + \cdots\right). \tag{6.201}$$

Dividing Eq. 6.196 by Eq. 6.195 gives us the following relation between the direction of the source in the two systems:

$$\tan\theta = \frac{\sin\theta'\sqrt{1 - V^2/c^2}}{\cos\theta' - V/c}. \tag{6.202}$$

This relation says that light emitted at a direction θ' in one frame will come from direction θ in the other frame. The phenomenon is called **aberration of light**. For instance, if the source is on the y' axis, the angle $\theta' = 90°$. But the angle θ in the S frame will be

$$\theta = -\tan^{-1}\left(\frac{\sqrt{1 - V^2/c^2}}{V/c}\right) \approx -\tan^{-1}\left(\frac{c}{V}\right). \tag{6.203}$$

6.8.4 Ives–Stilwell Experiment

In 1938 Ives and Stilwell conducted an experiment to verify the longitudinal Doppler effect. For technical reasons they could not perform the transverse Doppler effect which would require examination of the light emitted perpendicular to the emitting particle. In the Ives–Stilwell experiment canal rays (which are rays of positive ions) from a hydrogen gas discharge were accelerated to high speeds, as shown in Fig. 6.27. The ions then recombined with electrons to produce light.

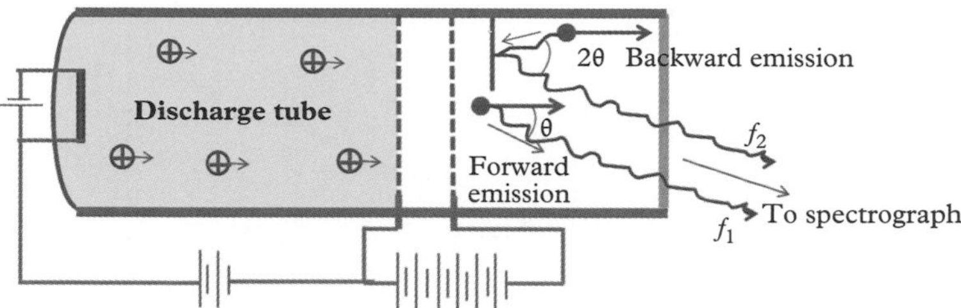

Fig. 6.27 *A schematic picture of the apparatus used by Ives and Stilwell to investigate Doppler shift.*

Ives and Stilwell detected the hydrogen emission in the blue, the Hβ line, for both forward and backward emitted light with the arrangement of a mirror in the tube. By varying the accelerating voltage they could control the speed of the emitted ion. They found the emitted light contained light of original wavelength 4849.32 Å and two symmetrically displaced wavelengths, one at a higher wavelength and the other at a lower wavelength. The magnitude of displacement of the wavelength depended upon the accelerating voltage. From the theory of the Doppler effect, if we ignore v^2/c^2 compared to v/c, then from Eq. 6.198 we expect

$$f \approx f_0\,[1 - (V/c)\cos\theta]\,,$$

which can be written for the wavelength as

$$\frac{c}{\lambda} \approx \frac{c}{\lambda_0}\,[1 - (V/c)\cos\theta]\,.$$

From this equation we can deduce the fractional shift $\Delta\lambda$ in the wavelength,

$$\left|\frac{\Delta\lambda}{\lambda_0}\right| = \frac{V}{c}\,|\cos\theta|\,. \tag{6.204}$$

Now, the speed of the light emitting particle can be related to the accelerating voltage $\mathcal{E}$,

$$e\mathcal{E} = \frac{1}{2}MV^2, \tag{6.205}$$

where e is the electronic charge and M is the mass of the emitting particle, which was either H_2 or H_3 in their experiment. From Eqs. 6.204 and 6.206 we see that

$$\left|\frac{\Delta\lambda}{\cos\theta}\right| = \text{const} \times \sqrt{\mathcal{E}}. \tag{6.206}$$

In the Ives–Stilwell experiment, rays were examined at 7°. A plot of $\Delta\lambda$ adjusted by dividing by cos 7° versus $\mathcal{E}$ from the paper by Ives and Stilwell is shown in Fig. 6.28. The data shows a good fit to prediction up to the linear approximation and confirms the predictions of the relativistic Doppler effect.

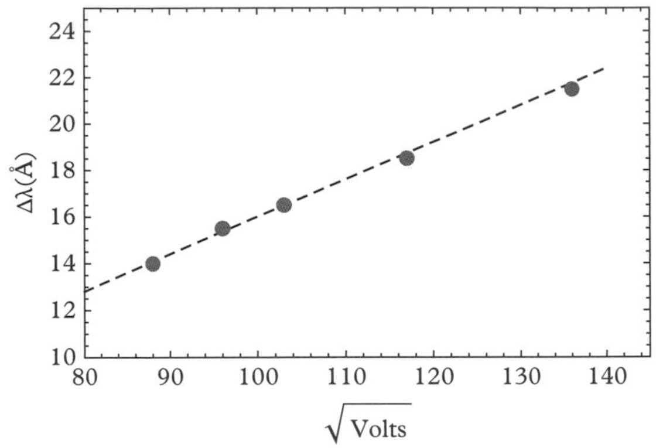

Fig. 6.28 *Doppler shift is plotted versus the square root of the voltage, which is proportional to the speed of the light-emitting particle. The dashed line is the theoretical prediction in the linear approximation.*

EXERCISES

(6.1) A harmonic traveling wave on a string is given by the following transverse displacement y as a function of x and t:

$$y(x, t) = 0.05\cos(2\pi x - 200\pi t),$$

where x and y are in m, and t is in sec. Find the values of the following quantities.

(a) Amplitude of the wave.
(b) The displacement of a particle of the string at (i) $x = 0$ and (ii) $x = 0.1$ m at an arbitrary instant t.
(c) Wavenumber.
(d) Angular frequency and frequency.
(e) Wavelength.
(f) Period.
(g) Phase velocity.
(h) Speed of a particle at $x = 1$ m at $t = 1/600$ sec.
(i) Maximum speed of any particle of the string and compare that to the wave speed.
(j) Minimum speed of any particle of the string.
(k) If mass per unit length of the string is 0.01 kg/m, what is the tension in the string?

(6.2) A harmonic transverse wave is traveling on a stretched string of tension 100 N and mass per unit length 0.01 kg/m with a wave amplitude of 4 cm and wavelength of 0.5 m.

(a) What is the wave function if the wave is moving towards the positive z-axis?
(b) What is the wave function if the wave is moving towards the negative z-axis?

(6.3) A harmonic longitudinal wave moves in a slinky of mass density 20 g/cm. The wave is given by the displacement $\psi(z, t) = 2\cos(0.8\pi z + 28\pi t)$, where ψ is in cm, z in cm, and t in sec.

(a) Which way in space is the wave traveling?
(b) What is the speed of the wave?

(c) What is the tension in the slinky?

(d) What is the average power transported by the wave?

(6.4) A harmonic transverse wave on a taut string is given by $y(x, t) = 3\cos(200\pi x + 400\pi t)$, where x and y are in cm and t is in sec.

(a) What is the speed of the particle at $x = 0$ at $t = 0.001$ sec?

(b) What is the phase difference in the vibration of the particle at $x = 0$ during a time interval of $\Delta t = 0.002$ sec?

(c) What is the phase difference in the vibration of the particles at $x = 0$ and at $x = 1$ cm at the same instant $t = 0$?

(6.5) In the text we studied a semi-infinite string vibrating at one end. Consider now a very long string of mass density μ [mass/length] along the x-axis, extending from $-\infty$ to $+\infty$, kept at tension T. Suppose we vibrate the string at $x = 0$ by displacing the point at $x = 0$ along the y-axis such that the displacement of the string at $x = 0$ is given by

$$y(x = 0, t) = y_0 \cos \omega t.$$

This vibration will send waves on both sides of $x = 0$. One wave will travel towards $x = +\infty$ and the other towards $x = -\infty$.

(a) Find the expressions of the two waves.

(b) Find (i) the force applied at $x = 0$ and (ii) the power supplied by this force.

(6.6) Recall that we can write a standing wave as a superposition of two traveling waves,

$$\text{Standing wave: } A\cos(kx)\cos(\omega t) = \frac{A}{2}\cos(kx - \omega t) + \frac{A}{2}\cos(kx + \omega t).$$

(a) Write the following traveling wave as a superposition of standing waves: $A\cos(kx - \omega t)$.

(b) Write the following superposition wave of two traveling waves: $\psi(x, t) = A\cos(kx - \omega t) + B\cos(kx + \omega t)$, as a superposition of two standing waves.

(6.7) **Traveling wave on a transmission line.** Consider a co-axial cable placed along the x-axis with an AC source at $x = 0$, $V(t) = V_0\cos(\omega t)$, as shown in Fig. 6.29. The other end of the co-axial cable is infinitely far away so that we do not have a reflected wave, just one wave traveling away from the source. Let C and L be the capacitance and inductance per unit length of the co-axial cable. This lossless (since resistance R = 0) transmission line is modeled by series inductors and parallel capacitors placed periodically with repeating units separated by $\Delta x = a$.

(a) Consider the circuit between arbitrary points x and $x + a$ where $x = (j - 1)\,a$ for some integer j. Examine the charging of the capacitor and show that in the limit $a \to 0$ it gives

$$\frac{\partial I}{\partial x} = -C\frac{\partial V}{\partial t}. \qquad (1)$$

(b) Examine the voltage across the inductor and show that

$$\frac{\partial V}{\partial x} = -L\frac{\partial I}{\partial t}. \qquad (2)$$

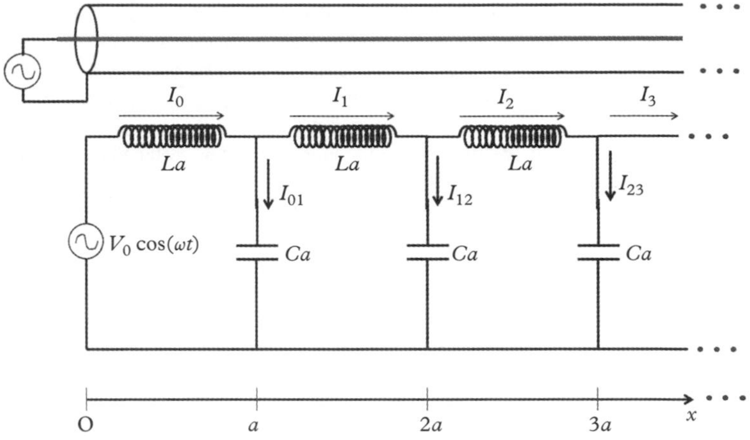

Fig. 6.29 *Exercise 6.7.*

(c) From (1) and (2) deduce the following transmission line wave equations:

$$\frac{\partial^2 I}{\partial x^2} = LC\frac{\partial^2 I}{\partial t^2}, \quad \frac{\partial^2 V}{\partial x^2} = LC\frac{\partial^2 V}{\partial t^2}. \quad (3)$$

(d) Verify that $I(x,t)$ and $V(x,t)$ satisfy the following traveling wave solution:

$$I(x,t) = I_0 \cos(kx - \omega t), \quad V(x,t) = V_0 \cos(kx - \omega t),$$

where $k = 2\pi/\lambda$ with λ the wavelength, and find the relation between I_0 and V_0.

(e) Find the characteristic impedance Z of the transmission line, defined by $V(x,t) = Z\,I(x,t)$.

(f) Find the average power the source puts into the transmission line.

(6.8) A harmonic current wave of amplitude 10 mA and frequency 100 MHz on a transmission line moves at a speed 1×10^8 m/s. Suppose the transmission line is laid out along the z-axis.

(a) Write the wave function of the wave moving towards the positive z-axis.

(b) What will be the phase change [in rad] of the wave at $z = 0$ in a time duration of $\Delta t = 2.0$ nanosec?

(c) How far apart are two points in space at which the phase difference is 0.5 rad?

(d) If the transmission line has a capacitance of 0.1 μF/m, what is the inductance per m?

(e) Find the characteristic impedance of the transmission line.

(6.9) **Wave in coupled pendulums.** Consider identical pendulums of mass m, length l coupled by identical springs with spring constant K, as shown in Fig. 6.30. Note: We will use K for spring constant and reserve k for the wave number.

The displacement of the first mass which is at $x = 0$ at equilibrium is controlled by the vibrator and is given by

$$\psi_0 = A\cos(\omega t). \quad (1)$$

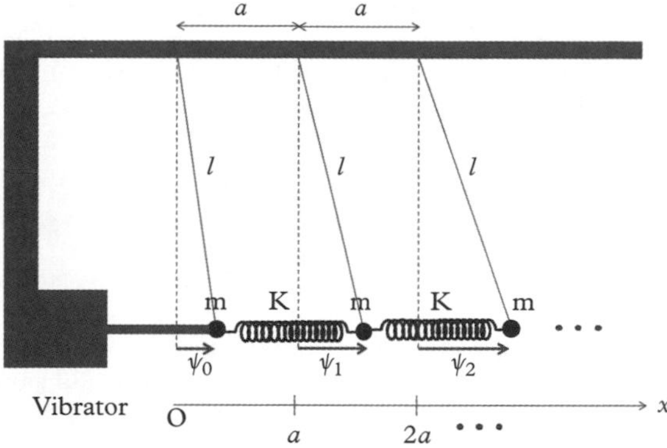

Fig. 6.30 *Exercise 6.9.*

(a) Examine the forces on the i-th mass and deduce the following equation of motion of this mass:

$$m\frac{d^2\psi_i}{dt^2} = \frac{mg}{l}\psi_i + K\left(\psi_{i+1} - 2\psi_i + \psi_{i-1}\right).$$

(b) Suppose the wavelength λ of the wave propagated through the system is much larger than a so that the collection of displacements $\{\psi_i(t)\}$ can be treated as a continuous function of x and written as $\psi(x,t)$. In the continuum limit we can replace the ratio of a finite difference over a distance a by the corresponding derivative. From the equation of motion for the i-th mass in the sequence, show that in the continuum limit the displacement obeys the following wave equation:

$$\frac{\partial^2\psi}{\partial t^2} = -\frac{g}{l}\psi + a^2\frac{K}{m}\frac{\partial^2\psi}{\partial x^2}. \quad (2)$$

(c) Find the dispersion relation.

(d) Find the phase and group velocities.

(e) Solve Eq. (2) with the condition in Eq. (1) to obtain a traveling wave down the chain of coupled pendulums.

(6.10) A very long spring with spring constant K, total mass M, and length L is tied at both ends so that the tension in the spring is T. The spring is then set in motion in the lowest longitudinal mode.

(a) Find the wave function for a standing wave of the lowest frequency mode.

(b) Find the time-averaged kinetic energy per unit length of the spring.

(c) Find the time-averaged potential energy per unit length of the string.

(6.11) A transverse wave $y(x,t)$, is riding on a taut string with speed $v = 50$ cm/s, which is independent of frequency. At $t = 0$ the wave has the following profile:

$$y(x,0) = 2\cos(10\pi x) + 3\cos(15\pi x),$$

where x and y are in cm. Find the wave function at an arbitrary instant t if the wave is moving towards the positive x-axis.

(6.12) A transverse wave $y(x, t)$ is generated on a taut string which is along the x-axis. The wave moves with speed $v = 50$ cm/s, which is independent of frequency. The particle of the string at $x = 0$ vibrates with the following vibration as the wave moves past this point while traveling in the positive x-axis direction:

$$y(0, t) = 2\cos(10\pi t) + 3\cos(15\pi t),$$

where y is in cm and t is in sec. Find the wave function at an arbitrary place x and arbitrary instant.

(6.13) A diffraction grating is made by etching saw teeth on a flat surface. The saw teeth "look" like a sawtooth wave pattern, as shown in Fig. 6.31. Suppose the saw teeth are along the x-axis as shown in the figure.
(a) Write the given sawtooth plot as a periodic function of x.
(b) Find the Fourier series representation of the periodic function.

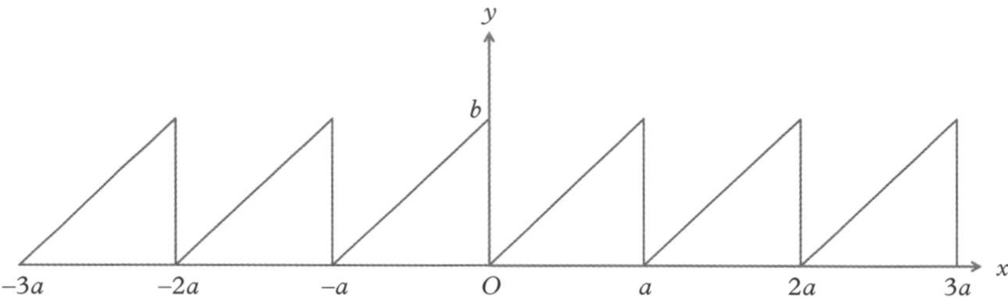

Fig. 6.31 *Exercise 6.13.*

(6.14) A very long string of linear density μ is tied to a mechanical vibrator at one end and the other end is fixed at a faraway point such that there is a tension T throughout the string. Let the string be along the positive x-axis with the vibrator at $x = 0$.

At $t = 0$ the vibrator gives a square pulse by suddenly pushing the string up to a displacement $y = b$, holding the end of the string in that position for a time duration τ, and then pulling back to the equilibrium position thereafter. The pulse so generated moves towards the other end. You can ignore the time in which the string changes from $y = 0$ to $y = b$ and from $y = b$ to $y = 0$.
(a) Find the state of the string at an instant t before the pulse reaches the end, assuming non-dispersive behavior.
(b) Describe what will happen to the pulse if the dispersion relation has the following form: $\omega^2 = a^2k^4$. Give an expression, which can be in an integral form, of the wave at an arbitrary time t.

(6.15) Suppose an infinitely long string of mass per unit length μ and tension T is placed along the x-axis and is driven by an oscillator at $x = 0$. The resulting vibrations at $x = 0$ have a saw-tooth pattern similar to the figure in Fig. 6.31, except that the abscissa in the figure is t instead of x.
(a) Assuming a non-dispersive behavior for the string find the expressions for the wave function $y(x, t)$.
(b) What is the power output of the wave generator?

(6.16) A wave packet for a one-dimensional wave moving towards the positive x-axis has the following profile in the wave number space:

$$A(k) = \begin{cases} 1, & k_0 < k < k_0 + \Delta k \\ 0, & \text{otherwise} \end{cases}$$
$$B(k) = 0,$$

where $A(k)$ and $B(k)$ are the cosine and sine transforms.

(a) Find the wave profile $\psi(x)$ in the x-space.

(b) What is the wave function at arbitrary time if the speed of the wave is v, which is independent of frequency.

(c) Plot $\psi(x)$ for the following values of k_0 and Δk and make observation about the relation between Δk and Δx: (i) $k_0 = 1,\ \Delta k = 1$, (ii) $k_0 = 1,\ \Delta k = 0.5$, (iii) $k_0 = 1,\ \Delta k = 0.1$.

(6.17) **Lorentzian Wave.** Suppose a transverse wave on a taut string of tension T in the shape of the Lorentzian function is traveling towards the positive x-axis,

$$y(x,t) = \frac{bw^2}{(x-vt)^2 + w^2}, \quad -\infty < x < \infty,$$

where $b > 0$ is the value of y at the peak.

(a) Plot $y(x,t)$ versus x at $t = 0$ for $w = 1$ and $w = 2$ and comment on what you see.

(b) Show that the width of $y(x,0)$ at half-height is $2w$.

(c) What is the velocity of a particle at $x = x$ at instant t?

(d) Let $w = 1$ and $b = 1$. Plot the wave profiles at $t = -0.5$ and $t = 0.5$ and the velocity profile at $t = 0$ on the same graph. Does the velocity plot make sense? Why, or why not?

(e) Find the energy contained in the pulse at $t = 0$.

(6.18) Two transverse pulses with opposite displacements ride in a taut string of tension T. The pulses have the shape of the Lorentzian function. The net displacement of the string at arbitrary time is given by

$$y(x,t) = \frac{bw^2}{(x-vt+a)^2 + w^2} + \frac{bw^2}{(x+vt-a)^2 + w^2},$$

with $-\infty < x < \infty,\ a \gg w > 0$.

(a) Show that at $t = 0$ the two pulses are centered about $x = -a$ and $x = a$ with the pulse on the negative x-axis moving towards the positive x-axis and the pulse on the positive x-axis moving towards the negative x-axis.

(b) Sketch the velocity profile of the wave at $t = 0$ and compare it to the wave profile.

(c) Find the time when the two pulses will completely overlap each other and the string will have a zero displacement everywhere.

(d) What happens after this time? Explain, your finding. Hint: Look at the velocity profile. You may experiment with numerical values and make sketches to get an idea about what may be happening here.

(6.19) A car horn at rest sounds a frequency of 400 Hz.

(a) You are running towards the horn at a speed of 10 m/s, what frequency will you hear?

(b) What frequency would you hear if you were running away from the horn? Assume the speed of sound in air to be approximately 343 m/s.

(6.20) A car horn at rest sounds a frequency of 400 Hz.

(a) If the car is moving towards you at speed 10 m/s, what frequency will you hear?

(b) If the car is moving away from you at speed 10 m/s, what frequency will you hear? Assume the speed of sound in air to be approximately 343 m/s.

(6.21) While sitting on your porch you hear a fire engine siren at a frequency 1650 Hz approaching you, and a frequency 1550 Hz when receding from you. From this data, determine the speed of the fire engine assuming it to be constant and the frequency of the sound generated by the fire engine. Use 343 m/s for the speed of sound.

(6.22) A humming bird is flying towards you with speed 10 m/s with respect to the ground while you are driving towards it at 20 m/s with respect to the ground. You hear sound of frequency 450 Hz. What is the frequency of the sound that the humming bird is making? Use 343 m/s for the speed of sound.

(6.23) A prominent emission of light from an excited hydrogen atom occurs in the Lyman series line L_α. Light corresponding to the L_α line has a wavelength of 121.6 nm when the emitting atom has a low speed and is considered to be at rest. The L_α line in the light from some star is observed to have a wavelength of 486.4 nm. Suppose the star is moving directly away from Earth, what is the speed of the star with respect to Earth?

(6.24) (a) Find an approximate relativistic Doppler effect formula when $v << c$ keeping terms up to linear in v/c.

(b) A police car parked at the side of the road sends out electromagnetic waves of frequency 5 GHz. What will be the frequency of the reflected wave from a car that is approaching the police car at a speed of 100 mph? Assume the wave is almost parallel to the velocity of the car.

(6.25) A galaxy is moving away from Earth at a speed of $0.9c$. (a) What will be the wavelength observed on Earth corresponding to a radio wave emitted at wavelength 21 cm from the galaxy? (b) The red-shift z of a galaxy is defined by $z = \Delta\lambda/\lambda_0$, where $\Delta\lambda$ is the change in the wavelength and λ_0 is the emitted wavelength. Find the red-shift of the galaxy.

(6.26) The red-shift z of a galaxy is defined by $z = \Delta\lambda/\lambda_0$, where $\Delta\lambda$ is the change in the wavelength and λ_0 is the emitted wavelength. A quasar is found to have a red-shift of 20. Supposing the quasar is moving away from Earth, what is the speed of the quasar with respect to Earth?

Waves in Three-Dimensional Space

7

Chapter Goals

In this chapter we will study waves in three dimensions. The concepts related to three-dimensional waves will be illustrated by using electromagnetic and matter waves. You will learn about propagation vectors, plane waves, mixed harmonic waves, and polarization of waves.

To keep the mathematics simpler we have worked largely with waves in one-dimensional space. Most waves of interest, whether electromagnetic, sound, or matter waves, actually travel in a three-dimensional space. Although most of the fundamental properties of waves from one-dimensional examples can be extended to higher-dimensional cases, waves in two and three dimensions have additional properties and we will discuss them here.

Many interesting phenomenon associated with waves can be illustrated by sound and light waves. Maxwell showed that light waves are actually electromagnetic waves. Therefore, we will learn about the formulation of light as electromagnetic waves first. Other examples of waves in three dimensions, such as sound and matter waves, will be described using the language we will develop for electromagnetic waves.

7.1 Waves in Three Dimensions

7.1.1 Harmonic Waves

Generalization of the one-dimensional classical wave equation describes many types of waves, in particular, the sound wave and the electromagnetic wave. Let $\psi(x, y, z, t)$ be the displacement of a property that obeys the classical wave equation for a wave of speed v. In the case of the sound wave ψ stands for the difference in pressure at (x, y, z) from the ambient pressure, while in the case of electromagnetic waves ψ stands for one of the Cartesian components of the electric or magnetic field at (x, y, z). The classical wave equation in three dimensions is

$$\frac{\partial^2 \psi}{\partial t^2} = v^2 \left(\frac{\partial^2 \psi}{\partial x^2} + \frac{\partial^2 \psi}{\partial y^2} + \frac{\partial^2 \psi}{\partial z^2} \right). \tag{7.1}$$

A First Course in Vibrations and Waves. First Edition. Mohammad Samiullah.

The harmonic wave solution of this equation refers to the harmonic time variation. That is, the solution can be taken to have the following form, using complex notation for simplicity:

$$\psi(x, y, z, t) = f(x, y, z)e^{-i\omega t}, \tag{7.2}$$

where ω is the frequency of the wave. Then, the two time derivatives will give rise to $-\omega^2\psi$. Therefore, the space part of harmonic waves will obey

$$\frac{\partial^2 f}{\partial x^2} + \frac{\partial^2 f}{\partial y^2} + \frac{\partial^2 f}{\partial z^2} = -\frac{\omega^2}{v^2} f. \quad \text{(harmonic wave)} \tag{7.3}$$

In one dimension, function f was a function of only one coordinate, which we had taken to be the x-axis. Consequently, this equation required boundary conditions at the two ends of the one-dimensional medium. In two dimensions f will be a function of two coordinates and the boundary will be a curve bounding the medium in a two-dimensional surface. In three dimensions f will be a function of all three coordinates and the boundary will be a surface enclosing the volume of the medium.

In previous chapters we had found that a wave in one dimension can be either a standing wave or a traveling wave—if the medium is bounded in a finite length we get standing waves and if the medium is unbounded we get traveling waves. The situation is more complicated in three dimensions with more options than a wave being either a standing wave or a traveling wave. In three dimensions you can have the medium bounded in all three directions, bounded in just two directions, or bounded in only one direction, or completely unbounded in all directions. Therefore, we can get standing waves in all three directions and no traveling wave, standing waves in two directions and a traveling wave in one direction, a standing wave in one direction and traveling waves in two directions, or traveling waves in all directions. For instance, when sound is sent through a rectangular pipe, the sound in the pipe consists of a traveling wave in the direction of the pipe and standing waves in the two other directions perpendicular to the axis of the pipe.

7.1.2 Plane Traveling Harmonic Waves in Three Dimensions

A particularly simple solution of the three-dimensional classical wave equation, Eq. 7.1, takes the following form in complex notation:

$$\psi(x, y, z, t) = Ce^{i(k_x x + k_y y + k_z z - \omega t)}, \tag{7.4}$$

where k_x, k_y, and k_z are constants, and C is a complex constant, whose magnitude A is the amplitude of the wave and whose phase ϕ is the phase constant of the wave,

$$C = Ae^{i\phi}. \tag{7.5}$$

The real part of this complex solution is the physical solution, which would be

$$\psi(x, y, z, t) = A\cos(k_x x + k_y y + k_z z - \omega t + \phi). \tag{7.6}$$

Inserting the solution from Eq. 7.4 into Eq. 7.1 we find that the classical wave equation has the following dispersion relation:

$$\omega^2 = v^2\left(k_x^2 + k_y^2 + k_z^2\right). \tag{7.7}$$

We notice that the constants k_x, k_y, and k_z are the Cartesian components of a vector whose magnitude is the wave number k of the wave. Therefore we define a **wave number vector**, $\vec{k}$, also called the **propagation vector** whose Cartesian components are k_x, k_y, and k_z,

$$\vec{k} = k_x\hat{x} + k_y\hat{y} + k_z\hat{z}, \tag{7.8}$$

where $\hat{x}$, $\hat{y}$, and $\hat{z}$ are unit vectors towards the positive Cartesian axes,

$$\hat{x}\cdot\hat{x} = \hat{y}\cdot\hat{y} = \hat{z}\cdot\hat{z} = 1, \tag{7.9}$$

$$\hat{x}\cdot\hat{y} = \hat{x}\cdot\hat{z} = \hat{y}\cdot\hat{z} = 0, \tag{7.10}$$

$$\hat{x}\times\hat{y} = \hat{z}, \quad \hat{y}\times\hat{z} = \hat{x}, \quad \hat{z}\times\hat{x} = \hat{y}. \tag{7.11}$$

Note that we are using $\hat{x}, \hat{y}, \hat{z}$ in place of $\hat{i}, \hat{j}, \hat{k}$ since we are reserving $\vec{k}$ for the propagation vector.

The magnitude of vector $\vec{k}$ is the wave number k which is related to the wavelength λ of the wave by the usual relation,

$$\lambda = \frac{2\pi}{k}, \quad k = \sqrt{k_x^2 + k_y^2 + k_z^2}. \tag{7.12}$$

The direction of the vector $\vec{k}$ is the direction towards which the wave travels. For instance, if the wave is traveling towards the positive x-axis we will have

$$k_x = k, \quad k_y = 0, \quad k_z = 0, \tag{7.13}$$

and the wave function will be

$$\psi(x, y, z, t) = Ae^{i\phi}e^{ikx}e^{-i\omega t} = Ae^{i(kx-\omega t+\phi)}. \tag{7.14}$$

The real part of this complex wave function is the wave function that we need here. This gives the following for the wave function of a plane harmonic wave traveling towards the positive x-axis.

$$\psi(x, y, z, t) = A\cos(kx - \omega t + \phi). \tag{7.15}$$

We see that the wave function of a plane harmonic wave is like that of a one-dimensional wave, although it travels in three-dimensional space. This occurs because the wave amplitude is independent of the points perpendicular to the wave direction.

7.1.3 Wavefront and Phase Velocity

The harmonic plane wave given in Eq. 7.6 is a function of a cosine (you could alternatively write this as a sine). The argument of sine or cosine here is called the **phase** of the wave. The phase $\varphi(x, y, z, t)$ at a particular location at some instant of time has the following value:

$$\varphi(x, y, z, t) = k_x\, x + k_y\, y + k_z\, z - \omega t + \phi \equiv \vec{k}\cdot\vec{r} - \omega t + \phi. \tag{7.16}$$

Now, note that for a fixed t and fixed φ, this equation is an equation of a plane,

$$k_x x + k_y y + k_z z = \text{constant}. \tag{7.17}$$

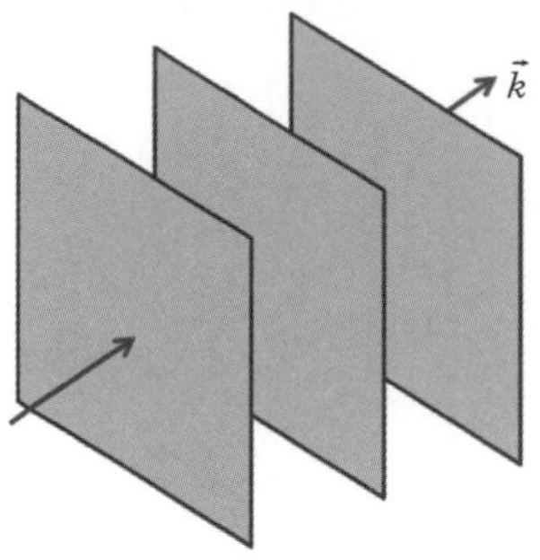

(a) Plane wavefronts

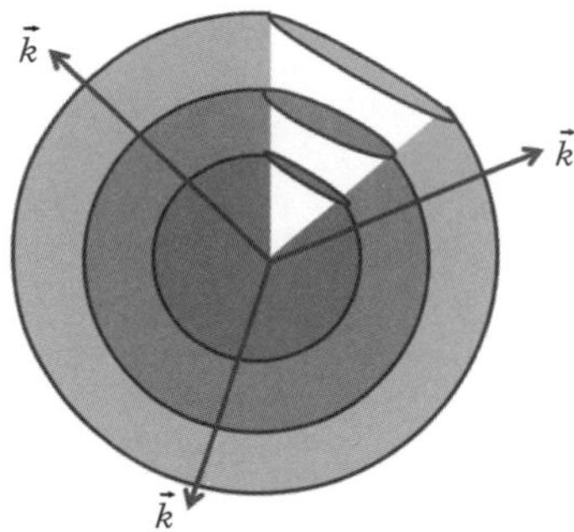

(b) Spherical wavefronts

Fig. 7.1 *(a) Plane wavefronts. The propagation vector is pointed towards a unique direction for plane waves. (b) Spherical wavefronts. The propagation vector points in different directions at different points of the wavefront for a spherical wave.*

Consider two points (x, y, z) and $(x+dx, y+dy, z+dz)$ in this plane. Since vector $\vec{k}$ is constant in space, Eq. 7.17 says that $\vec{k}$ is perpendicular to the displacement vector $d\vec{r} = dx\hat{x}+dy\hat{y}+dz\hat{z}$,

$$\vec{k} \cdot d\vec{r} = 0. \tag{7.18}$$

These planes are called **wavefronts**. For plane harmonic waves, the wavefronts are planar, as illustrated in Fig. 7.1(a), but for other types of waves, they may have other shapes. For instance, for spherical waves, the kind of waves that are emitted from a source isotropically in all directions, the wavefronts are spherical, as illustrated in Fig. 7.1(b).

Suppose we follow the motion of a planar wavefront corresponding to a particular value of the phase φ. During an interval dt the wavefront will move a displacement $d\vec{r}$ in the direction of $\vec{k}$ such that no change in phase occurs. Setting $d\varphi = 0$ in Eq. 7.16 we find

$$d\varphi(x, y, z, t) = \vec{k} \cdot d\vec{r} - \omega\, dt = 0. \tag{7.19}$$

Let the distance in space in the direction of $\vec{k}$ be dl such that $\vec{k} \cdot d\vec{r} = kdl$. Then, this says that

$$kdl - \omega\, dt = 0, \quad \frac{dl}{dt} = \frac{\omega}{k}, \tag{7.20}$$

where k is the magnitude of the propagation vector. The quantity dl/dt is called the **phase velocity** v_p; actually it is the magnitude of the phase velocity, whose direction is in the direction of the propagation vector $\vec{k}$. For the classical wave equation we see from above that

$$v_p = \frac{dl}{dt} = \frac{\omega}{k} = v, \tag{7.21}$$

where v is the velocity parameter in the classical wave equation, which depends on the elastic and inertial properties of the medium. In a non-dispersive medium, such as the string or vacuum for electromagnetic waves, the parameter v is independent of ω (or k). In these media, the dispersion relation is a linear relation between ω and k of the wave, and the phase velocity v_p is not a function of k or ω,

$$\omega = v_p\, k. \quad \text{(Non-dispersive medium)} \tag{7.22}$$

On the other hand, in a dispersive medium, v is a function of ω (or equivalently of k). For instance, v for electromagnetic waves in a dielectric is a function of ω, which is responsible for different amounts of bending of light rays depending upon the frequency when they pass through glass or water. This will make v_p a function of ω (or equivalently of k). Let us write the phase velocity in a dispersive medium as a function of k as $v_p(k)$. This means that in a dispersive medium the dispersion relation is a more complicated relation between ω and k of the wave,

$$\omega = v_p(k)\, k. \quad \text{(Dispersive medium)} \tag{7.23}$$

In a dispersive medium a more important velocity is the group velocity $\vec{v}_g$. You may recall from the last chapter that the group velocity for one-dimensional waves is given by

$$v_g = \frac{d\omega}{dk}. \quad \text{(one-dimensional case)} \tag{7.24}$$

The three-dimensional generalization of this will be the vector group velocity,

$$\vec{v}_g = \hat{x}\,\frac{\partial \omega}{\partial k_x} + \hat{y}\,\frac{\partial \omega}{\partial k_y} + \hat{z}\,\frac{\partial \omega}{\partial k_z}. \tag{7.25}$$

7.1.4 Spherical Traveling Harmonic Wave

Suppose a wave transmitter is at the origin and it is sending out waves isotropically in all directions. Such a traveling wave will depend only on the radial coordinate r of the spherical coordinate system, which is related to the x, y, and z of a point by

$$r = \sqrt{x^2 + y^2 + z^2}. \tag{7.26}$$

The radially outward moving spherical wave of frequency ω and wave number k is given by the following wave function $\psi(r,t)$:

$$\psi(r,t) = \frac{A}{r} e^{i(kr - i\omega t + \phi)}. \tag{7.27}$$

You can verify that this is a solution of the three-dimensional classical wave equation, Eq. 7.1 if you cast the equation in spherical coordinates. I leave that calculation for you to do. Here is the spherical coordinate form of Eq. 7.1,

$$\frac{1}{r^2}\frac{\partial}{\partial r}\left(r^2 \frac{\partial \psi}{\partial r}\right) + \frac{1}{r^2 \sin^2\phi}\frac{\partial^2 \psi}{\partial \theta^2} + \frac{1}{r^2 \sin\phi}\frac{\partial}{\partial \phi}\left(\sin\phi \frac{\partial \psi}{\partial \phi}\right) = \frac{1}{v^2}\frac{\partial^2 \psi}{\partial t^2}, \tag{7.28}$$

where ϕ is measured from the positive z-axis and θ is in the xy plane with respect to the positive x-axis.

The wave function in Eq. 7.27 has a spherical shape for constant phase, that is, the wavefronts are spherical about the origin. The wave function Eq. 7.27 give the displacement of the medium from equilibrium at a point a distance r away. If the wave transmitter is not at the origin, you can replace r accordingly by the distance from the source. We find that the amplitude of the wave decreases with distance from the source, unlike the plane wave whose amplitude is independent of position.

7.1.5 Mixed Harmonic Waves

A wave guided by a rectangular pipe such as shown in Fig. 7.2 is a traveling type of wave along the pipe and standing waves in the two perpendicular directions. Consider a wave that obeys the classical wave equation with wave speed v, guided through a pipe that has a rectangular cross-section with z-axis along the pipe and x- and y-axes parallel to the rectangular sides. Suppose the rectangular cross-section is in the area enclosed by $0 \leq x \leq a$ and $0 \leq y \leq b$.

Then, the wave traveling in the pipe will be standing waves in x- and y-directions and a traveling wave in the z-direction.

$$\psi(x,y,z,t) = A \sin(k_x x)\sin(k_y y)\cos(k_z z - \omega t + \phi), \tag{7.29}$$

where k_z will be a continuous variable and k_x and k_y will have discrete values given by

$$k_x = n\pi/a, \quad n = 1, 2, 3, \ldots,$$
$$k_y = m\pi/a, \quad m = 1, 2, 3, \ldots.$$

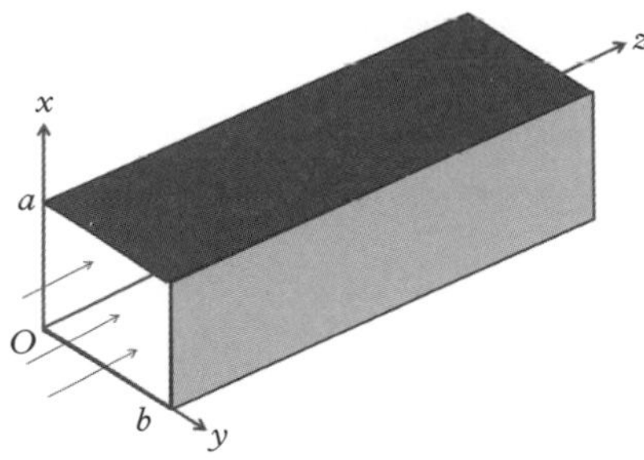

Fig. 7.2 *Wave through a rectangular wave guide has a traveling wave along the guide and standing waves in the directions perpendicular to the guide.*

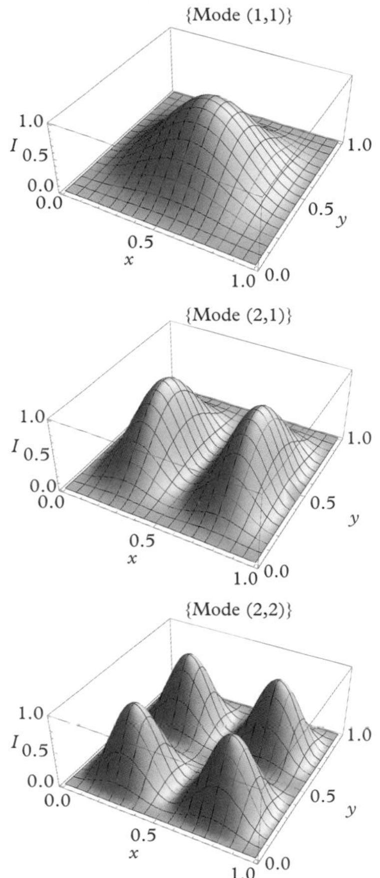

Fig. 7.3 *Relative intensity across the cross-section for modes (1,1), (2,1), and (2,2) in a rectangular wave guide.*

By substituting the wave function in the classical wave equation you can verify that the dispersion relation of the wave is

$$\omega^2 = v^2\left(k_x^2 + k_y^2 + k_z^2\right) = v^2\left(\frac{n^2\pi^2}{a^2} + \frac{m^2\pi^2}{b^2} + k_z^2\right). \tag{7.30}$$

Here $|k_z|$ is the wave number of the propagation direction and therefore, the wavelength of the traveling wave is $2\pi/|k_z|$. For a wave traveling towards the positive z-axis $k_z > 0$ and that traveling towards the negative z-axis, $k_z < 0$. We drop the subscript z from k_z and consider only the wave moving towards the positive z-axis. Each mode of the wave can be labeled with the indices n and m of the standing waves along the x- and y-axes respectively. For instance, for mode $n = 1, m = 1$, the dispersion relation is

$$\omega_{11}^2 = v^2\left(\frac{\pi^2}{a^2} + \frac{\pi^2}{b^2} + k_{11}^2\right), \tag{7.31}$$

and the dispersion relation for mode $n = 1, m = 2$ is

$$\omega_{12}^2 = v^2\left(\frac{\pi^2}{a^2} + \frac{4\pi^2}{b^2} + k_{12}^2\right). \tag{7.32}$$

The lowest frequency that will travel through the wave guide will be when $k_z = 0$. We see that this will happen for the $n = 1$, $m = 1$ case, since for all other cases the frequency will be higher when $k_z = 0$ in the dispersion relation for the corresponding mode. This is the lowest frequency cutoff for the wave guide,

$$\omega_{\text{cutoff}} = v\sqrt{\frac{\pi^2}{a^2} + \frac{\pi^2}{b^2}}. \tag{7.33}$$

The intensity of the plane wave along the wave guide varies across the cross-section of the guide but does not vary along the longitudinal direction. The variation across the cross-section is different for different modes.

Figure 7.3 shows the variation for modes (1,1), (2,1), and (2,2) in the case of a square cross-section. We notice that the lowest mode, (1,1), has the intensity peaked at the center while mode (2,1) has two peaks with zero intensity in the center, and mode (2,2) has four peaks. Similar modes exist for light traveling through an optical fiber. Since optical fibers are usually cylindrical in shape, the functions there are Bessel functions in place of the sine and cosine functions.

7.2 Acoustic Waves in Fluids

Acoustic waves are traveling vibrations in a medium, which can be solid, liquid, or gas. Acoustic waves in fluids (gas or liquid) are particularly simple since fluids can support only longitudinal stress if viscosity can be ignored. Therefore, only longitudinal acoustic waves propagate in fluids. The acoustic wave in air with frequency in the audible range, 20 Hz to 20,000 Hz, is also called sound, while the waves of frequency lower than 20 Hz are called infrasound, and those above 20,000 Hz are called ultrasound. The speed of acoustic waves in a medium depend on the properties of the medium, such as bulk modulus and density, as we have seen for waves on a taut string. This will become clear with our derivation of the wave equation for acoustic waves.

7.2.1 Acoustic Wave Equation in Fluid

Since hydrostatic pressure is the only stress in a fluid, the restoring force for vibrations in a fluid can only be sustained in the direction of the wave motion. That is, we have only longitudinal waves in a fluid. In an earlier chapter, Section 4.5.2, we studied the longitudinal vibration of a rod. Here we need to extend that study to the longitudinal vibration of a fluid in an arbitrary direction.

First note that when the fluid is at rest, the forces from pressure from various directions on any element of the fluid are balanced. For instance, consider an element of the fluid, shown in Fig. 7.4. As shown, there will be forces on all six faces of the box by the fluid elements outside the box due to the pressure in the fluid.

When the fluid does not have a wave in it, the pressure will only be the ambient pressure, p_0, which will vary with depth, here, along the y-axis. Let us ignore the variation of p_0 with depth and take p_0 to be uniform. Thus, at equilibrium we will have

$$F_x(\text{on left face}) + F_x(\text{on right face}) = 0, \tag{7.34}$$

$$F_y(\text{on top face}) + F_y(\text{on bottom face}) = 0, \tag{7.35}$$

$$F_z(\text{on front face}) + F_z(\text{on back face}) = 0, \tag{7.36}$$

where the force on each face is pressure times the area of that face,

$$F_i(\text{on a face}) = \pm p_0 \times \text{area of the face}, \tag{7.37}$$

where the sign is plus(+) when the force is pointed towards the positive axis and minus(-) when pointed towards the negative axis. Now, when a wave travels through the fluid, the fluid will be out of equilibrium. Pressure and density will differ from the equilibrium pressure and density and will be functions of position and time. Let

$$p(x,y,z,t) = p_0 + \delta p(x,y,z,t), \quad \rho(x,y,z,t) = \rho_0 + \delta\rho(x,y,z,t). \tag{7.38}$$

The change in pressure causes an incremental stress in the fluid, which is related to the **volume strain,** δV, which is the fractional change in volume,

$$\delta V = \frac{\Delta V}{V}. \tag{7.39}$$

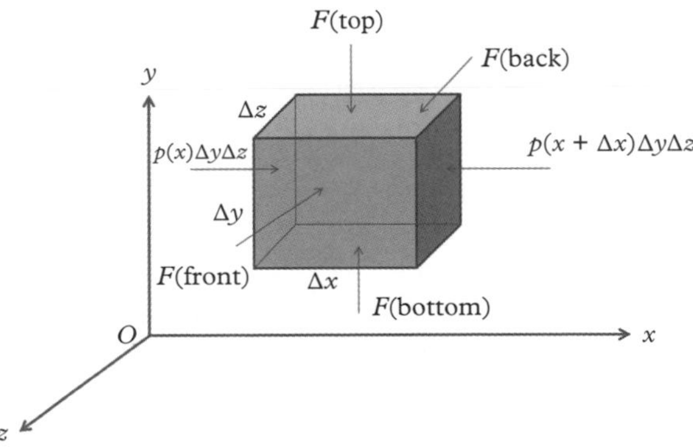

Fig. 7.4 *Stress forces acting through the surface of an element of the fluid of volume $\Delta x\Delta y\Delta z$ located at point (x, y, z). The force on any side is equal to the pressure at that side and the area of that side. The strain of the volume element is defined by the fractional change in volume as a result of compression or rarefaction.*

At a place where the pressure is higher than the ambient pressure, the volume is a little compressed and at a place where the pressure is less than the ambient pressure, the volume is a little expanded. The incremental change in the volume can be expressed in terms of displacements of the fluid particles. Consider the change in volume of the element $\Delta x \Delta y \Delta z$ at (x, y, z) to $\Delta x \Delta y \Delta z + \delta V$ and let the x-, y-, and z-components of the displacements of the particles in this volume be δx, δy, δz, respectively. Then, the volume strain will be

$$\delta V = \frac{\left(\Delta x + \frac{\partial \delta x}{\partial x}\Delta x\right)\left(\Delta y + \frac{\partial \delta y}{\partial y}\Delta y\right)\left(\Delta z + \frac{\partial \delta z}{\partial z}\Delta z\right) - \Delta x \Delta y \Delta z}{\Delta x \Delta y \Delta z} \tag{7.40}$$

$$= \frac{\partial \delta x}{\partial x} + \frac{\partial \delta y}{\partial y} + \frac{\partial \delta z}{\partial z}. \tag{7.41}$$

We can write this in a more compact vector notation using the vector operator $\vec{\nabla}$ and the displacement vector $\delta\vec{r} = \hat{i}\delta x + \hat{j}\delta y + \hat{k}\delta z$,

$$\vec{\nabla} = \hat{i}\frac{\partial}{\partial x} + \hat{j}\frac{\partial}{\partial y} + \hat{k}\frac{\partial}{\partial z}. \tag{7.42}$$

Using $\vec{\nabla}$, the volume strain can be written as

$$\text{volume strain, } \delta V = \vec{\nabla} \cdot \delta\vec{r}. \tag{7.43}$$

According to Hooke's law, as long as stress and strain are not too large ($\delta V \ll 1$), we will have a linear relation between them,

$$\delta p = -B\delta V, \tag{7.44}$$

where B is the **bulk modulus**. This gives the following relation between the incremental pressure and the displacement:

$$\delta p = -B\vec{\nabla} \cdot \delta\vec{r}. \tag{7.45}$$

To find the wave equation we will need the equation of motion of the fluid, which can be obtained by considering an element of the fluid at an instant when it is not in the equilibrium state. Then, the two time derivatives of the displacement vector $\delta\vec{r}$ will give the acceleration of the element and the net force will be from the imbalance of the incremental pressure δp at different faces of the fluid element. In Fig. 7.4 we see that the x-component of the net force will be

$$F_x = [-p(x + \Delta x, y, z, t) + p(x, y, z, t)]\,\Delta y \Delta z \tag{7.46}$$

$$= \frac{[-\delta p(x + \Delta x, y, z, t) - \delta p(x, y, z, t)]}{\Delta x}\,\Delta x \Delta y \Delta z \tag{7.47}$$

$$\longrightarrow -\frac{\partial \delta p}{\partial x}\,\Delta x \Delta y \Delta z. \tag{7.48}$$

Similarly you can show the y- and z-components of the force are

$$F_y = -\frac{\partial \delta p}{\partial y}\,\Delta x \Delta y \Delta z, \tag{7.49}$$

$$F_z = -\frac{\partial \delta p}{\partial z}\,\Delta x \Delta y \Delta z. \tag{7.50}$$

Let Δm be the mass in the volume element,

$$\Delta m = \rho_0 \Delta x \Delta y \Delta z, \tag{7.51}$$

where ρ_0 is the density of the fluid. Therefore, the equation of motion of the particles in the box will be

$$F_x = ma_x \implies -\frac{\partial \delta p}{\partial x} = \rho_0 \frac{\partial^2 \delta x}{\partial t^2}, \tag{7.52}$$

$$F_y = ma_y \implies -\frac{\partial \delta p}{\partial y} = \rho_0 \frac{\partial^2 \delta y}{\partial t^2}, \tag{7.53}$$

$$F_z = ma_z \implies -\frac{\partial \delta p}{\partial z} = \rho_0 \frac{\partial^2 \delta z}{\partial t^2}. \tag{7.54}$$

In vector notation, the equation of motion will be

$$-\vec{\nabla}\delta p = \rho_0 \frac{\partial^2 \delta \vec{r}}{\partial t^2}. \tag{7.55}$$

We can eliminate $\delta\vec{r}$ from this equation by making use of the Hooke's law equation, Eq. 7.45. You can work with x-, y, and z-components separately, or work in vector notation with all components in one place. I will work in vector notation here. Taking the divergence of both sides of Eq. 7.55 we get

$$-\vec{\nabla}\cdot\vec{\nabla}\delta p = \rho_0 \frac{\partial^2}{\partial t^2}\left(\vec{\nabla}\cdot\delta\vec{r}\right), \tag{7.56}$$

where I have pushed the space derivative through the time derivative on the right side. The left side is just the second derivatives,

$$\vec{\nabla}\cdot\vec{\nabla} = \frac{\partial^2}{\partial x^2} + \frac{\partial^2}{\partial y^2} + \frac{\partial^2}{\partial z^2} \equiv \nabla^2. \tag{7.57}$$

Now, using Eq. 7.45 here, we get

$$\nabla^2 \delta p = \frac{\rho_0}{B}\frac{\partial^2 \delta p}{\partial t^2}. \tag{7.58}$$

This is classical wave equation for the incremental pressure with speed,

$$v = \sqrt{\frac{B}{\rho_0}}, \tag{7.59}$$

which is the speed of the sound wave in the fluid. Equation 7.58 is a pressure wave equation suitable for studying a sound wave in a fluid such as air or water. We can write the acoustic wave equation in standard form by replacing the constants on the right side by the speed of the wave,

$$\nabla^2 \delta p = \frac{1}{v^2}\frac{\partial^2 \delta p}{\partial t^2}. \tag{7.60}$$

The bulk modulus B depends on the thermodynamic properties of the material. Previously we have noted that oscillation of the particles in a sound wave occurs much too fast for the process to be isothermal (i.e. constant temperature); instead, the process is mostly adiabatic (no exchange of heat, i.e. constant entropy). In adiabatic processes, pressure and volume change on an iso-entropic adiabatic curve. Using the adiabatic properties of air we can show that speed of wave can be obtained from temperature of air (see Eq. 7.68 below).

Example 7.1 Speed of Sound in Air

Assuming air to be an ideal gas of molecular weight 28 g/mole and $c_p/c_V = \frac{7}{5}$, which is suitable for diatomic molecules such as nitrogen and oxygen at room temperature, find the speed of sound in air at $T = 300$ K.

Solution

In an adiabatic process for an ideal gas, pressure p varies with volume V according to the following relation:

$$pV^\gamma = \text{constant} \quad \text{(ideal gas, adiabatic process)}, \tag{7.61}$$

where γ is the ratio of the specific heat at constant pressure and at constant volume,

$$\gamma = c_p/c_V. \tag{7.62}$$

Therefore, the incremental pressure δp and the corresponding volume change ΔV will be related by

$$V^\gamma \delta p + \gamma p V^{-\gamma-1}\Delta V = 0, \quad \text{(ideal gas, adiabatic process)} \tag{7.63}$$

Therefore,

$$\frac{\delta p}{\Delta V} = -\gamma\frac{p}{V}. \quad \text{(ideal gas, adiabatic process)} \tag{7.64}$$

With $\delta V = \Delta V/V$ in Eq. 7.44, we find that the bulk modulus for air will be

$$B_{\text{air}} = \gamma p. \tag{7.65}$$

Since the deviation in pressure is assumed to be small compared with the ambient pressure, p_0, we can replace $p(= p_0 + \delta p)$ here by p_0,

$$B_{\text{air}} = \gamma p_0. \tag{7.66}$$

Therefore, the speed of the sound wave in air will be

$$v_{\text{air}} = \sqrt{\frac{B}{\rho_0}} = \sqrt{\frac{\gamma p_0}{\rho_0}}. \tag{7.67}$$

By using the ideal gas equation, $p = RT\rho/\mu$, where μ is the molecular weight and R the universal gas constant, we can express the speed of sound in air in terms of the temperature of air,

$$v_{\text{in air}} = \sqrt{\gamma RT/\mu}, \tag{7.68}$$

where T is to be expressed in the absolute (Kelvin) scale. For a diatomic molecule such as nitrogen at room temperature $\gamma = 7/5$. Supposing the air to be all nitrogen, we have $\mu = 0.028$ kg/mole. We get the following for the speed of sound at room temperature of 27 °C = 300 K:

$$v_{\text{in air}} = \sqrt{\frac{(7/5) \times 8.3\ [\text{J K}^{-1}\text{mole}^{-1}] \times 300\ [\text{K}]}{0.028\ [\text{kg mole}^{-1}]}} = 353\ \text{m/s}. \tag{7.69}$$

7.2.2 Plane Acoustic Wave

Acoustic wave equation, Eq. 7.60, has the same form as the classical wave equation. Therefore, solutions of the acoustic wave equation will be similar. A particularly simple wave is the plane wave. Let x-axis be the direction of the wave, i.e. the wave is moving either towards the positive x-axis or towards the negative x-axis. A plane acoustic wave will then have pressure variations δp only as a function of x and t and will not depend on y or z. In that case, the wave equation Eq. 7.60 simplifies to

$$\frac{\partial^2 \delta p}{\partial x^2} = \frac{1}{v^2}\frac{\partial^2 \delta p}{\partial t^2}. \tag{7.70}$$

Harmonic Plane Waves

Harmonic plane waves can be produced by the vibrations of a flat plate moving back and forth perpendicular to the plate. The plate can vibrate in any arbitrary way or vibrate at a particular frequency. When the plate vibrates harmonically at a particular frequency ω, it will produce a harmonic plane traveling wave of that frequency and the incremental pressure will vary harmonically in space and in time. Let k be the wave number of the wave. Then, for a plane harmonic wave traveling towards the positive x-axis, it would be given by the real part—we use complex notation here for simplicity in calculation—of

$$\delta p(x, t) = p_m e^{i(kx-\omega t)}, \tag{7.71}$$

where p_m is the maximum pressure change at any point. For a wave moving towards the negative x-axis, the kx term in this equation will be negative.

Dispersion Relation

Putting the harmonic wave solution, Eq. 7.71 into Eq. 7.70 we obtain the dispersion relation of a sound wave in a fluid,

$$\omega^2 = v^2 k^2, \tag{7.72}$$

which for positive ω gives

$$\omega = vk. \tag{7.73}$$

Since ω is a linear function of k, sound waves in fluids are non-dispersive with group velocity and phase velocity equal,

$$v_p = \frac{\omega}{k}\omega = v, \quad v_g = \frac{d\omega}{dk} = v. \tag{7.74}$$

Characteristic Impedance

The characteristic impedance of the medium is defined by the ratio of the incremental pressure (δp) and the displacement velocity ($|d\delta\vec{r}/dt|$). If δp is harmonic so would the displacement $\delta\vec{r}$ be, as you can see from Eq. 7.55,

$$\delta\vec{r} = \vec{r}_m e^{i(kx-\omega t+\phi)}, \tag{7.75}$$

where ϕ is the phase of the displacement compared to the pressure and $\vec{r}_m$ is the amplitude of the displacement vector wave. From the plane wave solution in Eq. 7.71 for δp, the displacement velocity has only the x-component non-zero. Using the plane harmonic solution of Eq. 7.71 in Eq. 7.55 we find that

$$x_m = \frac{k}{\omega^2\rho_0}p_m, \quad \phi = \pi/2, \tag{7.76}$$

where x_m is the x-component of the displacement amplitude $\vec{r}_m$, whose y- and z-components are zero here. Therefore, the displacement wave for the longitudinal displacements along x-axis is

$$\delta x = \frac{kp_m}{\omega^2\rho_0}e^{i(kx-\omega t+\pi/2)}. \tag{7.77}$$

The velocity wave will be

$$\frac{d\delta x}{dt} = \frac{-ikp_m}{\omega\rho_0}e^{i(kx-\omega t+\pi/2)} = \frac{kp_m}{\omega\rho_0}e^{i(kx-\omega t)}, \tag{7.78}$$

where I have used $-i = e^{-i\pi/2}$, which cancels out the phase constant. Thus, the velocity wave is in phase with the pressure wave. The ratio of the amplitudes of the pressure wave and velocity wave gives us the characteristic impedance Z of the medium,

$$Z = \frac{p_m}{kp_m/\omega\rho_0} = \frac{\omega\rho_0}{k} = \rho_0 v. \tag{7.79}$$

Since $v = \sqrt{B/\rho_0}$ we can also write the characteristic impedance in terms of the bulk modulus and density of the medium,

$$Z = \rho_0 v = \sqrt{B\rho_0}. \tag{7.80}$$

For air treated as an ideal gas this would be

$$Z = \sqrt{\gamma p_0\rho_0} = \rho_0\sqrt{\gamma RT/\mu}. \tag{7.81}$$

7.2.3 Intensity of Acoustic Waves

The intensity of a wave gives the time-averaged power flow per unit area. Let us find the expression by adding up the kinetic and potential energy flows for a plane wave. The density of the kinetic energy in the wave will be given by the kinetic energy of each particle in a unit volume. Thus,

$$\text{KE/volume} = \frac{1}{2}\rho_0 \left| Re\left[\frac{d\delta\vec{r}}{dt}\right]\right|^2. \tag{7.82}$$

In the present example, we get

$$\text{KE/volume} = \frac{1}{2}\rho_0 \left(\frac{kp_m}{\omega\rho_0}\right)^2 \cos^2(kx-\omega t). \tag{7.83}$$

Writing $\omega/k = v$ in this expression we can simplify to

$$\text{KE/volume} = \frac{1}{2}\frac{p_m^2}{\rho_0 v^2}\cos^2(kx-\omega t). \tag{7.84}$$

For the potential energy, recall that in the case of Hooke's law force, the force varies linearly with deformation, as in the mass/spring system. In the mass/spring system, this leads to the potential energy being equal to the product of half the force and the displacement,

$$U = \frac{1}{2}kx^2 = \frac{1}{2}(kx)x = \frac{1}{2}Fx. \tag{7.85}$$

A similar analysis of the stress–strain due to the acoustic wave from the incremental pressure δp accompanying strain $\delta V \equiv \Delta V/V$ will give the potential energy per unit volume,

$$\text{PE/volume} = \frac{1}{2}\delta p\delta V = \frac{1}{2}\frac{(\delta p)^2}{B}. \tag{7.86}$$

For the plane harmonic wave in Eq. 7.71 we get

$$\text{PE/volume} = \frac{1}{2}\frac{(Re[\delta p])^2}{B} = \frac{1}{2}\frac{p_m^2}{B}\cos^2(kx-\omega t). \tag{7.87}$$

With $v^2 = B/\rho_0$ we can also write this as

$$\text{PE/volume} = \frac{1}{2}\frac{p_m^2}{\rho_0 v^2}\cos^2(kx-\omega t). \tag{7.88}$$

We notice that the potential energy density is equal to the kinetic energy density. This is the same conclusion as we found for the wave on a string. Both of these examples are examples of elastic waves in a non-dispersive medium. In these medias the wave carries, on average, equal amounts of kinetic and potential energies. The total energy density will just be the sum of the potential and kinetic energies,

$$\text{E/volume} = \text{KE/volume} + \text{PE/volume} = \frac{p_m^2}{\rho_0 v^2}\cos^2(kx-\omega t). \tag{7.89}$$

The energy contained in volume $A\lambda$, where $\lambda = 2\pi/k$ is the wavelength and A is the area of cross-section, will pass in a time given by λ/v. Therefore, the energy flux per unit time, or the power flux at any time at any instant will be

$$\begin{aligned} \text{P/area} &= \frac{\lambda}{(\lambda/v)} \frac{p_m^2}{\rho_0 v^2} \cos^2(kx - \omega t) \\ &= \frac{p_m^2}{\rho_0 v} \cos^2(kx - \omega t). \end{aligned} \tag{7.90}$$

Averaging this over time, which can be done by averaging over one period, gives us the average power flow per unit time, which is also called the intensity of the wave,

$$\begin{aligned} \text{Intensity, } I &= \langle \text{P/area} \rangle \\ &= \frac{\omega}{2\pi} \int_0^{2\pi/\omega} \frac{p_m^2}{\rho_0 v} \cos^2(kx - \omega t)\, dt \\ &= \frac{p_m^2}{2\rho_0 v}. \end{aligned} \tag{7.91}$$

Writing impedance Z for $\rho_0 v$, we can get a compact expression for the intensity of the acoustic wave,

$$I = \frac{1}{2} \frac{p_m^2}{Z}. \tag{7.92}$$

Recall that this is similar to the power flux in an AC circuit, which has $P_{rms} = (1/2)\, V_m^2/Z$. In AC circuits, the power can also be written in terms of the current. Similarly, the power flux in the acoustic wave can also be written in terms of the amplitude of the velocity wave, which is

$$\text{Amplitude of the velocity wave } = \text{Amplitude of } \left[\frac{d\delta x}{dt}\right] = Z p_m^2. \tag{7.93}$$

Let us denote this by v_m,

$$v_m \equiv \text{Amplitude of } \left[\frac{d\delta x}{dt}\right] = Z p_m^2. \tag{7.94}$$

Therefore, the intensity of acoustic waves in terms of pressure, amplitude, and velocity is

$$I = \frac{1}{2Z} p_m^2 = \frac{1}{2} Z v_m^2. \tag{7.95}$$

The student should verify that the SI unit of intensity will be W/m^2.

7.2.4 Pressure and Displacement Waves

We have found that acoustic waves can be considered as pressure waves or displacement or strain waves. Equation 7.45 shows that the pressure variation δp at any point is related to the strain $\delta\vec{r}$ by

$$\delta p = -B\vec{\nabla}\cdot\delta\vec{r}. \tag{7.96}$$

Therefore, if $\delta\vec{r}$ is a cosine of space coordinates, then δp will be a sine of space, and vice versa. Thus, at a point where displacement vanishes, such as at a surface of reflection, the δp will be a maximum and where δp vanishes, the displacement $\delta\vec{r}$ will be a maximum. That is, the pressure waves would be at an antinode when displacement wave was a node, and vice versa.

7.2.5 Standing Acoustic Waves in One Dimension

Consider a very wide tube of length L with both ends closed. For simplicity, let us look at sinusoidal waves that travel along the length of the tube which is oriented along the x-axis. We know that standing waves would be set up by two waves of the same frequency, one moving to the right and the other moving to the left. For simplicity, let the left and right moving waves have the same amplitude as well. The net wave will be a superposition of the two waves,

$$\delta p = p_m e^{i(kx-\omega t)} + p_m e^{i(-kx-\omega t)}. \tag{7.97}$$

Taking $p_m e^{-i\omega t}$ common we find

$$\delta p = 2p_m\cos(kx)e^{-i\omega t}. \tag{7.98}$$

This is a standing wave. The corresponding displacement wave is

$$\delta x = -\frac{kp_m}{\rho_0\omega^2}\sin(kx)e^{-i\omega t}, \tag{7.99}$$

and the velocity wave is

$$\frac{d\delta x}{dt} = \frac{ikp_m}{\rho_0\omega}\sin(kx)e^{-i\omega t}. \tag{7.100}$$

The real parts of δp, δx, and $d\delta x/dt$ give the wave functions in the tube. For δp and δx, we get

$$\delta p = 2p_m\cos(kx)\cos(\omega t),\quad \delta x = -\frac{kp_m}{\rho_0\omega^2}\sin(kx)\cos(\omega t), \tag{7.101}$$

and for $d\delta x/dt$ we get

$$\frac{d\delta x}{dt} = \frac{kp_m}{\rho_0\omega}\sin(kx)\cos(\omega t-\pi/2). \tag{7.102}$$

Since the tube is closed at the two ends, the displacement of particles at the ends must be zero. Let L be the length of the tube. Vanishing of the displacement wave at the two ends will give the normal mode frequencies of the air column. this gives the following condition:

$$\sin(kL) = 0, \tag{7.103}$$

whose solution gives (ignoring $k = 0$ and discarding the negative k)

$$k_n = \frac{n\pi}{L}, \quad n = 1, 2, 3, \ldots, \tag{7.104}$$

as we have found before for an air column closed at two ends. The corresponding resonant frequencies are

$$\omega_n = vk_n = n\frac{\pi}{L}\sqrt{\frac{B}{\rho_0}}, \quad n = 1, 2, 3, \ldots. \tag{7.105}$$

7.2.6 Standing Acoustic Waves in Three Dimensions

Standing waves can be set up in an enclosed volume. For simplicity, let us consider a rectangular box-shaped enclosure filled with air at room temperature, whose edges are parallel to the Cartesian coordinates. Let the inside of the box occupy the space with $0 \leq x \leq a$, $0 \leq y \leq b$, $0 \leq z \leq c$. Since the space is an enclosed space in all three dimensions, we will have standing waves along each axis. The displacement at the boundary will vanish, which means that a pressure wave will have its maximum there. That means we can try the following solution to the wave equation:

$$\delta p(x, y, z, t) = p_m \cos(k_x x)\cos(k_y y)\cos(k_z z)e^{-i\omega t}. \tag{7.106}$$

This form automatically satisfies the conditions at the walls of $x = 0$, or $y = 0$, or $z = 0$ where δ_x will be zero but δ_p will be maximum. To satisfy the boundary condition at $x = a$, we will need

$$\cos(k_x a) = 1. \tag{7.107}$$

This means

$$k_x = n_x\frac{\pi}{a}, \quad n_x = 0, 1, 2, 3, \ldots, \tag{7.108}$$

where I have kept only positive solutions, as we have done for the one-dimensional case above. Similarly, we will get

$$k_y = n_y\frac{\pi}{b}, \quad n_y = 0, 1, 2, 3, \ldots, \tag{7.109}$$

$$k_z = n_z\frac{\pi}{c}, \quad n_z = 0, 1, 2, 3, \ldots. \tag{7.110}$$

The indices n_x, n_y, and n_z label various modes with frequency of modes denoted by $\omega_{n_x n_y n_z}$. Note that, at most, two of these can be zero for any mode since if all three were zero simultaneously you would get a solution which does not vary with space and hence cannot be a wave. Using the solution given in Eq. 7.106 in the acoustic wave equation we get the frequency of a mode to be

$$\omega_{n_x n_y n_z} = \pi v \sqrt{(n_x/a)^2 + (n_y/b)^2 + (n_z/c)^2}, \tag{7.111}$$

where v is the speed of the wave. Suppose $a > b > c$. Then, the lowest frequency mode will be

$$\omega_{100} = \pi v/a. \tag{7.112}$$

The mode ω_{100} will have the highest pressure on the walls and a pressure node in the center of the tube, where the pressure will be equal to the ambient pressure. The mode ω_{110} will have nodes at $x = a/2$ and $y = b/2$. The nodes will divide the space in the box into four parts, where the adjacent parts will differ in phase. Similarly, the mode ω_{111} will have nodes at $x = a/2$, $y = b/2$, and $z = c/2$. The nodes will divide the space in the box into eight parts, where the adjacent parts will differ in phase.

7.3 Electromagnetic Waves

7.3.1 Maxwell's Equation in a Vacuum and the Electromagnetic Wave Equation

According to Maxwell's equations of electricity and magnetism, the oscillating electric field induces an oscillating magnetic field, which in turn induces an oscillating electric field so that the propagation of electromagnetic waves is self-sustaining and they can propagate in a vacuum. Maxwell also showed that the electric and magnetic fields of an electromagnetic wave in a vacuum are perpendicular to each other, and that they in turn are perpendicular to the direction of the propagation of the wave.

To understand how this comes about, let us start with Maxwell's equations in a vacuum. Let $\vec{E}(x, y, z, t)$ and $\vec{B}(x, y, z, t)$ be the electric and magnetic field at a space point P(x, y, z) at instant t. Suppose there is nothing at point P. Then, Maxwell's equations give the following equations, called Maxwell's equations in a vacuum:

$$\vec{\nabla} \cdot \vec{E} = 0, \tag{7.113}$$

$$\vec{\nabla} \cdot \vec{B} = 0, \tag{7.114}$$

$$\vec{\nabla} \times \vec{E} = -\frac{\partial \vec{B}}{\partial t}, \tag{7.115}$$

$$\vec{\nabla} \times \vec{B} = \mu_0 \epsilon_0 \frac{\partial \vec{E}}{\partial t}. \tag{7.116}$$

Here $\vec{\nabla}$ is the vector operator with the following form in Cartesian coordinates:

$$\vec{\nabla} = \hat{x}\frac{\partial}{\partial x} + \hat{y}\frac{\partial}{\partial y} + \hat{z}\frac{\partial}{\partial z}. \tag{7.117}$$

Equations 7.113–7.116 can be shown to give rise to both $\vec{E}$ and $\vec{B}$ obeying the classical wave equation. Let us see how. Let us start by differentiating Eq. 7.116 with respect to time t,

$$\mu_0 \epsilon_0 \frac{\partial^2 \vec{E}}{\partial t^2} = \vec{\nabla} \times \frac{\partial \vec{B}}{\partial t}. \tag{7.118}$$

Now, we can us Eq. 7.115 and replace $\partial\vec{B}/\partial t$ by $-\vec{\nabla}\times\vec{E}$ to obtain

$$\mu_0\epsilon_0\frac{\partial^2\vec{E}}{\partial t^2}=-\vec{\nabla}\times\left(\vec{\nabla}\times\vec{E}\right). \tag{7.119}$$

Now, we need a mathematical jujitsu in the form of a vector identity from vector calculus. For any arbitrary vector field $\vec{A}$ we have the following identity:

$$\vec{\nabla}\times\left(\vec{\nabla}\times\vec{A}\right)=\vec{\nabla}\left(\vec{\nabla}\cdot\vec{A}\right)-\left(\vec{\nabla}\cdot\vec{\nabla}\right)\vec{A}. \tag{7.120}$$

Using this identity in Eq. 7.119 gives

$$\mu_0\epsilon_0\frac{\partial^2\vec{E}}{\partial t^2}=\vec{\nabla}\left(\vec{\nabla}\cdot\vec{E}\right)-\left(\vec{\nabla}\cdot\vec{\nabla}\right)\vec{E}. \tag{7.121}$$

Now, from Eq. 7.113 we have $\vec{\nabla}\cdot\vec{E}=0$. Also, recall that the $\vec{\nabla}$ is a vector operator and its dot product with itself is

$$\vec{\nabla}\cdot\vec{\nabla}=\frac{\partial^2}{\partial x^2}+\frac{\partial^2}{\partial y^2}+\frac{\partial^2}{\partial z^2}. \tag{7.122}$$

This is also written more compactly as ∇^2,

$$\nabla^2\equiv\vec{\nabla}\cdot\vec{\nabla}=\frac{\partial^2}{\partial x^2}+\frac{\partial^2}{\partial y^2}+\frac{\partial^2}{\partial z^2}. \tag{7.123}$$

Therefore, we find that the electric field obeys the classical wave equation,

$$\frac{\partial^2\vec{E}}{\partial t^2}=\frac{1}{\mu_0\epsilon_0}\left(\frac{\partial^2\vec{E}}{\partial x^2}+\frac{\partial^2\vec{E}}{\partial y^2}+\frac{\partial^2\vec{E}}{\partial z^2}\right)=\frac{1}{\mu_0\epsilon_0}\nabla^2\vec{E}, \tag{7.124}$$

with the velocity v given by

$$v=\frac{1}{\sqrt{\mu_0\epsilon_0}}. \tag{7.125}$$

It is customary to write this speed using the symbol c, the speed of light in vacuum.

$$c=\frac{1}{\sqrt{\mu_0\epsilon_0}}. \tag{7.126}$$

To recap, we have shown that Maxwell's equations in a vacuum give rise to classical wave equations for an electric field. Similar calculations can be performed to eliminate $\vec{E}$. That would show that Maxwell's equations in a vacuum also give rise to classical wave equations for magnetic field $\vec{B}$. This task is left for the student to complete. Writing wave equations for the electric and magnetic fields in a compact form we have

$$\text{E-field wave:} \quad \frac{\partial^2 \vec{E}}{\partial t^2} = c^2 \nabla^2 \vec{E}, \tag{7.127}$$

$$\text{B-field wave:} \quad \frac{\partial^2 \vec{B}}{\partial t^2} = c^2 \nabla^2 \vec{B}. \tag{7.128}$$

Each of these two vector wave equations is actually three wave equations, one for each Cartesian component of the two fields,

$$\frac{\partial^2 E_x}{\partial t^2} = c^2 \nabla^2 E_x, \quad \frac{\partial^2 E_y}{\partial t^2} = c^2 \nabla^2 E_y, \quad \frac{\partial^2 E_z}{\partial t^2} = c^2 \nabla^2 E_z, \tag{7.129}$$

$$\frac{\partial^2 B_x}{\partial t^2} = c^2 \nabla^2 B_x, \quad \frac{\partial^2 B_y}{\partial t^2} = c^2 \nabla^2 B_y, \quad \frac{\partial^2 B_z}{\partial t^2} = c^2 \nabla^2 B_z. \tag{7.130}$$

7.3.2 Plane Harmonic Electromagnetic Wave

Transverse Electromagnetic Waves

We seek a solution of the wave equations, Eqs. 7.129 and 7.130, that also satisfies Maxwell's equations, Eqs. 7.113–7.116. Let us seek a wave with wave number k moving along the z-axis with the components of the propagation vector given by

$$k_x = 0, \quad k_y = 0, \quad k_z = \pm k, \tag{7.131}$$

where a + sign will be for the wave moving towards the positive z-axis and a – sign for moving towards the negative z-axis. Thus,

$$k_x x + k_y y + k_z z = \pm kz. \tag{7.132}$$

For the sake of simplicity, we will study only plane waves. That is, plane electric and magnetic field waves will only be functions of z and t and will have the following mathematical form:

$$\vec{E}(x, y, z, t) = \hat{x}\, E_x(z, t) + \hat{y}\, E_y(z, t) + \hat{z}\, E_z(z, t), \tag{7.133}$$

$$\vec{B}(x, y, z, t) = \hat{x}\, B_x(z, t) + \hat{y}\, B_y(z, t) + \hat{z}\, B_z(z, t). \tag{7.134}$$

Note that in these expressions the subscript x, y, and z refers to the vector components and the z within parentheses to the z-coordinate of the space point. Now, let us plug these into Maxwell's equation and see if there are any other restrictions on these waves. When we put $\vec{E}$ from Eq. 7.133 into Maxwell's equation Eq. 7.113 we find that

$$\vec{\nabla} \cdot \vec{E} = \frac{\partial E_z(z, t)}{\partial z} = 0. \tag{7.135}$$

This says that the z-component of the electric field is independent of the z-coordinate of the space point. That means E_z can only be a function of t, namely $E_z(t)$. Now, let us look at the z-component of Maxwell's equation, Eq. 7.116,

$$\mu_0 \epsilon_0 \frac{\partial E_z}{\partial t} = \left[\vec{\nabla} \times \vec{B}\right]_z = \frac{\partial B_x}{\partial y} - \frac{\partial B_y}{\partial x} = 0. \quad \text{(since no } x \text{ or } y \text{ dependence)} \tag{7.136}$$

Thus, we find that E_z is also not a function of t either. Since E_z is not a function of x, y, z, or t, it is at most a constant. Since a wave arises due to a displacement from equilibrium and constant E_z will not be a wave, therefore, we will set $E_z = 0$. The student is encouraged to perform similar calculations on the magnetic field to show that $B_z = 0$,

$$E_z = 0, \quad B_z = 0. \quad \text{(wave along } z\text{-axis)} \tag{7.137}$$

These calculations show that if a plane wave travels along the z-axis, it will only have the E_x, E_y, B_x, and B_y waves. This proves that the plane electromagnetic wave is a transverse wave. Now, we ask: Are the four waves of E_x, E_y, B_x, and B_y independent of each other? The answer is no, since Maxwell's equations, Eqs. 7.115 and 7.116 relate these components of electric and magnetic fields.

Independent Waves

To obtain the relations among the four waves, E_x, E_y, B_x, and B_y, let us work out the x- and y-components of Eqs. 7.115 and 7.116 by setting $\vec{E} = E_x(z,t)\hat{x} + E_y(z,t)\hat{y}$ and $\vec{B} = B_x(z,t)\hat{x} + B_y(z,t)\hat{y}$ in these equations. We find the following from Eqs. 7.115:

$$\mu_0\epsilon_0 \frac{\partial E_x}{\partial t} = -\frac{\partial B_y}{\partial z}, \quad \mu_0\epsilon_0 \frac{\partial E_y}{\partial t} = -\frac{\partial B_x}{\partial z}, \tag{7.138}$$

and from Eqs. 7.115,

$$\frac{\partial B_x}{\partial t} = \frac{\partial E_y}{\partial z}, \quad -\frac{\partial B_y}{\partial t} = \frac{\partial E_x}{\partial z}. \tag{7.139}$$

These equations show that E_x and B_y are not independent of each other, and E_y and B_x are similarly not independent of each other. The four waves $\{E_x, E_y, B_x, B_y\}$ split into two subsets $\{E_x, B_y\}$ and $\{E_y, B_x\}$. In each subset only one wave is independent. Usually, we take E_x and E_y as the independent waves and determine B_y and B_x from them respectively using the relations in Eq. 7.138 or 7.139.

Harmonic Plane Electromagnetic Waves and Polarization

Now, we will focus on harmonic plane electromagnetic waves of (angular) frequency ω and wave number k. Let the direction of the wave be towards the positive z-axis. The independent plane waves of E_x and E_y can have different amplitudes and different phase constants. Therefore, general expressions for these waves can be given as follows:

$$E_x = E_{0x}\cos(kz - \omega t + \phi_x), \tag{7.140}$$

$$E_y = E_{0y}\cos(kz - \omega t + \phi_y). \tag{7.141}$$

When the phase constants are equal, $\phi_x = \phi_y$, we call the electromagnetic wave a **linearly polarized** wave. When $\phi_x \neq \phi_y$ we call the wave **elliptically polarized**. The corresponding magnetic field waves will be

$$B_x = B_{0x}\cos(kz - \omega t + \phi_y), \tag{7.142}$$

$$B_y = B_{0y}\cos(kz - \omega t + \phi_x). \tag{7.143}$$

Note from Eq. 7.138 or 7.139 the phase of the B_x wave will be equal to the phase of the E_y wave and the phase of the B_y wave will equal that of the E_x wave. The amplitudes of the related electric and magnetic field waves are related also, as you can show from Eqs. 7.138 or 7.139, for plane waves traveling towards the positive z-axis

$$\frac{B_{0x}}{E_{0y}} = -\frac{k}{\mu_0 \epsilon_0 \omega} = -\frac{k}{\omega} = -\frac{1}{c}, \quad \frac{B_{0y}}{E_{0x}} = \frac{1}{c}. \tag{7.144}$$

Linearly Polarized Harmonic Electromagnetic Wave

For simplicity, we choose $x = 0$ and $t = 0$ such that we have $\phi_x = \phi_y = 0$ for the waves under study. That is, we have the following linearly polarized independent traveling waves:

$$E_x = E_{0x} \cos(kz - \omega t), \quad B_y = \frac{E_x}{c}, \tag{7.145}$$

$$E_y = E_{0y} \cos(kz - \omega t), \quad B_x = -\frac{E_y}{c}. \tag{7.146}$$

We can write these waves more compactly in vector notation,

$$\vec{E} = \vec{E}_0 \cos(\vec{k} \cdot \vec{r} - \omega t), \quad \vec{B} = \vec{B}_0 \cos(\vec{k} \cdot \vec{r} - \omega t),$$
$$\vec{B}_0 = \frac{1}{\omega} \vec{k} \times \vec{E}_0, \quad \frac{k}{\omega} = \frac{1}{c}. \tag{7.147}$$

From the direction aspects of the cross-product of vectors we find that vectors $\vec{E}_0$, $\vec{B}_0$, and $\vec{k}$ are perpendicular to each other such that the propagation direction is towards the cross-product of $\vec{E}$ and $\vec{B}$,

$$\vec{E}_0 \perp \vec{B}_0, \quad \vec{E}_0 \times \vec{B}_0 = \text{direction of } \vec{k}. \qquad E_{0x} = c\, B_{0y}, \quad E_{0y} = -c\, B_{0x}. \tag{7.148}$$

Figure 7.5 shows a schematic picture of such a wave with the propagation vector towards the positive z-axis. It is shown that if the electric field vector is along the x-axis (sometimes pointed towards the positive x-axis and at other times towards the negative x-axis), then the magnetic field would be along the y-axis for an electromagnetic wave traveling along the z-axis.

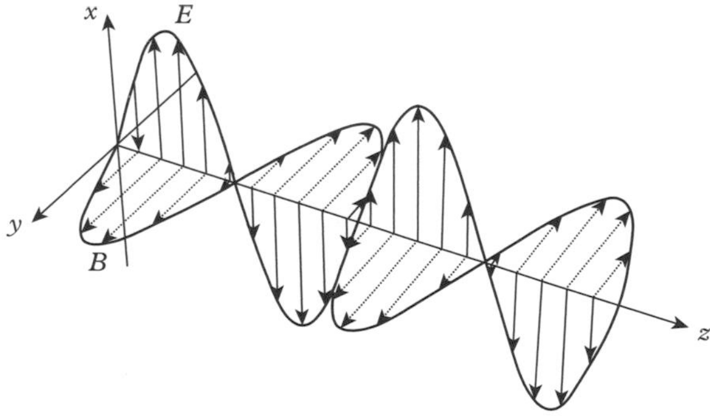

Fig. 7.5 *Plane electromagnetic wave. At each point in space, the directions of the electric and magnetic fields are perpendicular to each other, and the direction of propagation is perpendicular to the plane of the electric and magnetic vectors. The wave depicted here is moving towards the positive z-axis with electric field oscillating along the x-axis and the magnetic field along the y-axis.*

7.3.3 Intensity of Electromagnetic Wave

We have shown before that a traveling wave on a taut string transports energy from one place to another. You can also think of an electromagnetic wave in terms of the flow of electromagnetic energy in space. Consider a linearly polarized plane electromagnetic wave in a vacuum traveling towards the positive z-axis given by the following expressions:

$$E_x = E_0 \cos(kz - \omega t), \quad E_y = 0, \tag{7.149}$$

$$B_x = 0, \quad B_y = \frac{E_0}{c} \cos(kz - \omega t). \tag{7.150}$$

From electromagnetic theory, we know that the electric and magnetic fields have energy, given by the following formulas for energy per unit volume or **energy density** (u) of each:

$$\text{Energy density in electric field, } u_E = \frac{1}{2}\epsilon_0 \vec{E} \cdot \vec{E}, \tag{7.151}$$

$$\text{Energy density in magnetic field, } u_B = \frac{1}{2\mu_0} \vec{B} \cdot \vec{B}. \tag{7.152}$$

Therefore, the total energy density u in the electromagnetic wave would be the sum of the energy densities in the electric and magnetic field waves,

$$\begin{aligned} u = u_E + u_B &= \frac{1}{2}\epsilon_0 E_0^2 \cos^2(kz - \omega t) + \frac{1}{2\mu_0}\left(\frac{E_0^2}{c^2}\right)\cos^2(kz - \omega t) \\ &= \epsilon_0 E_0^2 \cos^2(kz - \omega t) = \frac{1}{2}\epsilon_0 E_0^2 \left[1 + \cos(2kz - 2\omega t)\right]. \end{aligned} \tag{7.153}$$

Here c is the speed of light in vacuum. Thus, the energy density moves with the wave and oscillates at twice the frequency of the wave itself. The electromagnetic energy density is also a wave of half the wavelength and twice the frequency of the electromagnetic wave. We can figure out the rate of transport of energy at any location by calculating the net energy passing through a cross-sectional area, perpendicular to the direction of the wave, in a small time interval Δt. As shown in Fig. 7.6, all energy in the volume of the cylinder of area A and length $c\Delta t$ will pass through area A in time interval Δt.

The energy contained in a box of length $c\Delta t$ and cross-sectional area A will pass through the cross-sectional plane in time Δt,

$$\text{Energy passing through area } A \text{ in time } \Delta t = u \times \text{volume} = uAc\Delta t. \tag{7.154}$$

The energy per unit area per unit time passing through the plane perpendicular to the wave is called the **power flux** and denoted by S,

$$S = \frac{\text{Energy passing through area } A \text{ in time } \Delta t}{A\Delta t} = uc. \tag{7.155}$$

The flux of power at any place also fluctuates in time, as can be seen by substituting u from Eq. 7.153,

$$S = c\epsilon_0 E_0^2 \cos^2(kz - \omega t). \tag{7.156}$$

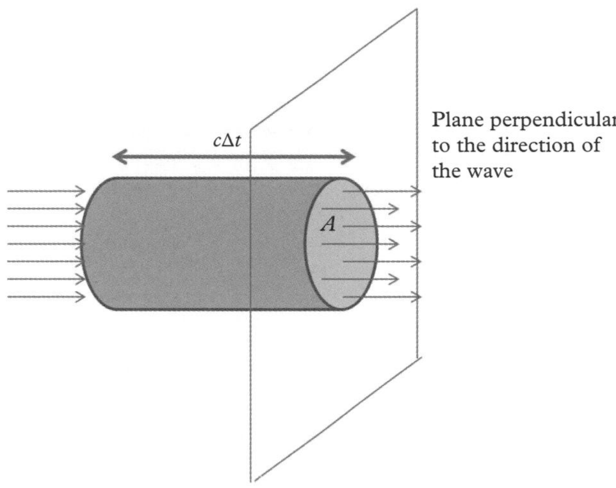

Fig. 7.6 *Energy $uAc\Delta t$ contained in volume $Ac\Delta t$ will pass area A in time Δt.*

Since the frequency of visible light is very high (of the order of 10^{14} cycles per second), the power flux will be an extremely rapidly varying quantity. Our eyes as well as most measuring devices cannot respond this fast and only an average over many cycles is observed. The quantity obtained by time-averaging the power flux is called the **intensity** or **irradiance** I of the electromagnetic wave,

$$\text{Intensity, } I = \langle S \rangle_{\text{time-average}}. \tag{7.157}$$

Time-averaging the expression for the power flux given above is easily performed by noting the following mathematical result:

$$\frac{1}{\text{one period}} \int_{\text{one period}} \cos^2(kx - \omega t)\, dt = \frac{1}{2}.$$

Hence, the intensity of light moving at speed c in a vacuum will be

$$I = \frac{1}{2} c\epsilon_0 E_0^2. \tag{7.158}$$

This can also be written in terms of the amplitude of the magnetic field in the wave,

$$I = \frac{1}{2} c\epsilon_0 E_0^2 = \frac{c}{2\mu_0} B_0^2. \tag{7.159}$$

If the wave is not in a vacuum but in some other linear medium such as glass or air, we will need to replace the speed c by the speed v in the medium, the permittivity of the vacuum ϵ_0 by the permittivity of the medium ϵ, and the permeability of the vacuum μ_0 by the permeability of the medium μ,

$$I = \frac{1}{2} v\epsilon E_0^2 = \frac{v}{2\mu} B_0^2. \quad \text{(Linear medium)} \tag{7.160}$$

Note that the amount of energy crossing a given area also depends upon the orientation of the area relative to the direction of light. To take the direction into account we introduce a power flux vector $\vec{S}$, called the **Poynting vector**, by the following defining equation:

$$\vec{S} = \frac{1}{\mu_0} \vec{E} \times \vec{B}. \tag{7.161}$$

The cross-product of $\vec{E}$ and $\vec{B}$ makes it evident that the Poynting vector $\vec{S}$ points in the direction of the propagation of the electromagnetic wave. The power crossing any arbitrary surface is then obtained by dividing the surface into small patches, calculating the power through each patch, and then summing the contributions. Let $\vec{S}$ be the average Poynting vector over a patch of area vector $\Delta\vec{A}$, then the power crossing area $\Delta\vec{A}$ will be given by the projection of $\vec{S}$ on $\Delta\vec{A}$ times the area, which is equal to the scalar product of $\vec{S}$ and $\Delta\vec{A}$,

$$\text{Power moving past the area element } = \vec{S} \cdot \Delta\vec{A}. \tag{7.162}$$

Here the direction of the area vector is perpendicular to the surface of the area element. The total power crossing any surface is equal to the sum of the contributions from all patches. The sum over infinitesimal patches leads to the following integral formula:

$$\text{Power crossing any surface } = \iint \vec{S} \cdot d\vec{A}. \tag{7.163}$$

Example 7.2 Intensity of an Isotropic Source

The intensity from an isotropic source such as the Sun falls off from the source in a particular way when we apply the principle of the conservation of energy to the energy flow. In this example we find the relation between intensities at two distances from an isotropic source. An isotropic source emits light in all directions equally.

Solution

Consider a point source that emits light (which is an electromagnetic wave) in a spherical symmetric way, e.g. in a spherical wave, as shown in Fig. 7.7. Energy passing the spherical surface with radius r_1 in an interval Δt must equal the energy passing through the surface with radius r_2 in the same interval of time.

Let the intensities at r_1 and r_2 be denoted by I_1 and I_2 respectively. As the product of intensity and surface area equals the total energy passing the surface per unit time, we obtain the following equality based on the energy conservation in each unit time interval:

$$4\pi r_1^2 I_1 = 4\pi r_2^2 I_2 \implies \frac{I_2}{I_1} = \frac{r_1^2}{r_2^2}. \tag{7.164}$$

This shows that the intensity of an isotropic source drops off as the inverse square of the distance from the source.

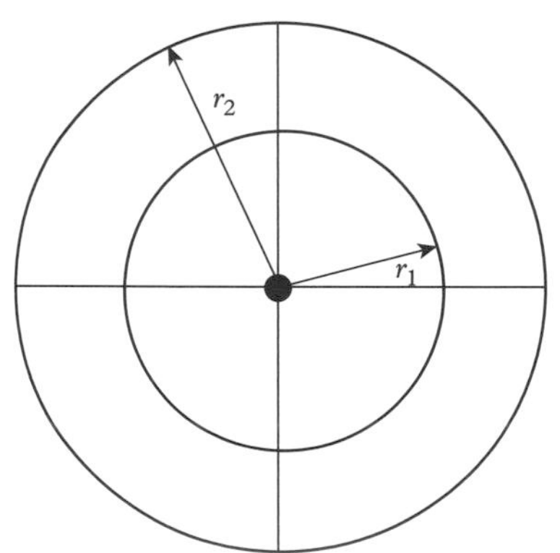

Fig. 7.7 *Spherical wavefronts at two distances from a point source at the center.*

Example 7.3 Electric and Magnetic Fields in Light

Consider a plane wave light source that delivers 3 mW of power over a 0.5 cm^2 area. Find the (a) intensity, and (b) amplitudes of electric and magnetic fields.

Solution

(a) The intensity is power per unit area,

$$I = \frac{3 \times 10^{-3}\ \text{W}}{0.5 \times 10^{-4}\ \text{m}^2} = 60\ \text{W/m}^2.$$

(b) The intensity is proportional to the square of the electric field amplitude

$$E_0 = \sqrt{\frac{2I}{c\epsilon_0}} = \frac{\sqrt{2 \times 60\ \text{W/m}^2}}{\sqrt{3 \times 10^8\ \text{m/s} \times 8.85 \times 10^{-12}\ \text{N.m}^2/\text{C}^2}} = 213\ \text{N/C}.$$

The magnetic and electric fields in an electromagnetic wave are related as follows:

$$B_0 = \frac{E_0}{c} = \frac{213\ \text{N/C}}{3 \times 10^8\ \text{m/s}} = 7.09 \times 10^{-7}\ \text{T}.$$

This magnetic field is approx 1.5% of the magnetic field of the Earth.

7.3.4 Electromagnetic Momentum and Radiation Pressure

Imagine an electromagnetic wave incident on a material. The electrons of the surface molecules are accelerated by the electric field in the wave. In addition they also experience a force from the magnetic field. Consequently, there would be a net force on the surface normal to the direction of the wave. This would result in a pressure on the material. We call the pressure by the electromagnetic wave **radiation pressure,** $\mathcal{P}$.

Maxwell showed that an electromagnetic wave also carries momentum. The momentum density, i.e. the momentum of an electromagnetic field per unit volume can be shown to be

$$\vec{p} = \epsilon_0 \vec{E} \times \vec{B}, \tag{7.165}$$

which can be expressed in terms of the Poynting vector as follows:

$$\vec{p} = \epsilon_0 \mu_0 \vec{S} = \frac{\vec{S}}{c^2}. \tag{7.166}$$

Since, the magnitude of the Poynting vector is related to the energy density u by $|\vec{S}| = uc$, we get

$$|\vec{p}| = \frac{u}{c}.$$

Thus, for a harmonic plane electromagnetic wave traveling towards the positive z-axis, the momentum per unit volume carried by the wave is given by

$$\vec{p} = \hat{k}\,\frac{\text{Energy per unit volume}}{c} = \hat{k}\,\frac{u}{c}. \tag{7.167}$$

Consider an area element A normal to the direction of the wave. In time interval Δt the total momentum ΔP_z passing will be

$$\Delta P_z = \frac{u \times A \times c\Delta t}{c}. \tag{7.168}$$

Therefore, the force applied by the electromagnetic field over area A per unit area will be

$$\frac{1}{A}\frac{\Delta P_z}{\Delta t} = u. \tag{7.169}$$

This is the radiation pressure at that location. Thus the radiation pressure $\mathcal{P}$ at a point in space where the energy density is u equals the energy density,

$$\mathcal{P} = u. \tag{7.170}$$

Do not confuse $\mathcal{P}$ with power. Let's see if the units are right;

$$[u] = \text{J/m}^3 = \text{N.m/m}^3 = \text{N/m}^2 = \text{unit of pressure!}$$

Equation 7.170 gives the instantaneous pressure exerted on a perfectly absorbing surface. But, since the energy density oscillates rapidly, we are usually interested in a time-averaged quantity, which can be written in terms of intensity,

$$\mathcal{P}_{\text{ave}} = u_{\text{ave}} = \frac{I}{c}. \quad \text{(perfectly absorbing)} \tag{7.171}$$

If a surface is perfectly reflecting, then the force on the surface will be twice as great, since the direction of the momentum of the incoming wave would be reversed upon reflection. Therefore the average pressure will be twice as much,

$$\mathcal{P}_{\text{ave}} = 2\frac{I}{c}. \quad \text{(perfectly reflecting)} \tag{7.172}$$

Example 7.4 Radiation Pressure

A laser light of power 3 W is spread evenly across the cross-section of the beam of diameter 2 mm. When the laser light is incident on a perfectly reflecting surface of a spherical particle of diameter 1.5 mm in a vacuum, the particle is observed to be suspended in space. Find the density of the particle.

Solution

Here, the force due to the radiation pressure is able to balance the force of gravity. Let ρ be the desired density of the spherical particle and R its radius. Let P be the radiation pressure at the site of the particle. The balancing of forces yields the following relation:

$$\mathcal{P}A_{\text{cross-section}} = (\rho V_{\text{particle}})g.$$

Here pressure acts on the cross-sectional area of the particle, as shown in Fig. 7.8, which is simply a circle of radius R, and the volume of the particle is just the volume of a sphere,

$$\mathcal{P}\pi R^2 = \left(\rho\frac{4}{3}\pi R^3\right)g.$$

Hence the required density is

$$\rho = \frac{3\mathcal{P}}{4gR}.$$

Now, we put in the numerical values given here:

$$R = 0.75 \text{ mm} = 7.5 \times 10^{-4} \text{ m}$$

$$g = 9.81 \text{ m/s}^2$$

$$\mathcal{P} = 2\frac{I}{c} = 2 \times \frac{3 \text{ W}}{\pi(1 \times 10^{-3} \text{ m})^2} \times \frac{1}{3 \times 10^8 \text{ m/s}} = 2 \times 10^{-3} \text{ Pa}.$$

Hence, $\rho = 0.2$ kg/m^3. It is interesting to compare the density found to the density of air at standard temperature and pressure, which is approximately 1.2 kg/m^3. Clearly, one will need a very powerful laser to suspend an ordinary material.

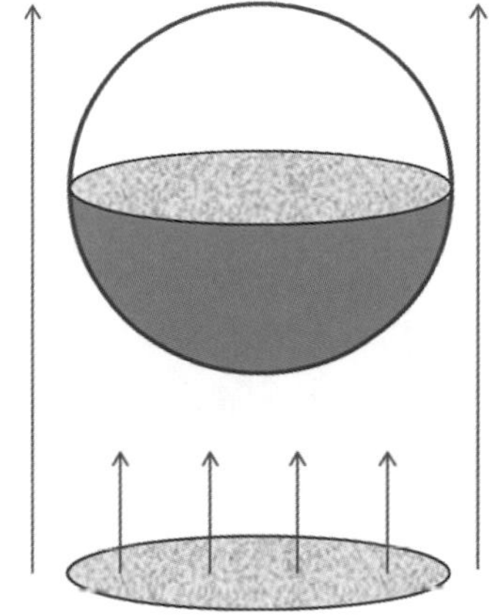

Fig. 7.8 *Example 7.4.*

7.3.5 Polarization of an Electromagnetic Wave

Describing Polarization

The polarization of a wave refers to the direction of oscillations in the medium when a wave travels through it. For instance, when a sound wave travels through the air the particles of air vibrate along the line of travel of the sound wave. We say that a sound wave in air is longitudinal or longitudinally polarized. A transverse wave on a string was seen to have oscillations in the string that were perpendicular to the direction of the wave, which was along the string. If the string is along the z-axis, then there are two independent directions for transverse vibrations, along the x- and y-axes. We say that the transverse wave on a string has transverse polarization.

The two types of polarization, longitudinal and transverse, can be illustrated by the motion of a slinky, as shown in Fig. 7.9. Suppose the slinky is along the z-axis, then for a longitudinal wave in the slinky, the string would remain along the z-axis and the displacements would travel down the slinky, as seen by the compressions moving from one place to the next on the slinky. You can generate a longitudinal wave in a slinky by compressing a part of the slinky and letting go. The compression would travel down the slinky with no displacements along the x- or y-axis. To set up a transverse wave you would oscillate one end of the slinky in the perpendicular direction, here in the xy-plane. You would find a transverse polarized wave traveling down the slinky.

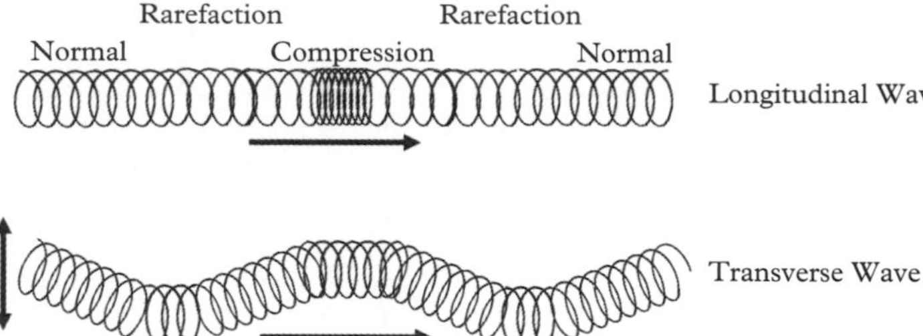

Fig. 7.9 *Longitudinal and transverse waves in a slinky.*

Linear and Elliptical Polarization

In the case of electromagnetic waves, the **polarization** of the wave is indicated by the direction of electric field vector $\vec{E}$, and the direction of the magnetic field $\vec{B}$ is deduced from the direction of $\vec{E}$ and the propagation vector $\vec{k}$. Note that we could equally well use the direction of the magnetic field vector for this purpose. However, traditionally, polarization of an electromagnetic wave is indicated by the direction of the electric field rather than the direction of the magnetic field.

It has been shown previously that the electromagnetic wave is transversely polarized. This would mean that if an electromagnetic wave is traveling towards the positive z-axis, then the polarization will be pointed in some direction in the xy-plane. Since an arbitrary vector in a plane can be constructed from its two components, there are only two independent polarization states for electromagnetic waves. For an electromagnetic wave traveling along the z-axis the z-component of the electric field is zero, $E_z = 0$, and the amplitudes E_{0x} and E_{0y} and phase constants ϕ_x and ϕ_y of E_x and E_y waves can be independently assigned,

$$E_x(z,t) = E_{0x}\cos(kz - \omega t + \phi_x), \tag{7.173}$$

$$E_y(z,t) = E_{0y}\cos(kz - \omega t + \phi_y). \tag{7.174}$$

If the two phase constants are equal, the two independent waves move together such that the net electric field always points along the same line in the xy-plane, as illustrated in Fig. 7.10. Such waves are called **linearly polarized**. Linearly polarized waves play a very important role in the study of optics since they are very simple to produce, manipulate, and analyze.

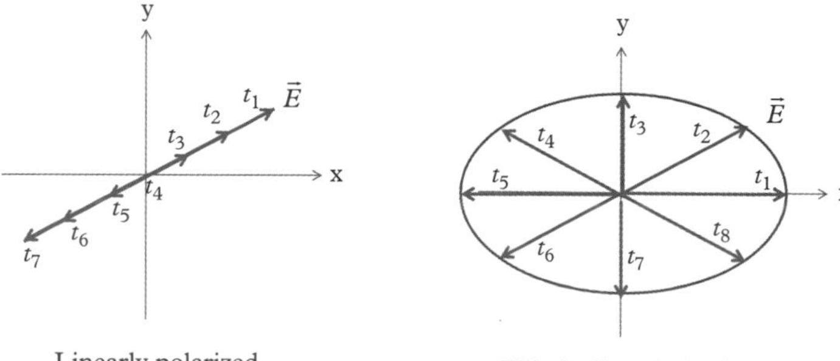

Fig. 7.10 *Linearly and elliptically polarized states. In the linearly polarized state, the electric field vector at a particular point points in a definite direction and the direction opposite to it and the magnitude changes with time. In the elliptically polarized wave, the direction of the electric field vector at a particular point rotates with time.*

If the two phase constants of E_x and E_y waves are not equal, the electric field vector will rotate from one direction at a particular z to another direction at another z, as illustrated in Fig. 7.10. Such waves are called **elliptically polarized,**

$$\text{Linear Polarization: } \phi_x = \phi_y \tag{7.175}$$

$$\text{Elliptic Polarization: } \phi_x \neq \phi_y. \tag{7.176}$$

The polarization direction of an elliptically polarized wave rotates as the wave travels through space, as shown in Fig. 7.11. When observed from the side of space towards which the wave is going, the polarization direction of the elliptically polarized wave will appear to rotate clockwise or counter-clockwise, depending upon whether $\phi_x < \phi_y$ or $\phi_x > \phi_y$ respectively.

A special case of an elliptically polarized wave occurs when the two amplitudes are equal $E_{0x} = E_{0y}$ and the phase constants differ by $\frac{\pi}{2}$ radians, $|\phi_x - \phi_y| = \frac{\pi}{2}$. These more restricted waves are called **circularly polarized** waves,

$$\text{Circular Polarization: } E_{0x} = E_{0y} \text{ and } |\phi_x - \phi_y| = \frac{\pi}{2}. \tag{7.177}$$

In the case of circularly polarized light, when $\phi_y = \phi_x - \pi/2$, we find that the wave rotates clockwise. This wave is called a **right-circularly polarized** wave. On the other hand when

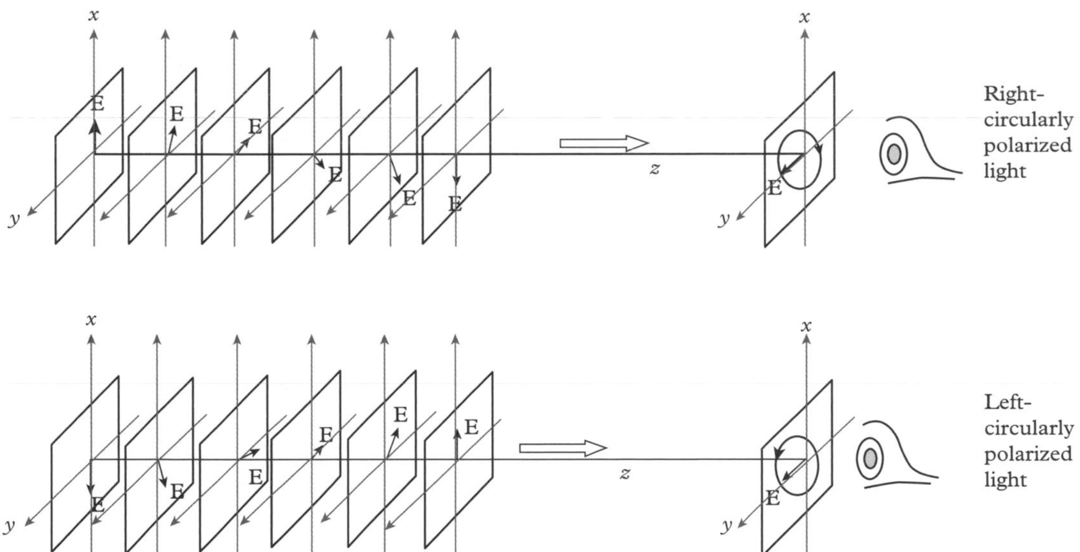

Fig. 7.11 *Right and left circularly polarized electromagnetic waves. When we look at the incoming wave, the electric field vector rotates clockwise for a right-circularly polarized wave and counterclockwise for a left-circularly polarized wave.*

$\phi_y = \phi_x + \pi/2$, we find that the wave rotates counterclockwise. This wave is called a **left-circularly polarized wave**,

$$\text{Right-Circular Polarization: } E_{0x} = E_{0y} \text{ and } \phi_y - \phi_x = -\frac{\pi}{2}, \tag{7.178}$$

$$\text{Left-Circular Polarization: } E_{0x} = E_{0y} \text{ and } \phi_y - \phi_x = +\frac{\pi}{2}. \tag{7.179}$$

Malus's Law

Natural light such as the light from the Sun is unpolarized. Sunlight is generally a random mixture of various waves with different polarizations and frequencies. Any device that produces polarized light is called a **polarizer**. When you pass an unpolarized light through a **polarizer** the transmitted light is polarized.

A **linear polarizer** is a device that lets through linearly polarized light that has the polarization parallel to the polarization axis of the polarizer and blocks any wave that oscillates at 90° to the axis. If incident light is neither polarized along the preferred axis nor at right-angles to it, then only the vector component of the electric field along the preferred axis passes through the linear polarizer. Therefore, the transmitted light will have a lesser intensity than the incident light when the incident light is either unpolarized or has its polarization direction not parallel to the axis of the linear polarizer.

What would be the intensity of the transmitted wave compared to the intensity of the incident wave? To find this, we start with a linearly polarized wave of known polarization and intensity and pass this wave through a linear polarizer at a different angle to the polarization direction. Therefore, the set-up for this experiment requires two polarizers, one to produce the wave of known polarized state and the other to analyze what happens at different angles. The first polarizer is called the **polarizer** and the second is called the **analyzer**. Figure 7.12 shows the polarizer/analyzer experiment arrangement, where the axis of the analyzer to the incoming linearly polarized light makes an angle θ to the latter.

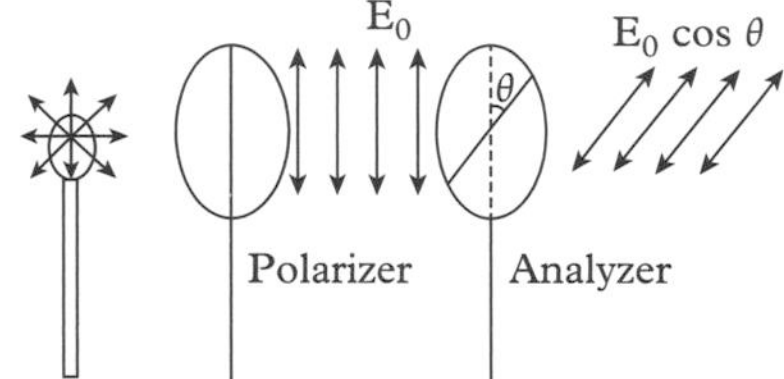

Fig. 7.12 *The polarizer and analyzer in a study of the polarization of light.*

If the amplitude of the electric field before the analyzer is E_0, then after the analyzer only the component along the axis of the polarizer, $E_0 \cos\theta$, will be transmitted. We know that the intensity of an electromagnetic wave is proportional to the square of the amplitude of the electric field,

$$I \propto (\text{Amplitude of Electric Field})^2. \tag{7.180}$$

Therefore the intensity of the electromagnetic wave will drop from E_0^2 to $E_0^2 \cos^2\theta$ when polarized light passes through the analyzer according to the following formula, that is,

$$\frac{I_{\text{after analyzar}}}{I_{\text{before analyzar}}} = \cos^2\theta. \tag{7.181}$$

This experimental observation is called **Malus's law**. According to this experimental observation, when a light wave is sent through two plane polarizers whose axes are 90° to each other, then no light would emerge at the other end since, in this case, we have $\cos 90° = 0$.

7.4 Matter Waves

Quantum mechanics is based on the idea that, at a fundamental level, matter is a wave. Louis de Broglie hypothesized that we can think of a particle of momentum p in terms of a wave of wavelength λ if the momentum and wavelength are related by the following fundamental relation:

$$\lambda = \frac{h}{p}, \tag{7.182}$$

where h is the Planck constant,

$$h = 6.6 \times 10^{-34}\ \text{J.s.}$$

Based on this proposal by de Broglie, Schrödinger introduced a dynamical equation for the matter wave function, $\psi(x, y, z, t)$. Unlike the wave functions we have studied previously, which were all real-valued functions, the matter wave is a complex function. For a particle of mass m moving in a potential $V(x, y, z)$ the wave function obeys Schrödinger's equation, given by

$$i\hbar\frac{\partial\psi}{\partial t} = -\frac{\hbar^2}{2m}\left(\frac{\partial^2\psi}{\partial x^2} + \frac{\partial^2\psi}{\partial y^2} + \frac{\partial^2\psi}{\partial z^2}\right) + V\psi, \tag{7.183}$$

where

$$\hbar = \frac{h}{2\pi}.$$

A free particle will have no potential acting on it and the $V\psi$ term will be zero,

$$i\hbar\frac{\partial\psi}{\partial t} = -\frac{\hbar^2}{2m}\nabla^2\psi, \tag{7.184}$$

where $\nabla^2\psi$ is a shorthand for the space derivatives,

$$\nabla^2\psi = \frac{\partial^2\psi}{\partial x^2} + \frac{\partial^2\psi}{\partial y^2} + \frac{\partial^2\psi}{\partial z^2}.$$

From the very beginning of quantum mechanics it has been difficult to understand the meaning of the complex wave function. Max Born proposed that the square of the absolute value of the complex function gives the probability density of finding the particle. Thus, the probability that the particle will be found in a volume $dxdydz$ at the point whose coordinate is (x, y, z) is given by

$$\text{Probability} = |\psi(x, y, z, t)|^2 dxdydz. \tag{7.185}$$

Since the probability that the particle is somewhere in space would be 1, an integral over all space should give the value 1,

$$\int_{\text{all space}} |\psi(x, y, z, t)|^2 dxdydz = 1. \tag{7.186}$$

7.4.1 Free particle in Open Space

Consider a particle moving towards the positive z-axis in open space, i.e. not constrained in a box. The matter wave function of this particle will be a traveling wave solution of Eq. 7.184, since the medium is unbounded. The simplest solution is a harmonic traveling wave with wave number k and frequency ω.

$$\psi(x, y, z, t) = Ae^{ikz}e^{-i\omega t}. \tag{7.187}$$

When we plug this wave function into Eq. 7.184 we obtain the following dispersion relation for the matter wave of a free particle:

$$\hbar\omega = \frac{\hbar^2 k^2}{2m}. \tag{7.188}$$

To relate this wave to the momentum and energy of the particle we first replace the wave number k by $2\pi/\lambda$ and use de Broglie's relation, Eq. 7.182, to write λ as h/p,

$$\hbar\omega = \frac{p^2}{2m}. \tag{7.189}$$

We note that the right side is just the expression for the energy E of the free particle. Therefore, the energy of the particle is related to the frequency of the wave,

$$E = \hbar\omega. \tag{7.190}$$

From the dispersion relation of the matter wave, Eq. 7.188, we note that the matter wave of a free particle is a dispersive wave with phase velocity

$$v_p = \frac{\omega}{k} = \frac{\hbar k}{2m}, \tag{7.191}$$

which is half of the velocity of the particle if we replace $\hbar k$ by p,

$$v_p = \frac{p}{2m}. \tag{7.192}$$

The group velocity of the wave is, however, equal to the velocity of the particle,

$$v_g = \frac{d\omega}{dk} = \frac{\hbar k}{m} = \frac{p}{m}. \tag{7.193}$$

This is consistent with our understanding that the energy of a wave moves at group velocity and not at phase velocity. The plane wave function in Eq. 7.187 has a definite energy $E = \hbar\omega$ and a definite momentum $p = \hbar k$. We say that this wave function represents a quantum state in which the particle has a definite energy and definite momentum. Note that in the dispersion relation, another state, $p = -\hbar k$, has the same energy as this state. The two states are said to be **degenerate** in energy.

Although the state represented by the plane wave function in Eq. 7.187 has definite energy and momentum, the position of the particle is completely uncertain, as we can show by the following calculation. Recall Born's interpretation of the wave function given above in which the square of the absolute value of the wave function gave the probability density and multiplication by a volume element gave the probability of finding the particle in that volume. Here, the absolute value of the wave function is constant $|A|$,

$$\text{Probability} = |\psi(x,y,z,t)|^2 dxdydz = |A|^2 dxdydz. \tag{7.194}$$

The probability density is independent of position. Therefore, the particle is equally likely to be found anywhere in space. This is a special application of Heisenberg's **uncertainty principle,** which states that the product of uncertainty in momentum and position cannot be less than a particular number. Let Δp be the uncertainty in momentum and Δx the uncertainty in position, then by Heisenberg's uncertainty principle,

$$\Delta x \Delta p \geq \frac{\hbar}{2}. \tag{7.195}$$

Since the momentum of the particle is known exactly here, we have $\Delta p = 0$. Therefore, we must have $\Delta x = \infty$, meaning the position of the free particle in open space will not be known with any precision. Now, this calculation is only for an idealized situation, in which we had assumed that the particle has a definite momentum.

In reality, we will not know the momentum precisely. In that case, the wave function will not be a harmonic traveling wave but a wave packet whose momentum will be centered about the most likely momentum and a width corresponding to the uncertainty in momentum. A wave packet travels as a ball in free space whose central position moves with v_g of the wave packet and whose width spreads out with time since the wave is dispersive.

7.4.2 Free Particle in a Confined Space

Let us consider a one-dimensional matter wave example in a confined geometry. Consider a free particle confined in a one-dimensional box of width a on the z-axis between $z = 0$ and $z = a$. The wave equation for this particle will be

$$i\hbar\frac{\partial\psi}{\partial t} = -\frac{\hbar^2}{2m}\frac{\partial^2\psi}{\partial z^2}, \tag{7.196}$$

with boundary conditions

$$\psi(z=0,t) = \psi(z=a,t) = 0. \tag{7.197}$$

Physically, the problem is similar to that of oscillations of a string tied at both ends. Therefore, we will get standing waves. The standing wave solutions that satisfy the boundary conditions in Eq. 7.197 are given by

$$\psi_n(z,t) = A_n \sin(k_n z)e^{-i\omega_n t}, \tag{7.198}$$

with

$$k_n = n\frac{\pi}{a}, \quad n = 1, 2, 3, \ldots, \tag{7.199}$$

and the dispersion relation

$$\omega_n = \frac{\hbar k_n^2}{2m}. \tag{7.200}$$

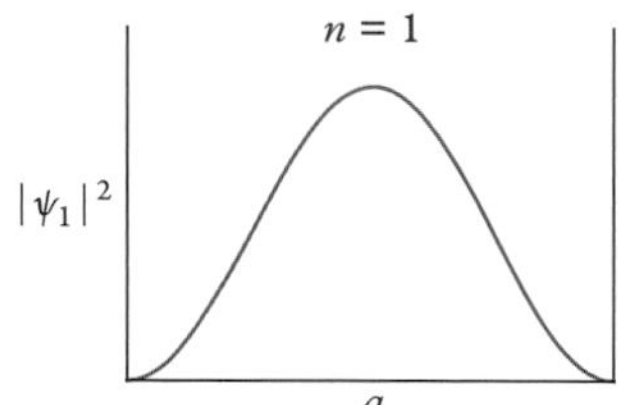

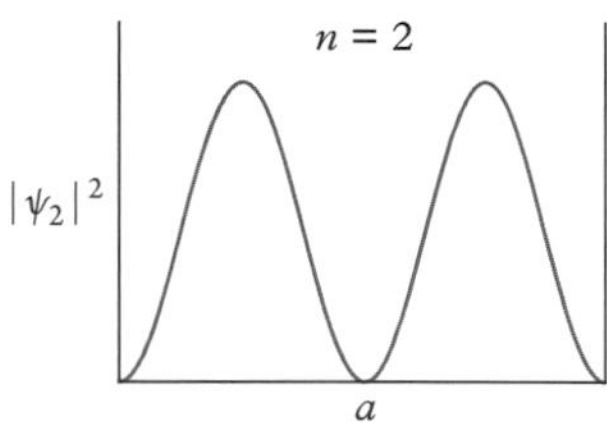

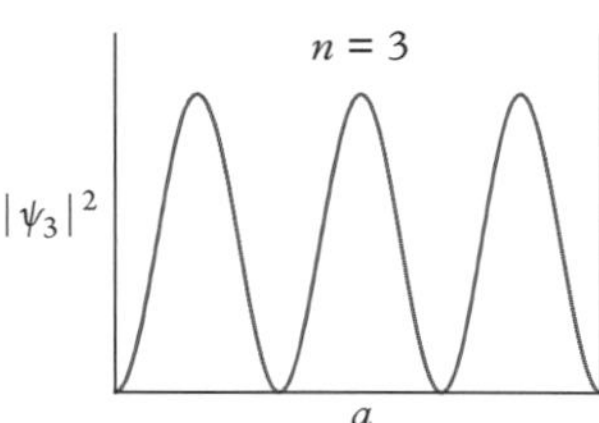

Fig. 7.13 *The probability densities of the three lowest energy states of a particle in a box. The horizontal axis is the x-axis here. Recall that the probability of finding the particle between x and x + dx is given by* $|\psi|^2 dx$.

These standing waves are called **stationary states** of the particle. They correspond to states in which the particle is in a state of definite energy, given by

$$E_n = \hbar\omega_n = \frac{\hbar^2 k_n^2}{2m} = n^2\frac{\hbar^2\pi^2}{2ma^2} = n^2\frac{h^2}{8ma^2}. \tag{7.201}$$

Figure 7.13 shows the probability densities corresponding to the three lowest energy states of a particle in a box.

The lowest energy state, called the **ground state**, has no nodes between the walls. That is, there is no place inside the box where the probability of finding the particle is zero. The ground state is also symmetric about the center of the box. The state with the next higher energy is the first excited state. The wave function of the first excited state has one node between the walls and hence the probability density has a value zero at one point at an inside point of the box. The second excited state has two nodes and each successively higher energy state has one more node. An important point of the energy levels of the particle in a box is that the separation between the states varies as n^2 of the states. Thus, the separation of the energy levels $n = n_1$ and $n = n_2$ will be

$$E_{n_2} - E_{n_1} = \left(n_2^2 - n_1^2\right) E_1. \tag{7.202}$$

Example 7.5 Energy Levels of a Particle in a Box

Consider an electron in a one-dimensional box of size equal to two times the Bohr radius, which is 52.9 pm. What will be the energies of the lowest three states?

Solution

Let us first calculate E_1 and then use $E_n = n^2 E_1$ to calculate the energies of other states,

$$E_1 = \frac{h^2}{8ma^2} = \frac{(6.63\times10^{-34}\text{J.s})^2}{8\times9.11\times10^{-31}\text{kg}\times(2\times52.9\times10^{-12}\text{m})^2} = 5.39\times10^{-18}\text{J}.$$

Therefore, E_2 and E_3 are

$$E_2 = 4E_1 = 2.16\times10^{-17}\text{J}, \qquad E_3 = 9E_1 = 4.85\times10^{-17}\text{J}.$$

Example 7.6 Position of Particle in the Box

Suppose the particle is in the ground state, what is the probability that the particle will be in the left one-third of the box?

Solution

The probability that the particle is between x and $x + dx$ is given by

$$dP = |\psi_1(x)|^2 dx.$$

Since the particle has to be somewhere in the box, the constant A_1 in Eq. 7.198 will be $\sqrt{2/a}$. Therefore, $\psi_1(x)$ will be

$$\psi_1(x) = \sqrt{\frac{2}{a}}\sin\frac{\pi x}{a}.$$

The left one-third of the box is between $x = 0$ and $x = a/3$. Therefore, we should integrate from $x = 0$ to $x = a/3$ to obtain the required probability.

$$P(\text{in left one-third of box}) = \int_0^{a/3} \frac{2}{a}\sin^2\left(\frac{\pi x}{a}\right) dx = \frac{1}{a}\int_0^{a/3}\left[1-\cos\left(\frac{2\pi x}{a}\right)\right] dx$$

$$= \frac{1}{a}\left[\frac{a}{3} - \frac{a}{2\pi}\sin\left(\frac{2\pi}{3}\right)\right] = \frac{1}{3} - \frac{\sqrt{3}}{4\pi} \approx 0.21.$$

The probability is not one-third. The particle in the ground state is more likely to be in the middle-third than in the thirds near the walls. Question to student: What will be the probability of the particle being in the middle-third of the box when in the ground state? Note: no additional calculation is necessary. Ans: 0.58.

Example 7.7 Position of Particle When in the First Excited State

Suppose the particle is in the first excited state, what is the probability that the particle will be in the left one-third of the box?

Solution

I have done the probability calculation for the ground state. In this problem we need to do the same steps for the first excited state ψ_2. The probability that the particle is between x and $x + dx$ is given by

$$dP = |\psi_2(x)|^2 dx.$$

The left one-third of the box is between $x = 0$ and $x = a/3$. Therefore, we should integrate from $x = 0$ to $x = a/3$ to obtain the required probability,

$$P(\text{left 1/3 box}) = \int_0^{a/3} \frac{2}{a}\sin^2\left(\frac{2\pi x}{a}\right) dx = \frac{1}{a}\int_0^{a/3}\left[1-\cos\left(\frac{4\pi x}{a}\right)\right] dx$$

$$= \frac{1}{a}\left[\frac{a}{3} - \frac{a}{4\pi}\sin\left(\frac{4\pi}{3}\right)\right] = \frac{1}{3} + \frac{\sqrt{3}}{4\pi} \approx 0.47.$$

The probability is more than one-third. The particle in the first excited state is more likely to be in the left-third and the right-third than in the middle third. Question to student: What will be the probability of the particle being in the middle-third of the box when in this state? Note: no additional calculation is necessary. Ans: 0.06.

7.5 Shock Waves

So far we have studied waves that are generated by the vibration in which vibrating particles have speeds less than the speed of the wave in the medium. For instance, A vibrating plate in air generates a plane sound wave in air if the plate has a speed less than the speed of sound in air. When you swim in water you set up a wave in the water which emanate smoothly from your motion.

What will happen if you shoot a bullet into water that has a higher speed than the speed of the acoustic wave in water? Near the moving particle the disturbance is very complicated and not described by the classical wave equation. However, away from the particle we see

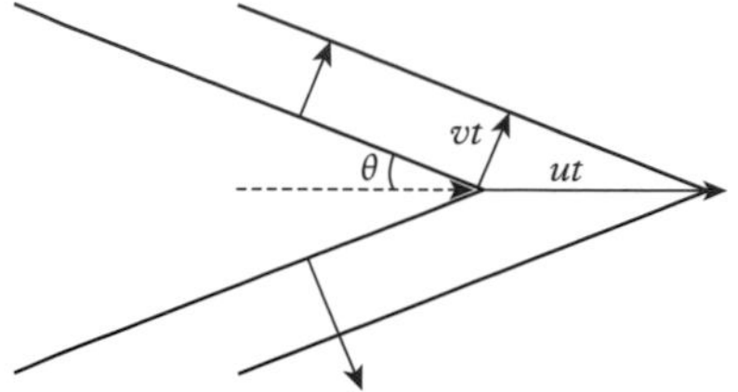

Fig. 7.14 *The wake of a shock wave behind a moving object whose speed u is greater than the speed of the elastic wave v in the material. The wave front moves a distance vt in the time the displacement of the particle is ut. The angle between the wave vector and particle displacement vector is 90° −θ. This gives the relation,* $v = u \sin\theta$.

waves moving such that there is a cone-shaped wake of wave that intersects at the moving bullet, as shown in Fig. 7.14, with the wave fronts parallel to the wake. These waves are called **shock waves**.

We can deduce an expression for the angle of the shock cone by the observation that the plane wave generated by the moving particle will move perpendicularly to the wake at speed v. In time t the wavefront will move a distance vt and the bullet will move a distance ut. From the figure we see that the projection of the displacement vector $\vec{u}t$ in the direction of the wave must equal vt. Therefore,

$$v = u \sin\theta. \tag{7.203}$$

A similar phenomenon occurs in the case of electromagnetic waves which are generated classically by accelerating charged particles. In a vacuum, charged particles cannot move faster than the speed of light, and therefore, we do not see any unusual activity—the electromagnetic wave is fully described by the classical wave equations for electric and magnetic fields. However, in a material medium, a charged particle can move faster than light in than medium since the speed of light in a medium, v, would be equal to the speed in a vacuum, c, divided by the refractive index n,

$$v = \frac{c}{n}.$$

Now, charged particles in accelerators and in nuclear reactions, can exceed v. In that case, the electromagnet waves come out in a cone whose angle is given by Eq. 7.203. These are called **Cherenkov radiation**. This radiation is responsible for the blue glow in water surrounding the nuclear fuel in a nuclear reactor.

EXERCISES

(7.1) A traveling pressure wave in a three-dimensional space is given by the following wave function:

$$\psi(x, y, z, t) = 0.05 \cos(2\pi x + \pi y - 2\pi z - 600\pi t),$$

where x, y, and z are in cm, t is in seconds, and ψ is in Pa. Determine the following quantities.

(a) Angular frequency and frequency.
(b) Propagation vector.
(c) Wave number and wavelength.
(d) Direction of wave motion.
(e) Wave velocity.
(f) Amplitude of the wave.
(g) The phase change at a particular point in space over a period of 3 msec.
(h) The phase difference at a particular instant between two points in space with coordinates (0,0,1) and (1,0,0).

(7.2) The electric field of an electromagnetic wave in a vacuum is given by the following vector function of space and time, where y is in meters and t is in seconds.

$$\vec{E} = \hat{i}\, 3 \times 10^5 (\text{N/C}) \cos\left(2\pi \times 10^{15}\, t + ky + \frac{\pi}{2}\right).$$

(a) Evaluate the frequency, wave number, and wavelength of the wave.
(b) What is the direction of the travel of the wave?
(c) What are the directions of the electric field and the magnetic field at the origin at $t = 0$?
(d) What are the directions of the electric field and the magnetic field at the origin at $t = 0.25 \times 10^{-15}$ seconds?
(e) What is the amplitude of the magnetic field wave at the following places and instants?
   (i) $x = z = 0, y = 0.03$ m, $t = 0$.
   (ii) $x = z = 0, y = 0.045$ m, $t = 0.05$ ns.
(f) Plot the electric field magnitude versus the y-coordinate of space at the following instants on the same graph.
   (i) $t = 0$.
   (ii) $t = 0.25 \times 10^{-15}$ sec.
   (iii) $t = 0.5 \times 10^{-15}$ sec.
   (iv) $t = 0.75 \times 10^{-15}$ sec.

(7.3) A plane harmonic elastic wave $\psi(x, y, z, t)$ is moving in the direction from $(1, 0, 0)$ to $(0, 1, 0)$ with amplitude 5 μm, wave number 1000 cm^{-1}, and frequency 2 MHz. Suppose the phase constant is zero. What will be the wave function of the wave?

(7.4) A light bulb radiates 10 W of light isotropically in space.
(a) What is the intensity of light 1 m from the bulb?
(b) What is the intensity of light at 2 m from the bulb?

(7.5) A spherical sound wave from a point source at the origin has an amplitude that drops with the radial distance r (cm) from the origin as follows:

$$\psi = \frac{A}{r} \cos(2\pi r/\lambda - 2\pi f t),$$

where $A = 5$ Pa.cm, $f = 10$ Hz, and $\lambda = 50$ cm is the wavelength.
(a) Find the wave number of the wave.
(b) What is the ratio of the intensity of the wave at $r = 100$ cm to the intensity at $r = 200$ cm?
(c) Plot the wave function at $t = 0$ from $r = \lambda$ to $r = 5\lambda$. On the same graph, plot the intensity and compare the two.

(7.6) A particular wave is a superposition of two monochromatic plane waves of different frequencies but the same speed v.

$$\psi(x, y, z, t) = A_1 \cos(k_1 x - \omega t) + A_2 \cos(k_2 x - 2\omega t).$$

What is the intensity of the wave at point $(x, 0, 0)$?

(7.7) A wave between two parallel plates parallel to the z-axis separated by a distance of $a = 20$ cm along the x-axis is given by

$$\psi(x, y, z, t) = A \sin\left(\frac{\pi x}{20\text{ cm}}\right) \cos\left(2\pi z/30\text{ m} - 2\pi f t\right).$$

This wave obeys a classical wave equation with speed $v = 2.0 \times 10^8$ m/s.
(a) Sketch the standing wave between the plates at $t = 0$ along the x-axis.
(b) Verify that the given mixed standing/traveling wave function satisfies the classical wave equation and find the value of the frequency of the wave.
(c) Show that the mixed standing/traveling wave can also be written as a superposition of two traveling waves.

(d) Where in the space between plates is the intensity strongest?
(e) Where in the space between plates is the intensity weakest?

(7.8) A wave inside a rectangular pipe of cross-sectional dimensions $a \times b$ is given by

$$\psi(x, y, z, t) = A \sin\left(\frac{2\pi x}{a}\right) \sin\left(\frac{2\pi y}{b}\right) \cos(k_z z - \omega t),$$

where $0 \leq x \leq a$, $0 \leq y \leq b$ and $0 \leq z < \infty$. This wave obeys a classical wave equation with speed v.

(a) Sketch the wave at $t = 0$ in the tube.
(b) (i) Find the dispersion relation of the wave, i.e. ω as a function of the propagation wave number k. (ii) From the dispersion relation, deduce the lowest frequency that will propagate in the tube.
(c) Show that the mixed standing/traveling wave can also be written as a superposition of two traveling waves.
(d) Where in the pipe is the intensity the strongest?
(e) Where in the pipe is the intensity the weakest?

(7.9) In the text we have studied sound waves in terms of two variables, the displacement δx and pressure incremental δp. Sometimes a sound wave is studied in terms of change in volume δV. Suppose sound is passing through an air tube of area of cross-section A. Then the change in volume at a point where the particle displacement is δx will be $\Delta V = A\delta x$. The rate of volume change will be related to speed of the particles as follows

$$\frac{\Delta V}{\Delta t} = A\frac{\Delta x}{\Delta t}, \quad \rightarrow \quad \frac{dV}{dt} = A\frac{dx}{dt}.$$

Show that the characteristic impedance, defined by the ratio of pressure and the volume flow rate, dV/dt,

$$Z = \frac{|p|}{|dV/dt|},$$

is equal to $\rho_0 v/A$, where v is the speed of sound and ρ_0 the density of the medium.

(7.10) A sound wave in air of density 1.225 kg/m^3 is given by

$$\delta p = 10 \text{ [Pa] } \cos(10z + \omega t),$$

where z is in m. Assume the speed of sound to be 350 m/s.

(a) What are the wavelength and frequency of the sound?
(b) What is the intensity of the wave?
(c) What is the expression of the particle displacement wave for the given pressure wave?

(7.11) The density and impedance of water at room temperature are 1000 kg/m^3 and 1.5×10^6 kg/m^2.s. A sound wave in water is given to be

$$\delta p = 10 \text{ [Pa] } \cos(10z + \omega t),$$

where z is in m.

(a) What will be the speed of sound in water at room temperature?
(b) What are the wavelength and frequency of the sound?
(c) What is the intensity of the wave?
(d) What is the expression of the particle displacement wave for the given pressure wave?

(7.12) A pulsating spherical ball (radius a) radiates a spherical acoustic wave in the medium outside the sphere, which we will take to be air. The radiating pressure wave outside the ball can be shown to have the following form:

$$p(r,t) = \frac{A}{r}\cos(kr - i\omega t), \quad (r > a)$$

where k is the wave number, ω the frequency, and A/a the amplitude at $r = a$. Clearly, the amplitude of the wave drops as $1/r$ from the source. Let ρ_0 be the density of air and v the speed of sound in air.

(a) Find the intensity of the wave at a radial distance r from the center of the ball.

(b) Find the total average power radiated by the ball.

(7.13) **Whiskey-bottle resonator.** (Adapted from *Waves*, by Frank Crawford, Jr., Berkeley Physics Course – Vol 3). When you blow across the mouth of a bottle you hear a tone, which is usually the lowest mode of the longitudinal oscillation of air in the bottle. If you think of the entire bottle as a tube of length L with one end open and the other end closed, the wavelength for this mode will be $4L$. For the speed of sound v, this would give the frequency of the tone to be $v/4L$. This turns out to give too low a frequency when compared to the actual sound produced. Helmholtz suggested another way to look at this system which gives much better agreement with the actual sound produced.

Suppose V_0 is the volume of the body of the bottle, l the neck length, and A the area of cross-section of the opening. Let ρ_0 be the density of air and p_0 the ambient pressure in the bottle. When the gas in the body expands into the neck to a distance x, the volume occupied by this air would be $V = V_0 + Ax$ and the pressure is p. The relation between (p_0, V_0) and (p, V) for an adiabatic process is

$$pV^\gamma = p_0 V_0^\gamma.$$

Helmholtz suggested thinking of the air in the bottle as a spring where the spring force is equal to $(p - p_0)A$ that acts on the particles of the air in the neck part only, as shown in Fig. 7.15.

(a) Let the x-axis be along the neck and assume $|p - p_0|/p_0 \ll 1$ to show that the x-component of the force on the mass of the air in the neck will be

$$F_x \approx -\left(\frac{\gamma\, p_0 A^2}{V_0}\right) x.$$

(b) Find the expression of the frequency of the mode generated. [Note: although your answer will come out in terms of l, the length of the neck, you would need to replace this l by l_{eff}, which is approximately $l + 2 \times 0.6R$, where R is the radius of the neck.]

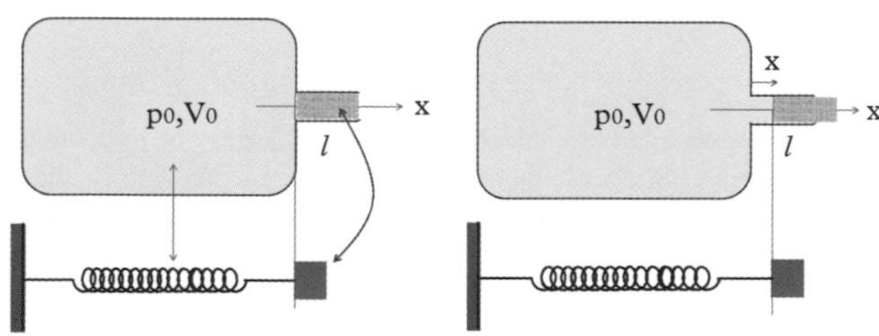

Fig. 7.15 *Exercise 7.13.*

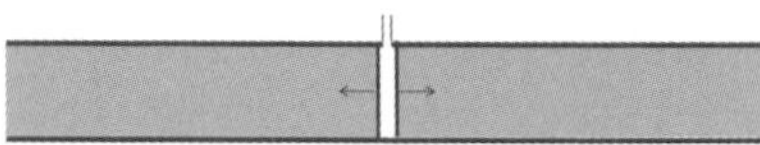

Fig. 7.16 *Exercise 7.14.*

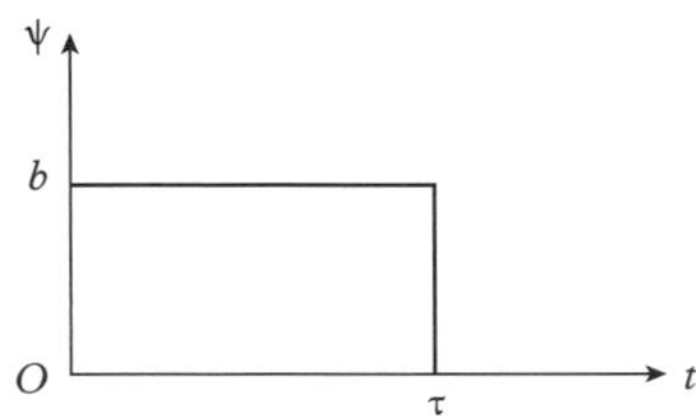

Fig. 7.17 *Exercise 7.14.*

(c) Evaluate the frequency for a bottle with $p_0 = 1$ atm, $V_0 = 1000\ \text{cm}^3$, $l = 4$ cm, and $A = 8\ \text{cm}^2$. Use $\gamma = 5/3$ and the density of air to be 0.001 g/cm^3. Make sure you convert units into the same system.

(7.14) **Simulating the sound wave in an explosion.** Consider an infinitely long tube of area of cross-section A, as shown in Fig. 7.16. Let us place a layer of explosive at $x = 0$, that goes off over a time duration τ, which results in a pressure wave towards $x > 0$ and $x < 0$.

Suppose the displacement ψ of the particles of air at $x = 0^+$ and $x = 0^-$ are given by the functions shown in Fig. 7.17.

(a) Express the pulse in time in terms of Fourier integrals by finding $A(\omega)$ and $B(\omega)$ in

$$\psi(0, t) = \int_0^\infty A(\omega)\cos(\omega t) + \int_0^\infty B(\omega)\sin(\omega t).$$

(b) Show that the intensity of the wave in any particular frequency is

$$I(\omega) \propto \left[\frac{\sin(\omega\tau/2)}{\omega\tau/2}\right]^2.$$

(c) Suppose air acts a non-dispersive medium and find the expression of the traveling waves in the two directions in the tube.

(7.15) We have studied the derivation of the electromagnetic wave equation in a vacuum. In a dielectric medium, the polarization of the medium also contributes to the electric field. Similarly, if we have magnetic material, the magnetic field has a contribution from magnetization as well. To include these effects, we introduce two additional fields, the electric displacement vector $\vec{D}$ and magnetic intensity $\vec{H}$, which are defined by the following relations to the electric field $\vec{E}$ and magnetic field $\vec{B}$:

$$\vec{D} = \epsilon_r\epsilon_0\vec{E}, \quad \vec{H} = \frac{\vec{B}}{\mu_r\mu_0},$$

where ϵ_r is called the relative permittivity or dielectric constant and μ_r the relative magnetic permeability. Maxwell's equations in matter are then given by the following four equations:

$$\vec{\nabla}\cdot\vec{D} = \rho_f, \tag{7.204}$$

$$\vec{\nabla}\times\vec{E} = -\frac{\partial\vec{B}}{\partial t}, \tag{7.205}$$

$$\vec{\nabla}\cdot\vec{B} = 0, \tag{7.206}$$

$$\vec{\nabla}\times\vec{H} = \vec{\mathcal{J}}_f + \frac{\partial\vec{D}}{\partial t}. \tag{7.207}$$

Here ρ_f is the density of free charge, i.e. not the charges due to polarization, and $\mathcal{J}_f$ is the free current. Suppose the medium is homogeneous with ϵ_r and μ_r constant. Let there be no free charges $\rho_f = 0$ or free currents $\vec{\mathcal{J}}_f = 0$. Starting from Maxwell's equations given here, prove the following aspects of electromagnetic waves in the medium.

(a) The velocity of a wave through the medium will be

$$v = \frac{c}{\sqrt{\epsilon_r \mu_r}},$$

where c is the speed of the wave in a vacuum.

(b) The impedance of the medium will be

$$Z = \frac{|\vec{E}|}{|\vec{H}|} = Z_0 \sqrt{\frac{\mu_r}{\epsilon_r}},$$

where $Z_0 = \sqrt{\mu_0/\epsilon_0}$ is the impedance of the vacuum.

(7.16) A plane electromagnetic wave in a vacuum is given by

$$\vec{E} = \hat{x}\, E_0 \cos(kz - \omega t). \tag{7.208}$$

(a) What will be the associated magnetic field wave?

(b) Find the Poynting vector $\vec{S}$ of the wave.

(c) Find the energy density u_E in the electric field wave. Do the same for the magnetic field wave and show that the energy density u_B of the magnetic field wave is equal to the energy density of the electric field wave.

(d) Show that the magnitude $|\vec{S}|$ of the Poynting vector can be expressed in terms of the energy density as follows:

$$|\vec{S}| = cu, \quad u = u_E + u_B.$$

(e) Find an expression for the time-averaged Poynting vector in terms of H_0 and the characteristic impedance $Z = \sqrt{\mu_0/\epsilon_0}$ of a vacuum, where $H_0 = B_0/\mu_0$, where B_0 is the amplitude of the magnetic field wave.

(7.17) Suppose a plane electromagnetic wave is incident on a plane *absorbing* surface at an angle θ with the normal of the surface.

(a) Show that the normal force exerted per unit area, i.e. pressure, will be

$$\mathcal{P} = \frac{F_n}{A} = u \cos^2 \theta,$$

where u is the energy density of the electromagnetic wave.

(b) Show that if the waves are incident from all angles with equal amplitude and the same frequency, i.e. if the averaging is done over the half-space on the side of the plate that the incident wave resides, the pressure will be

$$\mathcal{P} = \frac{1}{3} u.$$

(c) What will happen if the surface were a perfectly *reflecting* surface?

(7.18) A microscopic spherical dust particle of radius 2 μm and mass 10 μg is moving in outer space at a constant speed of 30 cm/sec. A wave of light strikes the particle from the opposite direction of its motion, and gets absorbed. Assuming the particle decelerates uniformly to zero speed in one second, what is the average electric field amplitude in the light wave?

(7.19) In the problem above, suppose light strikes the particle at 30° to the initial velocity, as shown in Fig. 7.18. What will be the particle's velocity (direction and magnitude) after 2 s? Assume complete absorption.

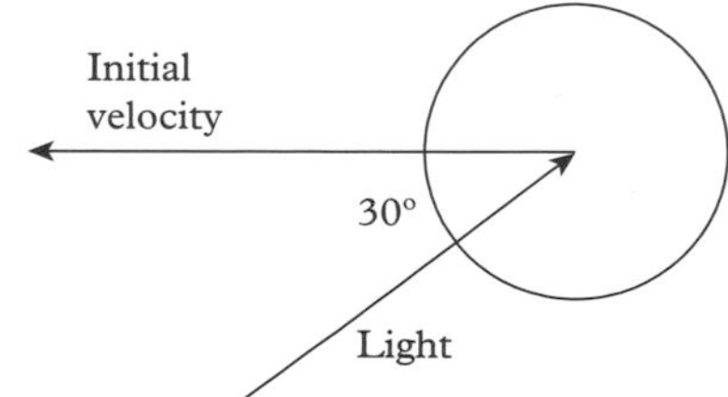

Fig. 7.18 *Exercise 7.19.*

(7.20) Radiation from the Sun has an intensity of 1.38 kW/m^2 at the surface of Earth. This is called **solar constant**. Other data you may need are: the radius of Earth = 6,371 km, the radius of the Sun = 695,800 km, the Earth–Sun distance = 149,600,000 km, the mass of Earth = 5.972×10^{24} kg, and the mass of the Sun = 1.989×10^{30} kg.

(a) Find the intensity of light at the surface of the Sun.

(b) Compare the force by the radiation pressure on Earth with the gravitational attraction of the Sun, assuming complete absorption of light by Earth.

(7.21) A Styrofoam spherical ball of radius 2 mm and mass 20 μg is to be suspended by radiation pressure in a vacuum tube in an Earth-based lab.

(a) How much intensity will be needed if completely absorbed by the ball?

(b) How much intensity will be needed if completely reflected by the ball?

(7.22) An astronaut is at a distance of $\frac{1}{4}R$ from the Sun, where R is the average Earth–Sun distance. He holds a perfectly reflecting mirror of area A perpendicular to the rays from the Sun. Find his acceleration due to the radiation pressure alone if his mass along with all the gear is M. Assume that the astronaut is completely behind the mirror so that light strikes the mirror only and the intensity of the sunlight at Earth is I_E. Check out the numerical value for $M = 100$ kg, $A = 100$ m^2, and $I_E = 1{,}400$ W/m^2.

(7.23) Use Malus's law to find the intensities I_1, I_2, and I_3 in terms of I_0 in Fig. 7.19.

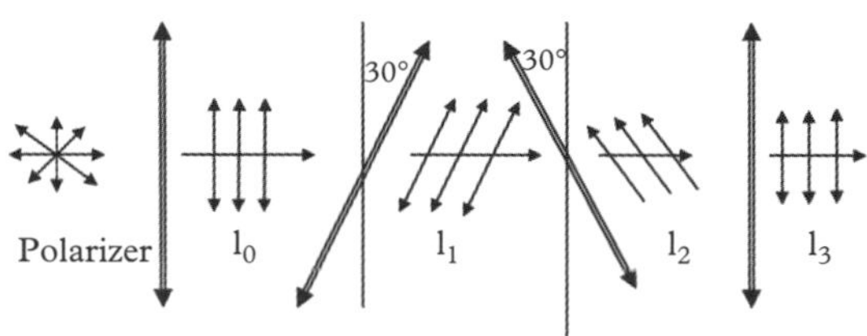

Fig. 7.19 *Exercise 7.23.*

(7.24) A polarizer is rotated at an angular speed between two identical stationary crossed polarizers, as shown in Fig. 7.20. Find the intensity of the emerging wave in terms of the intensity I_0 of the light after the first fixed polarizer.

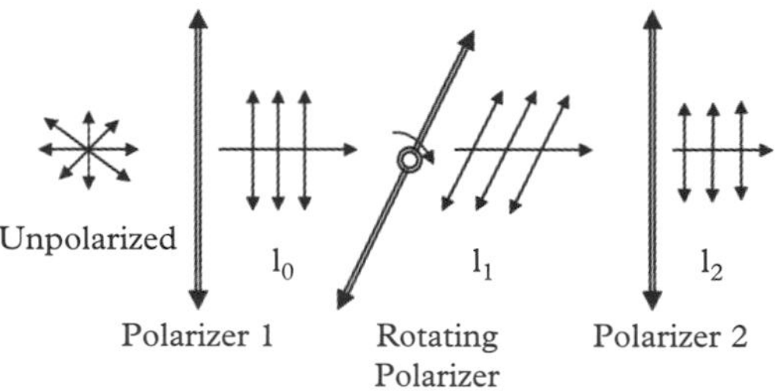

Fig. 7.20 *Exercise 7.24.*

(7.25) The following vectors give the electric field of an electromagnetic wave at the origin of a Cartesian coordinate system while traveling towards the positive z-axis. Draw these vectors for the following values of time t in units of the period of the waves: $t = 0, \frac{1}{4}, \frac{1}{2}, \frac{3}{4}, 1$.
(a) $\hat{x}\ \cos(2\pi t)$
(b) $\hat{y}\ \sin(2\pi t)$
(c) $\hat{x}\ \cos(2\pi t) + \hat{y}\ \sin(2\pi t)$
(d) $\hat{x}\ \cos(2\pi t) - \hat{y}\ \sin(2\pi t)$

(7.26) A circularly polarized light is often represented using a complex notation. There are two circularly polarized waves, one rotating clockwise and the other counterclockwise.

$$\vec{E}_1 = \left(\hat{x} + \hat{y}\, i\right) E_0 e^{i(kz-\omega t)}, \tag{7.209}$$

$$\vec{E}_2 = \left(\hat{x} - \hat{y}\, i\right) E_0 e^{i(kz-\omega t)}, \tag{7.210}$$

where $i = \sqrt{-1}$ and $\hat{x}$ and $\hat{y}$ are the unit vectors pointed towards the positive x-and positive y-axes respectively.
(a) Find the real parts of the given complex fields.
(b) Which one of these waves would appear to rotate clockwise and which one counterclockwise when viewed from the positive z-axis? Why? Note: The counterclockwise rotating polarization is also called right-circularly polarized and the clockwise rotating polarization is called the left-circularly polarized.
(c) Find the magnetic field waves corresponding to these electric field waves. Hint: Work out the x- and y-components separately and then put them together.

(7.27) A circularly polarized light has the following electric field:

$$\vec{E} = E_0 \left[\hat{x} \cos\left(2\pi f t - \frac{2\pi}{\lambda} z\right) + \hat{y} \sin\left(2\pi f t - \frac{2\pi}{\lambda} z\right)\right].$$

The wave passes through a linear polarizer whose polarizing axis is pointed 45° counterclockwise to the x-axis. Find the intensity of the emergent wave in terms of the intensity of the incident wave.

(7.28) No light can pass through two crossed polaroids whose axes are 90° to each other. Light however can get through two crossed polaroids if a third polaroid is inserted between the two in an appropriate way. What should be the angle between the axis of the polaroid in the middle and the axis of the first polaroid so that the intensity of the transmitted light is 5% of the intensity of unpolarized light incident on the first polaroid.

(7.29) A polaroid is rotated between two fixed crossed polaroids at a frequency of f Hz. Find the variation of emergent intensity with respect to time in terms of the incident intensity I_0 of the unpolarized light and the frequency f of rotation of the rotating polaroid.

(7.30) What are the wavelengths of the matter waves corresponding to the following objects:
(a) a 50 kg person running at 10 m/s,
(b) an electron ($m_e = 9.1 \times 10^{-31}$ kg) moving at (i) 10 m/s, (ii) 10^4 m/s, (iii) 10^8 m/s,
(c) a proton ($m = 1.67 \times 10^{-27}$ kg) moving at (i) 10 m/s, (ii) 10^4 m/s, (iii) 10^8 m/s?

(7.31) An electron's position in the hydrogen atom is known with an uncertainty of 10^{-10} m.

(a) What is the uncertainty in its momentum?

(b) What is the uncertainty in its speed?

(7.32) An electron is in a one-dimensional potential well of width a with infinite potential walls on the two sides. The electron is in the lowest energy state. What is the probability that the electron will be in

(a) the right half of the well,

(b) the right-hand quarter of the well,

(c) the right-hand fifth of the well?

(7.33) An electron is in a one-dimensional potential well of width a with infinite potential walls on the two sides. The electron is in the first excited state. What is the probability that the electron will be in

(a) the right half of the well,

(b) the right-hand quarter of the well,

(c) the right-hand fifth of the well?

Reflection and Transmission of Waves

8

Chapter Goals

In this chapter we will study what happens when a wave arrives at a discontinuity in the medium. The wave may be reflected, absorbed, and/or transmitted. We will also study structures that help to guide a wave.

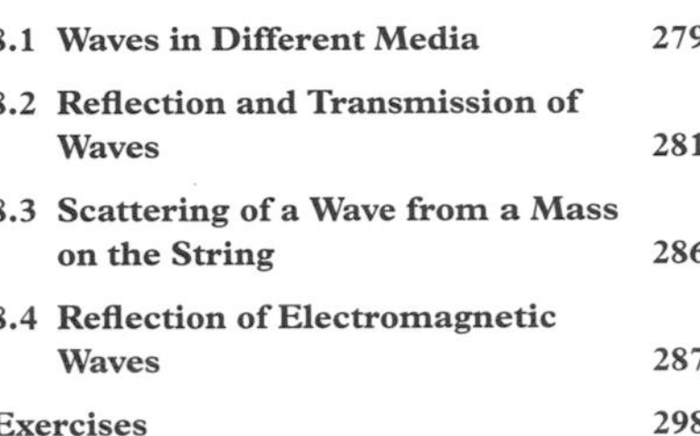

When we studied waves traveling in an open space we assumed that the medium in which the waves moved was infinite in extent. Realistically, no medium will be infinite in extent and the question arises as to what happens when the wave reaches the boundary of the medium in which it was generated? Will the wave be reflected back into the same medium or enter the next medium? What properties of the two media decide what will happen when the wave arrives at the boundary? We will address these questions in this chapter with examples from different areas of physics. To keep the mathematics simple and for the sake of continuity in our discussions we start with waves of transverse oscillations on strings. The results for one dimension can be applied directly without any modifications to plane waves in three dimensions with plane boundaries in which waves travel perpendicularly to the boundary.

8.1 Waves in Different Media

8.1.1 Wavelengths in Different Media

In the case of the waves on a string, the string is the medium of the wave. Suppose, we tie two strings of different mass densities μ_1 and μ_2 with a knot, and attach the ends to a vibrator at one end and a fixed post at the other, so that there is uniform tension T throughout the two strings, as shown in Fig. 8.1. Now, when we turn on the vibrator, a wave will be transmitted towards the knot. We call this wave the **incident wave**. When the wave arrives at the knot, the knot is vibrated by the incoming wave. As a result the knot acts as a wave transmitter sending waves in the two directions, one in the second string, called the **transmitted wave** and the other in the first string, called the **reflected wave**, which moves in the opposite direction to the wave coming from the vibrator. The reflected wave will come back and strike the vibrator.

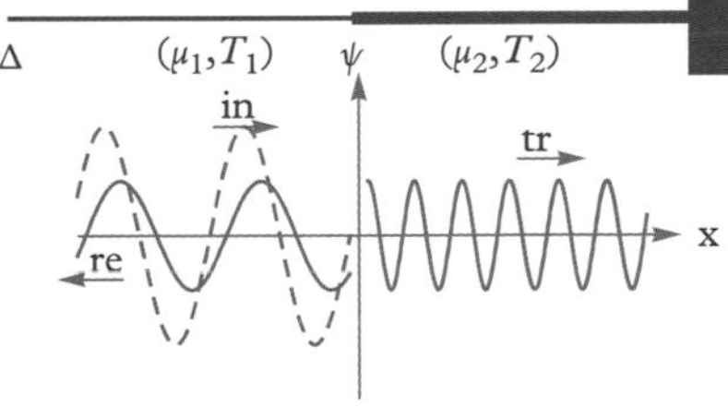

Fig. 8.1 *Harmonic waves on two connected strings with different mass densities and the same tension. The frequency of the wave is set by the frequency of vibrations of the wave generator. The wave has different wavelengths in the two media because the speed of the wave is different in the two media, while the frequency is the same. The configuration of the string on the left at a particular instant is obtained by the sum of the two waves on that part of the string at that instant.*

Suppose the vibrator is vibrating at a particular frequency ω. Then, the frequencies of the three waves, incident, reflected, and transmitted, will all be the same as ω. However, two of these waves, the incident and reflected, move in a medium of wave speed $v_1 = \sqrt{T/\mu_1}$, which is different from the transmitted wave, which moves in a medium of wave speed

A First Course in Vibrations and Waves. First Edition. Mohammad Samiullah.

$v_2 = \sqrt{T/\mu_2}$. As a result, the wave number k_1 of the incident and reflected waves will be different from the wave number k_2 of the transmitted wave,

$$k_1 = \omega/v_1 = \omega\sqrt{\mu_1/T}, \tag{8.1}$$

$$k_2 = \omega/v_2 = \omega\sqrt{\mu_2/T}. \tag{8.2}$$

Since the wave number k of a wave is related to the wavelength λ by $k = 2\pi/\lambda$, this says that the wavelengths of a wave of the same frequency will be different in the two media. Let λ_1 and λ_2 be the respective wavelengths, then we have

$$\frac{\lambda_1}{v_1} = \frac{\lambda_2}{v_2} = \frac{2\pi}{\omega}. \tag{8.3}$$

8.1.2 Boundary Conditions at the Junction of Two Media

Suppose the two strings in Fig. 8.1 are semi-infinite in extent, that is, we assume the vibrator and the post are far away from the knot at $x = 0$. Suppose the strings are tied together smoothly and the knot is considered to be *massless*. Let $\psi_1(x, t)$ and $\psi_2(x, t)$ be the transverse y-displacements of the string on the left of $x = 0$, i.e. when $x < 0$, and on the right of $x = 0$, i.e. when $x > 0$, respectively. Since the strings are continuous at the junction, the displacement of the string at the boundary at $x = 0$ must be the same whether we use ψ_1 or ψ_2 for the displacement at that point,

$$\text{Boundary Condition \#1: Continuity,} \quad \psi_1(0, t) = \psi_2(0, t). \tag{8.4}$$

With the knot being assumed to be massless, the mass times acceleration of the knot would be zero. This means that the net y-force on the knot will be zero. The y-force on the knot from each side would be equal to the corresponding tension times the slope of the strings on that side. This gives the second boundary condition at the junction of the two media:

$$\text{Massless junction:} \quad T\left(\frac{\partial \psi_1}{\partial x}\right)_{x=0^-} = T\left(\frac{\partial \psi_2}{\partial x}\right)_{x=0^+}. \tag{8.5}$$

Canceling the tension from the two sides, we obtain a simpler boundary condition when the tensions in the two strings are equal,

Boundary Condition #2: Massless junction (Case $T_1 = T_2$),

$$\left(\frac{\partial \psi_1}{\partial x}\right)_{x=0^-} = \left(\frac{\partial \psi_2}{\partial x}\right)_{x=0^+}. \tag{8.6}$$

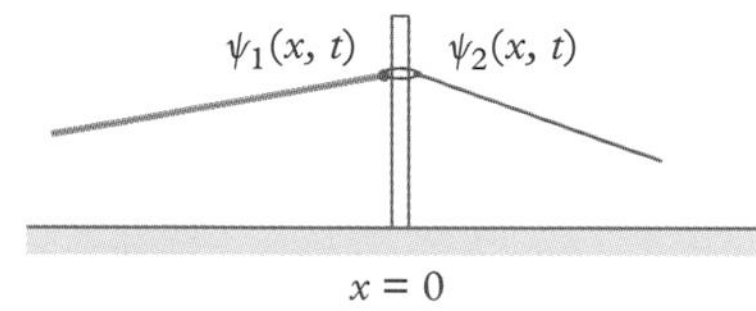

Fig. 8.2 *Two strings are tied on two sides of a massless ring which slides on a rod frictionlessly. The rod provides additional horizontal force so that the tensions on the two sides can be different.*

However, the strings could be fastened, not necessarily directly to each other, but through an intermediary such that tensions could be different in the two strings. For instance, if the strings are tied on the two sides of a ring which is mounted on a rod so that the ring slides frictionlessly, as shown in Fig. 8.2, then the rod can provide additional horizontal force, and tensions may not be equal on both sides. Let T_1 and T_2 be the tensions on the strings on the left and right sides of the junction. Then, the second boundary condition on the waves on the two strings would be

Boundary Condition #2: Massless junction (T_1, T_2 arbitrary),

$$T_1 \left(\frac{\partial \psi_1}{\partial x} \right)_{x=0^-} = T_2 \left(\frac{\partial \psi_2}{\partial x} \right)_{x=0^+}. \tag{8.7}$$

Equation 8.7 covers a more general scenario than Eq. 8.6 and will be used below before we specialize to the $T_1 = T_2$ case. Equations 8.4 and 8.7 are the boundary conditions that any wave function on the strings must satisfy.

8.2 Reflection and Transmission of Waves

8.2.1 Reflection and Transmission Coefficients

An important application of the boundary conditions discussed above is the study of reflection and transmission (or refraction) at the boundary. As an example, consider two very long strings of mass per unit length μ_1 and μ_2 respectively, fastened together at $x = 0$, as shown in Fig. 8.3. Let the knot at the junction be massless so that the tensions T_1 and T_2 in the two strings would be equal here.

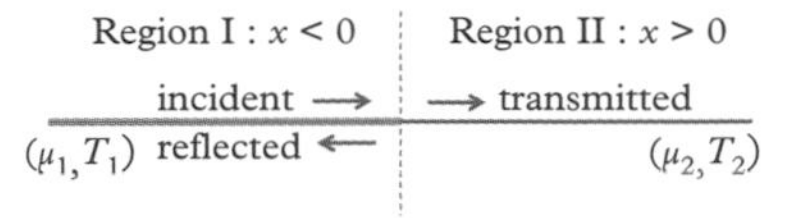

Let the wave be generated continuously in the left string by vibrating the end at $x = -\infty$ harmonically at frequency ω. The wave generated in the left string, that is traveling towards $x = 0$, is called the incoming or **incident wave** and will be denoted by ψ_{in}. Let us use complex number notation for waves since algebra with complex notation would be simpler,

Region I : $x < 0$ | Region II : $x > 0$
incident → | → transmitted
(μ_1, T_1) reflected ← | (μ_2, T_2)

Fig. 8.3 *Incident, reflected, and transmitted waves on two strings fastened at $x = 0$. The wave function ψ_1 on the left string is the sum of the incoming wave and the reflected wave, and the wave function ψ_2 on the string on the right is the transmitted wave.*

$$\psi_{\text{in}}(x, t) = A\, e^{i(k_1 x - \omega t)}, \quad (x < 0) \tag{8.8}$$

where

$$k_1 = \frac{\omega}{v_1} = \omega \sqrt{\frac{\mu_1}{T_1}}, \tag{8.9}$$

and A is the amplitude of the incident wave. When the incident wave arrives at the junction at $x = 0$, it vibrates the element at $x = 0$ continuously. In a sense the incoming harmonic wave acts as a vibrator at $x = 0$. Now, if you were to take the original two-string system and place an external vibrator at $x = 0$, you would find that such a vibrator would produce two waves, one going towards $x = -\infty$ and the other towards $x = \infty$. The same will happen as a result of vibration of the element at $x = 0$ caused by the incoming wave ψ_{in}. The wave going towards $x = -\infty$ is called the **reflected wave**, which will be denoted by ψ_{re}, and the one going towards $x = \infty$ is called the **transmitted wave**, which will be denoted by ψ_{tr}. Since the frequencies of the waves are set by the vibrations, which here is ω, the expressions for the two waves generated at the junction $x = 0$ would be given by the following expressions:

$$\psi_{\text{re}}(x, t) = B\, e^{i(k_1 x + \omega t)}, \quad (x < 0) \tag{8.10}$$

$$\psi_{\text{tr}}(x, t) = C\, e^{i(k_1 x - \omega t)}, \quad (x > 0) \tag{8.11}$$

where

$$k_2 = \frac{\omega}{v_2} = \omega \sqrt{\frac{\mu_2}{T_2}}. \tag{8.12}$$

Note that there are two waves on the left of $x = 0$, namely ψ_{in} and ψ_{re}. Therefore, in Region I ($x < 0$) the wave function will be a superposition of these two waves. In Region II ($x > 0$), there is only one wave, ψ_{tr}. Let us write the waves in the two regions now.

$$\text{Region I: } \psi_1(x,t) = \left(Ae^{ik_1x} + Be^{-ik_1x}\right)e^{-i\omega t}, \quad -\infty < x < 0 \tag{8.13}$$

$$\text{Region II: } \psi_2(x,t) = Ce^{ik_2x}e^{-i\omega t}, \quad 0 < x < \infty \tag{8.14}$$

$$\text{Boundary: } x = 0.$$

Note that the amplitudes A, B, and C may be complex with an amplitude and a phase constant. It is customary and also useful to introduce reflection and transmission coefficients ρ and τ by the ratio of the reflection and transmission amplitudes to the incident amplitude,

$$\rho = B/A, \quad \tau = C/A. \tag{8.15}$$

The boundary conditions here are the continuity of the wave function and the continuity of the slope of the wave function at $x = 0$. Applying them we get the following relations among the coefficients:

$$1 + \rho = \tau \quad \text{(continuity at } x = 0\text{)}, \tag{8.16}$$

$$iT_1k_1(1-\rho) = iT_2k_2\tau \quad \text{(zero y-force at } x = 0\text{)}. \tag{8.17}$$

Equations 8.16 and 8.17 can be solved for the reflection and transmission coefficients ρ and τ,

$$\rho = \frac{T_1k_1 - T_2k_2}{T_1k_1 + T_2k_2}, \tag{8.18}$$

$$\tau = \frac{2T_1k_1}{T_1k_1 + T_2k_2}. \tag{8.19}$$

Since the tensions in the two strings are equal here, we can cancel out the T's from these equations. This yields reflection and transmission coefficients solely in terms of the wave numbers,

$$\rho = \frac{k_1 - k_2}{k_1 + k_2}, \tag{8.20}$$

$$\tau = \frac{2k_1}{k_1 + k_2}. \tag{8.21}$$

It is possible to rewrite Eqs. 8.18 and 8.19 in terms of the wave-related properties of the two media, such as tension and wave speed, by dividing the numerator and denominator of Eqs. 8.18 and 8.19 by ω and using the dispersion relations $\omega^2 = v_1^2/k_1^2$ and $\omega^2 = v_2^2/k_2^2$ for waves on a taut string,

$$\rho = \frac{T_1/v_1 - T_2/v_2}{T_1/v_1 + T_2/v_2}, \tag{8.22}$$

$$\tau = \frac{2T_1/v_1}{T_1/v_1 + T_2/v_2}. \tag{8.23}$$

These relations are often written in terms of impedance of the two media. Let Z_1 and Z_2 be the mechanical impedance of the two strings. The impedance of a string was found to be

$$Z = \frac{T}{v} = \sqrt{\mu T}. \tag{8.24}$$

Therefore, Eqs. 8.22 and 8.23 can be written in terms of the impedance of the two media by using Eq. 8.24:

$$\rho = \frac{Z_1 - Z_2}{Z_1 + Z_2}, \tag{8.25}$$

$$\tau = \frac{2Z_1}{Z_1 + Z_2}. \tag{8.26}$$

These are more general equations than Eqs. 8.20 and 8.21 since the latter are only applicable when tensions in the strings are equal. Equation 8.25 shows that if the two media have equal impedance, then there is no reflection. This condition is called **impedance matching**. When two media have matched impedance all the energy of the incident wave simply moves on into the second medium.

Equation 8.25 shows that the reflection amplitude will have an opposite sign to that of the incident wave if $Z_2 > Z_1$. Therefore, if the wave is driven into a higher $Z = \sqrt{\mu T}$ string, then the reflected wave will be 180° out of step with the incoming wave at the boundary. That is, if the phase constant of the incident wave at the boundary is zero, then the phase constant of the reflected wave will be π radian,

$$\psi_{\text{in}}(x = 0, t) = A\cos(\omega t), \tag{8.27}$$

$$\psi_{\text{re}}(x = 0, t) = B\cos(\omega t - \pi) = -B\cos(\omega t). \quad (Z_2 > Z_1) \tag{8.28}$$

In Fig. 8.4 we plot the incident, reflected, and transmitted waves at 1/8-th periods. The figure also shows the sum of the incident and reflected waves, which would be the wave observed on the left string. Each particle of the string moves up and down as the wave travels along the horizontal direction. The figure shows the inverse phase relation between the incoming wave and the reflected wave for $Z_1 = 1$ and $Z_2 = 4$. The net wave on the left matches with the transmitted wave on the right, both in amplitude and slope, at the junction.

8.2.2 Perfect Reflection

Suppose, instead of tying to the right string, the string to which the wave generator is attached is tied to a fixed support. That is, we have one string which is being vibrated at some point, say, $x = -\infty$, and a wave is being sent down the string towards the positive x-axis, and let the string be attached to the fixed support at $x = 0$. We can think of the fixed support as if the second string in the example above has an infinite mass density, making $Z_2 = \infty$. Therefore, the large Z_2 limit of Eqs. 8.25 and 8.26 will give us a reflection from a massive string or a fixed post,

Reflection from fixed end:

$$\rho = \lim_{Z_2 \to \infty} \frac{Z_1 - Z_2}{Z_1 + Z_2} = -1, \tag{8.29}$$

$$\tau = \lim_{Z_2 \to \infty} \frac{2Z_1}{Z_1 + Z_2} = 0. \tag{8.30}$$

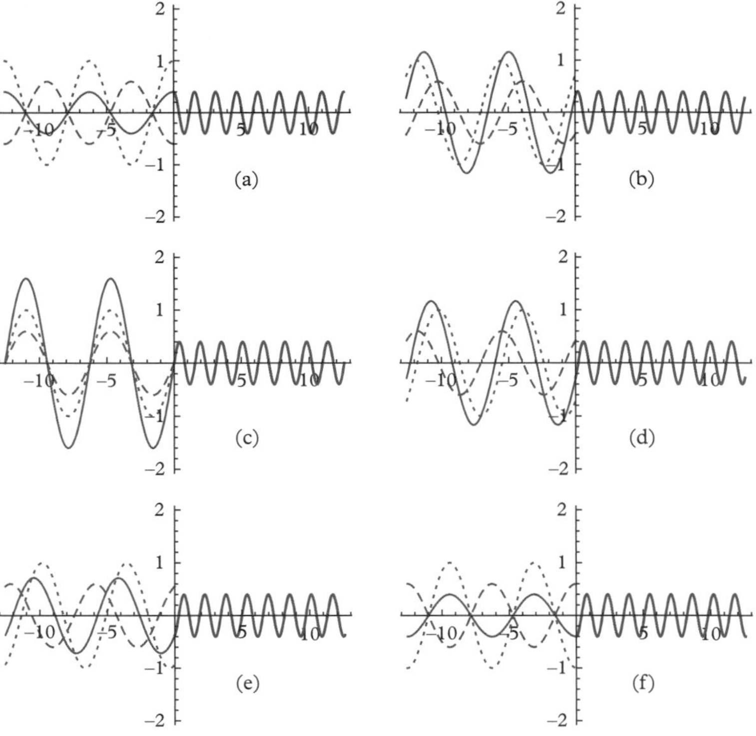

Fig. 8.4 *Incident, reflected, their sum, and transmitted waves on two strings ($z_1 = 1$, $z_2 = 4$) fastened at $x = 0$ at various times. (a) $t = 0$, (b) $t = 2T/8$, (c) $t = 4T/8$, (d) $t = 6T/8$, (e) $t = 7T/8$, (f) $t = 8T/8$, where $T = 2\pi/\omega$ is one period. The wave function ψ_1 on the left string, shown as a solid line on the left of $x = 0$, is the sum of the incoming wave (dotted) and the reflected wave (dashed), and the wave function ψ_2 shown as a solid line on the right of $x = 0$ on the string on the right is the transmitted wave.*

That is, the wave will be fully reflected with the same amplitude as the incident wave, but opposite phase and there will be no transmission for reflection from a fixed post. The right-moving wave and the left-moving wave add up to give a standing wave on the string,

$$A\cos(kx - \omega t) + (-1)A\cos(kx + \omega t) = 2A\sin(kx)\sin(\omega t). \tag{8.31}$$

8.2.3 Perfect Termination and Impedance Matching

From Eq. 8.25 we find that if impedance of the second medium is equal to the impedance of the first, i.e. when the impedances are matched, the reflection coefficient will be zero,

$$\rho = 0 \ \text{when} \ Z_1 = Z_2. \tag{8.32}$$

That is, the first string will not experience a recoil force and the wave will seamlessly enter the second string as if the boundary were not there. In this case the length of the second string does not matter since the second string just moves with the first string as the incident wave arrives at the junction. We say that the second string is a perfect terminator of the wave.

Example 8.1 Energy Conservation

In the last chapter we found that the energy in a wave over a distance of a wavelength λ for a sinusoidal wave of amplitude A, frequency ω, and wavelength λ traveling on a string of linear density μ is given by

$$E_\lambda = \frac{1}{2}\mu\lambda\omega^2 A^2. \tag{8.33}$$

This energy will pass through any point of the string in one period, $2\pi/\omega$. Therefore, the rate at which energy passes through any point of the string, i.e. the power, will be

$$P = \frac{E_\lambda}{2\pi/\omega} = \frac{1}{4\pi}\mu\lambda\omega^3 A^2 = \frac{1}{2}\mu\omega^2 v A^2. \tag{8.34}$$

We can arrive at this result also from a more general argument about power,

$$P = uv,$$

where u is the energy density, given by E_λ/λ. The expression for power in Eq. 8.34 can also be written more compactly in terms of impedance $Z = \sqrt{T\mu}$. By using $v = \sqrt{T/\mu}$ and after some algebra, you will find that

$$P = \frac{1}{2}Z\omega^2 A^2. \tag{8.35}$$

You can check that the unit of P is J/s. Use this expression for power in a wave to verify the conservation of energy when an incoming wave generates reflected and transmitted waves at the boundary of two media, discussed in the last section.

Solution

The incoming wave brings energy U_i with each cycle, since all energy contained in one wavelength of the incoming wave on the string on the left of the junction will arrive at the junction in one cycle,

$$U_i = P_i \times \frac{2\pi}{\omega} = \pi Z_1 \omega A^2. \tag{8.36}$$

Similarly, in each cycle, the energies in the reflected and transmitted waves are

$$U_r = \pi Z_1 \omega A^2 \rho^2, \tag{8.37}$$

$$U_t = \pi Z_2 \omega A^2 \tau^2. \tag{8.38}$$

The sum of the energy of outgoing waves from the junction at $x = 0$ must equal the energy of the incoming wave. Let us verify that this is indeed the case. We use Eqs. 8.25 and 8.26 to replace ρ and τ, the reflection and transmission coefficients.

$$\begin{aligned} U_r + U_t &= \pi Z_1 \omega A^2 \left(\rho^2 + \frac{Z_2}{Z_1}\tau^2\right) \\ &= \pi Z_1 \omega A^2 \left[\left(\frac{Z_1 - Z_2}{Z_1 + Z_2}\right)^2 + \frac{Z_2}{Z_1}\left(\frac{2Z_1}{Z_1 + Z_2}\right)^2\right] \\ &= \pi Z_1 \omega A^2 = U_i. \end{aligned}$$

8.3 Scattering of a Wave from a Mass on the String

When we discussed the reflection of a wave we assumed that the mass at the junction of the two regions was zero. What would happen if the junction were not massless? The simplest way we can investigate this question is to place a bead of mass m but of infinitesimal size on a string and send a traveling wave towards the bead, as shown in Fig. 8.5. For simplicity we will now work with the same Z on both sides of the junction where the bead is located.

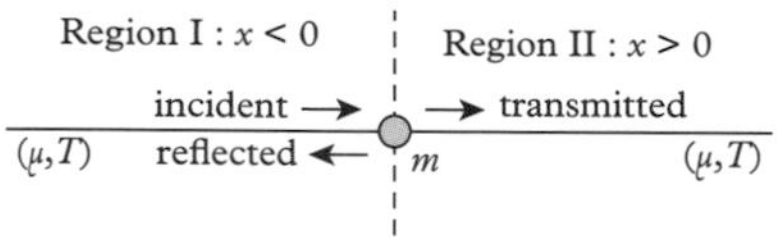

Fig. 8.5 *A bead on a string will reflect the wave on the string. The net force on the bead by the oscillations of the three waves at $x = 0$ gives rise to a net vertical acceleration of the bead. The calculation in the text shows that the reflected wave and the transmitted wave have a phase shift with respect to the phase of the incident wave.*

Just as above, there will be incident and reflected waves in the region I ($x < 0$) and a transmitted wave in region II ($x > 0$). Since the two strings have the same v for the waves $k_1 = k_2 \equiv k$,

$$\text{Region I: } \psi_1(x,t) = \left(Ae^{ikx} + Be^{-ikx}\right)e^{-i\omega t}, \quad -\infty < x < 0 \tag{8.39}$$

$$\text{Region II: } \psi_2(x,t) = Ce^{ikx}e^{-i\omega t}, \quad 0 < x < \infty \tag{8.40}$$

$$\text{Boundary: } x = 0. \tag{8.41}$$

Now, we impose the boundary conditions at $x = 0$. The first boundary condition comes from the continuity of the string at $x = 0$. This gives

$$A + B = C. \tag{8.42}$$

The second boundary condition in Eq. 8.7 requires a massless knot. Now, that is not the case. Therefore, we will not have that boundary condition any more. Instead, we will have $F_y = ma_y$ for the bead at $x = 0$ with

$$a_y = \frac{\partial^2 \psi(0,t)}{\partial t^2}, \tag{8.43}$$

and F_y given by

$$T\left[\frac{\partial \psi_2(x,t)}{\partial x}\right]_{x=0^+} - T\left[\frac{\partial \psi_1(x,t)}{\partial x}\right]_{x=0^-}. \tag{8.44}$$

Therefore, in place of Eq. 8.7 we will have the following equation:

$$T\left[\frac{\partial \psi_2(x,t)}{\partial x}\right]_{x=0^+} - T\left[\frac{\partial \psi_1(x,t)}{\partial x}\right]_{x=0^-} = m\,\frac{\partial^2 \psi(0,t)}{\partial t^2}. \tag{8.45}$$

Therefore, using ψ_2 for the right side we get

$$ikT(B - A + C) = -m\,\omega^2 C, \tag{8.46}$$

The same result will be obtained with ψ_1 on the right side. Simplifying, we get

$$A - B = \left(1 - i\frac{m\,\omega^2}{kT}\right)C. \tag{8.47}$$

Denoting $\frac{m\omega^2}{kT}$ by the symbol δ, we can write this equation as

$$A - B = (1 - i\delta)\, C, \quad \delta = \frac{m\,\omega^2}{kT}. \tag{8.48}$$

From Eqs. 8.42 and 8.48 we find the reflection and transmission coefficients to be

$$\rho = \frac{B}{A} = \frac{i\delta}{2 - i\delta}, \quad \tau = \frac{C}{A} = \frac{2}{2 - i\delta}. \tag{8.49}$$

These coefficients are complex. Let us write them in the polar form for complex numbers:

$$\frac{i\delta}{2 - i\delta} = |\rho|\, e^{i\theta}, \quad \frac{2}{2 - i\delta} = |\tau|\, e^{i\phi}, \tag{8.50}$$

with

$$|\rho| = \frac{\delta}{\sqrt{4 + \delta^2}}, \quad \theta = \frac{\pi}{2} + \tan^{-1}\frac{\delta}{2}, \tag{8.51}$$

$$|\tau| = \frac{4}{\sqrt{4 + \delta^2}}, \quad \phi = \tan^{-1}\frac{\delta}{2}. \tag{8.52}$$

Then we have B and C as

$$B = A|\rho|e^{i\theta}, \quad C = A|\tau|e^{i\phi}. \tag{8.53}$$

When we put these into the wave functions we get the following for the reflected and transmitted waves when the incident wave is $\psi_{\text{in}} = A\cos(kx - \omega t)$:

$$\psi_{\text{re}} = A\,|\rho|\,\cos(kx - \omega t + \theta), \tag{8.54}$$

$$\psi_{\text{tr}} = A\,|\tau|\,\cos(kx - \omega t + \phi). \tag{8.55}$$

The non-zero phase constants θ and ϕ tell us that the presence of the bead causes phase shifts in the reflected and transmitted waves. The student is encouraged to study the phase constants for $m = 0$ and $m \to \infty$ limits. As $m \to 0$, all the incident wave should be transmitted; nothing should be reflected since the two sides of $x = 0$ are strings with the same properties. You will also find that if you take $m \to \infty$, the phase of the reflected wave would be 180° and that of the transmitted wave $\phi = 90°$.

8.4 Reflection of Electromagnetic Waves

As an example of reflection and transmission in more than one dimension we will now work out the reflection of plane harmonic electromagnetic waves from a planar boundary. If the wave is incident normally on the boundary, then we do not need to do more than what we have done for the wave on a string. All the results will be the same since the electromagnetic wave can be thought of as one-dimensional wave. However, if the wave is incident at some angle to the interface, the reflection and transmission coefficients will also depend upon the angle of incidence.

It turns out that Maxwell's equation imposes boundary conditions on electric and magnetic fields that relate the components of electric and magnetic fields on one side of the interface to the corresponding values on the other side. We will denote the electric and magnetic fields in the two media by attaching subscripts 1 and 2 to the corresponding symbols as $\vec{E}_1$, $\vec{B}_1$, $\vec{E}_2$, and $\vec{B}_2$. Let $z = 0$ be the plane of interface between two dielectric media characterized by electric susceptibility and magnetic permeability (ϵ_1, μ_1) and (ϵ_2, μ_2); then Maxwell's equation can be used to deduce the following boundary conditions, assume no charges or currents at the interface.

Boundary conditions at $z = 0$:

$$\epsilon_1 E_{1x} = \epsilon_2 E_{2x}, \;\; \epsilon_1 E_{1y} = \epsilon_2 E_{2y}, \;\; E_{1z} = E_{2z}, \tag{8.56}$$

$$\frac{1}{\mu_1} B_{1x} = \frac{1}{\mu_1} B_{2x}, \;\; \frac{1}{\mu_1} B_{1y} = \frac{1}{\mu_1} B_{2y}, \;\; B_{1z} = B_{2z}. \tag{8.57}$$

We will now use these boundary conditions to derive the reflection and transmission coefficients for electric and magnetic field waves. To be concrete, let us consider a source of electromagnetic waves in a medium with electric and magnetic susceptibilities (ϵ_1, μ_1). Suppose the waves then strike the interface at an angle θ_1 with respect to the normal to the interface, as shown in Fig. 8.6. Let the reflected wave travel in the direction at an angle θ_1' with respect to the normal, and the transmitted wave in the second medium with electric and magnetic susceptibilities (ϵ_2, μ_2) moves in the direction of angle θ_2 with respect to the normal.

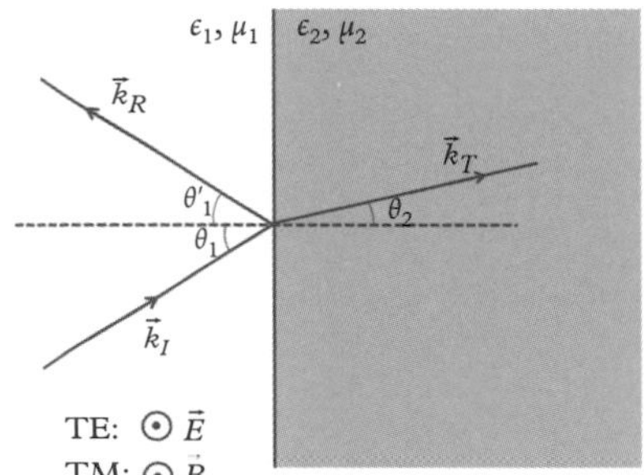

Fig. 8.6 *The geometry of incident, reflected, and transmitted waves. The propagation vectors $\vec{k}_I$, $\vec{k}_R$, and $\vec{k}_T$ are in the direction of the three waves respectively. In the transverse electric (TE) case, the electric field oscillates in the direction perpendicular to the plane and in the transverse magnetic (TM) case, the magnetic field oscillates in the direction perpendicular to the plane.*

The directions of the incident, reflected, and transmitted waves form a plane, called the **plane of incident**. The direction of $\vec{E}$ and $\vec{B}$ fields with respect to the plane of incidence gives us two special cases—transverse electric (TE) and transverse magnetic (TM)—separately. In the TE case, the electric field points perpendicular to the plane of incidence and the magnetic field points in a direction in the plane. In the TM case, the magnetic field points perpendicularly to the plane of incidence and the magnetic field points into the plane.

8.4.1 Transverse Electric (TE) Case

Figure 8.7 shows the electric and magnetic field directions for the incident, reflected, and transmitted waves for TE case. In drawing the vectors for the three waves I have chosen instants when the electric field vector at those points pointed up towards the positive y-axis. Beware that the overall signs in our formulas for reflection and transmission coefficients will depend on this arbitrary choice that we have made to perform the calculation since the calculations are done using components of these vectors.

Let E_0 be the amplitude of the incident electric wave and let us write $\rho_\perp E_0$ and $\tau_\perp E_0$ for the amplitudes for the reflected and transmitted waves respectively, where $\rho_\perp$ and $\tau_\perp$ are the reflection and transmission coefficients. I have attached a subscript $\perp$ to the symbols for the reflection and transmission coefficients to indicate that our results are tailored to the TE case in which the electric field is perpendicular to the plane of incidence. With the xz-plane being the plane of incidence and the $z = 0$ plane be the plane of interface between the two media, the electric waves will have the following components:

$$\text{Incident:} \quad E_{Ix} = 0, \;\; E_{Iy} = E_0 e^{i(k_{Ix}x + k_{Iz}z)} e^{-i\omega t}, \;\; E_{Iz} = 0 \tag{8.58}$$

$$\text{Reflected:} \quad E_{Rx} = 0, \;\; E_{Ry} = \rho_\perp E_0 e^{i(k_{Rx}x + k_{Rz}z)} e^{-i\omega t}, \;\; E_{Rz} = 0 \tag{8.59}$$

$$\text{Transmitted:} \quad E_{Tx} = 0, \;\; E_{Ty} = \tau_\perp E_0 e^{i(k_{Tx}x + k_{Tz}z)} e^{-i\omega t}. \;\; E_{Tz} = 0 \tag{8.60}$$

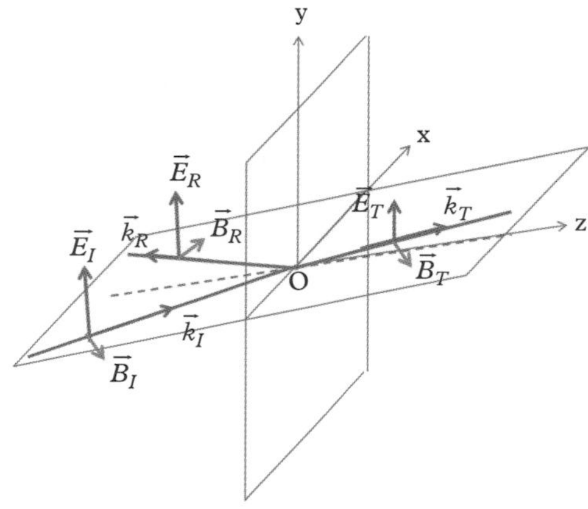

Fig. 8.7 *The directions of electric and magnetic field vectors for the TE mode reflection. The xz-plane is the plane of incidence and $\vec{E}$ is perpendicular to this plane.*

The associated magnetic field waves can be obtained from $\vec{k} \times \vec{E} = \omega\vec{B}$, with $\omega/k = v$, which is the speed of the wave in the appropriate medium. The magnetic field will have the y-component zero and x- and z-components non-zero. The magnetic field components are related to the electric field components as follows:

$$B_x = -\frac{k_z}{\omega}E_y, \ B_y = 0, \ B_z = \frac{k_x}{\omega}E_y. \tag{8.61}$$

Using Fig. 8.8 and $v_1 = \omega/k$ for waves in medium 1 and $v_2 = \omega/k$ for the transmitted wave in medium 2, with $v = 1/\sqrt{\epsilon\mu}$ for the corresponding medium, we obtain the following expressions for the magnetic field in the three waves. The components of the magnetic field are obtained as

Incident: $$B_{Ix} = -\frac{E_{Iy}}{v_1}\cos\theta_1, \ B_{Iy} = 0, \ B_{Iz} = \frac{E_{Iy}}{v_1}\sin\theta_1, \tag{8.62}$$

Reflected: $$B_{Rx} = \frac{E_{Ry}}{v_1}\cos\theta_1', \ B_{Ry} = 0, \ B_{Rz} = \frac{E_{Ry}}{v_1}\sin\theta_1', \tag{8.63}$$

Transmitted: $$B_{Tx} = -\frac{E_{Ty}}{v_2}\cos\theta_2, \ B_{Ty} = 0, \ B_{Tz} = \frac{E_{Ty}}{v_2}\sin\theta_2. \tag{8.64}$$

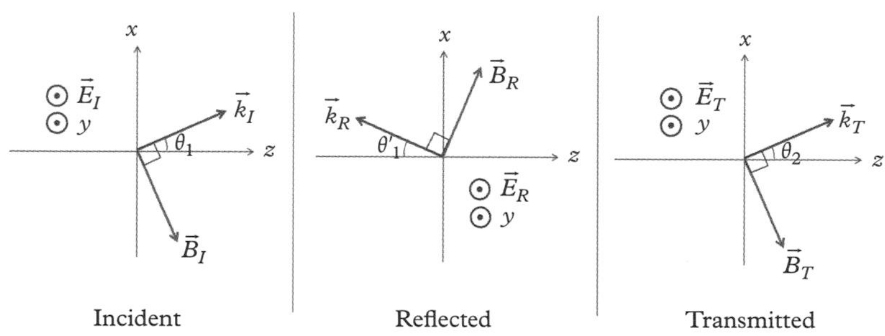

Fig. 8.8 *The directions of electric and magnetic field and propagation vectors for the TE mode reflection. The circle with a dot is an indicator for pointed-out-of-page direction.*

Satisfying Boundary Conditions on the Electric Field

For boundary conditions, we need the net electric field at the boundary on each side of the boundary. In medium 1, we need to add the incident and reflected fields to obtain the net field, while in medium 2 there is only the transmitted field. The boundary condition on the electric field yields the following equation after canceling out the common factor $E_0 e^{-i\omega t}$ and setting $z = 0$:

$$e^{ik_{Ix}x} + \rho_\perp e^{ik_{Rx}x} = \tau_\perp e^{ik_{Tx}x}. \tag{8.65}$$

This equation must be satisfied at all values of x. Therefore, we must have the exponents equal to each other,

$$k_{Ix} = k_{Rx} = k_{Tx}. \tag{8.66}$$

This yields the following relation for the reflection and transmission coefficients after canceling out the common factor:

$$1 + \rho_\perp = \tau_\perp. \tag{8.67}$$

Now, since we have a plane harmonic wave, the magnitude of the wave numbers in the two media will be related by the speeds in the two media, while the wave number for incident and reflected waves will be same. By working out the components of the wave number vectors in the figure it is readily seen that

$$k_1 \sin\theta_1 = k_1 \sin\theta_1' = k_2 \sin\theta_2. \tag{8.68}$$

From the first equality we get

$$\theta_1 = \theta_1', \tag{8.69}$$

that is, the angle of reflection is the angle of incidence, and from the second equality we get

$$k_1 \sin\theta_1 = k_2 \sin\theta_2. \tag{8.70}$$

Now, we replace the wave number in a medium by ω/v, where v is the speed of the wave in the medium,

$$\frac{\omega}{v_1} \sin\theta_1 = \frac{\omega}{v_2} \sin\theta_2. \tag{8.71}$$

Now, canceling out ω and multiplying by the speed of light c in a vacuum, we can rewrite this in terms of the refractive index n of the medium, which is defined by

$$n = \frac{c}{n}. \tag{8.72}$$

Thus, Eq. 8.71 is also written as

$$n_1 \sin\theta_1 = n_2 \sin\theta_2. \tag{8.73}$$

This relation is also called **Snell's law** of refraction.

Satisfying Boundary Conditions on the Magnetic Field

The continuity of $\vec{B}/\mu$ across the interface gives two equations here, one for the z-component and the other for the x-component. You can show that the z-component equation does not give a new relation, you get the same relation as Eq. 8.67. The x-component gives

$$-\frac{1}{v_1\mu_1}\cos\theta_1 + \frac{\rho_\perp}{v_1\mu_1}\cos\theta_1' = -\frac{\tau_\perp}{v_2\mu_2}\cos\theta_2. \tag{8.74}$$

Note that $1/\mu_1 v_1 = Z_1$, the characteristic impedance of medium 1, and similarly for medium 2, we can rewrite this equation in terms of Z_1 and Z_2,

$$(Z_1\cos\theta_1)(1-\rho_\perp) = (Z_2\cos\theta_2)\,\tau_\perp. \tag{8.75}$$

Fresnel Equations for the TE Case

Solving Eqs. 8.67 and 8.75 gives us the reflection and transmission coefficients for the TE case. To solve them, we can write Eq. 8.75 as

$$1-\rho_\perp = \beta\tau_\perp, \tag{8.76}$$

with

$$\beta = \frac{Z_2\cos\theta_2}{Z_1\cos\theta_1}. \tag{8.77}$$

From Eqs. 8.67 and 8.76 we get

$$\rho_\perp = \frac{1-\beta}{1+\beta} = \frac{Z_1\cos\theta_1 - Z_2\cos\theta_2}{Z_1\cos\theta_1 + Z_2\cos\theta_2}, \tag{8.78}$$

$$\tau_\perp = \frac{2}{1+\beta} = \frac{2Z_1\cos\theta_1}{Z_1\cos\theta_1 + Z_2\cos\theta_2}. \tag{8.79}$$

These relations are called Fresnel equations for the TE case. To emphasize that these equations refer to the geometry in which the waves are polarized perpendicular to the plane of incidence, we attach a subscript of $\perp$ to the quantities here,

$$\rho_\perp = \frac{Z_1\cos\theta_1 - Z_2\cos\theta_2}{Z_1\cos\theta_1 + Z_2\cos\theta_2}, \tag{8.80}$$

$$\tau_\perp = \frac{2Z_1\cos\theta_1}{Z_1\cos\theta_1 + Z_2\cos\theta_2}. \tag{8.81}$$

The magnetic susceptibility of dielectric materials does not vary by much. In that case, we can rewrite these relations in terms of refractive indices of the two media by employing the following relation for the impedances:

$$Z_1 = \frac{n_1}{c\mu},\quad Z_1 = \frac{n_2}{c\mu},\quad \text{using } \mu_1 = \mu_2 = \mu \tag{8.82}$$

This gives the following for the reflection and transmission coefficients:

$$\rho_\perp = \frac{n_1\cos\theta_1 - n_2\cos\theta_2}{n_1\cos\theta_1 + n_2\cos\theta_2}, \tag{8.83}$$

$$\tau_\perp = \frac{2n_1\cos\theta_1}{n_1\cos\theta_1 + n_2\cos\theta_2}. \tag{8.84}$$

8.4.2 Fresnel's Equations for the Transverse Magnetic (TM) Case

In the TM case the magnetic field is perpendicular to the plane of incidence and the electric field is pointing in a direction in the plane of incidence. We say that the electric field is parallel to the plane of incidence and denote reflection and transmission coefficients of the electric field waves by the subscript ∥, as in $\rho_\parallel$ and $\tau_\parallel$ respectively. Now, when we draw the electric and magnetic field vectors in the incident, reflected, and transmitted waves, as shown in Fig. 8.9, we find a different picture from what we had for the TE case. To work with components it is useful to draw out the field vectors in the plane of incidence. This is done in Fig. 8.10. It is seen that the x-component of the electric field vector is parallel to the interface. Writing the x-components of the electric field in terms of their magnitudes,

$$\text{Incident: } E_{0I}, \quad \text{Reflected: } E_{0R} = \rho_\parallel E_{0I}, \quad \text{Transmitted: } E_{0T} = \tau_\parallel E_{0I}, \tag{8.85}$$

and the angles that the propagation vector makes with the normal, we find the following expressions for the x-components of the amplitude of the waves:

$$\left.\begin{aligned} E_{Ix} &= E_{0I}\cos\theta_1, \quad E_{Rx} = -E_{0R}\cos\theta_1', \quad E_{Tx} = E_{0T}\cos\theta_2. \\ E_{Iz} &= -E_{0I}\sin\theta_1, \quad E_{Rz} = -E_{0R}\sin\theta_1', \quad E_{Tz} = -E_{0T}\sin\theta_2 \end{aligned}\right\} \tag{8.86}$$

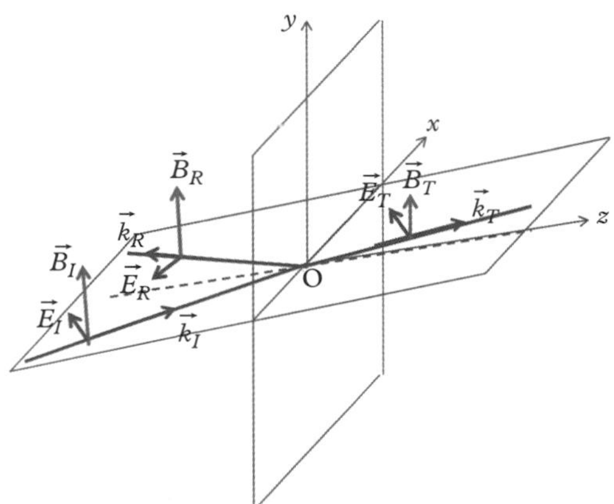

Fig. 8.9 *The directions of electric and magnetic field vectors for the TM mode reflection.*

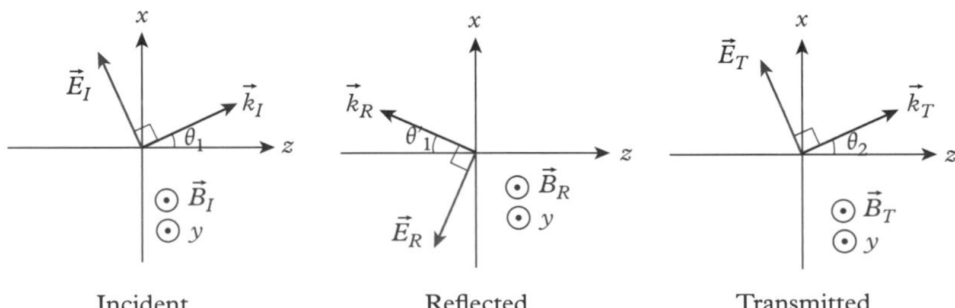

Fig. 8.10 *The directions of electric and magnetic field and propagation vectors for TM mode reflection.*

The continuity of the parallel components (here x- and y-components) of the electric field across the interface gives the following relation from the x-components:

$$\cos\theta_1 - \rho_\parallel \cos\theta_1' = \tau_\parallel \cos\theta_2. \quad (8.87)$$

Since $\theta_1' = \theta_1$, we can write this in the following form:

$$1 - \rho_\parallel = \frac{\cos\theta_2}{\cos\theta_1'}\tau_\parallel. \quad (8.88)$$

The continuity of the parallel component of $\vec{B}/\mu$ across the interface gives

$$\frac{1}{\mu_1}(B_{0I} + B_{0R}) = \frac{1}{\mu_2}B_{0T}. \quad (8.89)$$

The amplitudes of the magnetic field and electric field waves in each of the media are related by $B_0 = E_0/v$, where v is the speed of the electromagnetic wave in the medium. Using this we get the following relation:

$$\frac{1}{\mu_1 v_1}(E_{0I} + E_{0R}) = \frac{1}{\mu_2 v_2}E_{0T}. \quad (8.90)$$

Dividing out by E_{0I}, we write this equation as an equation for the reflection and transmission coefficients,

$$1 + \rho_\parallel = \frac{\mu_1 v_1}{\mu_2 v_2}\tau_\parallel. \quad (8.91)$$

Now, Eqs. 8.88 and 8.91 can be solved simultaneously for $\rho_\parallel$ and $\tau_\parallel$ to yield the following Fresnel's equations for the TM case:

$$\rho_\parallel = \frac{\mu_1 v_1 \cos\theta_1 - \mu_2 v_2 \cos\theta_2}{\mu_1 v_1 \cos\theta_1 + \mu_2 v_2 \cos\theta_2}, \quad (8.92)$$

$$\tau_\parallel = \frac{2\mu_2 v_2 \cos\theta_1}{\mu_1 v_1 \cos\theta_1 + \mu_2 v_2 \cos\theta_2}. \quad (8.93)$$

By substituting $Z = 1/\mu v$ for each medium we can rewrite these expressions in terms of the characteristic impedances,

$$\rho_\parallel = \frac{Z_2 \cos\theta_1 - Z_1 \cos\theta_2}{Z_2 \cos\theta_1 + Z_1 \cos\theta_2}, \quad (8.94)$$

$$\tau_\parallel = \frac{2Z_1 \cos\theta_1}{Z_2 \cos\theta_1 + Z_1 \cos\theta_2}. \quad (8.95)$$

For non-magnetic materials, $\mu_1 = \mu_2 = \mu$, we can write these relations in terms of the refractive indices of the two media,

$$\rho_\parallel = \frac{n_2 \cos\theta_1 - n_1 \cos\theta_2}{n_2 \cos\theta_1 + n_1 \cos\theta_2}, \quad (8.96)$$

$$\tau_\parallel = \frac{2n_1 \cos\theta_1}{n_2 \cos\theta_1 + n_1 \cos\theta_2}. \quad (8.97)$$

8.4.3 Consequences of Fresnel's Equations

It is interesting to explore how the coefficients of reflection and transmission vary with the incidence angle θ_1. We will use the expressions in Eqs. 8.83, 8.84, 8.96, and 8.97. Let us use Snell's law,

$$n_1 \sin\theta_1 = n_2 \sin\theta_2$$

and express the angle of refraction θ_2 in terms of the angle of incidence θ_1 and the ratio of the refractive indices of the two media,

$$\rho_\perp = \frac{n_1 \cos\theta_1 - \sqrt{n_2^2 - n_1^2 \sin^2\theta_1}}{n_1 \cos\theta_1 + \sqrt{n_2^2 - n_1^2 \sin^2\theta_1}}, \tag{8.98}$$

$$\tau_\perp = \frac{2n_1 \cos\theta_1}{n_1 \cos\theta_1 + \sqrt{n_2^2 - n_1^2 \sin^2\theta_1}}, \tag{8.99}$$

$$\rho_\parallel = \frac{n_2 \cos\theta_1 - (n_1/n_2)\sqrt{n_2^2 - n_1^2 \sin^2\theta_1}}{n_2 \cos\theta_1 + (n_1/n_2)\sqrt{n_2^2 - n_1^2 \sin^2\theta_1}}, \tag{8.100}$$

$$\tau_\parallel = \frac{2n_1 \cos\theta_1}{n_2 \cos\theta_1 + (n_1/n_2)\sqrt{n_2^2 - n_1^2 \sin^2\theta_1}}. \tag{8.101}$$

Normal Incidence

Before we discuss the behavior of these coefficients and their physical implications for all angles of incidence, let us look at the simple case of **normal incidence**. When a ray is incident normally, the angle of incidence will be zero and so would be the angles of transmission and reflection,

$$\text{Normal incidence: } \theta_1 = 0. \tag{8.102}$$

Taking the limit of $\theta_1 = 0$ in Eqs. 8.98–8.101 we get the following for the coefficients:

$$\rho_\perp = \frac{n_1 - n_2}{n_1 + n_2} = -\rho_\parallel, \tag{8.103}$$

$$\tau_\perp = \frac{2n_1}{n_1 + n_2} = \tau_\parallel. \tag{8.104}$$

The equality of the reflection and transmission coefficients for the TE and TM cases occurs because of the fact that when $\theta_1 = 0$, there is no plane of incidence since the propagation vectors $\vec{k}_I$, $\vec{k}_R$, and $\vec{k}_T$ are collinear. In the case of a light wave incident from air, refractive index $n_1 = 1.0$, on a glass plate such as a microscope slide of refractive index $n_2 = 1.5$, the reflection and transmission coefficients will be

$$\rho_\perp = -\frac{1}{5} = -\rho_\parallel,$$

$$\tau_\perp = \frac{4}{5} = \tau_\parallel.$$

General Incidence

Now, to study the behavior at all angles, we plot $\rho_\perp$, $\tau_\perp$, $\rho_\parallel$, and $\tau_\parallel$ versus the incidence angle θ_1 in the range $0 \le \theta_1 \le 90°$ for the air/glass case with $n_1 = 1.0$ and $n_2 = 1.5$. The plot in

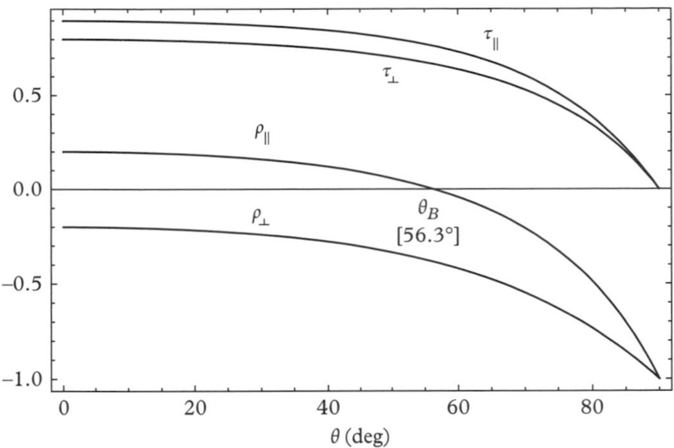

Fig. 8.11 *Plot of reflection and transmission coefficients of light at an air/glass interface with n(air) = 0 and n(glass) = 1.5. Note the vanishing of the reflected electric field for the TM case ($\rho_{\parallel}$) at 56.3°, which is the Brewster's angle for this interface. It is also important to note that at large angles, light is reflected off very efficiently in both the TE and TM cases.*

Fig. 8.11 shows that $\rho_{\perp}$ is always negative and $\tau_{\perp}$ and $\tau_{\parallel}$ are always positive, but $\rho_{\parallel}$ starts out being positive at $\theta_1 = 0$ and decreases to -1 at $\theta_1 = 90°$. At the angle labeled θ_B, called the **Brewster's angle** or **polarization angle**, the reflection coefficient in the TM case, $\rho_{\parallel} = 0$,

$$\rho_{\parallel}\Big|_{\theta_1=\theta_B} = 0. \tag{8.105}$$

From Eq. 8.100, this should happen when

$$n_2 \cos\theta_B = (n_1/n_2)\sqrt{n_2^2 - n_1^2 \sin^2\theta_B}. \tag{8.106}$$

Let $x = n_2/n_1$. Writing in terms of x we get

$$x\cos\theta_B = \sqrt{1 - \frac{\sin^2\theta_B}{x^2}},$$

which can be solved to yield

$$\sin\theta_B = \frac{x}{\sqrt{x^2+1}}, \quad \text{where } x = n_2/n_1.$$

We can write the Brewster's angle formula in a more compact form using a tangent as

$$\tan\theta_B = \frac{n_2}{n_1} \quad \text{(TM Waves)}. \tag{8.107}$$

Figure 8.11 shows that if the TM wave is incident at the Brewster's angle, i.e. if $\theta_1 = \theta_B$, the wave is completely transmitted and there is no reflected wave. Therefore, if you shine a beam of light on a planar interface at Brewster's angle, the reflected ray will have its electric field pointing only perpendicularly to the plane of incidence. Thus, we can use a ray with incidence at Brewster's angle to filter out TM waves from a mixture of TE and TM waves, as illustrated in Fig. 8.12. This is one way of obtaining a polarized light wave with known direction of electric field.

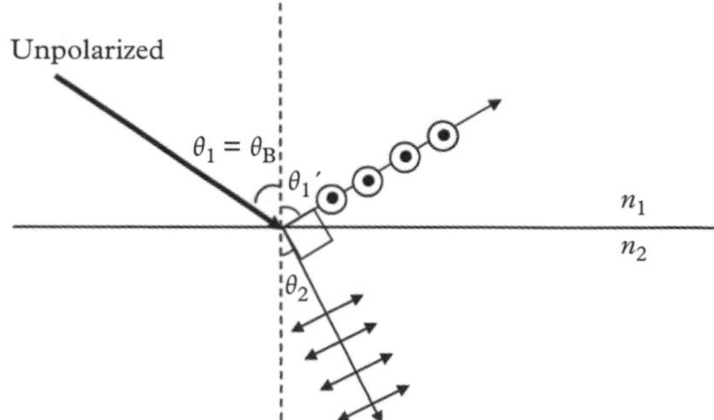

Fig. 8.12 *Reflection at the Brewster's angle gives rise to polarized light. The reflected light will be transverse electric (TE) since no TM wave will be reflected for light incident at the Brewster's angle.*

Example 8.2 Brewster's Angle

Find the angle of incidence of the sunlight so that the reflected light is fully polarized horizontal to the water surface.

Solution

The plane of incidence is the plane of the incident, reflected, and refracted rays. The polarization resulting from the refraction at Brewster's angle will be perpendicular to this plane, and hence parallel to the water surface. Therefore, we calculate the Brewster's angle using the following formula:

$$\tan\theta_B = \frac{n_2}{n_1}.$$

Putting $n_1 = 1$ and $n_2 = 4/3$, we find that the Brewster's angle is

$$\theta_B = \tan^{-1}\left(\frac{n_2}{n_1}\right) = \tan^{-1}\left(\frac{4}{3}\right) = 53°.$$

Therefore, the sunlight will be polarized upon reflection from the air/water interface if the reflected angle is 53°.

Example 8.3 Energy Conservation

Prove that the total energy in the reflected and transmitted waves is equal to the energy in the incident wave.

Solution

In a one-dimensional case, the energy of the incoming wave and the outgoing waves were relatively straightforward since the waves were all along one line. In the three-dimensional situation the three waves move in different directions and the energy flow from the incident wave to the reflected and transmitted waves occurs across a common area of cross-section whose normal is not in the directions of the waves. Let A be the area in the plane of the interface. The normal to this area makes an angle θ_1 with the incident wave. Therefore, the area of the cross-section of the beam that strikes an area A of the interface is $A\cos\theta_1$ as shown in Fig. 8.13.

Thus, if I_0 is the intensity of the incident wave, then the power flow through area A tilted at angle θ_1 with the direction of the wave will not be $I_0 A$, but rather $I_0 A\cos\theta_1$. That is, the energy per unit time arriving at the interface will be

$$P = I_0 A\cos\theta_1. \tag{8.108}$$

Here, the intensity is related to the wave amplitude E_0 and the properties, electric susceptibility ϵ_1 and wave speed v_1 of the medium in which the incident wave moves,

$$I_0 = \frac{1}{2}\epsilon_1 v_1 E_0^2. \tag{8.109}$$

Therefore, we have the following expression for the power arriving at the area A of the interface:

$$\text{Incident}: P_I = \frac{1}{2}\epsilon_1 v_1 E_0^2 A\cos\theta_1. \tag{8.110}$$

Similarly, the power in the reflected, and transmitted waves will be given by

$$\text{Reflected}: P_R = \frac{1}{2}\epsilon_1 v_1 \rho^2 E_0^2 A\cos\theta_1, \tag{8.111}$$

$$\text{Reflected}: P_T = \frac{1}{2}\epsilon_2 v_2 \tau^2 E_0^2 A\cos\theta_2, \tag{8.112}$$

where ρ and τ are the reflection and transmission coefficients in which I have suppressed the TE and TM indicator symbols. Energy conservation demands that we must have

$$P_I = P_R + P_T, \tag{8.113}$$

or,

$$\frac{P_R}{P_I} + \frac{P_T}{P_I} = 1. \tag{8.114}$$

The ratios P_R/P_I and P_T/P_I are called **reflectance** R and **transmittance** T respectively,

$$R = \frac{P_R}{P_I} = \rho^2, \tag{8.115}$$

$$T = \frac{P_T}{P_I} = \frac{\epsilon_2 v_2\cos\theta_2}{\epsilon_1 v_1\cos\theta_1}\tau^2. \tag{8.116}$$

We can use the Fresnell's Eqs. 8.83, 8.84, 8.96, and 8.97 to prove that in both TE and TM cases, the energy balance occurs, by proving that

$$R + T = \rho^2 + \frac{\epsilon_2 v_2\cos\theta_2}{\epsilon_1 v_1\cos\theta_1}\tau^2 = 1. \tag{8.117}$$

continued

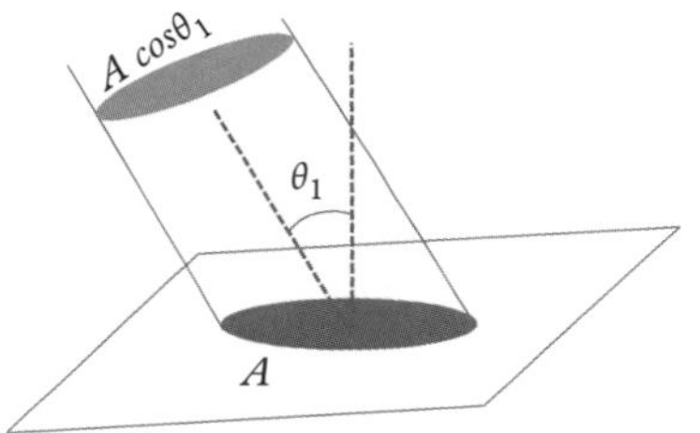

Fig. 8.13 *Example 8.3.*

Example 8.3 *continued*

For simplicity, let us check for the case of normal incidence and $\mu_1 = \mu_2$. In the case of $\mu_1 = \mu_2$, we have

$$\frac{\epsilon_2 v_2}{\epsilon_1 v_1} = \frac{\epsilon_2 \mu v_2}{\epsilon_1 \mu v_1} = \frac{1/v_2}{1/v_1} = \frac{c/v_2}{c/v_1} = \frac{n_2}{n_1}.$$

Setting $\theta_1 = 0 = \theta_2$ for the normal incidence case, we get

$$R + T = \left(\frac{n_1 - n_2}{n_1 + n_2}\right)^2 + \frac{n_2}{n_1}\left(\frac{2n_1}{n_1 + n_2}\right)^2$$

$$= \frac{(n_1 - n_2)^2 + 4n_1 n_2}{(n_1 + n_2)^2} = \frac{(n_1 + n_2)^2}{(n_1 + n_2)^2} = 1.$$

EXERCISES

(8.1) Two strings of mass densities μ_1 and μ_2 are tied together with a "massless" knot. The other end of the second string is tied to a fixed support far away and the free end of the combined string is pulled to generate a tension T throughout. Now, the free end of the combined string is vibrated harmonically so that a transverse harmonic wave of amplitude A and angular frequency ω travels with a speed $v_1 = \sqrt{T/\mu_1}$ towards the knot. When the incident wave arrives at the knot, the knot vibrates and generates two waves, one moving towards the vibrating end and the other towards the fixed end.

Thus, there would be two waves, the incident wave and the reflected wave, in the first string and one wave, the transmitted wave, in the second string. Denote the ratio μ_2/μ_1 by α and the speed in the second string by $v_2 = \sqrt{T/\mu_2}$. Suppose B is the amplitude of the reflected wave and C the amplitude of the transmitted wave. Also, let ϕ be the phase shift of the reflected wave compared to the incident wave, and δ the phase shift of the transmitted wave compared to the incident wave. Assume that the first string is in $-\infty < x < 0$ and the second string in $0 < x < \infty$.

(a) Find expressions for the amplitude ratios B/A and C/A and the phase shifts ϕ and δ.

(b) Find the average power carried by each of the waves and prove that the energy is conserved at the knot.

(c) Determine what happens to the average power of each wave when (i) $\alpha = 1$, and (ii) $\alpha = \infty$, and give a physical interpretation of these results.

(d) Find the displacement function that gives the vibration of the knot with time.

(8.2) Identical pendulums, of length l and mass m are connected by springs, as shown in Fig. 8.14, such that the spring constants of a spring in the $x < 0$ region is K_L and that of a spring in $x > 0$ region is K_R.

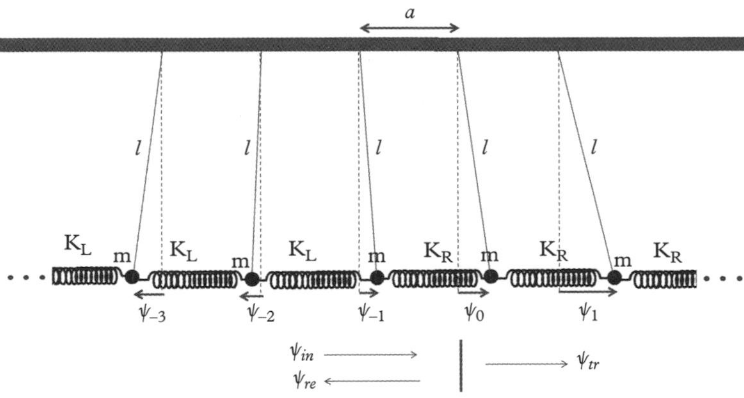

Fig. 8.14 *Exercise 8.2.*

(a) In the limit of large wavelength compared to the separation between the pendulums (the so-called continuum limit), the vibrations of pendulum bobs, denoted by the function $\psi(x,t)$, obey the following equation:

$$\frac{\partial^2\psi}{\partial t^2} = -\frac{g}{l}\psi + \frac{Ka^2}{m}\frac{\partial^2\psi}{\partial x^2}, \tag{8.118}$$

where $K = K_L$ when $x < 0$ and $K = K_R$ when $x > 0$. Derive this equation by examining the equation of motion of an arbitrary mass on either side of $x = 0$ in the discrete form of the problem shown in the figure. For instance, you can look at the equation of motion of a particle of index n, i.e. which is at $x = na$ at equilibrium. In this equation you can get the continuum limit by using the discrete form of the derivative with $\Delta x = a$.

(b) Let $\psi_L(x,t)$ be the solution for $x < 0$ and $\psi_R(x,t)$ for $x > 0$. Based on Eq. 8.118 find the relation between $\psi_L(x,t)$ and $\psi_R(x,t)$ at $x = 0$. You will get two relations, one between $\psi_L(0,t)$ and $\psi_R(0,t)$, and another between their derivatives with respect to x.

(c) Let there be an incident wave $\psi_{in} = A\cos(k_L x - \omega t)$ and a reflected wave $\psi_{re} = B\cos(-k_L x - \omega t + \phi)$ on the left part of $x = 0$, and a transmitted wave $\psi_{tr} = C\cos(k_R x - \omega t + \delta)$ on the right side of $x = 0$, where k_L and k_R are the wave numbers of the waves on the two sides of $x = 0$ and ϕ and δ are their phase shifts compared to the incident wave. Note that we are using capital letter K for the spring constant and small letter k for the wave number. Find expressions for the reflection and transmission coefficients.

(d) If $K_R = \frac{1}{2}K_L$, what is the percentage of incident energy that is transmitted in the $x > 0$ region?

(8.3) Identical pendulums of length l and mass m are connected by identical springs with spring constant K, as shown in Fig. 8.15. As shown in the figure, the springs are attached at different places on the left and right sides of $x = 0$. Suppose the springs in the $x > 0$ region are tied to the rigid rods of the pendulums at half the length of the rods.

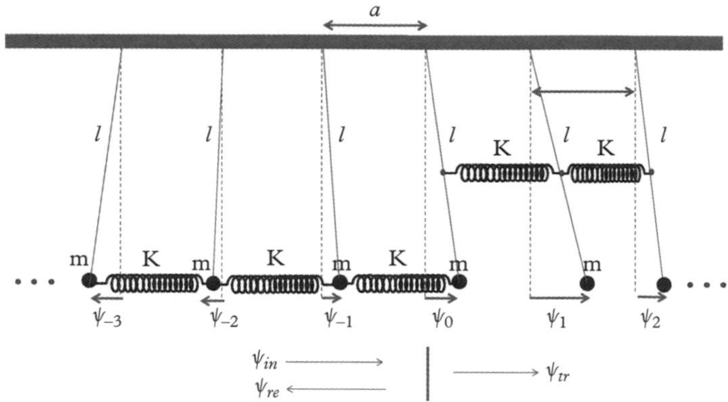

Fig. 8.15 *Exercise 8.3.*

(a) Deduce the wave equation for the propagation of vibrations in the regions $x < 0$ and $x > 0$.

(b) Let there be an incident wave $\psi_{in} = A\cos(k_L x - \omega t)$ and a reflected wave $\psi_{re} = B\cos(-k_L x - \omega t + \phi)$ on the left part of $x = 0$, and a transmitted wave $\psi_{tr} = C\cos(k_R x - \omega t + \delta)$ on the right side of $x = 0$, where k_L and k_R are the wave numbers of the waves on the two sides of $x = 0$. Note that we are using capital letter K for the spring constant and small letter k for the wave number. Find expressions for the reflection and transmission coefficients and phase shifts.

(8.4) **Reflection and transmission of waves in a transmission line.**
Consider a long two-conductor cable that has two types of wire connected at $x = 0$, such that the impedance for the transmission wire for $x < 0$ is Z_1 and that for $x > 0$ is Z_2, as shown in Fig. 8.16. An oscillating source of voltage is located at a far away point in the region $x < 0$.

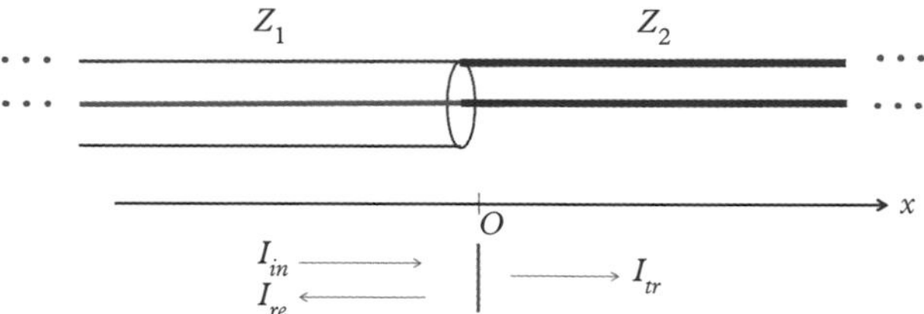

Fig. 8.16 *Exercise 8.4.*

When the oscillating source is turned on, it sets up current and voltage waves in the transmission line, which we will denote by the following expressions. The voltage waves will be

$$V_{in}(x,t) = V_0\, e^{i(k_1 x - \omega t)}, \quad x < 0$$

$$V_{re}(x,t) = V_{0R}\, e^{i(-k_1 x - \omega t + \phi_r)}, \quad x < 0$$

$$V_{tr}(x,t) = V_{0T}\, e^{i(k_2 x - \omega t + \phi_t)}. \quad x > 0$$

The current waves will be

$$I_{in}(x,t) = I_0\, e^{i(k_1 x - \omega t + \theta)}, \quad x < 0$$

$$I_{re}(x,t) = I_{0R}\, e^{i(-k_1 x - \omega t + \phi_r')}, \quad x < 0$$

$$I_{tr}(x,t) = I_{0T}\, e^{i(k_2 x - \omega t + \phi_t')}. \quad x > 0$$

Recall that current I and voltage V at a point on a transmission line are related by the following cable equations:

$$\frac{\partial I}{\partial x} = -C\frac{\partial V}{\partial t},$$

$$\frac{\partial V}{\partial x} = -L\frac{\partial I}{\partial t},$$

where C and L are the capacitance per unit length and inductance per unit length respectively at the point under consideration. Note that the wave numbers in the two parts of the transmission line have been denoted by k_1 and k_2, since the wave speeds in the two lines could be different, as given by

$$v_1 = \frac{1}{\sqrt{L_1 C_1}}, \quad v_2 = \frac{1}{\sqrt{L_2 C_2}}.$$

Also recall the relation of the impedance to the inductance and capacitance per unit length,

$$Z = \sqrt{\frac{L}{C}}.$$

(a) Prove the following relations.
   (i) $I_0 = \frac{V_0}{Z_1}, \quad \theta = \pi.$
   (ii) $I_{0R} = \frac{V_{0R}}{Z_1}, \quad \phi_r' = \phi_r.$
   (iii) $I_{0T} = \frac{V_{0T}}{Z_2}, \quad \phi_t' = \phi_t + \pi.$

(b) By matching the boundary conditions on I and V at $x = 0$ deduce the following:

$$V_{0R} = \frac{|Z_1 - Z_2|}{Z_1 + Z_2}, \quad \phi_r = \begin{cases} 0 & Z_2 > Z_1, \\ \pi & Z_1 > Z_2. \end{cases}.$$

$$V_{0T} = \frac{|2Z_2}{Z_1 + Z_2}, \quad \phi_t = 0.$$

(c) Verify that the average power coming into the junction at $x = 0$ is equal to the average power leaving the junction.

(8.5) **Reflection of waves in a transmission line.** In this exercise you will look at the reflection of the incoming wave in a transmission line, as shown in Fig. 8.17. Let Z_0 be the impedance of the transmission line and Z the impedance of the connection at the end point.

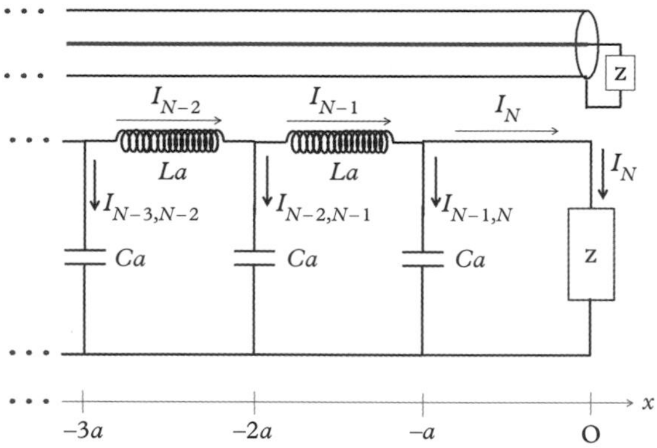

Fig. 8.17 *Exercise 8.5.*

We wish to study incoming and reflected waves that would ride on the transmission line in the region $-\infty < x < 0$. Let us write the incident and reflected waves of current and voltage as

$$I_{in}(x,t) = I_0\, e^{i(kx-\omega t)},$$

$$I_{re}(x,t) = \rho_i\, I_0\, e^{i(-kx-\omega t)},$$

$$V_{in}(x,t) = V_0\, e^{i(kx-\omega t)},$$

$$V_{re}(x,t) = \rho_v\, V_0\, e^{i(-kx-\omega t)},$$

where the coefficients ρ_i and ρ_v are complex

$$\rho_i = |\rho_i|\, e^{i\phi_i}, \quad \rho_v = |\rho_v|\, e^{i\phi_v}.$$

The net current and voltage at any point x on the transmission line is given by

$$I(x,t) = I_{in}(x,t) + I_{re}(x,t), \quad x < 0$$

$$V(x,t) = V_{in}(x,t) + V_{re}(x,t). \quad x < 0$$

The current and voltage at any point of the transmission line are related as given in the last problem. The current $I(0,t)$ and the voltage $V(0,t)$ across the terminating impedance Z at $x = 0$, as shown in the figure, are related as usual,

$$I(0,t) = \frac{V(0,t)}{Z}.$$

There are three cases of particular interest and we want to study them here.

(a) Suppose you connect the end at $x = 0$ with a short conducting wire such that $Z = 0$. We say that the end is shorted. Now, the voltage drop across the end will be zero since there is no impedance,

$$V(0, t) = 0 \text{ at the shorted termination.}$$

Find the amplitude and phase of the current in the shorting wire.

(b) Suppose you do not connect anything at the end at $x = 0$ so that the end is open. In that case $Z = \infty$. Now, the current at the open termination will have to be zero at all times.

$$I(0, t) = 0 \text{ at the open termination.}$$

Find the amplitude and phase of the voltage across the open terminal.

(c) Suppose you connect a circuit element whose impedance matches the impedance of the entire transmission line. That is, you pick $Z = Z_0$. (i) What will be the amplitude and phase of the current in the load? (ii) What is the average power delivered to the load?

(8.6) **Reflection of Sound.** A tube of length L is closed at one end and open at the other end. At the open end a tuning fork is vibrating at a frequency f such that you hear the fundamental mode of the air column in the tube. Let v be the speed of sound in air. Ignore the edge effect of the open end to analyze the situation here.

(a) What is the relation of f to the length of the tube?

(b) Let the positive x-axis go from the origin at the closed end towards the open end. Show that the standing wave in the tube consists of an incident wave and a reflected wave from the closed end.

(8.7) **Transmission and reflection of sound.** A tube of length L is filled with water so that the length of the air column now is h and below that is water. At the open end a tuning fork is vibrating at a frequency f such that you hear the fundamental mode of the air in the tube. Let v_1 be the speed of sound in air and v_2 the speed in water, and ρ_1 the density of air and ρ_2 the density of water. Suppose, the sound wave, which is due to the vibration of air particles driven by the tuning fork, is incident at 90° to the interface, i.e. at normal incidence. Let $\psi(x, t)$ be the displacement of particles of air or water from their equilibrium positions, then the boundary conditions on the $\psi(x, t)$ at the air/water interface will be

$$\psi(x, t)[\text{air side}] = \psi(x, t)[\text{water side}]$$
$$\rho_1 v_1^2 \frac{\partial \psi(x, t)}{\partial x}[\text{air side}] = \rho_2 v_2^2 \frac{\partial \psi(x, t)}{\partial x}[\text{water side}]$$

(a) Find the expressions for the reflection and transmission coefficients.

(b) Find expressions for the average power in the incident, reflected, and transmitted waves. Recall: the intensity of sound waves $= \frac{1}{Z} p_m^2$, where $Z = \rho v$, the characteristic impedance of the medium and p_m is the pressure amplitude, which is related to the displacement amplitude x_m by $kp_m = \omega^2 \rho x_m$, where k is the wave number and ω the angular frequency.

(c) Suppose $v_1 = 350$ m/s, $v_2 = 1500$ m/s, $\rho_1 = 1.3$ kg/m^3, and $\rho_2 = 1000$ kg/m^3, what is the percentage of energy in the sound wave that penetrates water?

(8.8) Find the Brewster's angle of a light ray incident from water on a water/glass interface. Use $\frac{4}{3}$ and $\frac{3}{2}$ as refractive indices for water and glass respectively.

(8.9) An electromagnetic wave of frequency $f = 4.5 \times 10^{14}$ Hz in glass is incident on a glass ($n = 1.5$)/air ($n = 1.0$) planar interface at an angle $\theta = 30°$.
(a) What are the wavelengths of the wave in the two media?
(b) Find the coefficient of reflection and transmission if the electric field is perpendicular to the incidence plane.
(c) What is the fraction of the incident energy that enters air?

(8.10) An electromagnetic wave of frequency $f = 4.5 \times 10^{14}$ Hz in glass is incident on a glass ($n = 1.5$)/water ($n = 1.3$) planar interface at an angle $\theta = 30°$.
(a) What are the wavelengths of the wave in the two media?
(b) Find the coefficient of reflection and transmission if the magnetic field is perpendicular to the incidence plane.
(c) What is the fraction of the incident energy that enters water?

Interference

9

Chapter Goals

In this chapter you will learn about interference of waves from two coherent point sources. We will derive the intensity of a combined wave from two waves overlapping in space which will show an interference pattern. You will also learn about applications of interference to reflection of light from dielectric films and instruments for measuring distances and frequencies.

So far we have studied waves generated by one source. When you have multiple sources of waves, the superposition of waves from different sources leads to the phenomenon of interference if the waves are coherent. Interference causes the intensity of the net wave to vary in space. The intensity is a maximum when waves interfere constructively and a minimum when they interfere destructively. The variation in intensity can be traced to different phase shifts of the waves from different sources, as we will find out below.

In this chapter we will study the interference of waves from two point sources. In the next chapter we will extend this study to the interference of waves from infinitely many sources.

9.1 The Superposition Principle

9.1.1 Linearity of Wave Equation

The wave equations obeyed by the wave functions $\psi(x, y, z, t)$ of the electromagnetic waves in a vacuum, the sound waves in air, shallow water waves, the waves on a string, and quantum mechanical matter waves, are linear in wave functions, meaning that each term of the equation contains only one power of the corresponding wave function. Because of the linearity of these wave equations, the solutions can be superposed to obtain new solutions. For instance, if $\psi_1(x, y, z, t)$ and $\psi_2(x, y, z, t)$ are two solutions, then any linear combination $\psi(x, y, z, t)$ of these solutions with constant coefficients will also be a solution of the wave equation,

$$\psi(x, y, z, t) = a\,\psi_1(x, y, z, t) + b\,\psi_2(x, y, z, t), \tag{9.1}$$

where a and b are constants, i.e. they do not depend on x, y, z, or t. This aspect of combining solutions to obtain new solutions can be applied to answer the question: what happens when two waves overlap? The **superposition principle** says that when two waves overlap, the combination of the two waves results in another wave whose amplitude is equal to the sum of the amplitudes of the constituent waves, i.e. $a = 1$ and $b = 1$ in Eq. 9.1,

$$\psi(x, y, z, t) = \psi_1(x, y, z, t) + \psi_2(x, y, z, t). \tag{9.2}$$

A First Course in Vibrations and Waves. First Edition. Mohammad Samiullah.

9.1.2 Intensity and Superposition Principle

The superposition principle of waves has important consequences for the intensity of the resulting wave due to the fact that the intensity of a wave is proportional to the square of the amplitude and not directly to the amplitude,

$$\text{Intensity, } I \propto \langle \psi^2 \rangle, \tag{9.3}$$

where $\langle \cdots \rangle$ refers to the time-averaging operation on the quantities enclosed within the angle brackets. Therefore, although, the amplitudes of waves will add simply according to Eq. 9.2, their intensities will add in a more complicated way:

$$\text{Intensity of combined wave} \quad I_{\text{net}} \propto \left\langle (\psi_1 + \psi_2)^2 \right\rangle. \tag{9.4}$$

Expanding the right side we obtain

$$I_{\text{net}} \propto \left\langle \psi_1^2 \right\rangle + \left\langle \psi_2^2 \right\rangle + 2\left\langle \psi_1 \psi_2 \right\rangle. \tag{9.5}$$

The first two terms on the right side are the intensities of the individual waves, which are both positive quantities, and the third term, called the **interference term**, can be either positive or negative depending upon the relative phases of the two waves. When the interference term is positive, it adds to the other positive terms and the net intensity at that point in space is more than a simple sum of the individual intensities of each of the waves. This is called **constructive interference**. When the interference term is negative, it subtracts from the other positive terms and the net intensity at that point in space is less than a simple sum of the individual intensities of each of the waves. This is called **destructive interference**.

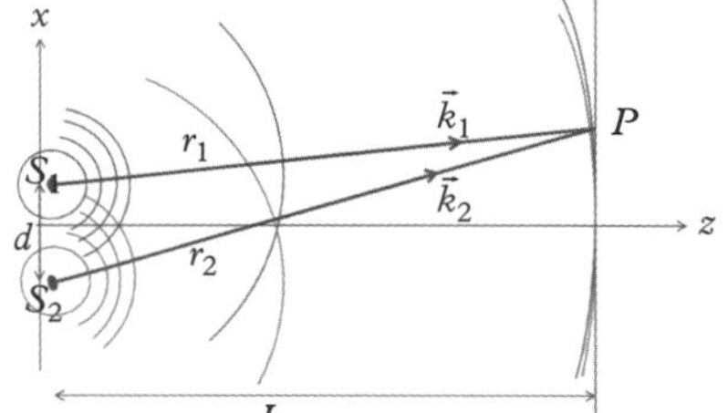

Fig. 9.1 *Interference of two identical sources of the same frequency. The two waves start out in phase at the two sources S_1 and S_2 and spread out from the sources. Here I have shown a few wavefronts for illustrative purposes; unlike the picture here, the wavefronts of monochromatic waves would be drawn equispaced. To reach point P, where a detector is placed, the two waves travel distances r_1 and r_2 and hence they undergo different phase changes. Therefore, when they reach point P, they may or may not be in phase even when they started out in phase at the source.*

9.2 The Interference Between Two Point Sources

9.2.1 The Derivation of Net Intensity at a Detector

As a simple example let us consider the interference at a point from two coherent point sources that emit waves at the same frequency, as shown in Fig. 9.1. We will discuss the concept of coherence below; it is sufficient to state here that the difference in the phases of two relatively coherent sources remains constant with time. A physical way of obtaining two coherent point sources would be by driving two identical speakers by the same oscillator circuit. Another example would be shining light from a monochromatic source through two infinitely narrow pin holes; the light emerging from the holes will be two coherent sources.

Another simplification we make is that we will place the detector at a point *far away* from the sources. This is to avoid having to deal with spherical waves in our calculations. As illustrated in Fig. 9.1, the wavefronts near the point sources will be spherical, but far away from the sources, which could be hundreds of wavelengths away, the curvature of the waves can be ignored and the waves can be treated as plane waves. The far region is also called **far-field**. One more technicality we wish to state without going into detail is that the separation d between the sources is much larger than the wavelength λ. Now, let us perform the calculations for the intensity at the detector.

Let the wave sources have frequency ω and have zero phase difference between the vibrations at the sources, so that the generated waves at S_1 and S_2, E_1 (at S_1) and E_2 (at S_2), respectively, are given by

$$E_1\,(\text{at } S_1) = E_{01}\cos(\omega t), \quad E_2\,(\text{at } S_2) = E_{02}\cos(\omega t), \tag{9.6}$$

where E_{01} and E_{02} are the amplitudes of the two waves. In the case of sound waves in air, the E's will be the vibrations of the air particles at the speakers, and in the case of light they can be thought of as one of the components of the electric or magnetic field at the source.

We want to know what will happen when the waves meet at point P. The two waves move in the directions given by their propagation vectors $\vec{k}_1$ and $\vec{k}_2$ and travel through distances $S_1P = r_1$ and $S_2P = r_2$ respectively. Since the two waves have the same frequency and speed, they will have the same wave number, which we can denote by k, and which is related to the wavelength λ as usual,

$$|\vec{k}_1| = |\vec{k}_2| = k = \frac{2\pi}{\lambda}. \tag{9.7}$$

At time t, the wave arriving at P from S_1 is emitted at the retarded time

$$t_1 = t - r_1/v,$$

and that from S_2 is emitted at the retarded time

$$t_2 = t - r_2/v,$$

where v is the wave speed,

$$v = \omega/k. \tag{9.8}$$

Therefore, the displacements at P at time t will be given by substituting t_1 and t_2 into Eq. 9.6,

$$E_1\,(\text{at } P) = E_{01}\ \cos(\omega[t - r_1/v]), \tag{9.9}$$

$$E_2\,(\text{at } P) = E_{02}\ \cos(\omega[t - r_2/v]). \tag{9.10}$$

Using Eq. 9.8 we can write these wave expressions in a more familiar form. Furthermore, since we will be working at point P only, we will leave out the qualifier (at P) from the expressions,

$$E_1 = E_{01}\cos(kr_1 - \omega t), \tag{9.11}$$

$$E_2 = E_{02}\cos(kr_2 - \omega t). \tag{9.12}$$

Now, applying the superposition principle, we claim that the net wave at point P will be

$$E = E_1 + E_2. \tag{9.13}$$

By definition, the net intensity will be proportional to the time-average of the square of the net wave function. In the case of light, the proportionality factor for intensity is the product of the speed of light and the permittivity of the medium. For other types of waves you would have some other multiplicative factors. To be general, we will keep the prefactors as some unspecified quantity α, since this quantity will be unimportant when we consider relative intensity on the screen. Putting in the prefactor in the formula for the intensity, we have the following for the intensity:

$$I = \alpha \langle E^2 \rangle_{\text{time average}}. \tag{9.14}$$

Using the net wave given in Eq. 9.13 we find that the net intensity at any point is equal to the sum of individual intensities I_1 and I_2 for the two waves separately and the **interference term**, which we will denote by I_{12},

$$I = I_1 + I_2 + I_{12}, \tag{9.15}$$

where

$$I_1 = \alpha \langle E_1^2 \rangle_{\text{time average}} = \frac{1}{2}\alpha E_{01}^2, \tag{9.16}$$

$$I_2 = \alpha \langle E_2^2 \rangle_{\text{time average}} = \frac{1}{2}\alpha E_{02}^2, \tag{9.17}$$

$$I_{12} = 2\alpha \langle E_1 E_2 \rangle_{\text{time average}}. \tag{9.18}$$

Evaluating the interference term explicitly we find the following:

$$I_{12} = \alpha\,(E_{10}E_{02})\,\cos\left[k\,(r_1 - r_2)\right]. \tag{9.19}$$

The argument of the cosine is the phase difference due to the different paths taken by the two waves to get to P, which we denote by Δ_{12},

$$\text{Phase difference in the two paths, } \Delta_{12} \equiv k\,(r_1 - r_2). \tag{9.20}$$

Thus, the net amplitude at point P is seen to be

$$I = I_1 + I_2 + 2\sqrt{I_1 I_2}\,\cos\left[k\,(r_1 - r_2)\right] \equiv I_1 + I_2 + 2\sqrt{I_1 I_2}\,\cos\Delta_{12}. \tag{9.21}$$

Since r_1 and r_2 would be different for different points in space, the interference term will yield different values depending on the location of point P, and the total intensity I may be equal to, greater than, or less than $I_1 + I_2$. The maximum intensity occurs at a point where the cosine factor is equal to 1, in which case, the point P is said to be a place of **total constructive interference**, or simply constructive interference,

$$I = I_{\text{max}} = I_1 + I_2 + 2\sqrt{I_1 I_2}, \tag{9.22}$$

when

$$\cos\left[k\,(r_1 - r_2)\right] = 1, \text{ or, equivalently, } \cos\Delta_{12} = 1, \tag{9.23}$$

which happens when

$$\Delta_{12} \equiv k\,(r_1 - r_2) = 0, \pm 2\pi, \pm 4\pi, \ldots. \tag{9.24}$$

We can write this condition in terms of wavelength λ also by writing $k = 2\pi/\lambda$,

$$r_1 - r_2 = 0, \pm\lambda, \pm 2\lambda, \ldots. \tag{9.25}$$

That is, if the path difference to P is an integral multiple of wavelength, then we will see total constructive interference. On the other hand, the minimum intensity will occur when the cosine factor is equal to –1, in which case, the point P is said to be a place of **total destructive interference** or simply destructive interference,

$$I = I_{\max} = I_1 + I_2 - 2\sqrt{I_1 I_2}, \tag{9.26}$$

when

$$\cos\left[k\left(r_1 - r_2\right)\right] = -1 = \cos\Delta_{12}, \tag{9.27}$$

which happens when

$$\Delta_{12} \equiv k\left(r_1 - r_2\right) = \pm\pi, \pm 3\pi, \pm 5\pi, \ldots. \tag{9.28}$$

In terms of wavelength λ this condition reads

$$r_1 - r_2 = \pm\frac{\lambda}{2}, \pm 3\frac{\lambda}{2}, \pm 5\frac{\lambda}{2} \cdots. \tag{9.29}$$

The conditions for total constructive and destructive interferences can be written more succinctly as

$$\text{Constructive Interference: } \Delta_{12} = 2m\pi \quad (m \text{ integer}), \tag{9.30}$$

$$\text{Destructive Interference: } \Delta_{12} = m'\pi \quad (m' \text{ odd integer}). \tag{9.31}$$

Or, equivalently, as

$$\text{Constructive Interference: } r_1 - r_2 = m\lambda \quad (m \text{ integer}), \tag{9.32}$$

$$\text{Destructive Interference: } r_1 - r_2 = m'\lambda/2 \quad (m' \text{ odd integer}). \tag{9.33}$$

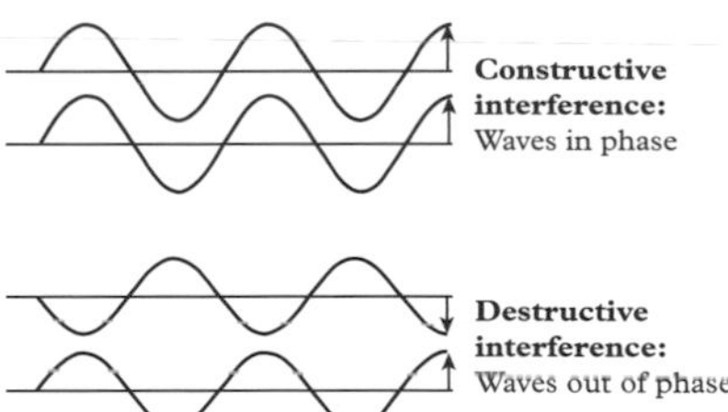

Fig. 9.2 *Constructive and destructive interference.*

The integers m and m' are called the **orders of constructive and destructive interference**, respectively. These conditions make sense when you look at what happens to the waves. They start out at the sources in step with each other, and when they travel different distances, they will tend to go out of step by varying amounts. They will be completely in step at places where path difference is an integral multiple of one wavelength, and they will be completely out of step, meaning if one is at the crest, then the other would be at the trough, if the difference in path is off by an odd integral multiple of half a wavelength as illustrated in Fig. 9.2.

9.2.2 Interference Conditions in Terms of Direction

Often we are interested in interference conditions with respect to direction from the sources. In the far-field case the distances S_1P and S_2P are much greater than the separation distance d of the slits, which in turn is large compared to the wavelength λ. In this case, the rays from S_1 and S_2 that make a small angle with respect to the horizontal direction will be almost parallel and we can assign the same angle θ to the two rays S_1P and S_2P, as shown in Fig. 9.3.

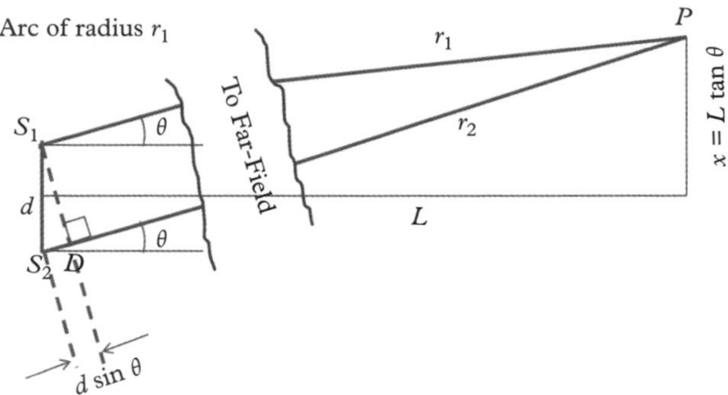

Fig. 9.3 *Geometry for calculation of the path difference $r_2 - r_1$ in terms of the angle θ to the observation point P and the separation d of the point sources. Drawing an arc of radius r_1 cuts the S_2P line at point D, which is very close to the normal line from point S_1 on line S_2P. The triangle ΔS_1S_2D gives $r_2 - r_1 = d\sin\theta$.*

Now, if you draw an arc with center at P and arc length r_1 in the figure, the difference in the paths will be given by the distance S_2D. From triangle S_1S_2D, the distance S_2D is

$$|r_2 - r_1| = d\sin\theta, \tag{9.34}$$

where d is the distance between the slits. Using this relation we can express the net intensity at point P, derived in Eq. 9.21, in terms of direction θ from the slits.

$$I = I_1 + I_2 + 2\sqrt{I_1 I_2}\,\cos(kd\sin\theta). \tag{9.35}$$

The constructive and destructive conditions are obtained by $\cos(kd\sin\theta) = \pm 1$. This gives the interference conditions for small angles as

$$d\sin\theta_m = \begin{cases} m\lambda & (m = 0, \pm 1, \pm 2, \cdots) \quad \text{Constructive} \\ m'\frac{\lambda}{2} & (m' = \pm 1, \pm 3, \pm 5, \cdots) \quad \text{Destructive.} \end{cases} \tag{9.36}$$

Often one places subscripts m and m' onto θ to indicate the special directions where these conditions hold. Note that because of the way angle θ has been defined in the figure, the positive m and m' here correspond to negative m and m' in Eqs. 9.32 and 9.33.

Since the sine of an angle cannot be greater than 1, d should be larger than $m'\lambda/2$ in the case of the destructive interference. Similarly, d must be larger than $m\lambda$ for constructive interference. Therefore, there are fringes possible only up to some values for m and m' for a given d and wavelength λ. These are the theoretical **maximum orders** of interference in a given experimental set-up,

$$\frac{m\lambda}{d} = \sin\theta \leq 1 \implies m_{\max} = \text{int}\left\lfloor \frac{d}{\lambda} \right\rfloor, \tag{9.37}$$

where int stands for the integer obtained when the floor function is evaluated on the ratio of d to λ. For instance, if the sources are separated by 3.5 times the wavelength, then you will observe up to seven constructive interference orders, corresponding to $m = 0, \pm 1, \pm 2, \pm 3$ and six destructive interferences for $m' = \pm 1, \pm 3, \pm 5$.

Interference on a Line Parallel to the S_1S_2 Line

The interference condition can be also expressed in terms of the position of point P along a line parallel to the S_1S_2 line. In Fig. 9.3 the position of point P is given by the x-coordinate with origin at the center line from the midway point between S_1 and S_2. The x-coordinate of P is given by

$$x = L\ \tan\theta. \tag{9.38}$$

For small angles, $\tan\theta \approx \sin\theta$,

$$x \approx L\ \sin\theta. \tag{9.39}$$

The intensity formula given above in terms of θ can now be written for points on a line on a screen, as shown in Fig. 9.4. The interference conditions for the coordinates of P along the x-axis on the screen in the figure will be

$$x = \begin{cases} m\lambda L/d & (m = 0, \pm 1, \pm 2, \ldots) \quad \text{Constructive} \\ m'\lambda L/2d & (m' = \pm 1, \pm 3, \pm 5, \ldots) \quad \text{Destructive.} \end{cases} \tag{9.40}$$

9.2.3 Identical Sources

Suppose the two sources S_1 and S_2 are identical with equal amplitudes $E_{01} = E_{02}$. Denoting the intensities I_1 and I_2 now by I_0, we find that the intensity at point P can be given by a simpler formula,

$$I = 2I_0 + 2I_0 \cos\Delta_{12} = 4I_0 \cos^2\left(\frac{\Delta_{12}}{2}\right). \tag{9.41}$$

Hence, the intensity varies from 0 to $4I_0$. Plotting the net intensity I versus the phase difference Δ_{12} shows an interference pattern of constructive and destructive interferences with the total constructive interferences when $\Delta_{12} = 0, \pm 2\pi, \pm 4\pi, \pm 6\pi$, etc., and the total destructive interferences at $\Delta_{12} = \pm\pi, \pm 3\pi, \pm 5\pi$, etc.

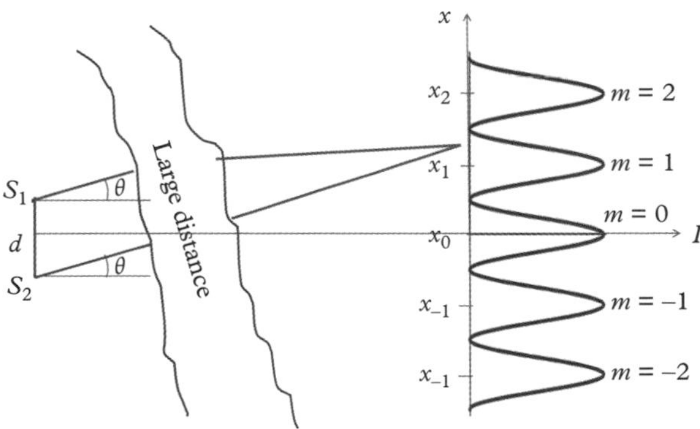

Fig. 9.4 *Intensity pattern on a line on the screen parallel to the $S_1 - S_2$ line. The points x_0 is the central peak, $x_{\pm 1}$, $x_{\pm 2}$ are the constructive interference points for first and second order respectively.*

9.2.4 Energy Conservation

The interference effect shows that the net intensity varies over space. For identical point sources, the intensity of the sources is $2I_0$, but the intensity varies between 0 and $4I_0$. Since intensity is the energy flow per unit time per unit area, it would appear that 1 + 1 can be any number between 0 and 4. This gives a false impression of energy becoming uncertain and perhaps energy conservation not being obeyed. The wave is however an extended object and we need to consider that the energy released at the source has spread out. You find that the principle of conservation of energy is not violated if you examine the space-average of intensity. The interference term goes to zero when you average over space since the average of a cosine will give zero,

$$[I_{12}]_{\text{space average}} = 0. \tag{9.42}$$

Therefore, the space-averaged net intensity is simply the sum of the intensities of the sources,

$$[I]_{\text{space average}} = I_1 + I_2. \tag{9.43}$$

That is, energy is conserved if you take into account the fact that energy spreads out in space with a higher concentration in some regions and a lesser in others.

9.2.5 Interference Conditions for Sources with Phase Difference

Some situations of practical interest, such as a the reflection of light from a soap bubble or oil layer on water, have point sources which do not oscillate in phase. Suppose the sources S_1 and S_2 have a constant phase difference $\Delta\phi = \phi_1 - \phi_2$ at the source so that the oscillations at the source have the following form:

$$E_1(\text{at } S_1) = E_{01}\cos(\omega t + \Delta\phi), \quad E_2(\text{at } S_2) = E_{02}\cos(\omega t), \tag{9.44}$$

where I have set $\phi_2 = 0$ as a reference and written ϕ_1 as $\Delta\phi$. How would the presence of a phase constant affect the conditions for constructive and destructive interference? After you carry out the calculation for the intensity at point P you will arrive at the following expression:

$$I = I_1 + I_2 + 2\sqrt{I_1 I_2}\cos\left(\Delta_{12} + \Delta\phi\right), \tag{9.45}$$

where

$$\Delta_{12} = k(r_1 - r_2) \approx kd\sin\theta. \tag{9.46}$$

Therefore, the constructive and destructive interference conditions will be

$$\text{Constructive:} \quad \Delta_{12} = 2m\pi - \Delta\phi \;\; (m \text{ integer}), \tag{9.47}$$

$$\text{Destructive:} \quad \Delta_{12} = m'\pi - \Delta\phi \;\; (m' \text{ odd integer}). \tag{9.48}$$

Of particular interest to applications is the case when the two sources are 180° out of step with each other, that is when

$$\text{Case: } \Delta\phi = \pi.$$

In this case, the conditions are

$$\text{Constructive:} \quad \Delta_{12} = m\pi \quad (m \text{ odd integer}), \tag{9.49}$$

$$\text{Destructive:} \quad \Delta_{12} = m'2\pi \quad (m' \text{ integer}), \tag{9.50}$$

which are the complete opposite of the conditions when the sources are in phase. In this case, the center of the screen will be a destructive interference since the path difference to the center from sources will be zero, but their phase difference will be 180°, which would mean destructive interference there.

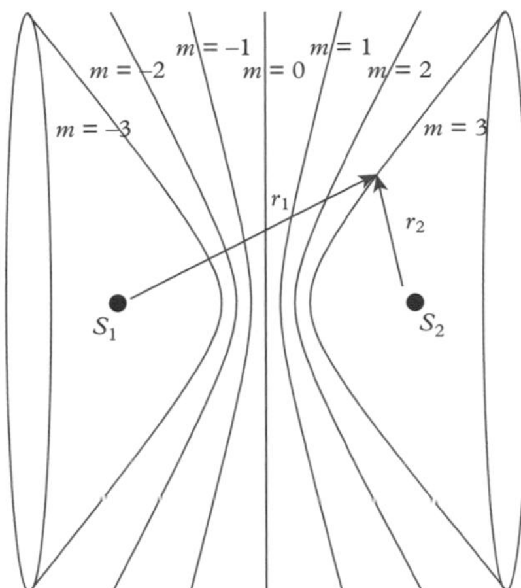

Fig. 9.5 *The hyperboloids of revolution for the constructive interference conditions $r_1 - r_2 = m\lambda$ for $m = 0, \pm 1, \pm 2, \pm 3$. Three-dimensional perspectives of only $m = \pm 3$ are shown here. The $m = 0$ is a plane surface equidistance from the two sources. A screen placed in the field will show interference fringes where the screen intersects with these hyperboloids.*

9.2.6 Interference Hyperboloids

The constructive and destructive interference conditions in space depend on the radial distances r_1 and r_2 from the sources. Let us write them for the case of in-phase oscillations of the source in the following form:

$$\text{Constructive:} \quad r_1 - r_2 = m\lambda \quad (m \text{ integer}), \tag{9.51}$$

$$\text{Destructive:} \quad r_1 - r_2 = m'\lambda/2 \quad (m' \text{ odd integer}). \tag{9.52}$$

Equation 9.51 locates points in space where $r_1 - r_2$ is a fixed value for a given m, and Eq. 9.52 locates points in space where $r_1 - r_2$ is a fixed value for a given m'. You may recall that the points in space satisfied by $r_1 - r_2$ = constant fall on a hyperboloid of revolution about the axis joining S_1 and S_2, which are at the foci. Thus, Eqs. 9.51 and 9.52, define a family of surfaces that are hyperboloids of revolution. Figure 9.5 illustrates some of these surfaces. If a screen is placed in this region, the screen will display constructive and destructive interference areas. In the case of light, these areas would appear as bright and dark lines, called **interference fringes**.

9.2.7 Coherence and the Interference Pattern

Since intensity is obtained by time averaging, the stability of the interference term to being a positive or a negative value at a particular location requires that the relative phases of the two waves are time-independent or nearly so. When the relative phase of the two waves ψ_1 and ψ_2 is constant in time, then we say that the two waves are **relatively coherent** or simply **coherent**.

Perfect Coherence

We can illustrate the idea of coherence by looking at the phase difference of two harmonic waves of the same frequency ω, two different propagation vectors $\vec{k}_1$ and $\vec{k}_2$, and two different phase constants ϕ_1 and ϕ_2, which are assumed to be constant in time,

$$\psi_1(x, y, z, t) = A_1 \cos(\vec{k}_1 \cdot \vec{r} - \omega t + \phi_1), \quad \psi_2 = A_2 \cos(\vec{k}_2 \cdot \vec{r} - \omega t + \phi_2). \tag{9.53}$$

Subtracting the two phases leads to the cancellation of the ωt term and we find that the relative phase does not depend on time, even when the two waves may move with different propagation vectors,

$$\Delta\varphi = \varphi_2 - \varphi_2 = (\vec{k}_2 - \vec{k}_1)\cdot\vec{r} + (\phi_2 - \phi_1). \tag{9.54}$$

Therefore, two waves having the same frequency and time-independent phase constants are coherent waves. The time averaging is completely controlled by the time of oscillations of the wave and we obtain the same intensity regardless of the interval over which we perform the time averaging as long as the interval is long enough to include many cycles of oscillations.

Partial Temporal Coherence and Interference

Two waves with different frequencies may still be coherent under the right conditions. Suppose, the two waves have frequencies, $\omega_0 - \frac{\Delta\omega}{2}$ and $\omega_0 + \frac{\Delta\omega}{2}$ respectively. Then, the relative phase of the two waves would be time dependent, given by

$$\Delta\varphi = \left[(\vec{k}_2 - \vec{k}_1)\cdot\vec{r} + (\phi_2 - \phi_1)\right] - \Delta\omega\, t. \tag{9.55}$$

This says that if the two waves had a zero phase difference at time $t = 0$ at some location, then at the same point in space, they would go completely out of step, meaning the phase difference will be π rad, in a time Δt given by

$$\Delta\omega\Delta t = \pi, \implies \Delta t = \pi/\Delta\omega. \tag{9.56}$$

Supposing that the "bandwidth" $\Delta\omega \ll \omega_0$ so that there are many cycles of the main oscillations at ω_0 within a time span of $\Delta t \sim \pi/\Delta\omega$. Now, the interference of the two waves may or may not give rise to an interference pattern depending on the time-averaging time t_{av} for the detector.

If the detector is fast and averages for a short time compared to $\pi/\Delta\omega$, then the detector will see the coherent source of the dominant oscillations at ω_0. In this case we will see an interference pattern,

$$\text{Coherent if: } t_{av} \ll \pi/\omega.$$

However, if the averaging is done over a time interval long compared with π/ω, the time dependence of $\Delta\varphi$ will wash out the signal and one will obtain only an average value. The time π/ω obtained in Eq. 9.56 for sources that have a spread $\Delta\omega$ in their frequencies is called the **temporal coherence time**. We will denote it by τ_c,

$$\tau_c = \pi/\Delta\omega. \tag{9.57}$$

We say that the two waves are **partially temporally coherent** for a time interval that is much smaller than the coherence time τ_c. The distance over which the wave will move in τ_c is called the **coherence length** l_c. Let v the wave speed, then the coherence length will be

$$l_c = v\tau_c. \tag{9.58}$$

Spatial Coherence and Interference for Extended Sources

Coherent point sources do not really exist. Even if you obtain a monochromatic light source, the source may not be point-like. In an extended source different points of the source may be uncorrelated. This would lead to waves with different phase constants coming from different parts of the source. In addition, the waves emitted at different parts of an extended source will reach the detector having traveled different distances. That is, the net wave will have a mixture of waves with different phases, the extent of mixture being a function of the size of the source. The interference between waves from different parts of the source may wash out the two-point-source interference if two extended sources are used in place of point sources.

The problem associated with an extended source was encountered early in the study of the interference of light from sunlight performed by Francesco Grimaldi in 1665. He let sunlight through two pinholes and projected the images on a white wall. If the two beams emitted from the other side of the pin holes were coherent he would have seen an interference pattern on the wall, but it turns out that the beams lacked spatial coherence as well as temporal coherence. The lack of spatial coherence was due to the fact that Grimaldi was using light waves that came from distant points of the Sun which were not coherent with each other.

To overcome the problem associated with spatial coherence of the sunlight, Thomas Young "spatially filtered" the sunlight first by passing the light through a pin hole before he let the light fall onto a screen that had two narrow slits to produce two coherent beams, as illustrated in Fig. 9.6. The light that came out behind the slits came from same wavefront behind the pin hole and therefore were relatively coherent. Young used these coherent light waves to prove that light beams interfered as waves.

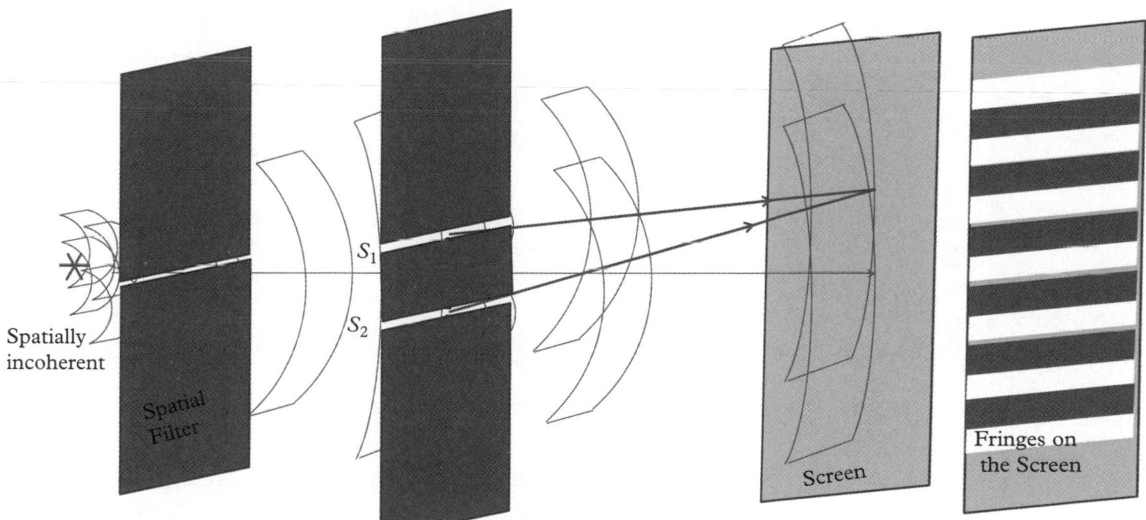

Fig. 9.6 *The use of a spatial filter to create spatially coherent waves in Young's double-slit experiment. The panel in the middle has two narrow slits that serve as two coherent sources whose waves interfere at the screen. Since the slits are horizontal, the interference patterns are horizontal bands spread out vertically.*

9.3 Interference Experiments

As we have discussed, the main difficulty with interference is obtaining coherent wave sources. For temporal coherence we need the source to be nearly monochromatic and for spatial coherence we need the source to be limited in size. There are basically two ways that spatially coherent sources are generated. We either use different points of the same wavefront or we use the same point of a wavefront, but split the amplitude of the wave into two waves; the former is called **wavefront-splitting** and the latter **amplitude-splitting**. Young's double-slit experiment is based on wavefront-splitting and double-beam and multiple-beam interference is based on amplitude-splitting. We will now discuss these techniques.

9.3.1 Wavefront-splitting and Young's Double Slit Experiment

Young's double-slit experiment was the first successful experiment that demonstrated the wave nature of light by showing that light beams could interact with each other just as waves. As illustrated in Fig. 9.6, an incoherent source of light such as sunlight is incident on a narrow slit, which acts as a spatial filter. Two parts of the same wavefront then pass through two slits S_1 and S_2 that lead to two cylindrical waves behind the slits.

If we suppose that the slits are infinitely long, then we have a cylindrical symmetry and we can perform the calculation in a single plane perpendicular to the length of the slits. In this plane, the slits become two single points and the waves to the right of the slits become circular. This calculation is similar to the calculation for two coherent sources done previously. Therefore, the relative intensity on the screen along the vertical direction in the figure can be stated easily. Let θ be the angle to a point on the screen, then the intensity there will be given by Eq. 9.21 for $I_1 = I_2$,

$$I = 2I_0 + 2I_0 \cos \Delta_{12}, \tag{9.59}$$

with

$$\Delta_{12} = kd \sin\theta = \frac{2\pi}{\lambda} d \sin\theta,$$

where d is the distance between the two slits. Because of the cylindrical nature of the problem, the interference pattern on the screen will be horizontal bands of constructive interference and destructive interference, which will occur in the following directions, as given in Eq. 9.36:

$$d \sin\theta = \begin{cases} m\lambda & (m = 0, \pm 1, \pm 2, \ldots) \quad \text{Constructive} \\ m' \frac{\lambda}{2} & (m' = \pm 1, \pm 3, \pm 5, \ldots) \quad \text{Destructive.} \end{cases} \tag{9.60}$$

Fig. 9.7 *Interference from a thin film. The colors of the fringes are observed in the colored version at the following website: http://en.wikipedia.org/wiki/Soap_film. (Picture courtesy of Wikicommons.)*

9.3.2 Amplitude-splitting and Double-Beam Interference

Bright fringes from oil spills and soap films are common examples of the phenomenon of interference due to the reflections from two sides of a dielectric film (see Fig. 9.7). Similar phenomena can be seen in other types of waves when a wavefront is split at an interface and the two parts come together later. In this section we will work out the physical conditions necessary for constructive and destructive interferences for reflections from a rectangular thin film. These results can be used to deduce the thickness of an oil slick or soap bubbles from the interference patterns observed in them.

For the sake of concreteness, we consider a light ray incident on a dielectric film of thickness d at an angle θ, as shown in Fig. 9.8. The amplitude of the original ray PA is split into a reflected part towards AD and a transmitted part towards AB. The refracted ray AB travels in the medium with refractive index n_2 until it encounters the n_2/n_3 interface at B. At that interface the amplitude splits again into a reflected part towards BC and a transmitted part, shown dashed in the figure, which we will ignore in the present discussion.

The reflected ray BC travels back to the n_1/n_2 interface and refracts into the first medium in the direction CC′, which travels parallel to the ray AD. The parallel rays are brought to be the point Q of the focal plane of a converging lens where they undergo constructive or destructive interference depending upon the difference in their phases.

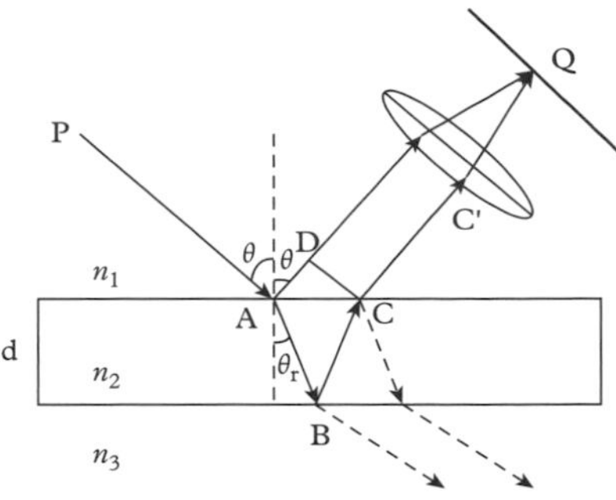

Fig. 9.8 *Interference from reflections off two sides of a planar film.*

Unlike the Young's double-slit experiment, the phase difference between two waves PADQ and PABCQ here arises due to two factors: (1) optical path difference and (2) any phase change due to reflections at A and B. Often, the phase difference is expressed as an optical path length difference. As we know that the phase of a wave changes by 2π radians when the wave travels one entire wavelength, the relation between the phase difference and the optical path length difference is simply

$$\text{Optical Path Length Difference} = \left[\frac{\text{Phase Difference}}{2\pi}\right]\lambda. \quad (9.61)$$

But since the wavelength depends on the refractive index of the medium, we will do our calculations using the phase difference and not bother with converting it into optical path length until the end, where we will write the interference condition in terms of the wavelength λ_2 in the dielectric film.

We have mentioned that the reflection of waves is also one of the sources of phase difference. It turns out that when a wave is reflected off from a medium of higher refractive index, the phase changes by 180° or π radians, as illustrated in Fig. 9.9. There is no phase change when reflecting from a medium of lower refractive index.

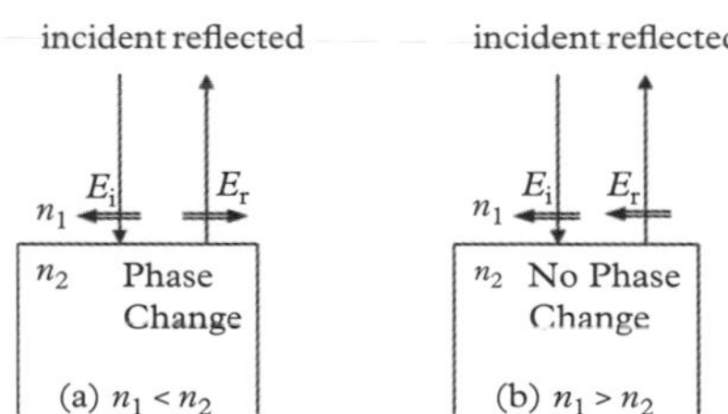

Fig. 9.9 *Reflection at an interface for $n_1 < n_2$ causes phase of the wave to change by π radians.*

The two rays resulting in a reflection from a dielectric film, actually reflect off at two different interfaces. Let us examine the most common case where the front and the back media are the same and their refractive indices are less than that of the film. This will be the case, for instance, when light is incident on a plastic film in air, or of a soap bubble which has air inside the bubble and outside the bubble.

Case: $n_1 = n_3 < n_2$ (example: air/plastic/air)

We can find the phase difference Δ_{12} between the two interfering waves by following the two rays in Fig. 9.8 starting from a point, such as point P, before the first reflection/refraction at A,

$$\begin{aligned}\Delta_{12} = &\ (\text{phase change due to reflection at A} + \text{phase change over AD}) \\ &- (\text{phase change over AB} + \text{phase change due to reflection at B} \\ &+ \text{phase change over BC}).\end{aligned}$$

In the case of $n_1 = n_3 < n_2$, the reflection at A causes a phase change of π radians, but the reflection at B does not have any phase change. The phase change for traveling a distance

AD in the first medium can be found by noting that the phase of a wave changes by 2π radians when it travels over a distance of one wavelength in the medium. We must be careful here and use the correct wavelength for each path since waves are traveling in different media. Let λ_1 and λ_2 denote the wavelength of light in the two media. Note that although the frequency of light is the same in the two media, the wavelength will be different in the two media because the speed of the wave depends on the refractive index,

$$\lambda_1 = \frac{v_1}{f} = \frac{c}{n_1 f}, \quad \lambda_2 = \frac{v_2}{f} = \frac{c}{n_2 f}, \quad \Longrightarrow \quad n_2\lambda_2 = n_1\lambda_1. \tag{9.62}$$

Therefore phase changes over the two paths are as follows:

$$\text{phase change over AD} = \frac{\text{AD}}{\lambda_1} \times 2\pi. \tag{9.63}$$

The phase changes over AB and BC paths are obtained by using the wavelength λ_2 in the second medium,

$$\text{phase change over AB} = \frac{\text{AB}}{\lambda_2} \times 2\pi, \tag{9.64}$$

$$\text{phase change over BC} = \frac{\text{BC}}{\lambda_2} \times 2\pi. \tag{9.65}$$

Therefore, the net phase difference Δ_{12} between the two waves will be

$$\Delta_{12} = \pi + \frac{2\pi}{\lambda_1}\left[\text{AD} - \frac{n_2}{n_1}(\text{AB} + \text{BC})\right]. \tag{9.66}$$

We would get a constructive interference if the phase difference Δ_{12} were an integer multiple of 2π, and a destructive interference if it were an odd multiple of π,

$$\Delta_{12} = \begin{cases} m \times 2\pi & m = 0, \pm 1, \pm 2, \dots \quad \text{Constructive} \\ m' \times \pi & m' = \pm 1, \pm 3, \dots \quad \text{Destructive.} \end{cases} \tag{9.67}$$

These interference conditions are more useful when written in terms of the angle of refraction in the film and the thickness of the film. To accomplish that, we use the geometry given in Fig. 9.8 and Snell's law at point A to rewrite the distances in terms of d and the angle of refraction θ_r. Notice the following relations among the sides of various triangles in the figure:

$$\text{AB} = \text{BC} = \frac{d}{\cos\theta_r}, \tag{9.68}$$

$$\text{AD} = \text{AC}\sin\theta = 2d\tan\theta_r\sin\theta = 2d\frac{n_2}{n_1}\frac{\sin^2\theta_r}{\cos\theta_r}, \tag{9.69}$$

where I have made use of Snell's law for the refraction at A,

$$n_1\sin\theta = n_2\sin\theta_r. \tag{9.70}$$

Substituting for AB, BC, and AD in the expression for Δ_{12} in Eq. 9.66 gives

$$\Delta_{12} = \pi + \frac{2\pi}{\lambda_2} \times 2d\cos\theta_r. \tag{9.71}$$

Hence, the interference conditions given in terms of the angle of refraction are as follows:

For $(n_1 = n_3 < n_2)$

$$2d\cos\theta_r = \begin{cases} \frac{|m|}{2}\lambda_2 & \text{constructive} \quad m = \pm1, \pm3, \pm5, \ldots \\ |m'|\lambda_2 & \text{destructive} \quad m' = \pm1, \pm2, \pm3, \ldots \end{cases} \tag{9.72}$$

We can rearrange this equation and write it in terms of the angle of incidence θ, but that would be a much more complicated expression.

For a nearly normal reflection, the angles of incidence and refraction will be zero, yielding $\cos\theta_r = 1$. This will give an interference condition that would depend only on the thickness of the film and the wavelength. Hence, depending upon the thickness, light of a particular wavelength will interfere constructively or destructively according to the following conditions.

Normal incidence for $(n_1 = n_3 < n_2)$:

$$2d = \begin{cases} \frac{|m|}{2}\lambda_2 & \text{constructive} \quad m = \pm1, \pm3, \pm5, \ldots \\ |m'|\lambda_2 & \text{destructive} \quad m' = \pm1, \pm2, \pm3, \ldots \end{cases} \tag{9.73}$$

Example 9.1 Colors from Soap Films

A soap film reflects blue light of wavelength 475 nm when viewed normally. Find the minimum thickness of the film if the soap water has a refractive index of 1.33.

Solution

To use the interference formula, first we need to evaluate the wavelength λ_2 inside the soap film, which is different from 475 nm in the air,

$$\lambda_2 = \frac{n_1}{n_2}\lambda_1 = \frac{1}{1.33} \times 475\text{ nm} = 375\text{ nm}.$$

We find the minimum thickness of the soap film by using $m = 1$ in the constructive interference condition,

$$d = \frac{1}{2}\frac{\lambda_2}{2} = \frac{1}{2}\frac{375\text{ nm}}{2} = 89.3\text{ nm}.$$

9.4 Practical Applications of Interference

Interference of light has been used to build instruments called interferometers that allow extremely precise and sensitive measurements of length, wavelength, and refractive index. Interferometers also play important roles in other applications such as optical communications. There are basically two types of instrument—two-beam and multiple-beam interferometers. In a two-beam optical interferometer, e.g. Michelson interferometer and Mach–Zehnder interferometer, a coherent light source, such as the light from a laser beam, is split into two parts by an optical device such as a partially silvered mirror, and the two partial beams are brought together after traveling through different paths. In multiple-beam optical interferometers, such as the Fabry–Perot interferometer, a coherent source of light undergoes multiple partial reflections at two sides of a medium and the partially transmitted beams from multiple refractions are made to interfere. We will study in detail the workings of the two most common interferometers. Interferometers for other waves have also been studied, but for the sake of concreteness we will look at optical devices only.

9.4.1 Michelson Interferometer

In a Michelson interferometer, shown in Fig. 9.10, light from an extended source is first split into two beams by a half-silvered beam splitter at 45° to the ray. The two resulting waves travel perpendicularly to each other, and reflect off mirrors M_1 and M_2. The reflected rays recombine resulting in interference which can either be viewed by the naked eye or by projecting it onto a screen with the use of a converging lens. The parallel rays from the extended source will converge as circles on the screen.

To compare the phases of the rays in the two arms we start from a common point of the two rays, say O′, when the two were together and go to a point, say E, when they have come together again. Therefore we calculate the phase difference between the paths $O'OBCM_1CBOE \equiv O'M_1E$ and $O'OAM_2AOE \equiv O'M_2E$.

The optical path length O′O is common to the two paths. A compensator is inserted in the path of ray OM_1O so that the optical path length (OPL) for BC+CB for the path on arm 1 is equal to the OPL for OA+AO for the path on arm 2. Therefore, the net optical path length difference between the two arms comes from the distances to the mirrors and the phase shift due to reflections at the beam splitter.

The ray $O'M_2E$ has one phase-changing reflection at M_2 that changes phase by π radians, while ray $O'M_1E$ has two such reflections, one at M_1 and the other at O. Therefore, the condition of destructive interference will result if the phase difference between the two rays is 0, 2π, 4π, etc., because the difference in phase flip due to reflections makes the phases of the waves in the two arms already off by π radians,

$$2 \times |OB + CM_1 - AM_2| \times \frac{2\pi}{\lambda_0} = m' \times 2\pi, \quad m' = 0, 1, 2, \ldots \text{ (Destructive)} \qquad (9.74)$$

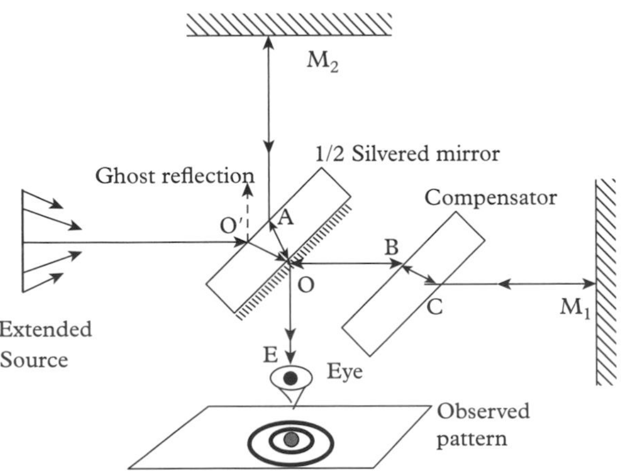

Fig. 9.10 *Michelson's interferometer. Wave from an extended source is split into two parts, which upon reflections from mirrors M_1 and M_2 are made to interfere at the detector.*

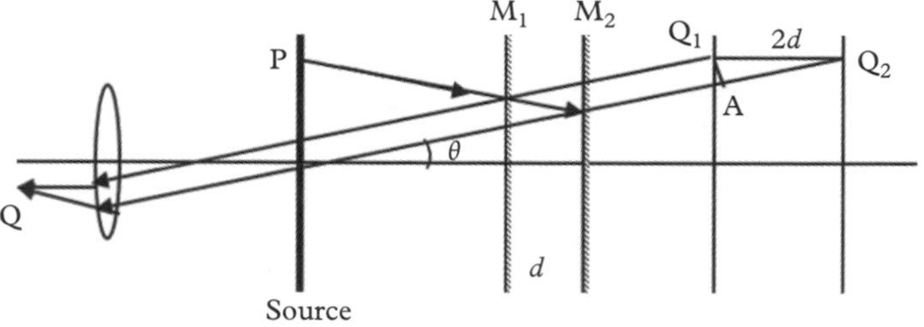

Fig. 9.11 *Effective geometry of interference of waves in a Michelson interferometer. A point P on the source has image Q_1 in mirror M_1 and image Q_2 in mirror M_2. The images serve as virtual sources which interfere at point Q at the detector.*

where λ_0 is the wavelength in air. Simplifying this equation, we find the following:

$$|OB + CM_1 - AM_2| = m'\frac{\lambda_0}{2}, \quad m' = 0,\ 1,\ 2, \ldots \text{ (Destructive)}. \tag{9.75}$$

But this only gives the condition for a ray coming horizontally from the source which will end up at the center point of the circles shown in Fig. 9.10. How do we find the interference conditions for rays at other angles? To address this question, notice that when you look at the incoming rays from the beam splitter, you are seeing reflections of the source in mirrors M_1 and M_2. In Fig. 9.11 we have redrawn a conceptual diagram of the Michelson interferometer, where we have displayed the images of the source in the mirrors, and how a ray from an off-axis point on the source must reflect to reach the screen, after converging from the lens at the site where the eye is shown in Fig. 9.10.

In Fig. 9.11, the point Q_1 is the image of the point P in mirror M_1 and the point Q_2 is the image in mirror M_2. To the eye at the screen, the rays appear to come from points Q_1 and Q_2. The interference of rays from Q_1 and Q_2 is observed by the eye. Rays from all points P on the circle on the source about the symmetry axis shown in Fig. 9.10 will meet on a circle at the screen that has point Q. Rays inclined at the same angle will either interfere constructively or destructively depending upon the difference in path, which is equal to AQ_2. We have already discussed the effects of reflection in the two rays. If the distance AQ_2 is an integer multiple of $\lambda_0/2$, then we would have destructive interference,

$$|AQ_2| = m'\frac{\lambda_0}{2}, \quad m' = 0,\ 1,\ 2, \ldots \text{ (Destructive)}. \tag{9.76}$$

If the difference between the distances to the mirrors is d, and the angle the rays make is θ, then the condition for destructive interference can be written in terms of d and θ as

$$2d\cos\theta = m'\frac{\lambda_0}{2}, \quad m' = 0,\ 1,\ 2, \ldots \text{ (Destructive)}. \tag{9.77}$$

This equation gives us the condition for destructive interference at points of a circle on the retina or screen that contain point Q. Since, the view of an observer is limited, one can only see circles made on the retina by rays that are inclined to a maximum angle. Various m' values in Eq. 9.77 refer to the corresponding angles of inclination for dark circles. Thus, $m' = 0$ interference is at the center, which occurs when $d = 0$. This is followed by a bright circle. The $m' = 1$ destructive interference happens at the following angle of inclination:

$$m' = 1: \quad \cos\theta = \frac{\lambda_0}{4d}. \tag{9.78}$$

Suppose you start with the distance to the mirror M_2 greater than the distance to M_1 by some distance d, and move mirror M_2 so that d decreases. That will cause $\cos\theta$ for the destructive condition to increase, which will imply that the condition for darkness is closer to the horizontal direction than before. This makes the $m' = 1$ circle on the screen smaller. As d gets smaller, the circle for $m' = 1$ also gets smaller, and when you have moved by a distance equal to $\lambda_0/2$, the $m' = 1$ circle disappears. All other fringes move correspondingly. The circle in space which used to be occupied by $m' = 1$, is now occupied by $m' = 2$, and the circle originally occupied by $m' = 2$ will now be occupied by $m' = 3$, and so on. Hence, if you focus on a particular place on the screen, a change in d can be determined by observing the number of fringes that pass. We multiply this number by half the wavelength to deduce the distance Δd moved,

$$\Delta d = N\frac{\lambda_0}{2}. \tag{9.79}$$

Since the wavelength of visible light is in the 0.4–0.7 μm range, this equation can be used for making precise measurements of sub-micrometer distances from a known source of light and the number of fringes that pass a reference point on the detector.

Example 9.2 Distance Measurements

A red laser light of wavelength 630 nm is used in a Michelson interferometer. While keeping mirror M_1 fixed, mirror M_2 is moved. The fringes are found to move past a fixed cross hair in the viewer. Find the distance mirror M_2 is moved for a single fringe to move past the reference line.

Solution

We use the result of the Michelson interferometer interference condition. For a 630 nm red laser light, for each fringe crossing ($N = 1$), the distance traveled by M_2 if you keep M_1 fixed will be

$$\Delta d = 1 \times \frac{630\text{ nm}}{2} = 315\text{ nm} = 0.315\ \mu\text{m}.$$

9.4.2 Fabry–Perot Interferometer

In France in 1899 Marie Fabry and Jean Perot built an interferometer which used multiple beams by coating the two parallel plates partially with a highly reflecting material. They found that multiple reflections resulted in much narrower interference fringes than with other interferometers. As a result of the narrow fringes the Fabry–Perot interferometer has a much higher resolving power. It is widely used for precision spectroscopy, and as a gas laser cavity, among other things.

A Fabry–Perot interferometer, shown in Fig. 9.12, essentially consists of two parallel plates of glass whose inner surfaces are polished to a very high degree of flatness, and then coated with a highly reflecting thin film of silver or aluminum, or some similarly good reflector. The outer surfaces of these plates are ground to make a wedge shape so that light escaping the cavity between the plates does not reflect back in. The distance d between

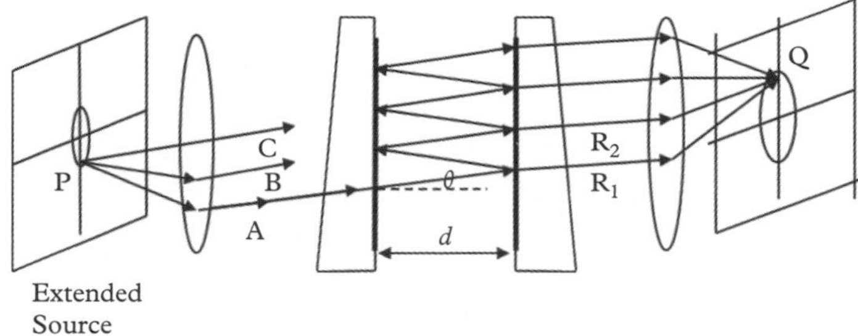

Fig. 9.12 *Fabry–Perot interferometer.*

the plates can vary from a few millimeters to a few centimeters. The light rays from an extended source enter the space between the reflecting plates, and emerge on both sides after multiple reflections. The rays emerging from the end opposite to the entrance are focused on a screen.

An extended light source, such as a mercury lamp is placed at the focal plane of a converging lens so that rays coming from a single point of the source emerge parallel on the other side of the lens at a particular angle of inclination, as shown in the figure. The transmitted rays from rays of the same angle of inclination enter the space between the two partial reflectors and focus on the same point at the focal plane of the second converging lens. For instance, the three rays A, B, and C from the source point P shown in the figure converge at the same point Q on the screen. The rays from points on the light source in a circle that contains P meet in a circle on the screen. If the condition at Q is met for constructive interference, we will see a bright ring, while if the condition there corresponds to destructive interference, there will be a dark ring there. The interference pattern on the screen will have alternating bright and dark rings. The reflectivity r of the coating can be adjusted to sharpen the rings, as illustrated in Fig. 9.13.

Let λ denote the wavelength of light in the medium between the reflecting plates, and n the refractive index of the medium. Each time a ray between the plates is incident on either the left or the right reflecting surface, it is partially reflected and partially transmitted. Let r denote the reflection coefficient of the reflecting surface. [Note: we have denoted this quantity by symbol ρ before.] The reflection coefficient tells us the factor by which the reflected amplitude of the electric field decreases when reflecting off a surface. Thus, if $r = 0.9$, the reflected electric field is 0.9 times the incident electric field.

We will also find it useful to define the square of the reflection coefficient, which we will denote by R,

$$R = r^2. \tag{9.80}$$

The quantity R will be related to the ratio of the intensity of the reflected wave to the intensity of the incident wave after a single reflection. A calculation (that you can find in a more advanced textbook on optics, such as Grant R. Fowles, *Introduction to Modern Optics*) of the intensity of light I_t at an arbitrary point Q on the screen in terms of the maximum intensity is given by an **Airy function** of the phase shift Δ_{12},

$$\frac{I_t}{(I_t)_{\max}} = \frac{1}{1 + F \sin^2 (\Delta_{12}/2)}, \tag{9.81}$$

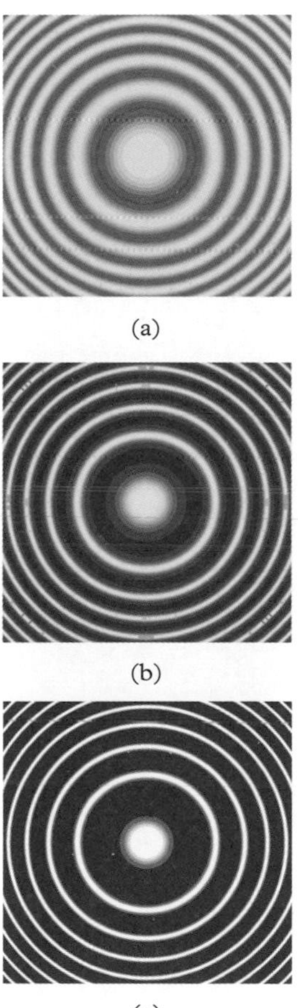

Fig. 9.13 *Calculated fringes for different reflectivities in a Fabry–Perot interferometer: (a) $r = 0.3$, (b) $r = 0.75$, (c) $r = 0.98$.*

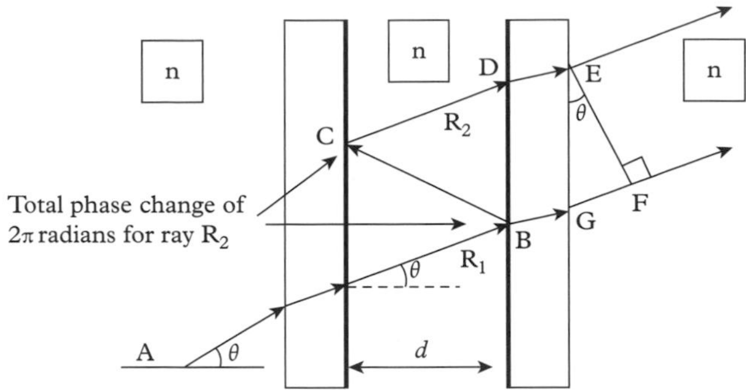

Fig. 9.14 *Geometry of Fabry–Perot for phase difference calculation between successively transmitted waves. The phase difference comes from the path difference of BCD and GF.*

where F is called the **finesse coefficient** or **quality factor**, which has the following relation to reflectance R:

$$F = \frac{4R}{(1-R)^2}, \tag{9.82}$$

and Δ_{12} is the phase difference between the optical paths of two successive transmitted waves, shown as rays R_1 and R_2 in Fig. 9.12. We redraw the figure with only two successive rays in Fig. 9.14. The phase difference in terms of angle θ, the distance d between the plates, and the wavelength of light λ between the plates is found to be

$$\Delta_{12} = \frac{2\pi}{\lambda}\, 2nd\cos\theta. \tag{9.83}$$

From Eq. 9.81 we find that the transmitted intensity is a maximum when $\sin^2(\Delta_{12}/2) = 0$. That means that constructive interference will take place if

$$\Delta_{12} = 0, \pm 2\pi, \pm 4\pi, \pm 6\pi, \ldots. \tag{9.84}$$

This yields the following condition for constructive interference in terms of the direction θ in physical space:

$$2nd\cos\theta = m\lambda \ \ \text{(Constructive)}, \tag{9.85}$$

where m is called the interference order which takes on values $= 0, \pm 1, \pm 2, \pm 3, \ldots$. The constructive maxima and minima of the intensity are plotted in Fig. 9.15. I have presented the plots against angle θ and phase difference Δ_{12}. Note that, while the plot in phase change has peaks at equal intervals of 2π radians, the same peaks are spread out differently in physical space.

Sharpness of Peaks and Reflectance

The interference pattern resulting from Eq. 9.81 is very sensitive to the value of the reflectance R and hence to the finesse coefficient F. A plot I_t/I_i versus Δ_{12} for a number of R values in Fig. 9.16 shows that fringes become sharper as the reflectivity R of the plates increases.

From Eq. 9.83, note that Δ_{12} is essentially a measure of the angular direction of the beams. For a given beam of fixed wavelength, increasing Δ_{12} corresponds to decreasing θ

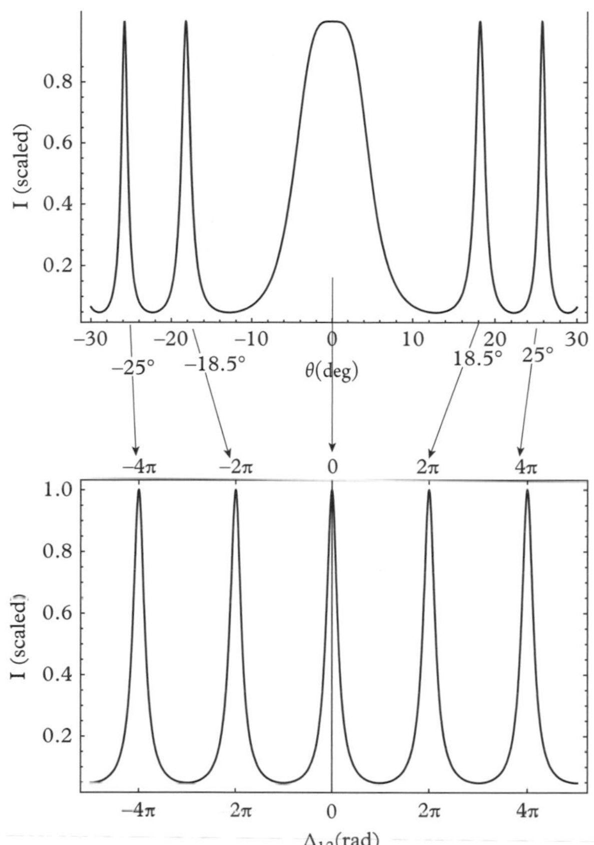

Fig. 9.15 *The transmitted intensity as a function of angle of transmission in a Fabry–Perot apparatus and as a function of the phase difference between successively reflected rays. Note that the maxima in Δ_{12} appear at equal intervals of 2π of Δ_{12} but they do not appear at equal intervals in the physical space of the angle. Also, the central peak in real space is much broader than it would appear from the plot in Δ_{12}.*

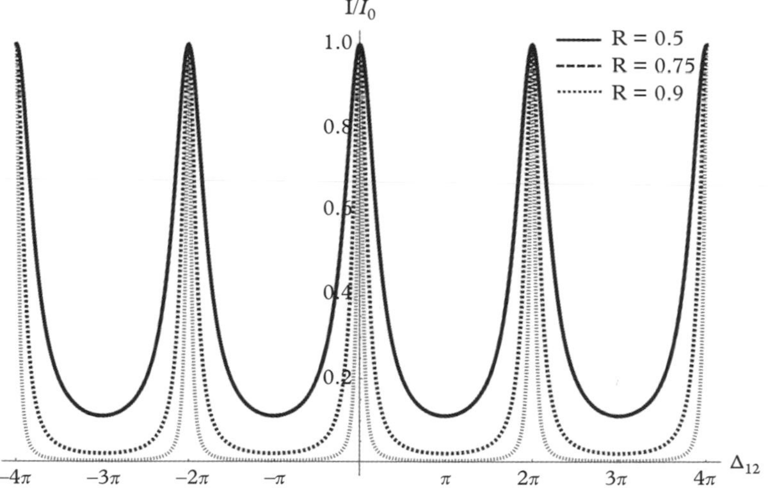

Fig. 9.16 *Sharpness of peaks increases with reflectance R.*

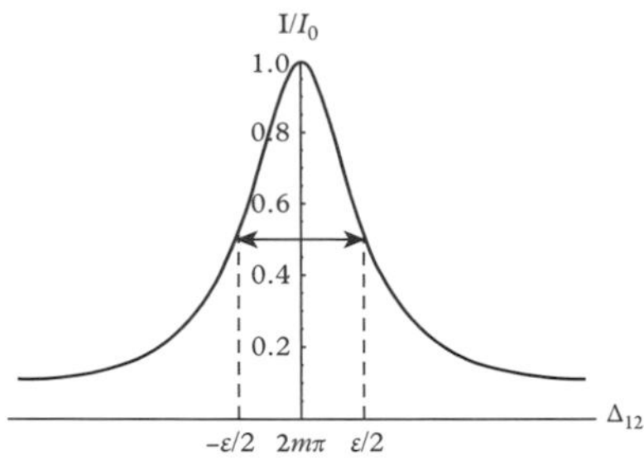

Fig. 9.17 *Defining the half width of a peak.*

due to the cosine on the right side. The constructive interference condition given in Eq. 9.84 shows that the fringes are equispaced with respect to Δ_{12}, but in real space where the direction is given by θ, the fringes will be bunched up as you go out from the center of the screen, with more separation near the middle than further out.

The width of a fringe at half height provides a standard way of describing the sharpness of the fringes. If the peak width is denoted by ϵ, as in Fig. 9.17, then, since phase Δ_{12} takes values $2m\pi$ at the peak maxima for the mth constructive interference, the intensity will have half the maximum at following values of the phase difference,

$$\Delta_{12} = 2m\pi \pm \frac{\epsilon}{2}. \tag{9.86}$$

Setting Δ_{12} from Eq. 9.86 into Eq. 9.81 for $I_t/I_i = 1/2$, we obtain

$$\frac{1}{2} = \frac{1}{1 + F\sin^2\left(m\pi \pm \frac{\epsilon}{4}\right)}. \tag{9.87}$$

One can solve this equation to obtain the width ϵ at half height. Since sharp peaks will have a very small ϵ, we can use the approximation $\sin(\epsilon/4) \approx \epsilon/4$ and $\cos(\epsilon/4) \approx 1$. After a simple algebra you can show that

$$\epsilon = \frac{4}{\sqrt{F}}. \tag{9.88}$$

It is customary to define a quantity called **finesse** (which should not to be confused with the finesse coefficient F defined earlier) which is useful for discussing the resolution of peaks,

$$\text{finesse} = \frac{\text{Separation of adjacent peaks}}{\text{Width at half height}}. \tag{9.89}$$

Here the separation of adjacent peaks in the phase difference is 2π radians. Therefore, the finesse is related to the finesse coefficient F as follows:

$$\text{finesse} = \frac{2\pi}{\epsilon} = \frac{\pi\sqrt{F}}{2}. \tag{9.90}$$

Equations 9.81–9.85 show that the interference conditions in the Fabry–Perot interferometer depend upon three physically controllable parameters—the refractive index n of the medium between the plates, the separation d of the plates, and the wavelength λ of light. Depending upon our particular purpose for using a Fabry–Perot interferometer, we may vary one of these parameters while keeping the other two quantities fixed, and study the change on the screen.

When the plates of the Fabry–Perot interferometer are fixed to a definite separation d, the arrangement is also called an **etalon**. For fixed n and d, the interference will depend upon the wavelength. So, if a light source consists of two wavelengths, say a red light of 650 nm and a green light of 530 nm, two separate rings for each interference order m are seen. We display the resolution of a mixture of two wavelengths in a black/white picture in Fig. 9.18, where the inner ring in each order corresponds to the smaller wavelength. Since the interference lines in a Fabry–Perot interferometer are very narrow, they are easier to resolve than is the case with other interferometers. For this reason, Fabry–Perot is used for spectroscopy.

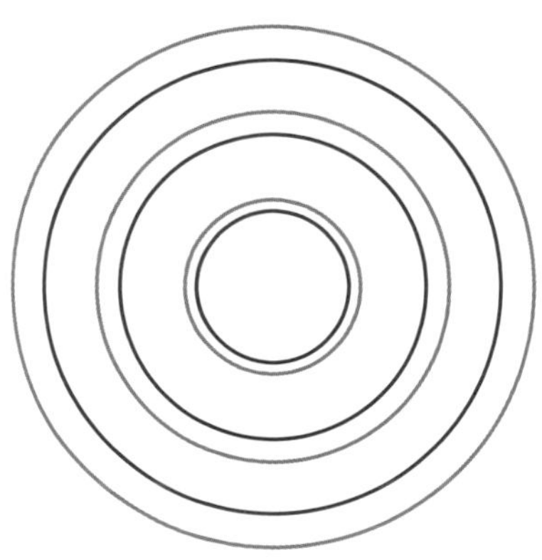

Fig. 9.18 *The schematics of separation of lights of different wavelength by a Fabry–Perot apparatus. The inner ring in each order corresponds to the smaller wavelength.*

EXERCISES

(9.1) Consider two coherent sources S_1 and S_2 of waves that travel towards a detector D as shown in Fig. 9.19. While source S_1 is fixed at $(x = 0, y = 0)$, source S_2 can be moved between $(x = 0, y = 1)$ cm and $(x = 0, y = +7)$ cm, and the detector D is fixed at $x = 100$ cm. Suppose the wavelength of the waves is 4 cm and assume the waves start in-phase at S_1 and S_2. Determine the phase difference of the two waves when they arrive at point D on the screen with the coordinate $(x = 100\text{ cm}, y = 0)$ when the y-coordinate of source S_2 is:
(a) $y = 1$ cm,
(b) $y = 3$ cm,
(c) $y = 5$ cm,
(d) $y = 7$ cm.

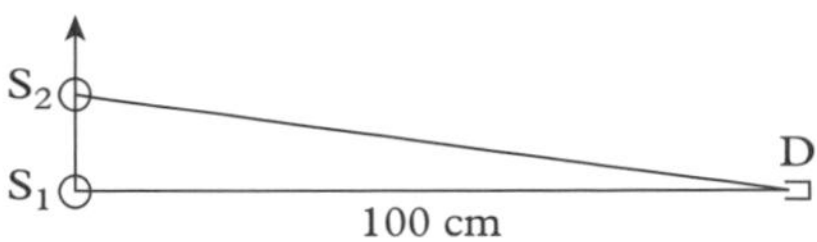

Fig. 9.19 *Exercise 9.1.*

(9.2) What are the coherence times and coherence lengths of the following waves?
(a) A radio station emitting a signal in a band of frequencies, between 1.5 MHz ± 10 kHz,
(b) A speaker emitting sound in air in the frequency range 2000 Hz to 2005 Hz,
(c) White light of wavelength between 400 nm and 700 nm.
(d) A laser emitting light in the wavelength range 590 nm ± 0.0004 nm.

(9.3) A laser light of wavelength 632.8 nm passes through two narrow slits separated by 10 μm and forms an interference pattern on a screen 1.5 m away.
(a) Suppose the positive y-axis points up from the central spot on the screen. Find the y-coordinates of the locations of three bright and two dark spots on the screen.
(b) What is the theoretical maximum order of constructive interference that could form on the screen?

(9.4) Three waves of identical wave amplitude ψ_0, and identical wavelength λ are in phase at the plane shown with a dashed line in Fig. 9.20. Find a formula for the intensity of light at the arbitrary point in terms of angle θ, assuming $\theta \ll 1$ when expressed in radians.

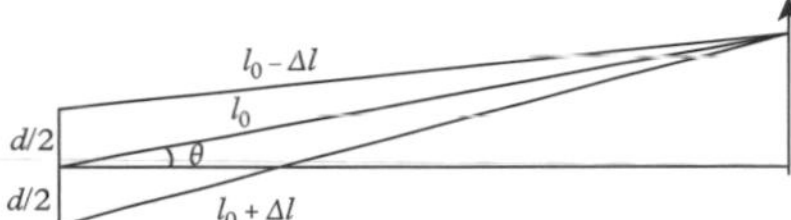

Fig. 9.20 *Exercise 9.4.*

(9.5) A speaker generates a sound of frequency 2000 Hz which strikes a wall with two narrow vertical openings separated by 0.66 m. The sound is detected by a microphone at a far-away place (say, five meters) behind the wall with the openings. Use speed of sound in air 340 m/s.
(a) Find the direction of three interference maxima and four interference minima from the openings.
(b) Find the angular width of the $m = 0$ peak, defined as the directions from which the intensity of the peak is more than half that at the center of the peak.

(9.6) Two light waves start from the same coherent source and come together after traveling the same physical distance L. While one wave travels in vacuum, there is a glass slide of thickness d and refractive index n in the path of the other wave, as shown in Fig. 9.21. What is the phase difference between the two waves if the wavelength has the wavelength λ_0 in a vacuum?

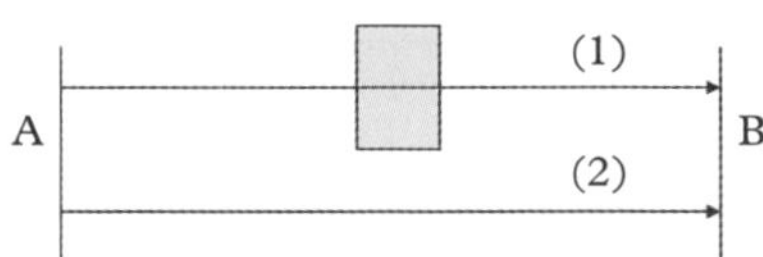

Fig. 9.21 *Exercise 9.6.*

(9.7) A coherent light of wavelength λ_0 in air $(n = 1)$ is incident on a double slit of separation d. One of the slits is covered with a glass slide of thickness h and refractive index n.
(a) What is the phase difference between the two rays in direction θ from the slits, assuming θ is small?
(b) Find the direction θ of the constructive interference of order m.

(c) What is the order of interference corresponding to the center direction, i.e. when $\theta = 0$ or nearest to the $\theta = 0$ direction, for $\lambda_0 = 500$ nm, $d = 1\ \mu$m, $h = 2\ \mu$m, and $n = 1.5$?

(9.8) Two radio antennas are 1 km apart. They broadcast 4 MHz signals in phase. A car moves at a constant speed v on a road parallel to the line joining the two antennas. The road is 20 km away from the antennas, i.e. the length of a line perpendicular to both the line joining the antennas and the road is 20 km. When the car radio is tuned to the signal at 4 MHz the signal increases and decreases periodically.

(a) State why you would observe such variation in the intensity.

(b) Determine five successive places on the road where you will get the strongest signal.

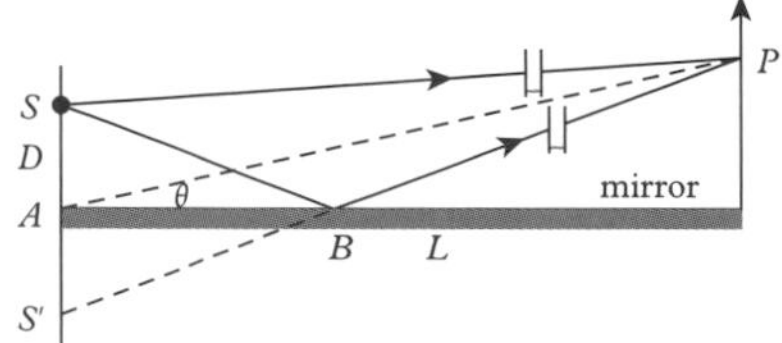

Fig. 9.22 *Exercise 9.9.*

(9.9) A **Lloyd's mirror** is a set-up that allows a double-slit experiment with only one source S by placing a mirror between the slit and the screen, as shown in Fig. 9.22. The image S' of the slit serves as the second slit. Use the fact that the reflection of light from the mirror adds an additional phase change of π radians.

(a) Find formulas for the constructive and destructive interferences on the screen if the screen is a distance L away and the slit is a distance D above the mirror. Assume $D \ll L$.

(b) If $L = 2$ m, $D = 2\ \mu$m, and the wavelength of light used is the yellow sodium D line of wavelength 589 nm, find two places on the screen you will find bright spots and two places where there will be dark spots.

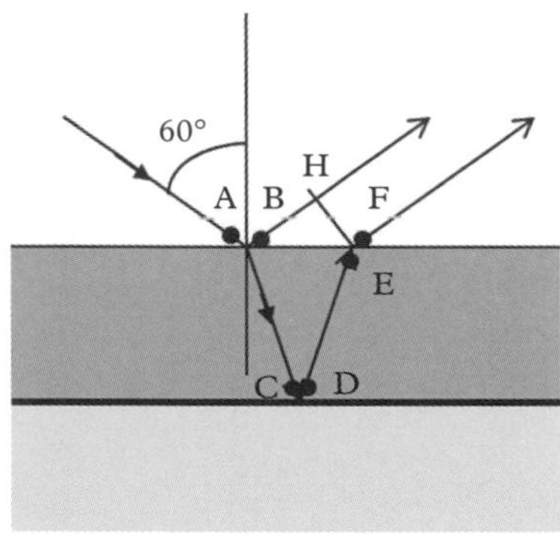

Fig. 9.23 *Exercise 9.10.*

(9.10) **Dielectric Thin Film**. Light reflects off a thin oil film over water, as shown in Fig. 9.23. Let h be the thickness of the film and n_{oil} be the refractive index of oil, which you can assume to be less than the refractive index n_{w} of water. Consider a ray of light of wavelength λ_0, incident at an angle θ_1 on the film from air. (You may use $h = 150$ nm, $n_{\text{oil}} = 1.2$, $n_{\text{w}} = 1.33$, $n_{\text{air}} = 1.0$, $\lambda_0 = 550$ nm, $\theta_1 = 60°$.)

(a) Find refraction angle θ_2 in the oil.

(b) Find the distances AC, AF, and BH.

(c) Determine the phase difference between the following points shown in the figure: (i) A and B, (ii) C and D, (iii) E and F, (iv) H and A, (v) F and A, and (vi) H and F.

(d) As you vary the angle of incidence, you will find that at some angles the light has a stronger intensity than at some other angles. Find a general relation among n_0, λ_0, θ_1, θ_2, and h corresponding to the constructive interference for the air/oil/water system.

(9.11) Two optically flat glass sheets of length l and refractive index n are used to form a wedge of air by placing a narrow wire of thickness h between them at one end. A monochromatic light of wavelength λ_0 in air is incident normally on the top glass plate. A constructive interference is observed in the light reflected at a distance $\frac{l}{2}$ from one end.

(a) How far from this point towards the edge with the wire would you need to move to observe the next constructive interference?

(b) What will be the numerical value of the distance if $l = 25$ cm, $h = 100\ \mu$m, $\lambda_0 = 550$ nm, $n(\text{air}) = 1$ and $n(\text{glass}) = 1.55$?

(9.12) In a Michelson interferometer, light of wavelength 632.8 nm from a He–Ne laser is used. When one of the mirrors is moved by a distance D, eight fringes move past the field of view. What is the value of the distance D?

(9.13) In one arm of a Michelson interferometer a glass container is placed, as shown in Fig. 9.24. The space inside the container is 2 cm wide. The inside of the glass container can be filled with a gas to desired density or pressure. Initially the container is empty. As gas is slowly filled into the container, it is observed that dark fringes of the interferogram move past a reference line in the field of observation. By the time the container is filled to the desired density of gas, 1220 fringes move past the reference line when light of wavelength 632.8 nm is used. Use the data given here to find the refractive index of the gas at the final density.

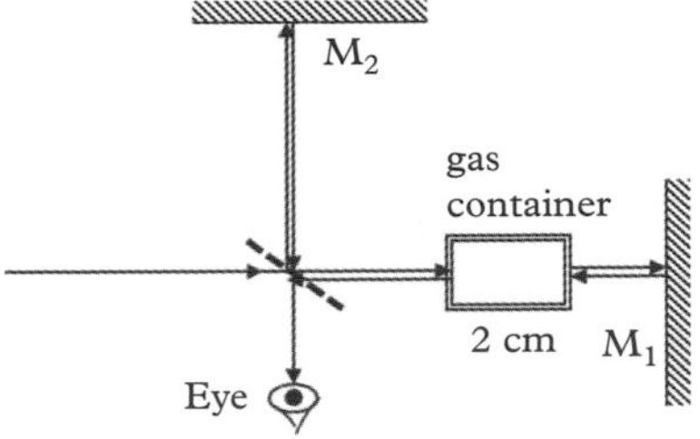

Fig. 9.24 *Exercise 9.13.*

10 Diffraction

Chapter Goals

In this chapter you will learn about the Huygens principle of propagation of waves. We will apply the Huygens principle to understand diffraction. You will learn to calculate the consequences of interference of infinitely many coherent sources and apply the results to understand diffraction in a variety of systems. You will also learn the limitation of imaging apparatus due to diffraction that limits the resolution.

You might have noticed that water waves slosh around a rock in their way and may sometimes form again and continue on. A siren from a car can be heard around the corner. A light wave coming out of a laser spreads out as it travels in space. It is a common property of all waves that when a wave encounters an opening or an obstacle, the wave is distorted and spreads out. These are examples of the property of **diffraction**.

The theoretical understanding of diffraction is due to **Augustin Fresnel** (1788–1827), who in 1819 proposed his theory of diffraction in a winning essay to the French academy and was instrumental in the acceptance of the idea of light as waves. The theory of diffraction is based on the superposition principle we discussed in the last chapter and Huygens' principle that we will state below. When a wave interacts with an object in its path, the interaction excites the particles of the object which creates additional waves. In the region near the obstacle, the net wave is quite complicated, but very far from the interaction the picture is somewhat simpler. Diffraction in the far-field region is also called **Fraunhofer diffraction**. In this chapter we will study diffraction through apertures in the far-field region to gain an understanding about fundamental aspects of diffraction.

10.1 Huygens–Fresnel Principle

The theoretical understanding of diffraction is based on a fundamental principle of wave propagation enunciated in 1690 by the Dutch physicist **Christian Huygens**. The **Huygens principle** states that:

> Every point on a primary wavefront serves as a point source for spherical secondary wavelets of the same frequency and speed as the original primary wave such that the **primary wavefront** at a later time is the envelope of these **secondary spherical wavelets**.

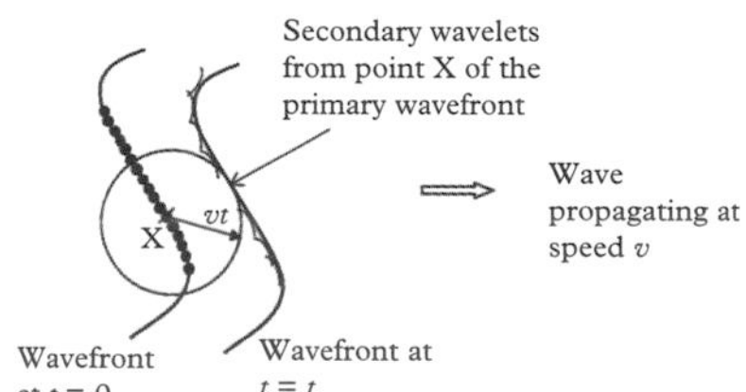

Fig. 10.1 *Traveling of waves according to Huygens' secondary spherical wavelets.*

In Fig. 10.1, I have illustrated Huygens' principle by sketching the time evolution of a wavefront. The wavefront at $t = t$ is at a distance from the wavefront at $t = 0$ for a wave propagating with speed v. Using pictures like these, Huygens could successfully explain the laws of reflection and refraction.

In spite of the successes of Huygens' principle, it suffers from several drawbacks. It does not tell us what to do with the backward traveling part of the secondary spherical wavelet. Furthermore, it only makes use of the envelope of the secondary wavelets, and does not tell us what to do with the rest of the wavelet. More importantly, it gives rise to the same wavefront regardless of the value of the wavelength. Thus, if we follow the prescription outlined above for waves of two different wavelengths, we will get exactly the same result. But we know from experiments that different diffraction patterns result for the same aperture, depending on the wavelength.

Augustin Jean Fresnel used the ideas of interference to modify the Huygens principle so that one can understand diffraction phenomena better. Fresnel hypothesized that:

> Every point of a primary wave could be thought of as producing secondary wavelets whose overlap and interference forms the primary wave at an instant later.

The modified version is called the **Huygens–Fresnel principle**. The main difference comes from the introduction of interference in Fresnel's modification in place of the envelope that determines the primary wave later. We will now apply the Huygens–Fresnel principle to simple cases of diffraction from single slit, double slit, circular aperture, double slit and diffraction grating to gain qualitative and quantitative understanding of the diffraction phenomena.

10.2 Diffraction through a Single Slit

10.2.1 Near-Field versus Far-Field

Consider a light source in front of a slit in an opaque material (Fig. 10.2). We place a screen behind the slit to observe the diffraction pattern there. If the screen is placed immediately behind the slit, we will find a **shadow of the slit** on the screen. When we move the screen further out, the shadow develops fringes which are sensitive to the distance from the slit. These patterns are called **Fresnel or near-field diffraction**.

Moving the screen considerably far away from the slit you reach a region where the patterns stabilize and although they spread out more as you increase the distance to the screen, the diffraction pattern itself remains the same. We call this pattern the **Fraunhofer or far-field diffraction**, after **Joseph von Fraunhofer**.

It turns out that the mathematics for near-field study is quite advanced for this book. I will recommend a more advanced book on optics such as Fowles, *Modern Optics* if you want to learn more. Here we will study only Fraunhofer diffraction.

10.2.2 Calculation of Intensity in the Far-Field Region

We will now calculate the intensity at a far away point from a wide slit. The set-up and coordinate axes for the calculation are shown in Fig. 10.3. Briefly, let a plane harmonic wave be incident on a rectangular slit, with one side being very long. According to the Huygens–Fresnel principle, the part of the wavefront in the area of the slit will serve as

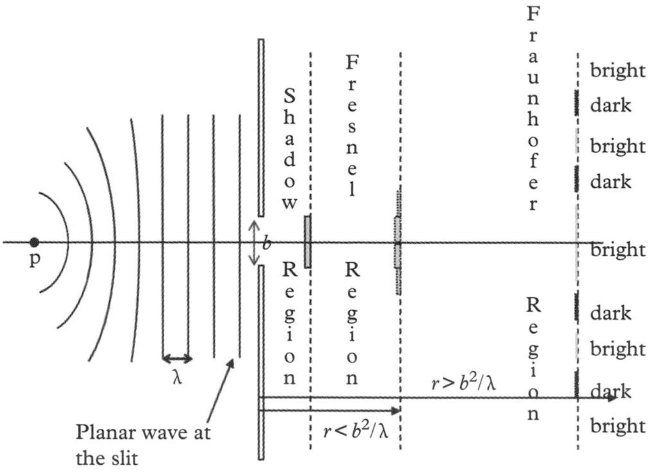

Fig. 10.2 *Shadow, Fresnel, and Fraunhofer diffraction regions and patterns from a horizontal slit in front of a point source placed far away from the slit of width b.*

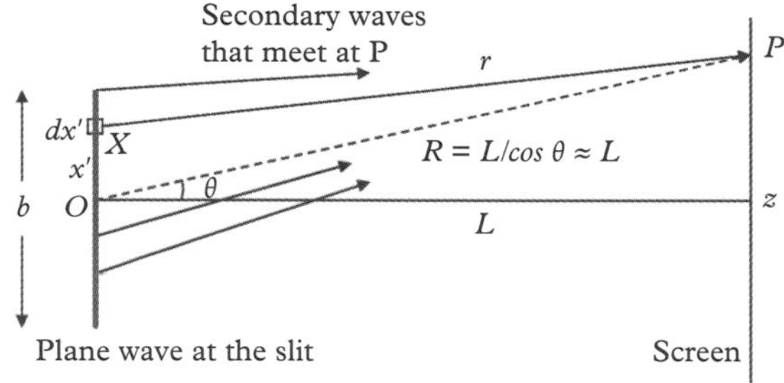

Fig. 10.3 *Geometry for the calculation of Fraunhofer diffraction from a plane wave at a slit. The slit is parallel to the y-axis. In this figure only the cross-section in the xz-plane is shown.*

a source of waves when the wave emerges behind the slit. Each point of the wavefront whose cross-section in the xz-plane is shown in Fig. 10.3 serves as a coherent point source of spherical wavelets which propagate to the screen. The wave amplitude at an arbitrary point P on the screen will be equal to the superposition of waves from every region of the wave segment on the area of the slit.

Since the wave segment on the screen is very long in the y-direction, the result at any y will be same as at any other y. Therefore, we will perform the calculation in the xz-plane only. In the xz-plane we will have a line source of length b along the x-axis, as shown in Fig. 10.3. We will use coordinate symbol x' to locate the points on the wavefront and symbol x to locate a point on the screen, as in the figure.

To perform the superposition calculation of spherical wavelets from points on the $O'x'$ axis, consider an infinitesimal segment between x' and $x' + dx'$. Let the amplitude of the vibration there be Adx', the frequency ω, wave number k, and the phase constant $\phi = 0$. The traveling spherical wavefront generated by the source at dx', which we can write as $d\psi$, will be given by the following spherical wave function:

$$d\psi\,(\text{at } P) = \frac{Adx'}{r}\cos(kr - \omega t), \tag{10.1}$$

where r is the distance from the element. Let L be the distance between the slit and the screen, x' the x-coordinates of the element, and x be the x-coordinate of the point P on the screen. Then,

$$r = \sqrt{(x-x')^2 + L^2}. \tag{10.2}$$

The contribution of the wave from the dx' element to the wave at P will be

$$d\psi\,(\text{at } P) = \frac{A dx'}{\sqrt{(x-x')^2 + L^2}} \cos\left(k\sqrt{(x-x')^2 + L^2} - \omega t\right). \tag{10.3}$$

Summing up contributions from every point of the wavefront at the slit amounts to performing an integration over the variable x' from $x' = -b/2$ to $x' = b/2$,

$$\psi\,(\text{at } P) = \int_{-b/2}^{b/2} \frac{A dx'}{\sqrt{(x-x')^2 + L^2}} \cos\left(k\sqrt{(x-x')^2 + L^2} - \omega t\right). \tag{10.4}$$

In general, this integration is hard to do. However, for far-field regions we can make some reasonable approximations. Let R be the distance of point P from the center of the slit and θ be the angle between the line from the center of the slit to P and the horizontal direction. Let us write the variable r in terms of R and θ rather than x, x', and L. Using the cosine law we note that

$$r = \sqrt{R^2 + x'^2 - 2Rx' \cos(90^\circ - \theta)} = \sqrt{R^2 + x'^2 - 2Rx' \sin\theta}. \tag{10.5}$$

Now, note that R is $\approx L$ since θ is small,

$$R = \frac{L}{\cos\theta} \approx L. \tag{10.6}$$

Therefore,

$$r \approx \sqrt{L^2 + x'^2 - 2Lx' \sin\theta}. \tag{10.7}$$

Note that x' here is the x-coordinate of a point on the slit, therefore $|x'| < b$. In the far-field region $L \gg b$, therefore we will have $L \gg x'$. Therefore, we can consider x'/L as a small value variable and expand r in a Maclaurin series in powers of x'/L. Keeping only the leading order in x'/L gives

$$\begin{aligned} r &= \left(L^2 + x'^2 - 2Lx' \sin\theta\right)^{1/2} = L\left[1 + \left(\frac{x'}{L}\right)^2 - 2\frac{x'}{L}\sin\theta\right]^{1/2} \\ &\approx L\left[1 - \frac{x'}{L}\sin\theta\right] = L - x'\sin\theta. \end{aligned}$$

Therefore,

$$r \approx L - x'\sin\theta. \tag{10.8}$$

Since $|x' \sin\theta| \ll L$ we may be tempted to drop $x' \sin\theta$ altogether. This turns out to be not a wise choice as it throws out the diffraction effect altogether. The distance r in Eq. 10.1 appears in the amplitude $[Adx'/r]$ and in the phase $[kr - \omega t]$. The kr in the phase has a large k and the result is acted on by a sine function and small changes in r tend to make a large impact on the sine of kr. Therefore, in the phase we must keep $x' \sin\theta$ along with L when approximating r. Dropping $x' \sin\theta$ compared to L in the amplitude $[Adx'/r]$ does not make much of a difference. Making these changes we get

$$d\psi\,(\text{at } P) \approx \frac{Adx'}{L} \sin\left[k(L - x' \sin\theta) - \omega t\right]. \tag{10.9}$$

Integration of this expression is now much easier to do with the following result:

$$\psi\,(\text{at } P) = \frac{bA}{L} \left[\frac{\sin\left(\frac{1}{2} kb \sin\theta\right)}{\frac{1}{2} kb \sin\theta} \right] \sin\left[kL - \omega t\right]. \tag{10.10}$$

The intensity at P is obtained by time averaging the square of this rapidly fluctuating wave. Since time averaging of $\sin^2(\omega t)$ or $\cos^2(\omega t)$ gives $\frac{1}{2}$, we obtain the intensity in direction θ from the slits to be

$$I(\theta) = \alpha \left(\frac{bA}{R}\right)^2 \left[\frac{\sin\beta}{\beta}\right]^2, \tag{10.11}$$

where α is a proportionality factor which is equal to $\epsilon_0 c$ for light, and β is the following function of slit width b, angle θ, and wave number k (or equivalently, wavelength λ),

$$\beta = \frac{1}{2} kb \sin\theta = \frac{\pi b \sin\theta}{\lambda}. \tag{10.12}$$

The quantity $\alpha \left(\frac{bA}{R}\right)^2$ is the intensity in direction $\theta = 0$, or when $\beta = 0$, which is also the maximum intensity. We may simplify the formula for intensity further by casting the general formula in terms of the intensity for $\theta = 0$. Then the intensity in direction θ is more compactly written as

$$I(\theta) = I_0 \left[\frac{\sin\beta}{\beta}\right]^2. \tag{10.13}$$

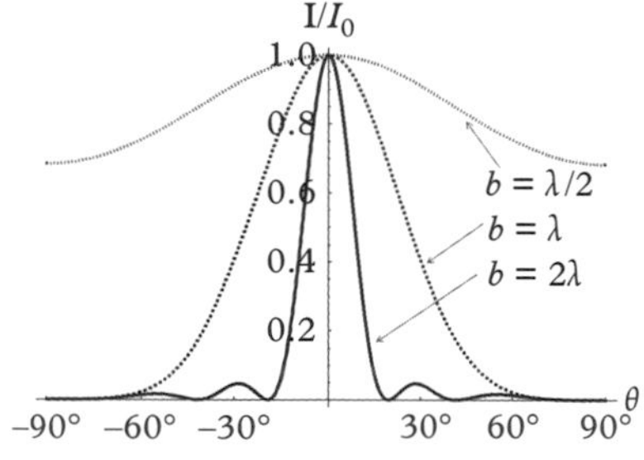

Fig. 10.4 *A plot of intensity at the screen various slit width. The intensity variation on the screen corresponds to the diffraction pattern at the screen.*

By expressing θ in terms of x of point P on the screen we can also write β as a function of x on the screen,

$$\frac{x}{L} = \tan\theta \approx \sin\theta, \implies \beta = \frac{\pi b x}{\lambda L}. \tag{10.14}$$

Figure 10.4 shows three plots of the intensity versus the direction of beams for width $b = \lambda/2, \lambda, 2\lambda$. We see that when the width of the slit $b < \lambda$ there is no diffraction minimum. Only when the width is larger than the wavelength do minima appear in the diffraction pattern. Overall, the intensity drops off away from the center. The second peak is about 20 times smaller than the central peak.

10.2.3 The Maxima and Minima of the Diffraction Pattern

The maxima and minima of a diffraction pattern refer to the maxima and minima of the intensity. Since intensity can never be negative, zero is the lowest value that intensity can have. From the formula for intensity in Eq. 10.13 it is easy to find when intensity will be zero,

$$I = 0 \ \text{when} \ \frac{\sin\beta}{\beta} = 0. \tag{10.15}$$

That is, when

$$\sin\beta = 0, \ \text{with} \ \beta \neq 0. \tag{10.16}$$

When $\beta = 0$, $\sin\beta/\beta$ is not zero, but 1. Therefore, we find that intensity will have minima at the following values of β:

$$\text{Minima:} \ \beta = m'\pi, \ \ m' = \pm1, \pm2, \pm3, \ldots. \tag{10.17}$$

The integers m' are called diffraction minima orders. We often place a subscript onto the parameter to indicate various orders,

$$\text{Minima:} \ \beta_{m'} = m'\pi, \ \ m' = \pm1, \pm2, \pm3, \ldots. \tag{10.18}$$

This can be written for direction angle θ as

$$\text{Minima:} \ b \sin\theta_{m'} = m'\lambda, \ \ m' = \pm1, \pm2, \pm3, \ldots, \tag{10.19}$$

and for the position on the screen,

$$\text{Minima:} \ b\, x_{m'} = m'\lambda L, \ \ m' = \pm1, \pm2, \pm3, \ldots. \tag{10.20}$$

Intensity maxima do not occur at points halfway between the minima. We can obtain the maxima of intensity I by treating it as a function of β and setting the derivative to zero,

$$\frac{dI}{d\beta} = 0. \tag{10.21}$$

This condition will include both the minima and maxima. Since we already know all the minima, we can identify the maxima by filtering out the minima from the result. Carrying out the derivation explicitly we obtain two conditions on β,

$$\sin\beta = 0, \tag{10.22}$$

$$\tan\beta = \beta. \tag{10.23}$$

The first condition gives extrema at

$$\beta = 0, \pm1\pi, \pm2\pi, \pm3\pi, \ldots. \tag{10.24}$$

All except $\beta = 0$ are the minima. Therefore, we have identified $\beta = 0$ as one of the maxima. The second equation, Eq. 10.23 gives us all the other maxima. This equation is a **transcendental equation** and can be solved graphically by plotting the left and right sides of the equation, $\tan\beta$ and β, against β, as shown in Fig. 10.5. The values of β where the two graphs intersect correspond to the solutions of the equation $\tan\beta = \beta$.

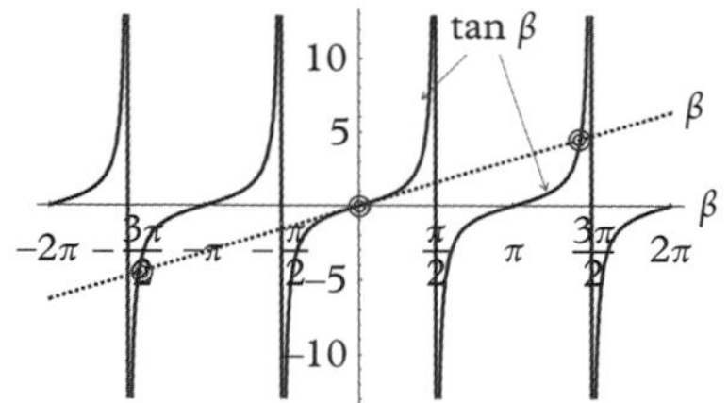

Fig. 10.5 *The graphical solution of Eq. 10.23. The dashed line is the plot of β versus β and the solid lines are $\tan\beta$ versus β. The intersection of the two plots gives the points where $\tan\beta = \beta$.*

The approximate values for the three solutions shown in Fig. 10.5 are $\beta = 0, \beta \approx \frac{3}{2}\pi, \beta \approx -\frac{3}{2}\pi$. A more precise and expanded graph gives $\beta = 0$, $\beta \approx \pm 1.4303\pi$, $\pm 2.4590\pi$, $\pm 3.4707\pi$, ... The locations of the diffraction maxima, $\beta = 0$, $\pm 1.4303\pi$, $\pm 2.4590\pi$, $\pm 3.4707\pi$, ... are referred to as constructive diffraction orders $m = 0, \pm 1, \pm 2, \pm 3, \ldots$.

Width of Diffraction Maxima

The central maximum occurs for $\theta = 0$, which is the same as $\beta = 0$. The peak widths in β and θ are related in a nonlinear way. For instance, if you plot intensity as a function of β, you would find that the central maximum is between $\beta = -\pi$ and $\beta = \pi$, while the other peaks are between $m\pi$ and $(m \pm 1)\pi$. That means that the central maximum is between $-\pi$ and π, which gives it a width of 2π, but the other maxima have a width of only π.

That the central maximum is twice as wide in the beta space does not carry over to the direction in real space. In real space, the angular width of the central peak, $\Delta\theta$, is related nonlinearly to $\Delta\beta$, as you can easily see from Eq. 10.12,

$$\beta = \frac{\pi b \sin\theta}{\lambda} \implies \Delta\beta = \frac{\pi\, b\, \cos\theta\, \Delta\theta}{\lambda}. \tag{10.25}$$

That is, the width of a peak in terms of the physical angle θ would depend upon the direction θ as well,

$$\Delta\theta = \frac{\lambda}{\pi\, b\, \cos\theta}\, \Delta\beta. \tag{10.26}$$

The central maximum has a width of $\Delta\beta = 2\pi$ rad in β-space. This peak will have the following width in real space:

$$(\Delta\theta)_{\text{central}} = \frac{\lambda}{\pi\, b\, \cos\theta}\, 2\pi = \frac{2\lambda}{b\, \cos\theta}.$$

For the $\theta \approx 0$ near the central peak, therefore, we have $\cos\theta = 1$, hence

$$(\Delta\theta)_{\text{central}} = \frac{2\lambda}{b}. \tag{10.27}$$

Example 10.1 Single-slit Fraunhofer Diffraction

Light of wavelength 550 nm passes through a slit of width 2 μm. Find the location of four minima on a screen about the central bright location (a) in terms of the angle subtended with the horizontal direction from the center of the slit and (b) the positions of the dark bands if the screen is 50 cm away.

Solution

(a) The minima are given by the following condition:

$$b \sin\theta = m\lambda \quad (m = \pm1, \pm2, \pm3, \ldots) \quad (\textit{Minima}).$$

Hence, the four minima around the central maxima will have $m = \pm1, \pm2$. The corresponding directions from the slit are

$$\theta_{\pm1} = \pm \sin\left(\frac{\lambda}{b}\right) = \pm16^\circ,$$

$$\theta_{\pm2} = \pm \sin\left(\frac{2\lambda}{b}\right) = \pm33^\circ.$$

Pictorially, the directions from the slit for the minima are given in Fig. 10.6.

(b) The location of the diffraction minima on the screen can be deduced from the right-angled triangles. Let the positive y-axis be pointed upwards on the screen, then we will have

$$y = (50 \text{ cm}) \tan\theta.$$

Denote the positions of the four minima by y_{+1}, y_{-1}, y_{+2} and y_{-2},

$$y_{\pm1} = (50 \text{ cm}) \tan\theta_{\pm1} = \pm14 \text{ cm},$$

$$y_{\pm2} = (50 \text{ cm}) \tan\theta_{\pm2} = \pm33 \text{ cm}.$$

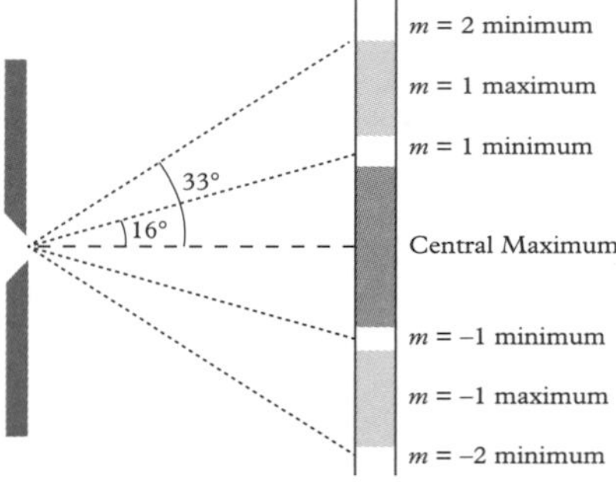

Fig. 10.6 *Example 10.1.*

10.3 Diffraction through a Circular Aperture

10.3.1 The Diffraction Pattern

When light goes through a circular hole or a lens we find that the beam spreads out. If the aperture has appropriate dimensions, you can also see a diffraction pattern on a screen behind the aperture with a central bright spot and alternating dark and bright rings. You can see the same type of diffraction pattern when you pass sound or any other type of waves through a circular aperture. The development of these diffraction patterns can be understood from application of the Huygens–Fresnel principle, just as for diffraction through a wide slit.

Suppose a plane harmonic wave of frequency ω and wave number $k = 2\pi/\lambda$ is incident on a circular aperture of radius a, as shown in Fig. 10.7. What will be the intensity on a screen a distance L away from the aperture? Calculation of the intensity is somewhat complicated and will not be presented here. We will simply cite the final expression for the intensity in direction θ from the center,

$$I(\theta) = I(0)\left[2\frac{\mathcal{J}_1(ka\sin\theta)}{ka\sin\theta}\right]^2, \tag{10.28}$$

where $\mathcal{J}_1(x)$ is the Bessel function of the first kind or order 1 and $I(0)$ is the intensity at $\theta = 0$. Note that for a small argument x the function $\mathcal{J}_1(x)$ has the following limit:

$$\lim_{x\to0} \frac{\mathcal{J}_1(x)}{x} = \frac{1}{2}. \tag{10.29}$$

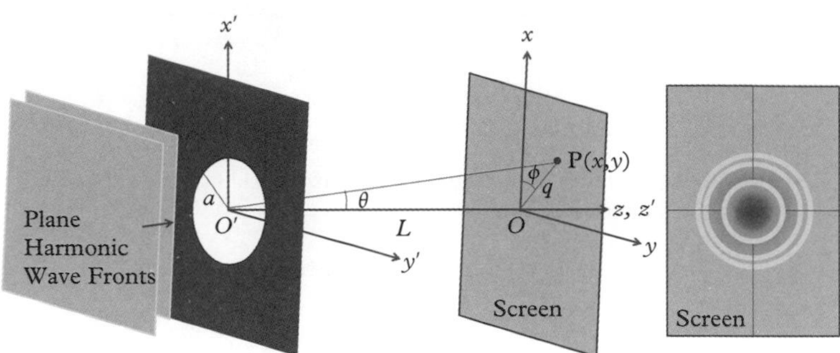

Fig. 10.7 *The circular aperture geometry.*

The intensity in direction θ is independent of direction ϕ about the axis joining the center of the aperture and the center of the screen. In terms of the position on the screen, we can write this as a condition on the radial distance of a polar coordinate at the screen. Let (q, ϕ) be polar coordinates at the screen, then

$$q = L\tan\theta \approx L\sin\theta, \quad \tan\phi = y/x. \tag{10.30}$$

Therefore, intensity is a function of the radial coordinate only, which predicts the diffraction pattern on the screen to be circular.

$$I(q) = 4I(0)\left[\frac{\mathcal{J}_1(kaq/L)}{kaq/L}\right]^2. \tag{10.31}$$

Figure 10.8 shows a plot of intensity as a function of the angle, the wave amplitude and the intensity at the screen. The central peak contains most of the signal with the second peak approximately 1.8% of the central peak. The zeros of the Bessel function, $\mathcal{J}_1(x) = 0$, give us the location of the diffraction minimum except for $x = 0$. The first few zeros of $\mathcal{J}_1(x)$ are $x = 0,\ 3.8317,\ 7.0156,\ 10.1735, \ldots$ Therefore there will be a dark circle of radius q_1 around a central filled circle with q_1 given by

$$kaq_1/L = 3.8317. \tag{10.32}$$

Solving for q_1 we get

$$q_1 = 3.8317\frac{L}{ka} = 1.22\frac{\lambda L}{D}, \tag{10.33}$$

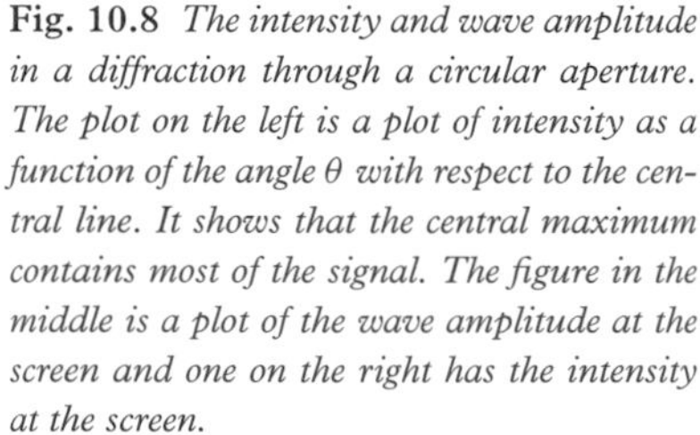

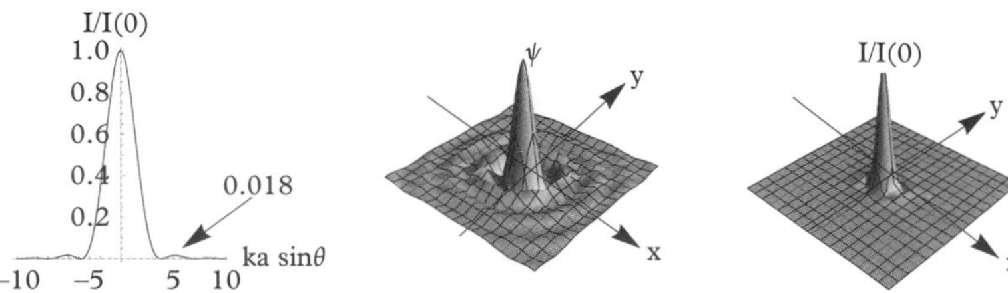

Fig. 10.8 *The intensity and wave amplitude in a diffraction through a circular aperture. The plot on the left is a plot of intensity as a function of the angle θ with respect to the central line. It shows that the central maximum contains most of the signal. The figure in the middle is a plot of the wave amplitude at the screen and one on the right has the intensity at the screen.*

where D is the diameter of the aperture,

$$D = 2a. \tag{10.34}$$

The angle θ subtended at the center of the aperture by the first circle will be

$$\theta_1 \approx \sin\theta_1 = 1.22\frac{\lambda}{D}. \tag{10.35}$$

This is an important formula for resolution of images limited by diffraction effects, as we will see below. Similarly, we can obtain the radius q_2 of the next dark circle,

$$q_2 = 7.0156\frac{L}{ka} = 2.23\frac{\lambda L}{D}, \tag{10.36}$$

and the angle θ_2 corresponding to this diffraction minimum,

$$\theta_2 \approx \sin\theta_2 = 2.23\frac{\lambda}{D}. \tag{10.37}$$

10.3.2 Limitations on Imaging Due to Diffraction

Diffraction through a circular aperture has many applications. The circular aperture may be just a circular hole in an opaque object. The circular aperture may also be a circular lens through which light could pass, such as in the viewing tube of a telescope or microscope. When light in a narrow beam passes through a circular aperture the beam will spread out in accordance with Eqs. 10.28 or 10.31. Therefore, rather than a point on the screen you will observe a bright circular disk on the screen surrounded by a dark circle. The pattern of bright and dark circles will repeat on the screen.

The circular images of two nearby point objects, such as two stars, may overlap making it difficult or even impossible to distinguish them. The resolvability of two objects is given by the **Raleigh criterion**, which states that two images are barely resolvable if the center of one is at the edge of the other circle (Fig. 10.9).

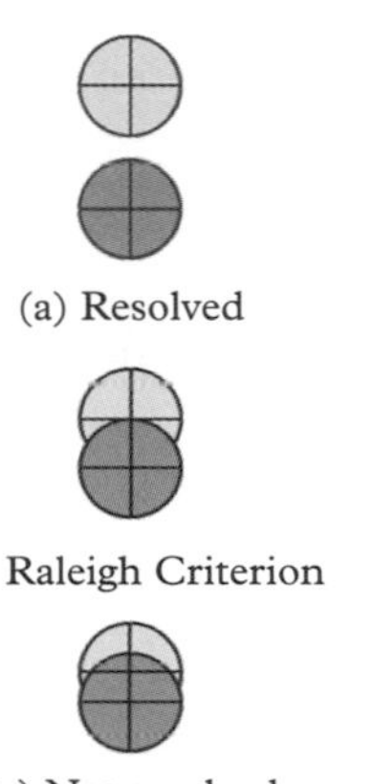

Fig. 10.9 *Raleigh criterion of resolution.*

In terms of the angular separation of the centers of the two images, the Raleigh criterion states that the angle of separation θ of the centers of the images, as shown in Fig. 10.10 must be greater than a minimum angle given by the first dark ring of the diffraction pattern of one of the objects,

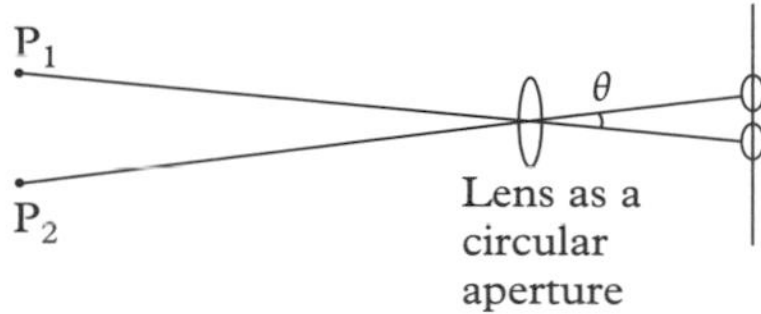

Fig. 10.10 *Angular separation of two images. For the stars to be resolvable the angle θ must be larger than θ_1 given by $1.22\lambda/D$, where D is the diameter of the lens serving as the aperture.*

$$\text{Angular separation, } \theta > \theta_1 = \frac{1.22\lambda}{D} \quad \text{(Raleigh Criterion)}. \tag{10.38}$$

Thus, according to the Raleigh criterion, two stars cannot be resolved by a telescope of aperture D operating at wavelength λ, if the stars are not far enough apart: they have to subtend an angle larger than angle θ_1. Since the loss of resolution due to diffraction effects cannot be eliminated by employing a better lens or adding additional optical elements, we say that the resolution is **diffraction limited**. An image that is diffraction limited may be improved by changing the aperture or observing in a different part of the electromagnetic spectrum.

Example 10.2 Resolution of Images Formed by a Lens

A circular converging lens of diameter 100 mm and focal length 50 cm is to be used to project the images of two faraway point sources of wavelength 632.8 nm. The image distance can be taken to be equal to the focal length and the image assumed to be on the focal plane of the lens. (a) Looking from the lens's center, what should be the minimum angular separation of the two objects so as to satisfy the Raleigh criterion? (b) How far apart will the central bright spots be on the screen at the Raleigh criterion?

Solution

(a) The figure for this example will be the same as Fig. 10.10. For the Raleigh criterion the angular separation θ must be larger than the angular width for the central peak, which is θ_1,

$$\theta \approx \tan\theta = \theta_1, \quad \text{where } \theta_1 = 1.22\frac{\lambda}{D}.$$

Putting in the numerical values appropriate for this problem we get

$$\theta_1 = \frac{1.22 \times 632 \times 10^{-9}\ \text{m}}{0.1\ \text{m}} = 7.7 \times 10^{-6}\ \text{rad}.$$

Note that we need to express both λ and D in the same units.

(b) Let one image be centered at point Q_1 with q being the radius of $m = 1$ minimum for Q_1 as shown in Fig. 10.11, The image Q_2 of the second object must be at the minimum distance of q from Q_1. That is the distance Q_1Q_2 will be equal to q of one of the images,

$$Q_1Q_2 = q.$$

The distance from the lens is equal to the focal length.

Therefore, the distance between the images will be

$$Q_1Q_2 = q = L\theta = f\theta.$$

Putting in the numbers we obtain

$$Q_1Q_2 = 50\ \text{cm} \times 7.7 \times 10^{-6}\ \text{rad} = 3.85\ \mu\text{m}.$$

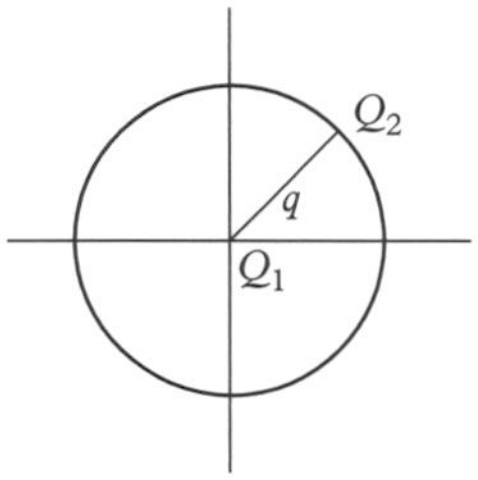

Fig. 10.11 *Example 10.2.*

10.4 Fraunhofer Diffraction through a Double Slit

10.4.1 The Diffraction Pattern

In the last chapter when we were studying interference in the Young's double-slit experiment we ignored the diffraction effect in each slit. We assumed that the slits were so narrow that on the screen you only saw the interference of light from just two point sources at the slits. If a slit width is smaller than the wavelength then Fig. 10.4 shows that there is just a spreading of light and no peaks or troughs on the screen from the diffraction effect. Therefore, it was reasonable to leave out the diffraction effect in the last chapter. Figure 10.4 also shows that, if the slit width is larger than a wavelength, you cannot ignore diffraction, since now the intensity will be affected by both interference of waves from the two slits and the interference of waves from the same slit.

To distinguish the interference pattern resulting from an interference of different parts of the same slit from the interference between two different slits, we call the former a **diffraction pattern**, and the latter the **interference pattern**, although there is really no difference in the fundamental way of distinguishing them since both effects arise from the interference of waves. Generally, interference from two finite number of waves is termed interference and interference from infinitely many waves is termed diffraction.

In this section we will study the complications of the double-slit experiment when you also need to take into account the diffraction effect of each slit. Figure 10.12 shows the basic arrangement of a double-slit experiment. The intensity on a point P on the screen is usually written as a function of the angle θ that the point makes at the symmetry point O, as shown in the figure.

The secondary spherical wavelets from each slit would interfere with other wavelets from the same slit, as well as the wavelets from the other slit. The resulting intensity in direction θ, which is derived below, would be a combined diffraction and interference effect and shows up as a product of the two effects,

$$I(\theta) = 4I_{\text{ref}} \left(\frac{\sin\beta}{\beta} \right)^2 \cos^2\alpha. \tag{10.39}$$

Here I_{ref} is the intensity of light from each slit treated as a point source. The constants β and α characterize the slit width and the slit separation respectively, and are defined by the following relations, as before:

$$\alpha \equiv \frac{\pi a \sin\theta}{\lambda} = \frac{ka\sin\theta}{2}; \quad \beta \equiv \frac{\pi b \sin\theta}{\lambda} = \frac{kb\sin\theta}{2}. \tag{10.40}$$

The factor $(\sin\beta/\beta)^2$ in Eq. 10.39 comes from the diffraction of the waves originating from the same slit and the factor $\cos^2\alpha$ arises from the interference of the waves from the two different slits. The diffraction pattern has a minimum whenever $I(\theta)$ becomes a minimum.

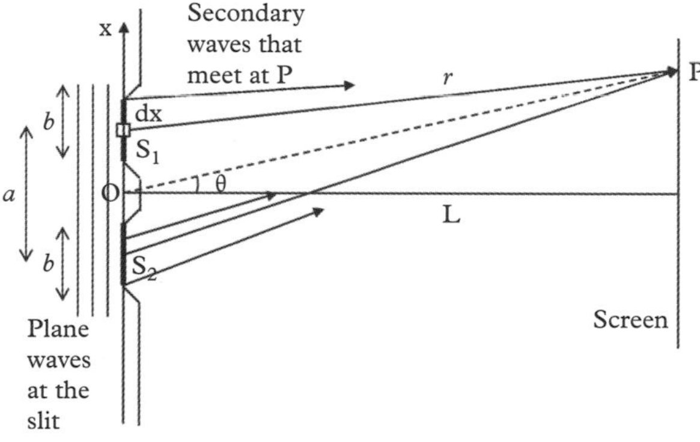

Fig. 10.12 *Fraunhofer diffraction through a double slit.*

Minima:

$$\alpha = \pm\frac{\pi}{2}, \pm 3\frac{\pi}{2}, \ldots \tag{10.41}$$

$$\beta = \pm\pi, \pm 2\pi, \pm 3\pi, \ldots \tag{10.42}$$

The minima due to α are called the **interference minima** and those due to β are called the **diffraction minima**. It is also useful to rewrite the minima conditions in terms of the wavelength λ and the slit dimensions a and b.

Minima:

$$\text{Interference:} \quad a\sin\theta = \pm\frac{\lambda}{2}, \pm 3\frac{\lambda}{2}, \ldots \tag{10.43}$$

$$\text{Diffraction:} \quad b\sin\theta = \pm\lambda, \pm 2\lambda, \pm 3\lambda, \ldots \tag{10.44}$$

The maxima of $\cos^2\alpha$ are called the maxima due to interference, or interference maxima. They occur when $\alpha = m\pi$ for integer m. These values of α correspond to the following values of the angle θ:

$$\text{Interference maxima:} \quad a\ \sin\theta = m\lambda, \quad m = 0, \pm 1, \pm 2, \ldots. \tag{10.45}$$

Some of these maxima will be missing from the pattern if the direction also corresponds to a diffraction minimum in Eq. 10.44. Thus, at an angle for which both Eqs. 10.45 and 10.44 are satisfied, we will have a zero intensity, even when a maximum intensity is expected, based on interference alone (Eq. 10.45). We refer to these missing peaks as **missing order**. One example diffraction pattern on the screen is presented in Fig. 10.13, the solid line with multiple peaks of various heights is the observed intensity on the screen. It is a product of the interference pattern from waves from separate slits and the diffraction of waves from within one slit. You can see that the $m = 3$ interference peak is suppressed since this interference maximum is co-incident with a diffraction minimum in the same direction.

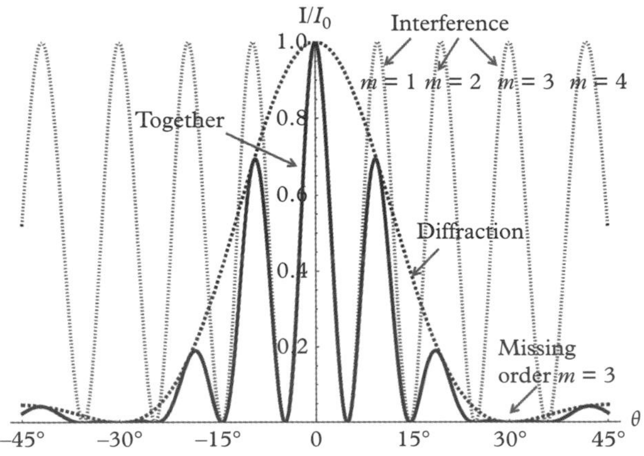

Fig. 10.13 *Fraunhofer diffraction from a double slit. The dashed line with same height peaks is from the interference of the waves from two slits, the dashed line with one big hump in the middle is the diffraction of waves from within one slit, and the solid line is the product of the two, which is the pattern observed on the screen. The plot shows the expected result for a slit width $b = 2\lambda$ and slit separation $a = 6\lambda$. The maximum of $m = \pm 3$ order for the interference is missing because the minimum of the diffraction occurs in the same direction.*

10.4.2 Derivation of the Intensity Formula

In case you are wondering how one can obtain the formula in Eq. 10.39 for the intensity cited above, here it is. Consider plane waves incident on an opaque shield with two long horizontal slits each with width a and center-to-center distance b, as shown in Fig. 10.12. Then the superposition of secondary wavelets from the plane wavefront at the two slits will proceed in a similar way as for a single slit, except now the integration over x will have two parts, one over each slit,

$$\begin{aligned}\psi(\text{at } P) &= \text{Contribution from the wave through slit 1}\\ &\quad + \text{Contribution from the wave through slit 2.}\end{aligned}$$

We have already worked out the math for one slit in detail when we studied the diffraction through a single slit. The same math is applied here with the range of x for the slits being: $\frac{a}{2} - \frac{b}{2}$ to $\frac{a}{2} + \frac{b}{2}$ for slit S_1 and $-\frac{a}{2} - \frac{b}{2}$ to $-\frac{a}{2} + \frac{b}{2}$ for slit S_2,

$$\begin{aligned}\psi(\text{at } P) &= \frac{A}{L}\left[\int_{a/2-b/2}^{a/2+b/2} \sin(kL - kx\sin\theta - \omega t)\,dx\right]_{\text{slit 1}}\\ &\quad + \frac{A}{L}\left[\int_{-a/2-b/2}^{-a/2+b/2} \sin(kL - kx\sin\theta - \omega t)\,dx\right]_{\text{slit 2}},\end{aligned}$$

where A is the amplitude at the slits. For simplicity we have assumed the same amplitude at the points of the two slits. The integrals are easy to do with the following result:

$$\psi(\text{at } P) = 2\frac{bA}{L}\left(\frac{\sin\beta}{\beta}\right)\cos\alpha\ \sin(kL - \omega t), \tag{10.46}$$

where α and β are same as defined above. The intensity is obtained by time-averaging the square of the field and then multiplying the result by an appropriate factor for that wave; for a light wave the factor is $\epsilon_0 c$, where ϵ_0 is the permittivity and c the speed of light. Let us use this factor here and tailor our calculation to light waves. At the end, we can write the formula for intensity that will apply to any type of wave. The time averaging of $\sin^2(kR - \omega t)$ gives $\frac{1}{2}$. Therefore, the intensity in the direction at an angle θ from the slits will be

$$I(\theta) = \left[2\epsilon_0 c\left(\frac{bA}{R}\right)^2\right]\left(\frac{\sin\beta}{\beta}\right)^2\cos^2\alpha, \tag{10.47}$$

where the quantity in the square bracket has units of intensity. As we are mostly interested in how the intensity changes on the screen and not in the absolute value, it will be nice to find a reference intensity I_{ref} in terms of which we can express $I(\theta)$. The intensity of light from one slit on the screen at the horizontal spot from the slit when the other slit is covered provides a nice reference intensity. This intensity is easily obtained from noting that the electric field has amplitude A at any point of the slit and accounting for the spreading out in a sphere of radius R when traveling to the screen,

$$I_{\text{ref}} = \frac{1}{2}\epsilon_0 c\left(\frac{bA}{R}\right)^2. \tag{10.48}$$

Therefore intensity $I(\theta)$ from two slits is written as follows:

$$I(\theta) = 4I_{\text{ref}} \left(\frac{\sin\beta}{\beta} \right)^2 \cos^2\alpha. \text{ (all waves)} \tag{10.49}$$

Note that when both slits are open, the intensity is four times the intensity of one slit rather than two times. This is because the amplitudes add giving 2 × the amplitude of one slit, but the intensity is proportional to the square of amplitudes, and therefore you get 4 × the intensity of one slit.

Example 10.3 Interference and Diffraction in a Double-Slit Diffraction

Find the angular positions of the interference maxima within the central peak of a double-slit diffraction for a monochromatic light of wavelength 628 nm on slits of width 1.5 μm separated by 4 μm.

Solution

First let us find the range of angle included within the central maximum of the diffraction by locating the first diffraction minima,

$$b \sin\theta = \pm\lambda.$$

Therefore, the central peak is between $-\sin^{-1}(\lambda/b)$ and $+\sin^{-1}(\lambda/b)$, which gives −23.3° to +23.3°. We need to find the interference maxima within this range. The condition for interference maxima is

$$a \sin\theta - m\lambda, \quad m - 0, \pm1, \pm2, \pm3, \ldots.$$

The value of m for which the edge of the central peak of diffraction is reached is obtained by setting the angle to 23.3°,

$$m = \frac{a \sin\theta}{\lambda} = \frac{4\ \mu\text{m}\ \sin(23.36\text{o})}{0.628\ \mu\text{m}} = 2.5.$$

Hence, the central peak will have five interference peaks, corresponding to $m = 0, \pm1, \pm2$.

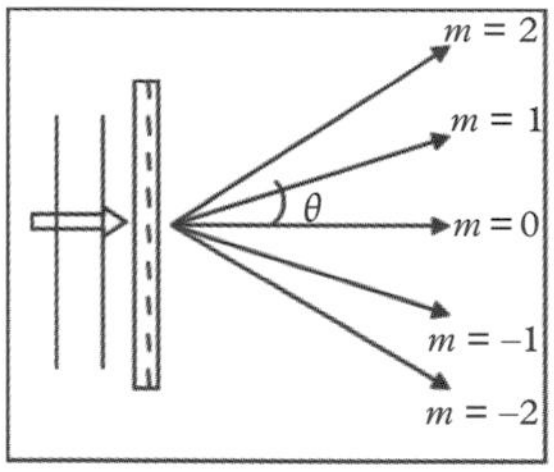

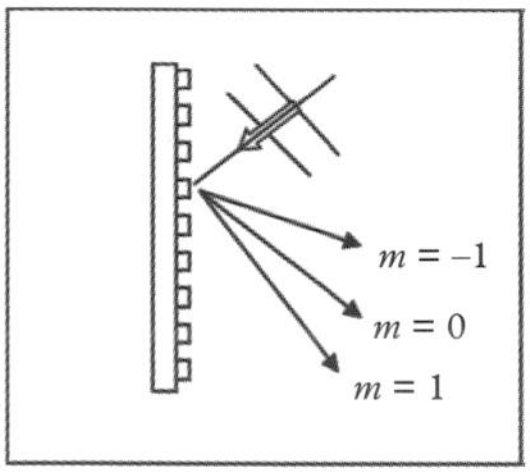

Fig. 10.14 *(a) Transmission grating and (b) reflection grating.*

10.5 Diffraction Grating

A material consisting of periodic reflectors or periodic transmission slits is called a diffraction grating. The diffraction gratings with periodic reflectors are more specifically called **reflection gratings** and those with periodic slits are called **transmission gratings** (See Fig. 10.14).

A transmission grating for light can be made from a glass plate by machining fine parallel lines onto it so that the spaces between the lines serve as slits. In a transmission grating, light is incident on one side of the slits and emerges on the other side having passed through the slits (Fig. 10.14a), so that each slit serves as a new source of a wave. The reflection gratings for light are made by etching parallel grooves onto a glass or metal plate from which light can be reflected (Fig. 10.14b). Music and computer CDs acts as reflection gratings for light waves. You may have noticed that white light is separated into component colors when reflected off a CD. This effect is due to the CD acting as a reflection grating. Diffraction gratings for sound waves are even easier to construct. For a transmission grating for sound you can tape or nail plastic or metal plates, usually 2 or 3 cm wide for audible sound, with the periodicity of grating required. To observe the effects of a diffraction grating of sound you can either vary the spacing in the grating and/or adjust the wavelength of the sound.

Let us examine a transmission grating, which is simply a system of many slits. Diffraction through multiple slits can be understood in much the same way that we have discussed diffraction through two slits previously. If a wave is incident on a diffraction grating such that the wave passes through N slits, each of width b, and slit separation distance a, then the superposition of the waves from these N slits will give the following wave amplitude:

$$\psi(\text{at}P) = \left[A\left(\frac{\sin\beta}{\beta}\right)\frac{\sin(N\alpha)}{\sin\alpha}\right]\cos(kL-\omega t). \tag{10.50}$$

The derivation of this result is left as an exercise for the student. Therefore, the intensity on the screen will be

$$I(\theta) = I(0)\left(\frac{\sin\beta}{\beta}\right)^2\left[\frac{\sin(N\alpha)}{\sin\alpha}\right]^2, \tag{10.51}$$

where $I(0)$ is the intensity at $\theta = 0$. Here, α and β are same as above.

$$\alpha \equiv \frac{\pi a\sin\theta}{\lambda} = \frac{ka\sin\theta}{2},$$
$$\beta \equiv \frac{\pi b\sin\theta}{\lambda} = \frac{kb\sin\theta}{2}.$$

The parameters α and β respectively control the effect of interference of waves from adjacent slits and the diffraction from secondary wavelets within the same slit.

In the case of narrow slits, i.e. when the slit widths are much smaller than the wavelength, we can assume $\beta \ll 1$ and make the approximation $\sin\beta \approx \beta$, which gives $\sin\beta/\beta \approx 1$. In these circumstances, one finds that a regular interference pattern is modulated in intensity with the intensity of the peaks in the center being the largest. When discussing the effect of the number of slits it is better to "turn off" the diffraction by using slits where the slit width $b \ll \lambda$ so that $\beta \ll 1$. In this case, we will have a simpler formula for the intensity,

$$I(\theta) = I(0)\left[\frac{\sin(N\alpha)}{\sin\alpha}\right]^2. \quad \text{(narrow slits)} \tag{10.52}$$

10.5.1 Principal Maxima

The diffraction pattern from multiple slits consists of major and minor peaks. The major peaks, also called **principal maxima**, occur when the denominator in Eq. 10.52 goes to zero,

$$\text{Principal maxima when: } \sin\alpha = 0, \tag{10.53}$$

which means that α has the following values at principal maxima:

$$\text{Principal maxima when: } \alpha = m\pi,\quad m = 0, \pm1, \pm2, \ldots. \tag{10.54}$$

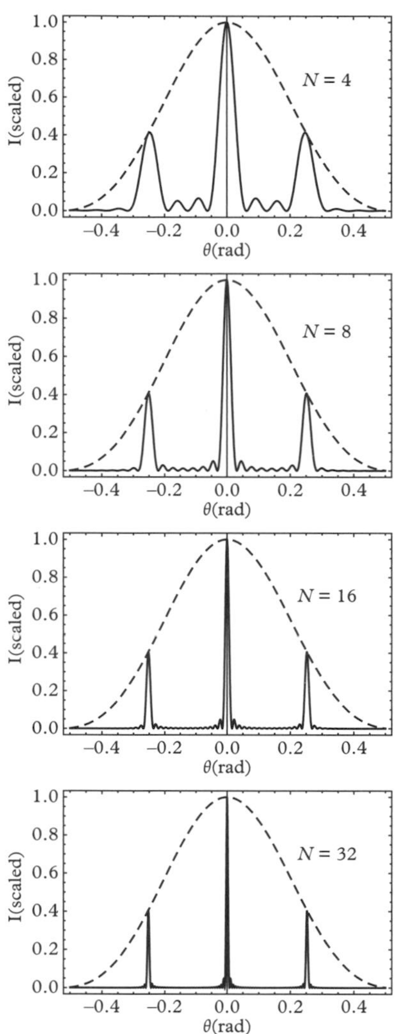

Fig. 10.15 *Multiple slits diffraction. Normalized intensity is plotted against θ with the slit width $b = \lambda/2$ and distance between slits $a = 4\lambda$ for four values of N, $N = 4$, 8, 16, 32. The dashed envelope is the diffraction of a single slit. The intensity was normalized by dividing the actual intensity by N^2.*

Note that when α has these values, the numerator of Eq. 10.52 will also be zero since N is an integer and $\sin(Nm\pi) = 0$ when m and N are integers. That is, at principal maxima, both the denominator and numerator go to zero. But, the intensity in Eq. 10.52 is well defined since the ratio of $\sin(N\alpha)/\sin\alpha$ for these values of α has a limit of $\pm N$ as $\alpha \to m\pi$. Let us set $\alpha = m\pi - \epsilon$ and then take $\epsilon \to 0$,

$$\lim_{\alpha\to m\pi}\frac{\sin(N\alpha)}{\sin\alpha} = \lim_{\epsilon\to 0}\frac{\sin(Nm\pi - N\epsilon)}{\sin(\pi-\epsilon)} = \lim_{\epsilon\to 0}\frac{(-1)^{Nm}\sin(N\epsilon)}{(-1)^m\sin(\epsilon)}$$

$$= (-1)^{(N-1)m}\lim_{\epsilon\to 0}\left[\frac{(N\epsilon) - \frac{1}{6}(N\epsilon)^3 + \cdots}{\epsilon - \frac{1}{6}\epsilon^3 + \cdots}\right]$$

$$= (-1)^{(N-1)m}\, N = \pm N.$$

In terms of angle θ of the direction from the center of the diffraction grating, the principal maxima occur at

$$\text{Principal Maxima when: } b\sin\theta = m\lambda,\quad m = 0, \pm 1, \pm 2, \ldots, \tag{10.55}$$

where I have used $\alpha = \pi a \sin\theta/\lambda$.

10.5.2 Angular Width of Principal Maxima

The sharpness of the principal maxima depends on N. In Fig. 10.15, diffraction patterns of a monochromatic light incident across N slits of a transmission grating for various values of N are shown. You can easily notice that the peaks become sharper with increasing N. The peaks also become more intense. To compare the patterns for different N I have divided the actual intensities by N^2. Thus, the peak of intensity 1 for $N = 32$ is 64 times more intense than the peak of intensity 1 for $N = 4$.

From the formula for intensity in diffraction for an N-slit grating, Eq. 10.51, we note that the direction and the width of the principal peaks are controlled by the following function:

$$f(\theta) = \left[\frac{\sin(N\alpha)}{\sin\alpha}\right]^2. \tag{10.56}$$

The principal peaks are at

$$\alpha = m\pi,\quad \text{or,}\quad a\sin\theta = m\lambda,\quad m = 0, \pm 1, \pm 2, \ldots.$$

To find the width of a particular peak, we need the directions on either side of that peak where intensity is a minimum. Let us illustrate the calculation of the width of a peak by working out the width of the central peak. The central maximum occurs at $\alpha = 0$, which is in the direction $\theta = 0$. From Eq. 10.56, we see that the nearest minimum of f, which is $f = 0$, about this direction is when $N\alpha = \pm\pi$. that is, the α for the minimum around the central peak are given by

$$N\alpha = \pm\pi. \tag{10.57}$$

This condition in terms of the angle θ is

$$\sin\theta = \pm\frac{\lambda}{Na}.$$

For small angles we can use the $\sin\theta = \theta$ approximation. Therefore, zeros of intensity occur in directions $\theta = -\lambda/Na$ and $\theta = +\lambda/Na$. The half-width is defined as the angle from the center to one of these minima, as shown in Fig. 10.16.

Hence, half-width θ_{hw} of the central peak will be

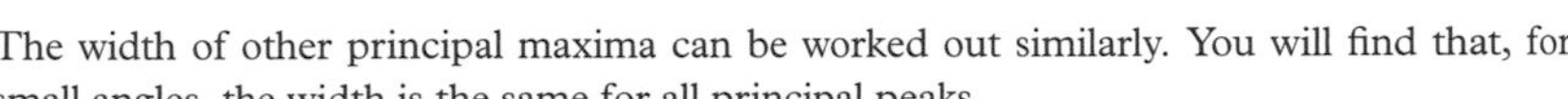

$$\theta_{\text{hw}} = \frac{\lambda}{Na} \quad (\theta_{\text{hw}} \text{ in radian}). \quad \text{(Central peak)} \tag{10.58}$$

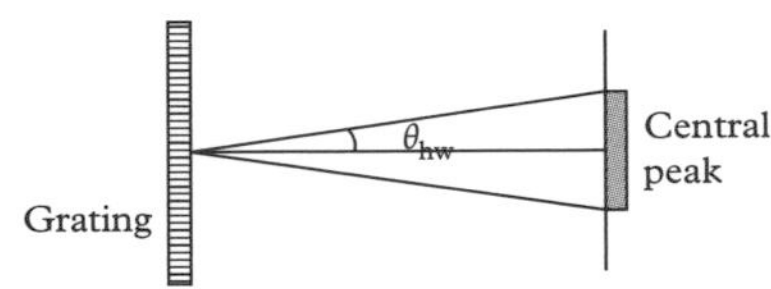

Fig. 10.16 *Half-width for central peak.*

The width of other principal maxima can be worked out similarly. You will find that, for small angles, the width is the same for all principal peaks.

Example 10.4 Diffraction Grating for Spectroscopy

Find the separation in angles of a light of wavelength 550 nm and 575 nm for the first order ($m = 1$) principal peak of the two waves from a diffraction grating with 800 lines per mm.

Solution

Since there are 800 lines per mm, the distance between adjacent slits is

$$a = \frac{1\text{ mm}}{800} = 1.25\ \mu\text{m}.$$

Hence, the directions for $m = 1$ principal peaks corresponding to the two wavelengths in the problem are

$$\theta_{550} = \sin^{-1}\left(\frac{0.550\ \mu\text{m}}{1.25\ \mu\text{m}}\right) = 26.1°$$

$$\theta_{575} = \sin^{-1}\left(\frac{0.575\ \mu\text{m}}{1.25\ \mu\text{m}}\right) = 27.4°$$

Therefore, $\theta_{575} - \theta_{550} = 1.3°$.

10.5.3 Resolving Power

A diffraction grating produces sharp lines in its diffraction pattern. The angular positions of the maxima depend upon slit separation and the wavelength of the light. Therefore a common use of diffraction gratings is in the separation of lights of different colors and measurement of their wavelengths and the corresponding intensities. However if the wavelengths of light to be separated are too close then their diffraction peaks may not be distinguishable. Once again, we use the Raleigh criterion to deduce the minimum separation for resolvability. The Raleigh criterion is based on the fact that the peaks of the diffraction pattern have a finite width and when two peaks overlap significantly we may not be able to tell whether there is one peak or two. To properly discuss the criterion of separability we use the concept called the half-width of the peak calculated above.

The ability of a diffraction grating to separate waves of different wavelength is often designated by its **resolving power**. A diffraction grating with higher resolving power produces narrower peaks and is therefore able to separate wavelengths that are closer in value. The resolving power is defined by making use of the smallest wavelength interval that can barely be resolved. Let $\Delta\lambda \equiv |\lambda_1 - \lambda_2|$ be the smallest wavelength interval resolvable so that λ_1 and λ_2 can both be seen at the screen. Then the **resolving power** R is defined as

$$R = \frac{\lambda_{\text{ave}}}{\Delta\lambda}, \tag{10.59}$$

where λ_{ave} is the average of the two wavelengths,

$$\lambda_{\text{ave}} = \frac{\lambda_1 + \lambda_2}{2}.$$

Using the Raleigh criterion, two peaks are barely resolvable if the peak of one is at the zero of the other peak. For wavelengths λ_1 and λ_2 to be resolvable by the Raleigh criterion their peaks will be separated by a half-width given in Eq. 10.58 (we work with small angles),

$$\theta_{\text{hw}} = \frac{\lambda_{\text{ave}}}{Na}. \tag{10.60}$$

We will equate this to the angular separation between peaks of λ_1 and λ_2. Now, using the condition for the maxima of the principal peaks we find the angular separation between peaks corresponding to two wavelengths separated by $\Delta\lambda = \lambda_1 - \lambda_2$,

$$a \sin\theta = m\lambda \implies (a\cos\theta)\Delta\theta = m\Delta\lambda \implies \Delta\theta = \frac{m\Delta\lambda}{a\cos\theta} \approx \frac{m\Delta\lambda}{a}. \tag{10.61}$$

Equating the quantities of Eqs. 10.60 and 10.61 we get

$$\frac{\lambda_{\text{ave}}}{Na} = \frac{m\Delta\lambda}{a} \implies \frac{\lambda_{\text{ave}}}{\Delta\lambda} = Nm. \tag{10.62}$$

Hence, the resolving power R of a diffraction grating with N slits spread over a distance Na is given by

$$R = \frac{\lambda_{\text{ave}}}{\Delta\lambda} = Nm. \tag{10.63}$$

Example 10.5 Resolving the Sodium Doublet

The D-line of sodium consists of two different wavelengths, 589.0 nm and 589.6 nm. A beam of light from a sodium lamp forms a beam of width 10 mm. The beam is incident perpendicularly on a diffraction grating that has 8000 rulings over a width of 10 mm. (a) Find the directions of the first-order principal peaks for the two wavelengths. (b) Decide if the two peaks are resolvable in the first order. (c) Find the number of lines of a grating so that peaks in the first order are barely resolved.

Solution

(a) We need the separation a between the slits to figure out the direction of the first-order principal peak,

$$a = \frac{10 \text{ mm}}{8000 \text{ lines}} = 1250 \text{ nm}.$$

For $m = 1$, the condition for the principal peak is $a \sin\theta = \lambda$. Let us denote the angle for wavelength 589.00 nm by θ and the angle for wavelength 589.59 nm by ϕ. Hence the directions of $m = 1$ peaks for the two wavelengths are

$$\theta_1 = \sin^{-1}\left(\frac{589.00 \text{ nm}}{1250 \text{ nm}}\right) = 0.49065 \text{ rad},$$

$$\phi_1 = \sin^{-1}\left(\frac{589.59 \text{ nm}}{1250 \text{ nm}}\right) = 0.49119 \text{ rad}.$$

(b) We can compare the resolution required to the resolving power of the grating to see if the peaks will be resolved. The resolving power required for resolving the two given wavelengths is

$$R_{\text{req}} = \frac{\lambda_{\text{ave}}}{\Delta\lambda} = \frac{589.295 \text{ nm}}{0.59 \text{ nm}} = 999.$$

The resolving power of the given grating for $m = 1$ peaks is

$$R_{\text{grating}} = Nm = 8000.$$

Since $R_{\text{grating}} > R_{\text{req}}$, the peaks will be resolved.

(c) Using the required resolving power, we can work backwards and deduce the number of lines required in 10 mm, the beam width, so that the grating can resolve the two given waves,

$$Nm = R_{\text{req}} \implies N \times 1 = 999 \implies N = 999.$$

Figure 10.17 shows the two peaks separated with a grating of $N = 999$.

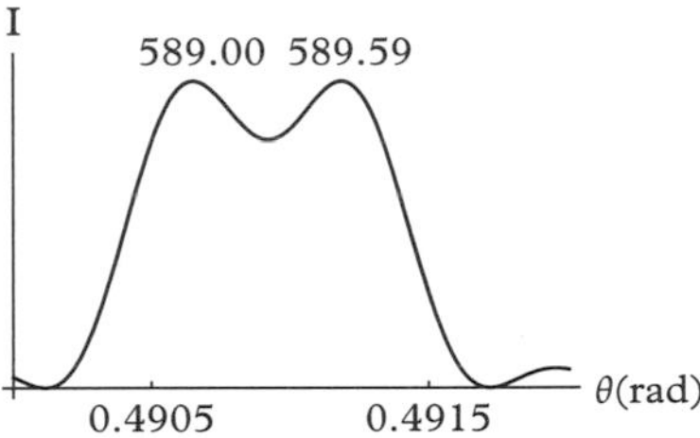

Fig. 10.17 *Example 10.5. Separation of $m = 1$ order peaks of the sodium doublet by a grating with $N = 999$.*

EXERCISES

(10.1) Monochromatic light of wavelength 530 nm passes through a horizontal single slit of width 1.5 μm in an opaque plate. A screen is 1.2 m away from the slit.
  (a) Which way is the diffraction pattern spread out on the screen?
  (b) What are the directions of the minima in terms of the angle from the slit and of the locations on the screen?
  (c) What are the directions of the maxima in terms of the angle from the slit and of the locations on the screen?
  (d) How wide in cm is the central bright fringe on the screen?
  (e) How wide in cm is the next bright fringe on the screen?
  (f) What is the intensity of the first side peak compared to the intensity of the central peak?

(10.2) Light of wavelength 630 nm passes through a vertical single slit of width 1.5 μm in an opaque plate. A screen of dimensions 2 m × 2 m is 1.2 m away from the slit.
   (a) Which way is the diffraction pattern spread out on the screen?
   (b) Where on the screen are the diffraction minima?
   (c) Where on the screen are the diffraction maxima?
   (d) What is the angular width of the central peak?

(10.3) A monochromatic light of wavelength 632.8 nm is incident on a single slit. The emerging light behind the slit is projected on a screen 1.6 m away where the central bright spot is found to have a width of 5 cm.
   (a) How wide is the slit?
   (b) What will be the width of the central bright spot if light of wavelength 540 nm is used?

(10.4) Solve the following equations graphically. (a) $x^2 = \sin(x)$, (b) $x\tan(x) = 1$.

(10.5) Two slits of width 2 μm each in an opaque material are separated by a center-to-center distance of 6 μm. A monochromatic light of wavelength 450 nm is incident on the double slit. One finds a combined interference and diffraction pattern on the screen.
   (a) How many peaks of the interference will be observed in the central maximum of the diffraction pattern?
   (b) How many peaks of the interference will be observed in the central maximum of the diffraction pattern if the slit width is doubled while keeping the distance between the slits the same?
   (c) How many peaks of interference will be observed in the central maximum of the diffraction pattern if the slits are separated by twice the distance, i.e. 12 μm, while keeping the width of the slits 2 μm?
   (d) What would happen in (a) if instead of 450-nm light another light of wavelength 680 nm were used?
   (e) What is the value of the ratio of the intensity of the central peak to the intensity of the next bright peak in (a)?
   (f) Does the ratio in (e) depend on the wavelength of the light?
   (g) Does the ratio in (e) depend on the width or separation of the slits?

(10.6) When a monochromatic light of wavelength 430 nm is incident on a double slit with slit separation 5 μm, there are 11 interference fringes in its central maximum. How many interference fringes will be in the central maximum of a light of wavelength 632.8 nm for the same double slit?

(10.7) A converging lens of diameter 8 cm and focal length 20 cm is used to focus the image of a star. Think of the lens as a circular aperture and find the radius of the central bright spot by using 550 nm for the wavelength of light from the star.

(10.8) What is the minimum distance between two points on the Moon that the 40-in refractor telescope at Yerkes observatory in Wisconsin can resolve if the resolution were limited to due to diffraction only? Use 540 nm as the wavelength of light and a distance of 4×10^8 m from the Earth to the Moon.

(10.9) Two lamps producing light of frequency 589 nm are fixed 1 meter apart on a wooden plank. How far can the plank be moved from the eye so that the eye can still resolve them if the resolution is affected solely by diffraction of light in the eye? Assume light enters the eye through a pupil of diameter 4.5 mm.

(10.10) A diffraction grating contains 500 lines per mm.
   (a) What is the separation between lines of the grating?
   (b) What will be the angle of separation between the $m = 1$ peaks of a yellow light of wavelength 590 nm and a red light of wavelength 630 nm?

(10.11) How many lines per mm are there in the diffraction grating if the second-order principal maximum for a light of wavelength 536 nm occurs at an angle of 24° with respect to the line from the grating to the center of the diffraction pattern?

(10.12) The spectrum of mercury has a pair of yellow orange lines at 576.959 nm and 579.065 nm. A diffraction grating with 200 lines per mm is available in the lab. Determine if this grating will do the job of resolving the two mercury yellow orange lines in $m = 1$ order if the beam diameter is 4 mm.

(10.13) A light beam of an unknown wavelength is incident over 5 mm of a diffraction grating that has 400 lines per mm, and the first-order peak is observed at an angle of 20°. Find (a) the wavelength, and (b) the half-width of the peak.

(10.14) Suppose a is the ratio of center-to-center distance between two slits and b is the slit width of each slit in a double-slit experiment. Let $p = \frac{a}{b}$ with $p > 1$.

(a) Prove that the number of interference peaks inside the central maximum of the diffraction peak is $2p + 1$.

(b) Find the orders of the interference maxima that will be missing if (i) $p = 3$, (ii) $p = \frac{3}{2}$, (iii) $p = \frac{5}{3}$, (iv) $p = \sqrt{5}$.

(10.15) A light ray of wavelength 461.9 nm emerges from a 2 mm circular aperture of a krypton ion laser. Because of diffraction the beam expands as it moves out. How large is the central bright spot at

(a) 1 m?

(b) 1 km?

(c) 1000 km? and

(d) the surface of the Moon at a distance of 400,000 km from the Earth?

(10.16) The Hubble telescope has an aperture diameter of 2.4 m and is approximately 600 km from the surface of the Earth. If the resolution is only limited by the diffraction of the telescope aperture, what is the smallest separation between two points on the Earth that the Hubble telescope can resolve? Use 550 nm for light.

(10.17) Deduce the condition for the minima when light is incident at an angle ψ rather than normally incident on the slit, as shown in Fig. 10.18.

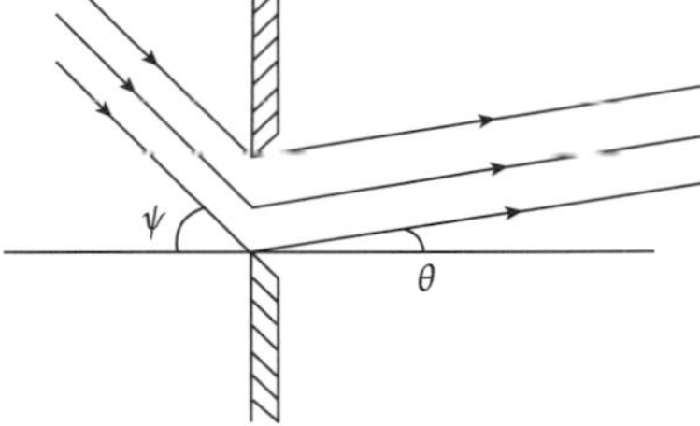

Fig. 10.18 *Exercise 10.17.*

Appendix
Solutions

Chapter 1

1.1 $v_x = 2b_2\tau - b_3 b_4 \sin(b_4\tau)$, $a_x = 2b_2 - b_3 b_4^2 \cos(b_4\tau)$.

1.2 $x = \frac{b_1}{2}\tau^2 + \frac{b_2}{12}\tau^4 - \frac{b_3}{b_4^2}[\cos(b_4\tau) - 1]$, $v = b_1\tau + \frac{b_2}{3}\tau^3 + \frac{b_3}{b_4}\sin(b_4\tau)$.

1.3 $\vec{r} = \frac{q}{2m}\vec{E}_0\tau^2$, $\vec{v} = \frac{q}{m}\vec{E}_0\tau$.

1.4 $b \sin D$.

1.5 (a) $x = \frac{c}{6m}\tau^3$, $v_x = \frac{c}{2m}\tau^2$; (b) $\frac{c^2}{2m}t^3$; (c) $\frac{c^2}{8m}\left(\frac{6mD}{c}\right)^{4/3}$.

1.6 (a) $\Delta v_x = g(t_2 - t_1) + \frac{g}{c}\left(e^{-ct_2} - e^{-ct_1}\right)$; (b) $\Delta x = \frac{g}{2}(t_2^2 - t_1^2) - \frac{g}{c^2}\left(e^{-ct_2} - e^{-ct_1}\right) - \frac{g}{c}(t_2 - t_1)$;
(c) Hint: Perform the integration in $< P > = \frac{1}{t_2 - t_1}\int_{t_1}^{t_2} F_x v_x$.

1.7 $u(1 + \ln M)(t_2 - t_1) + \frac{u}{\alpha}\left[(M - \alpha t_2)\ln|M - \alpha t_2| - (M - \alpha t_1)\ln|M - \alpha t_1|\right]$.

1.8 With origin at a fixed point to the left of the two masses let x_1 an x_2 be their coordinates. Then, equations of motion are: $m\, d^2x_1/dt^2 = -k(x_1 - x_2)$, $m\, d^2x_2/dt^2 = +k(x_1 - x_2)$.

1.9 $m\, d^2x/dt^2 = -(k_1 + k_2)x$.

1.10 $m\frac{d^2y}{dt^2} = -(k_1 + k_2)(l_1 - l_0)\frac{y}{l_1}$, with $l_1 = \sqrt{l_0^2 + y^2}$.

1.11 (a) $U = \frac{1}{2}(k_1 + k_2)(l_1 - l_0)^2$, with $l_1 = \sqrt{l_0^2 + y^2}$, (b) Hint: $F_y = -dU/dy$.

1.12 (a) Let $l_1 = \sqrt{(l_0 + x)^2 + y^2}$ and $l_2 = \sqrt{(l_0 - x)^2 + y^2}$, then the x-component of the equation of motion is

$$m\frac{d^2x}{dt^2} = -k_1(l_1 - l_0)\frac{l_0 + x}{l_1} - k_2(l_2 - l_0)\frac{l_0 - x}{l_2}.$$

(b) The y-component of the equation of motion will be

$$m\frac{d^2y}{dt^2} = -k_1(l_1 - l_0)\frac{y}{l_1} - k_2(l_2 - l_0)\frac{y}{l_2}.$$

1.13 (a) $U = \frac{1}{2}k_1\left[\sqrt{(l_0 + x)^2 + y^2} - l_0\right]^2 + \frac{1}{2}k_2\left[\sqrt{(l_0 - x)^2 + y^2} - l_0\right]^2$.

1.14 Hint: The vertical component of tension $\approx mg$ in the small angle approximation.

1.15 (a) The equations of motion are

$$m_1\frac{d^2x_1}{dt^2} = -\frac{m_1 g}{l}x_1 + k(x_2 - x_1 - l_0).$$

$$m_2\frac{d^2x_2}{dt^2} = -\frac{m_2 g}{l}x_2 - k(x_2 - x_1 - l_0).$$

(b) Hint: Set $x_2' = x_2 - l_0$ in the equations in (a).

1.16 (a) $\sqrt{2}e^{i\pi/4}$, (b) $5e^{i\pi}$, (c) $\sqrt{2}e^{i5\pi/4}$, (d) $2e^{i3\pi/2}$, (e) $\sqrt{7}e^{i\arctan(2/\sqrt{3})}$, (f) $\sqrt{7}e^{i\arctan(\sqrt{3}/2)}$.

1.17 (a) 1, (b) i, (c) -1, (d) $-i$, (e) 1, (f) $1 + i\sqrt{3}$.

1.18 (a) $-2 + i2\sqrt{3}$, (b) 4, (c) -4, (d) i, (e) $\sqrt{3} + i$, (f) $3 + i3\sqrt{3}$, (g) 8, (h) $i2$, (i) -3, (j) $\frac{5}{2}$.

1.19 (a) When $\theta_2 > 0$, the vector rotates counterclockwise and when $\theta_2 < 0$, the vector rotates clockwise, (b) When $a > 0$ the vector keeps its direction when $a < 0$, the vector rotates by 180°. When $|r < 1|$, the length of the vector shrinks and when $|r > 1|$, the length of the vector increases.

1.20 $z = \left(\frac{a}{a^2+b^2}\right) - i\left(\frac{b}{a^2+b^2}\right)$.

1.21 Hint: Use Euler's formula.

1.22 Hint: Use the polar form.

1.23 Hint: List the roots and use the fact that for any integer n we get $e^{i2\pi n} = 1$.

1.24 $z^{29} = 2^{28} - i2^{28}\sqrt{3}$, $z^{1000} = -2^{999} - i2^{999}\sqrt{3}$.

1.25 (a) $z = 0$, the only solution, (b) four solutions: $z = 1, i, -1, -i$, (c) four solutions: $z = e^{i\pi/8}, e^{i\pi/8+i\pi/2}, e^{i\pi/8+i\pi}, e^{i\pi/8-i\pi/2}$.

1.26 (a) $z = \sqrt{i} = \left(e^{i\pi/2 + i2n\pi}\right)^{1/2} = e^{i\pi/4 + in\pi}$. The rectangular forms: $z = \frac{1}{\sqrt{2}} + \frac{i}{\sqrt{2}}, -\frac{1}{\sqrt{2}} - \frac{i}{\sqrt{2}}$.
(b) $z = \left(e^{i\pi/2 + i2n\pi}\right)^i = e^{-(\pi/2 + 2n\pi)}$, $n = 0, \pm1, \pm2, \cdots$.

1.27 $\sqrt{z} = \sqrt{r}e^{i\theta/2 + in\pi} = \sqrt{r}\cos(\theta/2 + n\pi) + i\sqrt{r}\sin(\theta/2 + n\pi)$, with $n = 0, \pm1, \pm2, \cdots$.

1.28 Hint: Use a Taylor series of the exponential function.

1.29 Hint: Use Euler's formula.

1.30 (a) $f_{av} = 250$ Hz, (b) $f_{beat} = 20$ Hz, (c) 12.5 oscillations in 1 beat.

1.31 (a) $[1 + 2\cos(0.1\omega t)]\cos(1.1\omega t)$, (b) $T_{beat} = 20\pi/\omega$ since max amplitudes at $0.1\omega t = 0, 2\pi, 4\pi, \cdots$.

1.32 (a) $[\cos\omega t - \cos(\omega t - \epsilon t) - \cos[\omega t - (N+1)\epsilon t] + \cos(\omega t + N\epsilon t)]/\ 2(1 - \cos\epsilon t)$, (b) Writing the answer in (a) the form $A(t) \times$ Oscillating function we get the period of $A(t)$ to be $2\pi/\epsilon$, which would be the beat period here.

1.33 (a) $x = A\cos 5t + B\sin 5t$. (b) $e^{-3t}(A\cos 4t + B\sin 4t)$, (c) See text, (d) Hint: $\sin(A) = \cos(A - \pi/2)$.

1.34 Let $\omega = qB_0/m$. (a) $x = (v_0/\omega)\sin(\omega\tau)$, $y = (v_0/\omega)[\cos(\omega\tau) - 1]$, $z = 0$; $v_x = v_0 \cos(\omega\tau)$, $v_y = -v_0\sin(\omega\tau)$, $v_z = 0$. (b) $R = v_0/\omega$.

Chapter 2

2.1 We are given

$$m = 200 \text{ g} = 0.2 \text{ kg}, \quad k = 100 \text{ N/m}.$$

Therefore,

$$\omega = \sqrt{k/m} = \sqrt{100 \text{ [N/m]}/0.2 \text{ kg}} = 10\sqrt{5} \text{ sec}^{-1}$$
$$f = \omega/2\pi = \frac{5\sqrt{5}}{\pi} \text{ Hz}$$
$$T = 1/f = \frac{\pi}{5\sqrt{5}} \text{ sec}.$$

2.2 (a) Comparing the given oscillations with $x(t) = A\cos(\omega t - \phi)$ we can read off the various quantities in the given x_1 and x_2,

$$\omega_1 = 1 \text{ sec}^{-1}, \ A_1 = 2 \text{ cm}, \ \phi_1 = 0,$$
$$\omega_2 = 1 \text{ sec}^{-1}, \ A_2 = 2 \text{ cm}, \ \phi_2 = -\pi.$$

(b) Taking the time derivatives of x_1 and x_2 gives the velocities (or rather, the x component of velocities), which we denote by v_1 and v_2,

$$v_1 = -2\sin(t), \quad v_2 = -2\sin(t + \pi).$$

Setting $t = 0$ in x and v gives the position and velocity at $t = 0$,

$$x_1(0) = 2 \text{ cm}, \ v_1(0) = 0,$$
$$x_2(0) = -2 \text{ cm}, \ v_2(0) = 0.$$

(c) See Fig. A.1. Note that the two oscillations are 180° out of phase with each other. In this case, we cannot tell which one is "ahead" and which one "behind".

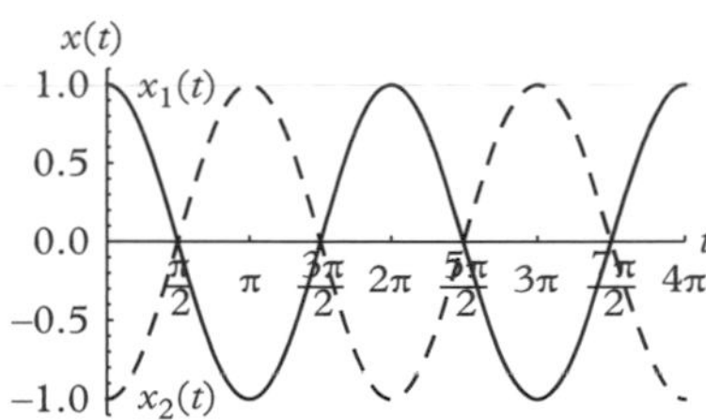

Fig. A.1 *Exercise 2.2(c).*

(d) We can calculate kinetic and potential energies from their definitions. For the potential energy, note that $k = m\omega^2$, and write $PE = \frac{1}{2}kx^2 = \frac{1}{2}m\omega^2 x^2$,

$$K_1(0) = 0, \ K_2(0) = 0,$$
$$U_1(0) = \frac{1}{2}m_1\omega_1^2 x_1(0)^2 = \frac{1}{2} \times 0.2 \text{ kg} \times (1 \text{ s}^{-1})^2 \times (0.02 \text{ m})^2 = 40 \ \mu\text{J},$$
$$U_2(0) = \frac{1}{2}m_2\omega_2^2 x_2(0)^2 = 80 \ \mu\text{J}.$$

2.3 The solution will have the form

$$x(t) = A\cos(\omega t) + B\sin(\omega t),$$

with $\omega = 40\pi$ rad/sec. The velocity will be

$$v(t) = \omega[-A\sin(\omega t) + B\cos(\omega t)].$$

We can now use the initial position and velocity to determine constants A and B. At $t = 0$, $x = 3$ cm, but $v = -10$ cm/s since the velocity is pointed towards the negative x-axis.

$$x(0) = A = 3 \text{ cm}, \text{ and } v(0) = B\omega = -10 \text{ cm/s}.$$

Therefore,

$$A = 3 \text{ cm}, \quad B = -\frac{1}{4\pi} \text{ cm}.$$

This gives the displacement

$$x(t) = (3 \text{ cm}) \cos(40\pi t) - \left(\frac{1}{4\pi} \text{ cm}\right) \sin(40\pi t).$$

If we wish to write this answer in the single cosine form, we will get

$$x(t) = (3.0 \text{ cm}) \cos(40\pi t - 0.027).$$

2.4 Expanding $A\cos(t - \phi)$ gives

$$A\cos(t - \phi) = (A\cos\phi)\cos(t) + (A\sin\phi)\sin(t).$$

Now, equating this to $x(t) = C_1 \cos(t) + C_2 \sin(t)$, we get

$$(C_1 - A\cos\phi)\cos(t) + (C_2 - A\sin\phi)\sin(t) = 0.$$

Since $\sin(t)$ and $\cos(t)$ are orthogonal functions, their coefficients must separately go to zero for their sum to be zero. Therefore,

$$C_1 = A\cos\phi, \ C_2 = A\sin\phi.$$

2.5 Let us get the v from the given $x(t)$,

$$v(t) = \frac{dx}{dt} = -\omega C_1 \sin(\omega t) + \omega C_2 \cos(\omega t).$$

Now, setting $t = 0$ in x and v gives

$$x(0) = C_1, \quad v(0) = \omega C_2.$$

Therefore,

$$C_1 = x_0, \quad C_2 = v_0/\omega.$$

2.6 Since the amplitude is A, the total energy E will be all potential energy when $x = A$, since at that instant $v = 0$. We can write k as $m\omega^2$ or $4\pi^2 mf^2$,

$$E = \frac{1}{2}kA^2 = 2\pi^2 mf^2 A^2.$$

(a) When $x = A/2$, the potential energy will be

$$U = \frac{1}{2}k(A/2)^2 = \frac{1}{4}\left[\frac{1}{2}kA^2\right] = \frac{E}{4}.$$

That is, at this instant, the potential energy is only 25% of the total energy. The potential energy will be given by the following expression:

$$U = \frac{1}{2}\pi^2 mf^2 A^2.$$

(b) Since $K + U = E$, the kinetic energy at this instant will be

$$K = \frac{3}{4}E, \quad \text{that is, 75\% of the total energy.}$$

The expression for K will be

$$K = \frac{3}{2}\pi^2 mf^2 A^2.$$

(c) The lowest speed will be $v = 0$ since the oscillator will come to rest momentarily at the turning points $x = \pm A$.

(d) The largest speed will be when $U = 0$. That happens when $x = 0$. At that instant, the kinetic energy would equal the total energy. This gives the following for the maximum speed:

$$K = E, \implies v_{\max} = 2\pi f A.$$

2.7 Let θ be the counterclockwise angle of rotation from the equilibrium. Then, the rotational equation of motion of the disk for small angles will be

$$I\frac{d^2\theta}{dt^2} = -\kappa\theta,$$

with the moment of inertia $I = \frac{1}{2}mR^2$. This equation of motion leads to harmonic motion with frequency given by

$$\omega = \sqrt{\kappa/I},$$

in analogy with the mass/spring system. The frequency f will be

$$f = \frac{1}{2\pi}\omega = \frac{1}{2\pi}\sqrt{\kappa/I} = \frac{1}{2\pi}\sqrt{2\kappa/mR^2},$$

which upon simplification gives

$$f = \frac{1}{2\pi R}\sqrt{2\kappa/m}.$$

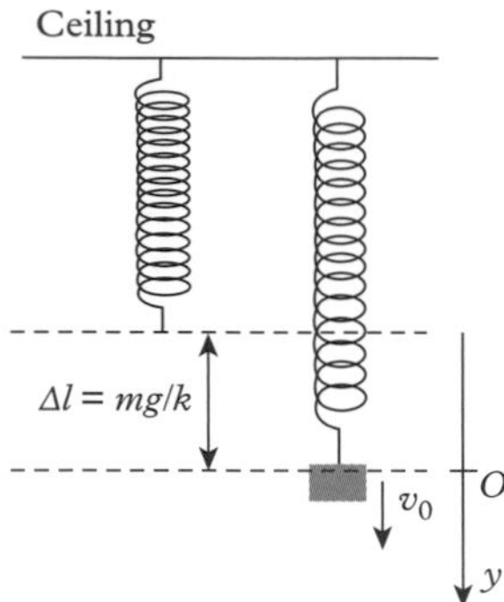

Fig. A.2 *Exercise 2.8.*

2.8 Figure A.2 shows the situation of the block immediately after it was hit with the hammer.

(a) When the block is hung from the spring, the spring stretches. At the static equilibrium, the weight of the block is matched by the spring force from the stretch Δl giving

$$\Delta l = \frac{mg}{k} = \frac{5 \text{ kg} \times 9.81 \text{ m/s}^2}{100 \text{ N/m}} = 0.49 \text{ m}.$$

Therefore, the block will be 50 + 49 = 99 cm from the ceiling and 101 cm from the floor.

(b) We will use the static equilibrium point as the reference for both potential energies, one due to gravity and the other due to the spring force. We will place the origin of the coordinates at the equilibrium place and point the positive y-axis downwards, as shown in the figure. Thus,

$$U_g = 0, \quad U_s = 0. \text{ (Choice of reference)}$$

(c) The kinetic energy

$$K = \frac{1}{2}mv^2 = \frac{1}{2}(5 \text{ kg})(0.4 \text{ m/s})^2 = 400 \text{ mJ}.$$

(d) The change in kinetic energy will be the work done by the applied force since other forces are balanced and their work will cancel out,

$$W = \Delta K = 400 \text{ mJ} - 0 = 400 \text{ mJ}.$$

(e) Since the spring force is the restoring force (gravity is a constant force, always acting downward, and therefore it is not a restoring force here), the frequency will be same as the mass/spring system,

$$f = \frac{1}{2\pi}\sqrt{\frac{k}{m}} = \frac{1}{2\pi}\sqrt{\frac{100 \text{ N/m}}{5 \text{ kg}}} = \frac{\sqrt{5}}{\pi} \text{ Hz}.$$

(f) With y-axis pointed down, the initial $v > 0$, actually $v_y > 0$, but we will use v in place of v_y. Since the motion will be simple harmonic, we will use $y = A\cos(2\pi ft - \phi)$. This would mean $v = -2\pi fA\sin(2\pi ft - \phi)$. Now we satisfy the given initial conditions,

$$y(0) = 0, \implies \cos(\phi) = 0, \implies \phi = \frac{\pi}{2}, -\frac{\pi}{2}.$$

Now, since $v(0) > 0$, $\sin(\phi) > 0$, which will mean that we will have $\phi = +\pi/2$. Therefore, the solution is

$$y = A\cos(2\pi ft - \pi/2) = A\sin(2\pi ft).$$

(g) The solution with y pointed down is $y(t) = A\sin(2\pi ft)$. From the initial velocity we get

$$v(0) = 40 \text{ cm/s} = 2\pi fA.$$

Therefore,

$$A = \frac{40\ \text{cm/s}}{2\pi f} = \frac{40\ \text{cm/s}}{2\sqrt{5}\ \text{s}^{-1}} = 8.9\ \text{cm}$$

2.9 (a) The free-body diagram of the block is shown in Fig. A.3.

Since both forces are in the same direction, they will add, giving the following equation of motion:

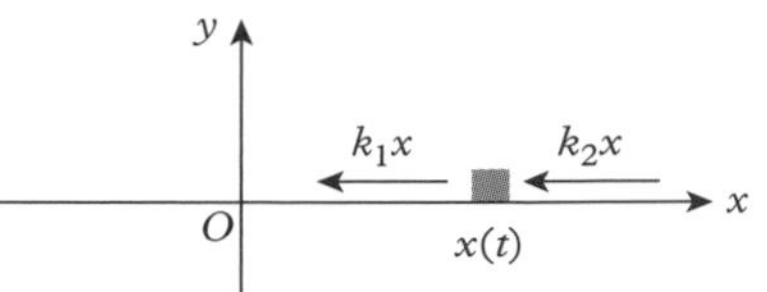

Fig. A.3 *Exercise 2.9.*

$$m\frac{d^2x}{dt^2} = -(k_1 + k_2)x.$$

(b) The equation of motion in (a) is just the equation for a mass attached to one spring with spring constant $k = k_1 + k_2$. Therefore, the angular frequency will be

$$\omega = \sqrt{\frac{k}{m}} = \sqrt{\frac{k_1 + k_2}{m}} = \sqrt{\frac{k_1}{m} + \frac{k_2}{m}} = \sqrt{\omega_1^2 + \omega_2^2}.$$

(c) The above argument also apples to parallel springs attached to a block since both springs will stretch or contract by the same amount x, as happened in this exercise. Therefore,

$$k_{\text{eff}} = k_1 + k_2.$$

(d) The energy of the mass/springs system in part (a) will be

$$E = \frac{1}{2}k_1x^2 + \frac{1}{2}k_2x^2 + \frac{1}{2}m\left(\frac{dx}{dt}\right)^2$$
$$= \frac{1}{2}(k_1 + k_2)x^2 + \frac{1}{2}m\left(\frac{dx}{dt}\right)^2$$

From this it is clear that the square of the angular frequency of oscillation will be

$$\omega^2 = \frac{k_1 + k_2}{m} = \frac{k_1}{m} + \frac{k_2}{m} = \omega_1^2 + \omega_2^2.$$

2.10 (a) The physical situation and the free-body diagrams of the block and glue are shown in Fig. A.4.

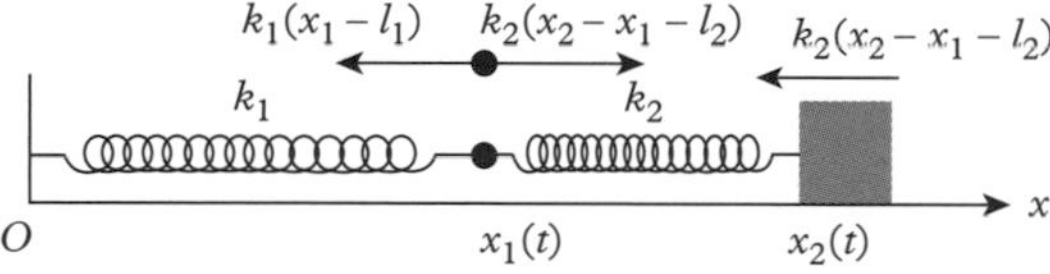

Fig. A.4 *Exercise 2.10. Here l_1 and l_2 are unstretched lengths of the two springs. All coordinates are from the origin at the fixed support.*

There is one force on the block and two forces on the glue. Let the position of the glue be x_1 and that of the block be x_2. Then, the stretching of the two springs are $x_1 - l_1$ and $x_2 - x_1 - l_2$, respectively. Since the position of the block is given by the coordinate x_2, the acceleration of the block will be equal to the second derivative of x_2. Therefore, the equation of motion of the block will be

$$m\frac{d^2x_2^2}{dt^2} = -k_2(x_2 - x_1 - l_2). \quad \text{(A.1)}$$

(b) Since the mass of the glue is negligible, we will get the net force to be zero. This gives

$$k_2(x_2 - x_1 - l_2) - k_1(x_1 - l_1) = 0. \tag{A.2}$$

(c) Solving Eq. A.2 for x_1 we get

$$x_1 = \frac{k_2 x_2 - k_2 l_2 + k_1 l_1}{k_1 + k_2}. \tag{A.3}$$

Put this in the right side of Eq. A.1 and you would get

$$k_2(x_2 - x_1 - l_2) = \frac{k_1 k_2}{k_1 + k_2} x_2 - \frac{k_1 k_2}{k_1 + k_2}(l_1 + l_2).$$

Therefore, the equation of motion of the block becomes

$$m\frac{d^2 x_2^2}{dt^2} = -\frac{k_1 k_2}{k_1 + k_2}(x_2 - l_1 - l_2). \tag{A.4}$$

The variable $x = x_2 - l_1 - l_2$ can be introduced that refers to the displacement of the block with respect to its equilibrium position. In variable x, the equation of motion of the block will be

$$m\frac{d^2 x_2^2}{dt^2} = -\frac{k_1 k_2}{k_1 + k_2} x. \tag{A.5}$$

This will give a harmonic motion of frequency ω whose square will be

$$\omega^2 = \frac{k_1 k_2}{k_1 + k_2} \times \frac{1}{m}.$$

If we write $k_1 = m\omega_1^2$ and $k_2 = m\omega_2^2$ we can show

$$\omega^2 = \frac{\omega_1^2 \omega_2^2}{\omega_1^2 + \omega_2^2}.$$

(d) The equation of motion, Eq. A.5 shows that the effective spring constant is

$$k_{\text{eff}} = \frac{k_1 k_2}{k_1 + k_2} \Longrightarrow \frac{1}{k_{\text{eff}}} = \frac{1}{k_1} + \frac{1}{k_2}.$$

(e) We need the expressions of the KE and PE at an arbitrary instant. For the potential energy we just need to add up the potential energies in the two springs. We will use Eq. A.3 to eliminate x_1 and express the potential energy in terms of x_2,

$$\begin{aligned} U &= \frac{1}{2}k_1(x_1 - l_1)^2 + \frac{1}{2}k_2(x_2 - x_1 - l_2)^2 \\ &= \frac{1}{2}\frac{k_1 k_2}{k_1 + k_2}(x_2 - l_1 - l_2)^2. \end{aligned} \tag{A.6}$$

The kinetic energy would be

$$K = \frac{1}{2}m\left(\frac{dx_2}{dt}\right)^2.$$

We can rewrite KE and PE in terms of the variable $x = x_2 - l_1 - l_2$ so that we study only the displacement with respect to the equilibrium. Thus,

$$K = \frac{1}{2}m\left(\frac{dx}{dt}\right)^2, \quad U = \frac{1}{2}\frac{k_1 k_2}{k_1 + k_2}x^2. \tag{A.7}$$

Clearly, KE and PE are the kinetic energy and potential energy of a simple harmonic oscillator and from the ratio of the constants in the PE and KE we get the square of the frequency to be

$$\omega^2 = \frac{k_1 k_2/(k_1 + k_2)}{m} = \frac{\omega_1^2 \omega_2^2}{\omega_1^2 + \omega_2^2},$$

which is the same as we found above.

2.11 (a) Figure A.5 shows the fluid levels at an arbitrary instant. Suppose the potential energy is zero for the configuration when the fluid is in equilibrium and the fluid is at the same level at both ends, shown dashed in the figure. The figure shows that fluid of mass $\rho A y$ has moved up a height y with respect to the equilibrium situation. Therefore, the potential energy at the instant shown in the figure will be

$$U = (\rho A y)gy = \rho A g y^2. \tag{A.8}$$

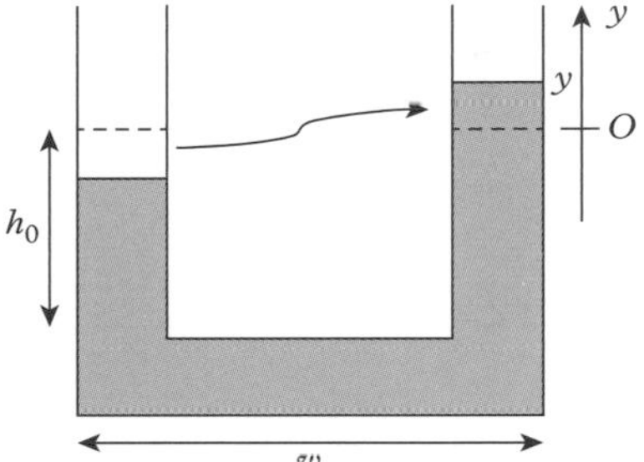

Fig. A.5 *Exercise 2.11.*

(b) Suppose the level on the right is rising at this instant. The speed of the rising level will be dy/dt. Actually, this will be the speed of every particle of fluid since the tube is uniform in cross-section. We have a total mass of fluid $\rho A(w + 2h_0)$ all moving with speed $|dy/dt|$. Therefore, the kinetic energy of the fluid will be

$$K = \frac{1}{2}[\rho A(w + 2h_0)]\left(\frac{dy}{dt}\right)^2. \tag{A.9}$$

(c) From the ratio of the coefficients of the speed squared in K (Eq. A.9) and the displacement squared in U (Eq. A.8) we get the square of the angular frequency ω,

$$\omega^2 = \frac{\rho A g}{\frac{1}{2}\rho A(w + 2h_0)} = \frac{2g}{w + 2h_0}.$$

Therefore, the frequency of oscillation will be

$$f = \frac{1}{2\pi}\sqrt{\frac{2g}{w + 2h_0}}.$$

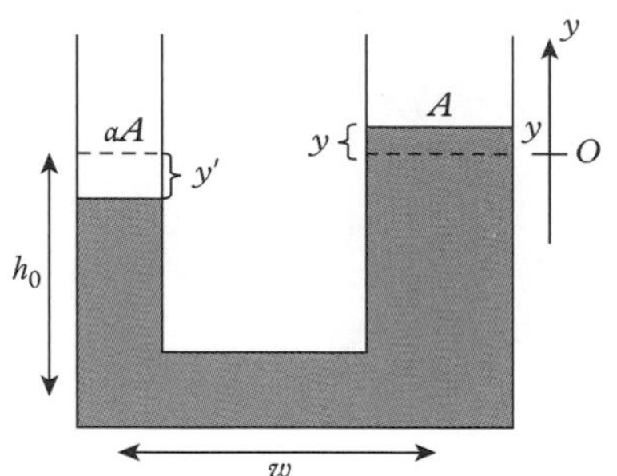

Fig. A.6 *Exercise 2.12.*

2.12 We will use Fig. A.6. Let y' denote the height of water moved on the side of area of cross-section αA when the water in the other arm of area A moves by y.

From the volume conservation of an incompressible fluid, we find that the height moved will be different in the two arms. The volume moved in arm A is Ay and that in the other arm $\alpha Ay'$. They must be equal,

$$(\alpha A)y' = Ay.$$

Therefore,

$$y' = \frac{y}{\alpha}.$$

The distance between the CM of heights moved in the two sides will be

$$\Delta h = \frac{y}{2} + \frac{y'}{2} = \frac{y}{2}\left(1 + \frac{1}{\alpha}\right) = \frac{y}{2}\left(\frac{1+\alpha}{\alpha}\right).$$

The potential energy with respect to zero when the fluid has the same level on the two sides will be

$$U = (\rho Ay)g\Delta h = \frac{\rho Ag(1+\alpha)}{2\alpha}\, y^2.$$

To find the kinetic energy, we note that the fluid on the right and bottom will move with speed $|dy/dt|$, but in the left arm with speed $|dy'/dt|$. We will also assume that $|y| \ll h_0$. Then, the kinetic energy will be

$$K = \frac{1}{2}\,[\rho A h_0]\,(dy/dt)^2 + \frac{1}{2}\,[\rho\alpha A(h_0 + w)]\,(dy'/dt)^2.$$

Expressing y' in terms of y and simplifying we get

$$K = \frac{1}{2}\rho A\left[\frac{w}{\alpha} + \left(\frac{1+\alpha}{\alpha}\right)h_0\right](dy/dt)^2.$$

Therefore, the energy of the fluid will be

$$E = \frac{\rho Ag(1+\alpha)}{2\alpha}\, y^2 + \frac{1}{2}\rho A\left[\frac{w}{\alpha} + \left(\frac{1+\alpha}{\alpha}\right)h_0\right](dy/dt)^2,$$

which has the form of a simple harmonic motion with the square of the angular frequency given by

$$\omega^2 = \frac{\text{coeff. of } y^2}{\text{coeff. of } (dy/dt)^2} = \frac{(1+\alpha)g}{w + (1+\alpha)h_0}.$$

Therefore, the frequency of oscillation will be

$$f = \frac{1}{2\pi}\sqrt{\frac{(1+\alpha)g}{w + (1+\alpha)h_0}}.$$

2.13 Figure A.7 shows the water levels at an arbitrary time. The level on the side with area $A = \pi R^2$ is up by y and that on the other side with area $A' = \pi b^2 R^2$ is down by y'.

From the conservation of volume of an incompressible fluid we must have

$$A'y' = Ay, \implies y' = \frac{1}{b^2}y.$$

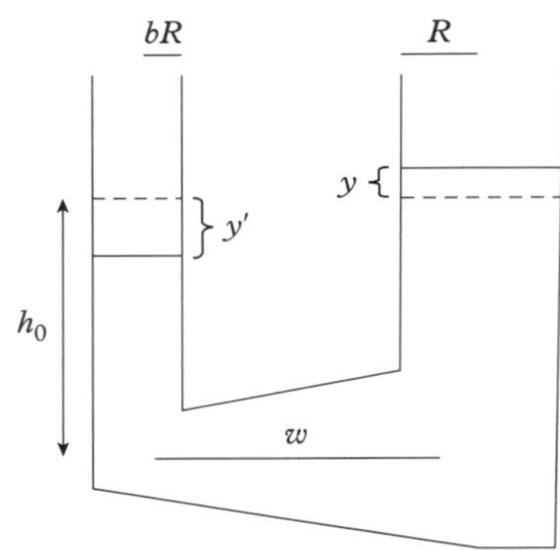

Fig. A.7 *Exercise 2.13.*

The difference between the CM of the fluid that used to be in the bR side and what is now in the R side gives the change in the potential energy with respect to the equilibrium. Taking the potential energy of the equilibrium state to be the reference zero, the potential energy of the fluid at the instant shown in the figure will be

$$U = (\rho\pi R^2 y)g\left(\frac{y}{2} + \frac{y'}{2}\right).$$

This gives

$$U = \left[\rho\pi R^2 g\left(\frac{1+b^2}{2b^2}\right)\right] y^2.$$

Now, we will find the kinetic energy of the fluid at this instant. We separate the calculation into three parts: K_R, K_{bR}, and K_{join} for the kinetic energy of the fluid in the arm with radius R, the arm with radius bR, and the base tube that joins the two arms. Assuming $y \ll h_0$ we can suppose that the length of fluid in the arms is h_0 for each. The speed of fluid in the two arms will be related by the relation between y and y' found above. Let us denote the speeds by the same subscripts,

$$v_{bR} = \frac{1}{b^2}v_R, \text{ with } v_R = dy/dt.$$

Then, ignoring the small problems at the corners and assuming $|y| < h_0$, the kinetic energies in the two arms will be

$$K_R = \frac{1}{2}(\rho\pi R^2 h_0)(dy/dt)^2,$$

$$K_{bR} = \frac{1}{2}(\rho\pi b^2 R^2 h_0)\left[\frac{1}{b^2}(dy/dt)\right]^2.$$

To find the expression for kinetic energy in the joining base, we note that the speed will depend on the location since the area of cross-section is changing. Let $r(x)$ be the radius at coordinate x and let $v(x)$ be the flow speed at that point, as shown in Fig. A.8.

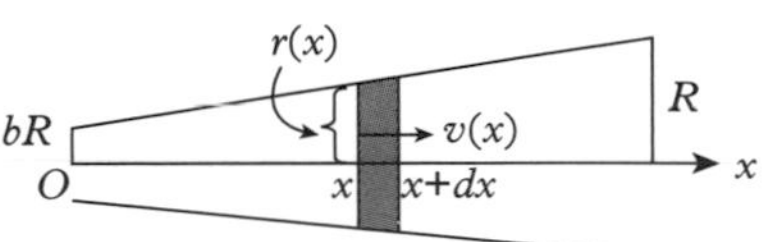

Fig. A.8 *Exercise 2.13.*

The radius $r(x)$ at distance x from the R-end will be

$$r(x) = R\left[b + (1-b)\frac{x}{w}\right].$$

Let $v(x)$ denote the velocity at a distance x from the R-end. Then, due to conservation of volume we will have

$$\pi r(x)^2 v(x) = \pi R^2 v_R.$$

Therefore,

$$v(x) = \frac{dy/dt}{[b + (1-b)x/w]^2}.$$

To find the kinetic energy K_{join} we first write the kinetic energy of a slice of width dx at a distance x from the bR-end,

$$dK_{\text{join}} = \frac{1}{2}(\rho\pi r(x)^2 dx)v(x)^2.$$

Writing this equation explicitly we find

$$dK_{\text{join}} = \frac{1}{2}\left[\rho\pi R^2\left(\frac{dy}{dt}\right)^2\right]\frac{dx}{[b + (1-b)x/w]^2}.$$

Integrating from $x = 0$ to $x = w$ gives the K_{join},

$$K_{\text{join}} = \frac{1}{2}\frac{\rho\pi R^2 w}{b}\left(\frac{dy}{dt}\right)^2.$$

The total kinetic energy will be

$$K = \frac{1}{2}\rho\pi R^2\left(\frac{h_0}{b^2} + h_0 + \frac{w}{b}\right)\left(\frac{dy}{dt}\right)^2.$$

Now, we combine all the kinetic energies and the potential energy to obtain the total energy,

$$E = \left[\frac{1}{2}\rho\pi R^2\left(\frac{h_0}{b^2} + h_0 + \frac{w}{b}\right)\right]\left(\frac{dy}{dt}\right)^2 + \left[\rho\pi R^2 g\left(\frac{1+b^2}{2b^2}\right)\right]y^2.$$

This expression is analogous to the expression for the energy of a simple harmonic oscillator. Therefore, the angular frequency can be readily obtained from the ratio of the coefficients of y^2 and v^2,

$$\omega = \sqrt{\frac{(1+b^2)g}{bw + (1+b^2)h_0}}.$$

You can check for $b = 1$ to see if the answer is the same as that obtained above when the two sides had the same radii.

2.14 (a) Figure A.9 shows a cross-section of the configuration of water in the tank at an arbitrary instant when the level on the right side is at height y_w above the equilibrium level. By symmetry, the level on the left side will be down by y_w, as shown.

The CM at the arbitrary instant can be calculated from the CM at P of the rectangular part CEGF and the CM at Q of the triangular part CDE. The dimension in the direction perpendicular to the figure is b. Let ρ be the density of water.

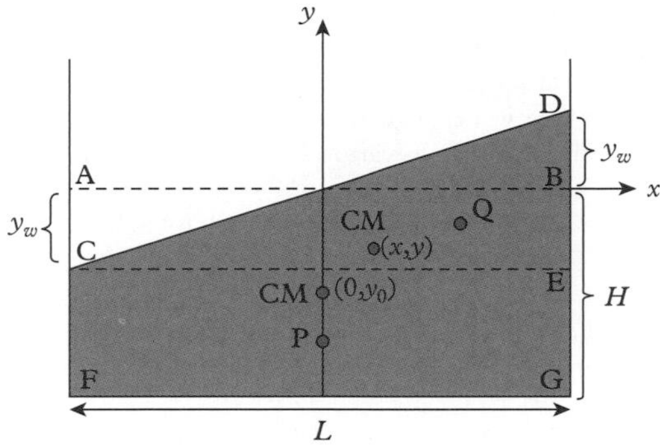

Fig. A.9 *Exercise 2.14. Here $y_0 = H/2$. The CM at an arbitrary instant is at (x, y), which is calculated from the CM of CEGF at P and CDE at Q.*

CEGF part: The mass will be

$$m_1 = \rho(Lb)(H - y_w).$$

The coordinates of the CM of this part will be at

$$x_1 = 0,\ y_1 = \frac{H + y_w}{2}.$$

CDE part: The mass will be

$$m_2 = \rho Lby_w,$$

and the coordinates of the CM of this part will be

$$x_2 = \frac{L}{2} - \frac{L}{3} = \frac{L}{6},\ y_2 = -\frac{1}{3}y_w.$$

Using these we get the (x, y) of the CM of the water in the tank at an arbitrary instant given in the figure to be at

$$x_{cm} = \frac{m_1 x_1 + m_2 x_2}{m_1 + m_2} = \frac{L}{6H}y_w,$$
$$y_{cm} = \frac{m_1 y_1 + m_2 y_2}{m_1 + m_2} = -\frac{1}{2}H + \frac{1}{6}\frac{y_w^2}{H} = -\frac{1}{2}H + \frac{6H}{L^2}x_{cm}^2.$$

(b) Let the zero of the potential energy be at the CM when the fluid is at equilibrium. Then, by the change in the height of the CM of the entire fluid in the tank we get the potential energy at the arbitrary instant to be

$$U = mgh,$$

with $m = \rho LbH$ and $h = y_{cm} - y_0$ with $y_0 = -H/2$,

$$U = \rho LbHg(y_{cm} - y_0) = \rho LbHg\frac{1}{6}\frac{y_w^2}{H} = \frac{1}{6}\rho bLgy_w^2. \tag{A.10}$$

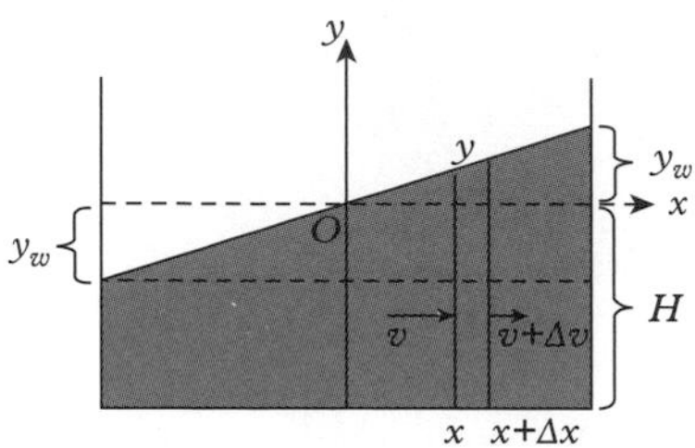

Fig. A.10 *Exercise 2.14.*

(c) Consider a slice between $x = x$ and $x = x + \Delta x$, as in Fig. A.10.

Let the x-component of the velocity at x be v and pointed into the slice and let the velocity at $x + \Delta x$ be $v + \Delta v$ and pointed away from the slice. The height $H + y$ of the slice will change by Δy during an interval Δt due to more liquid flowing into the slice than leaving it. Notice the following in time Δt:

$$\text{volume flowing in} = v\Delta t \times (H + y)b,$$
$$\text{volume flowing out} = (v + \Delta v)\Delta t \times (H + y)b.$$

Therefore, the volume accumulating in this slice can be obtained by subtracting the amount that left the slice from the amount that came into the slice,

$$v\Delta t(H + y)b - (v + \Delta v)\Delta t(H + y)b = \Delta v\Delta t(H + y)b \approx bH\Delta v\Delta t,$$

where I have assumed $y \ll H$. This should equal the volume change resulting in the rise Δy of the strip of width Δx, i.e.

$$b\Delta y\Delta x.$$

Equating the two volume changes gives

$$bH\Delta v\Delta t = b\Delta y\Delta x.$$

Now, dividing both sides by $H\Delta t\Delta x$, we get the desired result. Show the following relation:

$$\frac{\Delta v}{\Delta x} = -\frac{1}{H}\frac{\Delta y}{\Delta t}.$$

Taking $\Delta t \to 0$ limit will give us the following:

$$\frac{dv}{dx} = -\frac{1}{H}\frac{dy}{dt}. \tag{A.11}$$

(d) We suppose the water level at each point rises so that the top is slanted in a straight line. This will mean that dy/dt varies linearly with x, being zero when $x = 0$ and having the value dy_w/dt when $x = L/2$. This will give the following expression for dy/dt when $x = x$:

$$\frac{dy}{dt}(\text{when } x = x) = \frac{x}{L/2}\frac{dy_w}{dt}.$$

(e) Replacing dy/dt in Eq. A.11 and integrating from $x = 0$ to $x = x$ we get

$$v(x) - v(0) = -\left(\frac{1}{HL}\frac{dy_w}{dt}\right)x^2. \tag{A.12}$$

Note: while integrating with respect to x, the velocity dy_w/dt is fixed and comes out of the integral. Evaluating this for $x = L/2$ and setting $v(L/2) = 0$ as the velocity of the fluid particles must be zero since they are reflected at the wall and must switch direction with $v = 0$ at the wall, we get

$$0 - v(0) = -\left(\frac{1}{HL}\frac{dy_w}{dt}\right)(L/2)^2,$$

which gives

$$v(0) = \frac{L}{4H}\frac{dy_w}{dt}.$$

(f) The slice between x and $x + dx$ has width dx, height y, and length b, and moves with velocity $v(x)$. Therefore, the kinetic energy of the fluid particles in this slice will be

$$dK = \frac{1}{2}\left[\rho b(H + y)dx\right] v(x)^2.$$

With $y \ll H$, we can ignore y in comparison to H and get

$$dK = \frac{1}{2}\rho bHv(x)^2 dx,$$

and $v(x)$ is given in Eq. A.12,

$$dK = \frac{1}{2}\rho bH\left(\frac{L}{4H}\frac{dy_w}{dt}\right)^2\left(1 - \frac{8}{L^2}x^2 + \frac{16}{L^4}x^4\right)\, dx.$$

(g) To integrate dK along the width of the tank, we need the integral

$$\int_{-L/2}^{L/2}\left(1 - \frac{8}{L^2}x^2 + \frac{16}{L^4}x^4\right)\, dx.$$

This integral is easily done by changing $z = 2L/x$ to

$$L\int_0^1 \left(1 - 2z^2 + z^4\right)\, dz = \frac{8}{15}L.$$

Thus, integration of dK from $x = -L/2$ to $L/2$ will give the total kinetic energy,

$$K = \left[\frac{\rho bL^3}{60H}\right]\left(\frac{dy_w}{dt}\right)^2.$$

(h) Take the ratio of the coefficient of y_w^2 in potential energy in Eq. A.10 and the coefficient of (dy_w/dt) in kinetic energy gives us the square of the angular frequency,

$$\omega^2 = \frac{\frac{1}{6}\rho bLg}{\rho bL^3/(60H)} = \frac{10gH}{L^2}.$$

2.15 (a) Kater's pendulum is a physical pendulum. From our discussions in the chapter you know that the period of oscillations of the pendulum will depend on the moment of inertia about the suspension point. Let us recall the formula here,

$$T = 2\pi\sqrt{\frac{I}{Mgl}},$$

where l is the distance to the CM. Let I_0 be the moment of inertia of Kater's pendulum about its CM and I_1 and I_2 be the moments of inertia about O_1 and O_2. Let k be the radius of gyration. Then, we have the following formulas for the moments of inertia:

$$I_0 = Mk^2, \quad I_1 = I_0 + Ml_1^2, \quad I_2 = I_0 + Ml_2^2,$$

where l_1 and l_2 are the distance to the CM from O_1 and O_2, respectively. Therefore, the periods T_1 and T_2 will be

$$T_1 = 2\pi\sqrt{(I_0 + Ml_1^2)/Mgl_1}, \quad T_2 = 2\pi\sqrt{(I_0 + Ml_2^2)/Mgl_2}.$$

Using $I_0 = Mk^2$ and simplifying we get

$$T_1 = 2\pi\sqrt{(k^2 + l_1^2)/gl_1}, \quad T_2 = 2\pi\sqrt{(k^2 + l_2^2)/gl_2}.$$

(b) Equating T_1 to $T-2$ gives us the condition when the two periods will be equal. This would give us a relation between k and l_1 and l_2,

$$T_1 = T_2 \implies \frac{k^2 + l_1^2}{gl_1} = \frac{k^2 + l_2^2}{gl_2}.$$

Simplifying this we get

$$k^2 = l_1 l_2.$$

Putting this in either T_1 or T_2 will give the required period,

$$T = 2\pi\sqrt{\frac{k^2 + l_1^2}{gl_1}} = 2\pi\sqrt{\frac{l_1 l_2 + l_1^2}{gl_1}} = 2\pi\sqrt{\frac{l_1 + l_2}{g}}.$$

With $L = l_1 + l_2$ this is

$$T = 2\pi\sqrt{\frac{L}{g}}.$$

2.16 (a) To find the minima of the potential energy function, firstly we set the first derivative to zero and then test each extremum by the second derivative test to distinguish minima from maxima. With $U(x) = ax(x-1)^2$ we get

$$\frac{dU}{dx} = a(x-1)^2 + 2ax(x-1) = a(x-1)(3x-1).$$

Therefore $dU/dx = 0$ would be

$$a(x-1)(3x-1) = 0,$$

which has two solutions,

$$x = 1, \ \frac{1}{3}.$$

Now we find the expression for the second derivative to test each solution,

$$\frac{d^2U}{dx^2} = a(3x-1) + 3a(x-1).$$

We have $a > 0$ here. Therefore, we have

$$\left(\frac{d^2U}{dx^2}\right)_{x=1} = a(3-1) = 2a > 0,$$

$$\left(\frac{d^2U}{dx^2}\right)_{x=1/3} = a \times 3 \times \left(\frac{1}{3} - 1\right) = -2a < 0.$$

Therefore, $x = 1$ is a minimum and $x = \frac{1}{3}$ is a maximum.

(b) Expanding $U(x)$ in a Taylor series around $x = 1$ we get

$$\begin{aligned} U(x) &= U(1) + \left(\frac{dU}{dx}\right)_{x=1} (x-1) + \frac{1}{2!}\left(\frac{d^2U}{dx^2}\right)_{x=1} (x-1)^2 + \cdots \\ &= \frac{1}{2} 2a(x-1)^2 + \cdots . \end{aligned}$$

The potential energy near $x = 1$ is therefore the same as the potential energy of a simple harmonic oscillator with "spring constant" k given by

$$k = 2a.$$

Therefore, the frequency of oscillations will be

$$f = \frac{1}{2\pi}\sqrt{2a/m}.$$

(c) Yes. Since $x = \frac{1}{3}$ is a maximum, the x of the particle must not get less than $x = \frac{1}{3}$, otherwise the particle will roll down the hill. Therefore, the potential well around the $x = 1$ minimum for harmonic oscillations is between the following x values:

$$\frac{1}{3} < x < 1 + \frac{1}{3}.$$

2.17 In each of the given cases we will need to check ω_0 versus $\frac{\Gamma}{2}$ to decide whether the given system is under-damped, over-damped, or critically damped.

(a) The values are

$$\omega_0 = \sqrt{\frac{k}{m}} = \sqrt{\frac{10 \text{ N/m}}{0.2 \text{ kg}}} = \sqrt{50} \text{ sec}^{-1} \sim 7.1 \text{ sec}^{-1},$$

$$\Gamma = \frac{b}{m} = \frac{6 \text{ kg/s}}{0.2 \text{ kg}} = 30 \text{ sec}^{-1} \implies \frac{\Gamma}{2} = 15 \text{ sec}^{-1}.$$

Since $\omega_0 < \frac{\Gamma}{2}$, the oscillator is over-damped.

(b) The values are

$$\omega_0 = \sqrt{\frac{k}{m}} = \sqrt{\frac{20\text{ N/m}}{0.2\text{ kg}}} = 10\text{ sec}^{-1},$$

$$\Gamma = \frac{b}{m} = \frac{8\text{ kg/s}}{0.2\text{ kg}} = 40\text{ sec}^{-1} \implies \frac{\Gamma}{2} = 20\text{ sec}^{-1}.$$

Since $\omega_0 < \frac{\Gamma}{2}$, the oscillator is over-damped.

(c) The values are

$$\omega_0 = \sqrt{\frac{k}{m}} = \sqrt{\frac{32\text{ N/m}}{0.8\text{ kg}}} = \sqrt{40}\text{ sec}^{-1} \sim 6.3\text{ sec}^{-1},$$

$$\Gamma = \frac{b}{m} = \frac{8\text{ kg/s}}{0.8\text{ kg}} = 10\text{ sec}^{-1} \implies \frac{\Gamma}{2} = 5\text{ sec}^{-1}.$$

Since $\omega_0 > \frac{\Gamma}{2}$, the oscillator is under-damped.

(d) The values are

$$\omega_0 = \sqrt{\frac{k}{m}} = \sqrt{\frac{80\text{ N/m}}{0.8\text{ kg}}} = 10\text{ sec}^{-1}.$$

$$\Gamma = \frac{b}{m} = \frac{16\text{ kg/s}}{0.8\text{ kg}} = 20\text{ sec}^{-1} \implies \frac{\Gamma}{2} = 10\text{ sec}^{-1}.$$

Since $\omega_0 = \frac{\Gamma}{2}$, the oscillator is critically damped.

2.18 (a) We will illustrate the method for part (a) and you can carry out the others in a similar fashion. Since this oscillator is over-damped, we need to use the solution corresponding to the over-damped case,

$$x(t) = e^{-\frac{\Gamma}{2}t}\left(C_1 e^{-\alpha t} + C_2 e^{\alpha t}\right), \tag{A.13}$$

with

$$\omega_0 = \sqrt{50}\text{ sec}^{-1}, \quad \frac{\Gamma}{2} = 15\text{ sec}^{-1}, \quad \alpha = \sqrt{(\Gamma/2)^2 - \omega_0^2} = \sqrt{175}\text{ sec}^{-1}.$$

The velocity will be

$$v = \frac{dx}{dt} = -\frac{\Gamma}{2}e^{-\frac{\Gamma}{2}t}\left(C_1 e^{-\alpha t} + C_2 e^{\alpha t}\right) + \alpha e^{-\frac{\Gamma}{2}t}\left(-C_1 e^{-\alpha t} + C_2 e^{\alpha t}\right).$$

Now, we apply the initial conditions of x and v, i.e. $x(0) = 1$ and $v(0) = 0$. This gives

$$C_1 + C_2 = 1,$$

$$-\frac{\Gamma}{2}(C_1 + C_2) + \alpha(-C_1 + C_2) = 0.$$

Solving these equations yields

$$C_1 = \frac{\alpha - \Gamma/2}{2\alpha},$$
$$C_2 = 1 - C_1 = \frac{\alpha + \Gamma/2}{2\alpha}.$$

Therefore, we have the two constants in the solution in Eq. A.13. Using the numerical values of Γ and α gives the required numerical values in the solution,

$$C_1 = \frac{\sqrt{175} - 15}{2\sqrt{175}} = -0.067, \quad C_2 = \frac{\sqrt{175} + 15}{2\sqrt{175}} = 1.067.$$

Therefore, the displacement $x(t)$ will be

$$x(t) = e^{-15t}\left(-0.067e^{-13.23t} + 1.067e^{13.23t}\right).$$

(b) Try $x(t) = e^{-\frac{\Gamma}{2}t}(C_1 e^{-\alpha t} + C_2 e^{\alpha t})$.

(c) Try $x(t) = e^{-\frac{\Gamma}{2}t}(C_1 \cos(\omega_1 t) + C_2 \sin(\omega_1 t))$, with $\omega_1 = \sqrt{\omega_0^2 - (\Gamma/2)^2}$.

(d) Try $x(t) = e^{-\frac{\Gamma}{2}t}(C_1 + C_2 t)$.

2.19 We can read off $\Gamma/2$ from the exponent of the exponential and ω_1 from the argument of the cosine. Since $\omega_1 \gg \Gamma$, we will set $\omega_0 = \omega_1$. From ω_0 and Γ, we get Q by

$$Q = \frac{\omega_0}{\Gamma}.$$

In (a) we have

$$\omega_0 = 2\pi \text{ sec}^{-1}, \quad \frac{\Gamma}{2} = 0.1 \text{ sec}^{-1} \implies \Gamma = 0.2 \text{ sec}^{-1}.$$

Therefore, $Q = \omega_0/\Gamma = 10\pi \approx 31$.
Similarly, Q for other parts can be obtained. They are (b) 25, (c) 400π, (d) 209,440.

2.20 (a) We plot five cycles of oscillations. in Fig. A.11. We see that the oscillations decay with time as expected for an underdamped oscillator.

(b) In Fig. A.11 the envelope of the positive and negative amplitudes is shown. They are plots of $2e^{-0.1t}$ and $-2e^{-0.1t}$. Either can be used to determine the time to drop by e^{-1},

$$e^{-0.1t} = e^{-1}.$$

Therefore, the time for the envelope to drop by a factor of $1/e$ is $t = 10$ sec.

(c) Setting $e^{0.1t} = e^{-2}$ gives $t = 20$ sec.

(d) From the expression of $x(t)$ we have $\omega_1 = 2\pi \text{ sec}^{-1}$ and $\Gamma = 2 \times 0.1 = 0.2 \text{ sec}^{-1}$. Since $\omega_1 \gg \Gamma/2$, we will approximate $\omega_0 \approx \omega_1$. Therefore, the Q of the oscillator is

$$Q = \frac{\omega_0}{\Gamma} = 10\pi \approx 31.$$

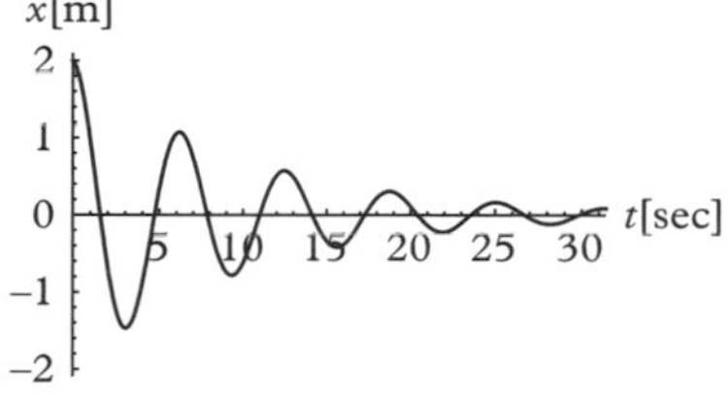

Fig. A.11 *Exercise 2.20. The solid line is the plot of x(t) and the dashed lines constitute the envelope.*

(e) The rate of dissipation of energy is proportional to the energy at that point in time,

$$\frac{dE}{dt} = -\Gamma E.$$

At $t = 0$, the displacement $x(0) = 2$ m and the velocity is

$$v(0) = \left.\frac{dx}{dt}\right|_{t=0} = -0.2 \text{ m/s}.$$

Therefore, the total energy at $t = 0$ will be

$$E(0) = \frac{1}{2}mv(0)^2 + \frac{1}{2}kx(0)^2.$$

Putting in the numerical values we get

$$E(0) = \frac{1}{2} \times 0.25 \text{ kg} \times (-0.2 \text{ m/s})^2 + \frac{1}{2}[0.25 \text{ kg} \times (2\pi \text{ s}^{-2})^2](2 \text{ m})^2 \approx 2\pi^2 \text{ J}.$$

Therefore,

$$\left.\frac{dE}{dt}\right|_{t=0} = -\Gamma E(0) = -0.4\pi^2 \text{ J/s}.$$

(f) From $x(t)$ we already know $\omega_1 = 2\pi$ sec^{-1}. Therefore, the frequency of oscillation is

$$f = \frac{\omega_1}{2\pi} = 1 \text{ Hz}.$$

(g) Since the energy of the oscillator drops in time exponentially as $E \sim e^{-\Gamma t}$, 90% of energy will be lost in time for the t obeyed by

$$e^{-\Gamma t} = 0.9.$$

Take the natural log of both sides to get

$$t = \frac{-\ln(0.9)}{\Gamma} = 0.53 \text{ sec}.$$

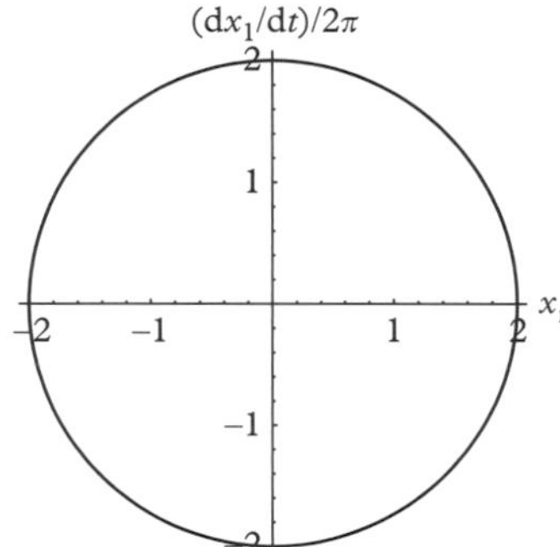

Fig. A.12 *Exercise 2.21 (a).*

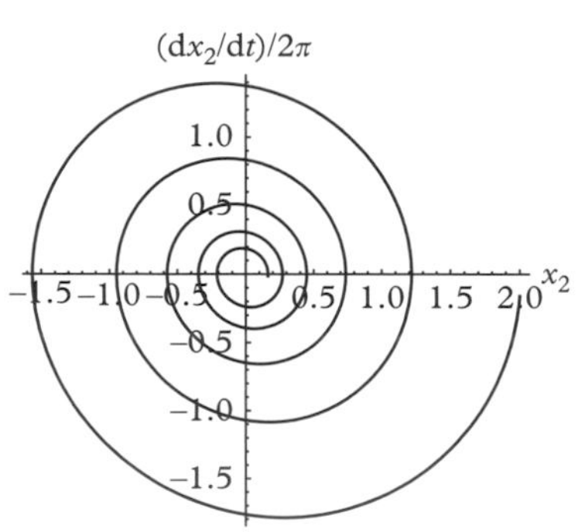

Fig. A.13 *Exercise 2.21 (b).*

2.21 (a) Figure A.12 shows the parametric plot of position versus velocity. The curve is closed and repeats with time without decaying to zero velocity since there is no damping in this oscillator.

(b) Figure A.13 shows the parametric plot of position versus velocity. The curve is not closed but rather spirals in and will end up at $x = 0$, $dx/dt = 0$ as $t \to \infty$ due to damping.

(c) The two plots in Figs. A.12 and A.13 show that the trajectory in the (x, v) plane for the damped case spirals to the origin while the trajectory for the ideal case stays stable. Thus, when there is no damping, the oscillations continue for ever, but when there is damping, the oscillations will eventually stop.

2.22 (a) From the given figure we read 10 oscillations take 1.0 sec. Therefore, the period T is 0.1 sec.

(b) Reading the envelope of the oscillations shows that only 20% of the amplitude is left after 1 sec. Therefore, we must have

$$e^{-\frac{\Gamma}{2}\times 1\,\text{sec}} = 0.2.$$

Therefore,

$$-\frac{\Gamma}{2} = \ln(0.2)\ \text{sec}^{-1} \implies \Gamma = 3.2\ \text{sec}^{-1}.$$

(c) From T we can get ω_1,

$$\omega_1 = \frac{2\pi}{T} = 20\pi\ \text{sec}^{-1}.$$

Since $\omega_1 \gg \Gamma$, we can set $\omega_0 \approx \omega_1$. Therefore, the Q of the oscillator will be

$$Q = \frac{\omega_0}{\Gamma} = \frac{20\pi}{3.2} = 19.6.$$

(d) Energy decays as $\sim e^{-\Gamma t}$. Therefore, set this to e^{-1} to obtain the time,

$$e^{-\Gamma t} = e^{-1},$$

which gives

$$t = \frac{1}{\Gamma} = \frac{1}{3.2\ \text{sec}^{-1}} = 313\ \text{msec}.$$

This is different than the decay rate of $x(t)$ since energy goes as x^2.

2.23 (a) Figure A.14 shows the situation for the problem. The left side of the figure is the equilibrium situation. Let us mark the level of the block as dashed and we follow the motion of the block by locating the position of the dashed line. This way, we

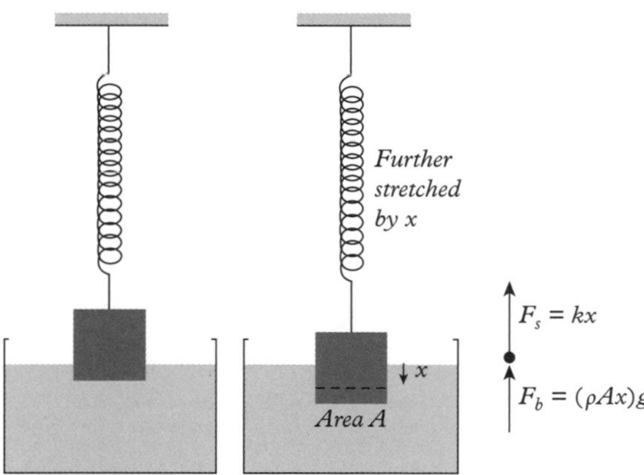

Fig. A.14 *Exercise 2.23.*

would not need to worry about gravity since its effect would already be included. Let us take the level of fluid to be the origin and point the positive x-axis facing down. Then, the stretch of the spring with respect to the equilibrium will be given by the x-coordinate of the dashed line on the block.

From the free-body diagram we get the following equation of motion for the block:

$$m\frac{d^2x}{dt^2} = -kx - \rho Axg = -(k + \rho Ag)\, x.$$

This is the equation of motion of a simple harmonic motion with angular frequency,

$$\omega_0 = \sqrt{\frac{k + \rho Ag}{m}}.$$

(b) When there is a drag during the motion of the block we will get

$$m\frac{d^2x}{dt^2} = -kx - \rho Axg - b\frac{dx}{dt}.$$

If $\omega_0 > (b/2m)$, then the new frequency will be

$$\omega_1 = \sqrt{\left(\frac{k + \rho Ag}{m}\right)^2 - \left(\frac{b}{2m}\right)^2}.$$

Chapter 3

3.1 For the given problem the positions of the two pendulums A and B can be written in terms of the normal mode frequencies ω_1 and ω_2 for this system,

$$\omega_1 = \omega_0 = 2\ \text{sec}^{-1},\quad \omega_2 = \sqrt{\omega_0^2 + 2\omega_c^2} = \sqrt{4 + 2} = \sqrt{6}\ \text{sec}^{-1}.$$

The general solution will be a superposition of the normal modes. Let us write the general solution in the form of sines and cosines,

$$x_A = [C_1 \cos(\omega_1 t) + D_1 \sin(\omega_1 t)] + [C_2 \cos(\omega_2 t) + D_2 \sin(\omega_2 t)],$$
$$x_B = [C_1 \cos(\omega_1 t) + D_1 \sin(\omega_1 t)] - [C_2 \cos(\omega_2 t) + D_2 \sin(\omega_2 t)].$$

The corresponding velocities will be

$$v_A = \omega_1[-C_1 \sin(\omega_1 t) + D_1 \cos(\omega_1 t)] + \omega_2[-C_2 \sin(\omega_2 t) + D_2 \cos(\omega_2 t)],$$
$$v_B = \omega_1[-C_1 \sin(\omega_1 t) + D_1 \cos(\omega_1 t)] - \omega_2[-C_2 \sin(\omega_2 t) + D_2 \cos(\omega_2 t)].$$

Now, we use the given initial conditions: $x_A(0) = 0$, $v_A = 0$, $x_B(0) = 1$, and $v_B(0) = 0$, to obtain the following relations for the coefficients:

$$C_1 + C_2 = 0,$$
$$\omega_1 D_1 + \omega_2 D_2 = 0,$$
$$C_1 - C_2 = 1,$$
$$\omega_1 D_1 - \omega_2 D_2 = 0.$$

Solving these equations simultaneously we get the values of the coefficients in cm to be

$$C_1 = \frac{1}{2}\ \text{cm},\ \ C_2 = -\frac{1}{2}\ \text{cm},\ \ D_1 = 0,\ \ D_2 = 0.$$

Therefore, the positions of the two pendulums at the instant t will be

$$x_A = \frac{1}{2}\ \text{cm}\ \cos(2t) - \frac{1}{2}\ \text{cm}\ \cos(\sqrt{6}t),$$

$$x_B = \frac{1}{2}\ \text{cm}\ \cos(2t) + \frac{1}{2}\ \text{cm}\ \cos(\sqrt{6}t),$$

where t is in seconds.

3.2 Figure A.15 shows the configuration of the two pendulums at an arbitrary instant.

Let the spring be attached at length fl. In the given problem $f = \frac{1}{2}$, but I want to be more general at first. From the figure it is clear that the change in length of the spring will be $f(x_B - x_A)$. Therefore, the x-components of the equations of motion of the two pendulum bobs are

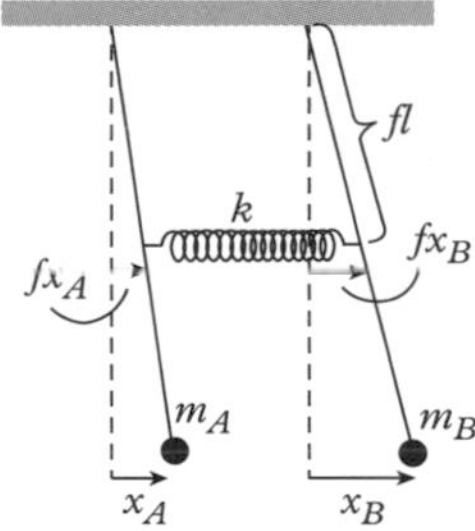

Fig. A.15 *Exercise 3.2. Here f is the fraction of the total length l where the spring is attached. In the problem $f = \frac{1}{2}$.*

$$m_A \frac{d^2x_A}{dt^2} = -\frac{m_A g}{l}\, x_A + k\,[f(x_B - x_A)],$$

$$m_B \frac{d^2x_B}{dt^2} = -\frac{m_B g}{l}\, x_B - k\,[f(x_B - x_A)].$$

We have $m_A = m_B = m$. Let

$$\omega_0^2 = \frac{g}{l},\ \ \omega_c^2 = \frac{fk}{m}.$$

Rewriting the equations of motion in terms of ω_0 and ω_c we get

$$\frac{d^2x_A}{dt^2} = -(\omega_0^2 + \omega_c^2)x_A + \omega_c^2 x_B,$$

$$\frac{d^2x_B}{dt^2} = -(\omega_0^2 + \omega_c^2)x_B + \omega_c^2 x_A,$$

which have the same form as the equations of two identical pendulums when they are connected directly to the bobs. We will use the analogy to get answers to various parts.

(a) The normal modes will be

$$\omega_1 = \omega_0 = \sqrt{g/l},$$

$$\omega_2 = \sqrt{\omega_0^2 + 2\omega_c^2} = \sqrt{\frac{g}{l} + \frac{2fk}{m}}.$$

Setting $f = 1/2$ we get the mode frequencies when the spring is connected at half the length,

$$\omega_1 = \sqrt{g/l},$$

$$\omega_2 = \sqrt{\frac{g}{l} + \frac{k}{m}}.$$

The normal mode amplitudes will be

$$(B/A)_1 = +1, \quad (B/A)_2 = -1,$$

which are independent of f.

(b) Modes are the same, but the frequencies are different now. The effective coupling is fk here versus k when spring is attached to masses. Hence, the bobs are less strongly coupled now.

(c) In this case $f = 0$. This would make $\omega_2 = \omega_0 = \omega_1$. There will be only one normal mode, the mode of a single pendulum, since each pendulum is independent of the other.

(d) In this case $f = a/l$.

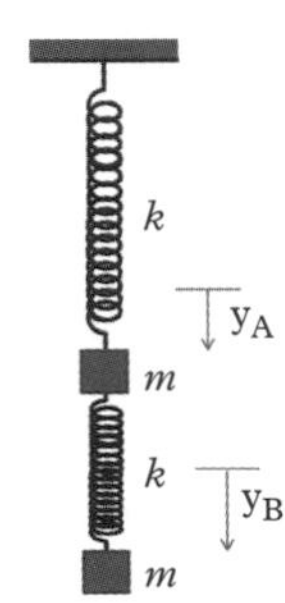

Fig. A.16 *Exercise 3.3.*

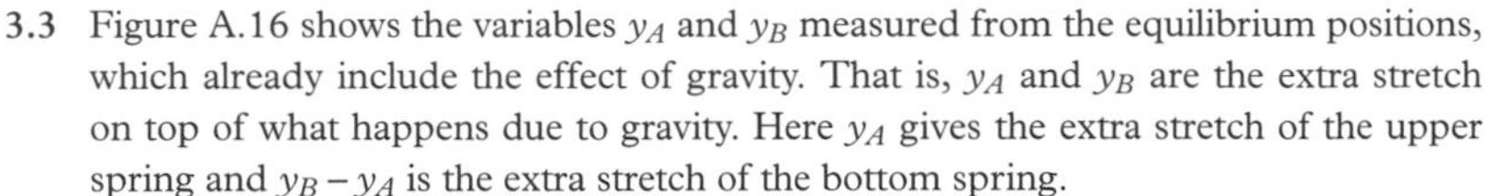

3.3 Figure A.16 shows the variables y_A and y_B measured from the equilibrium positions, which already include the effect of gravity. That is, y_A and y_B are the extra stretch on top of what happens due to gravity. Here y_A gives the extra stretch of the upper spring and $y_B - y_A$ is the extra stretch of the bottom spring.

(a) The equations of motion of the two blocks will be

$$m\frac{d^2 y_A}{dt^2} = -ky_A + k(y_B - y_A),$$

$$m\frac{d^2 y_B}{dt^2} = -k(y_B - y_A).$$

(b) Let us rewrite the equations of motion using

$$\omega_0^2 = \frac{k}{m}.$$

We now have a simplified form of the equations of motion,

$$\frac{d^2 y_A}{dt^2} = -\omega_0^2(2y_A - y_B),$$

$$\frac{d^2 y_B}{dt^2} = -\omega_0^2(y_B - y_A).$$

To find the normal modes, we will seek a solution of the form,

$$y_A = A\cos(\omega t), \quad y_B = B\cos(\omega t).$$

These convert the equations of motion into equations for A, B, and ω,

$$-\omega^2 A = -\omega_0^2(2A - B),$$
$$-\omega^2 B = -\omega_0^2(B - A).$$

Let $\lambda = \omega^2/\omega_0^2$, and rewrite these equations to obtain

$$(\lambda - 2)A + B = 0, \tag{A.14}$$

$$A + (\lambda - 1)B = 0. \tag{A.15}$$

Eliminating A and B we get

$$\lambda^2 - 3\lambda + 1 = 0,$$

whose solutions are

$$\lambda_1 = \frac{3-\sqrt{5}}{2}, \quad \lambda_2 = \frac{3+\sqrt{5}}{2}.$$

The mode frequencies for these modes will be

$$\omega_1 = \left(\frac{3-\sqrt{5}}{2}\right)^{1/2} \omega_0,$$

$$\omega_2 = \left(\frac{3+\sqrt{5}}{2}\right)^{1/2} \omega_0.$$

The ratios of amplitudes for oscillations in the two modes can be obtained by using λ_1 and λ_2 in either Eq. A.14 or A.15 respectively,

$$\left(\frac{B}{A}\right)_1 = 2 - \lambda_1 = \frac{1+\sqrt{5}}{2} > 0,$$

$$\left(\frac{B}{A}\right)_2 = 2 - \lambda_2 = \frac{1-\sqrt{5}}{2} < 0.$$

(c) In mode 1 both masses move in the same direction and in mode 2 they move in opposite directions. The ratios B/A give the ratio of their amplitudes. For instance, for a motion in mode 1, if $y_A = \cos(\omega_1 t)$, then $y_B = (B/A)_1 \cos(\omega_1 t) = \frac{1+\sqrt{5}}{2}\cos(\omega_1 t)$. Similarly, for a motion in mode 2, if $y_A = \cos(\omega_2 t)$, then $y_B = (B/A)_2 \cos(\omega_2 t) = \frac{1-\sqrt{5}}{2}\cos(\omega_2 t)$. The sketch is left for the student to complete.

(d) Let us write the solutions in the form of sines and cosines and use a single symbol to denote the amplitude ratios. Let

$$\alpha_1 \equiv \left(\frac{B}{A}\right)_1, \quad \alpha_2 \equiv \left(\frac{B}{A}\right)_2.$$

Then, the general solution will be

$$y_A = C_1 \cos(\omega_1 t) + D_1 \sin(\omega_1 t) + C_2 \cos(\omega_2 t) + D_2 \sin(\omega_2 t),$$
$$y_B = \alpha_1 \left[C_1 \cos(\omega_1 t) + D_1 \sin(\omega_1 t)\right] + \alpha_2 \left[C_2 \cos(\omega_2 t) + D_2 \sin(\omega_2 t)\right].$$

The corresponding velocities will be

$$v_A = \omega_1[-C_1 \sin(\omega_1 t) + D_1 \cos(\omega_1 t)] + \omega_2[-C_2 \sin(\omega_2 t) + D_2 \cos(\omega_2 t)],$$
$$v_B = \omega_1\alpha_1[-C_1 \sin(\omega_1 t) + D_1 \cos(\omega_1 t)] + \omega_2\alpha_2[-C_2 \sin(\omega_2 t) + D_2 \cos(\omega_2 t)].$$

Now, we use the initial conditions at $t = 0$ to obtain the following:

$$y_A(0) = C_1 + C_2 = 0,$$
$$y_B(0) = \alpha_1 C_1 + \alpha_2 C_2 = 1,$$
$$v_A(0) = \omega_1 D_1 + \omega_2 D_2 = 0,$$
$$v_B(0) = \alpha_1\omega_1 D_1 + \alpha_2\omega_2 D_2 = \frac{1}{2}.$$

These can be solved for the coefficients,

$$C_1 = -C_2 = 2\omega_1 D_1 = -2\omega_2 D_2 = \frac{1}{\alpha_1 - \alpha_2} = \frac{1}{\sqrt{5}}.$$

These coefficients fix the values in the y_A and y_B and we have definite values of y_A and y_B at an arbitrary instant t.

3.4 (a) The equations of motion will be

$$m_A \frac{d^2 x_A}{dt^2} = -kx_A - k(x_A - x_B),$$
$$m_B \frac{d^2 x_B}{dt^2} = k(x_A - x_B) - kx_B.$$

Let us introduce the following symbols:

$$\alpha^2 \equiv \frac{k}{m_A}, \quad \beta^2 \equiv \frac{k}{m_B}.$$

Then, the equations of motion are

$$\frac{d^2 x_A}{dt^2} = -2\alpha^2 x_A + \alpha^2 x_B,$$
$$\frac{d^2 x_B}{dt^2} = \beta^2 x_A - 2\beta^2 x_B.$$

To find the normal modes we will set

$$x_A = A\cos(\omega t), \quad x_B = B\cos(\omega t).$$

This turns the equations of motion into

$$-\omega^2 A = -2\alpha^2 A + \alpha^2 B, \tag{A.16}$$
$$-\omega^2 B = \beta^2 A - 2\beta^2 B. \tag{A.17}$$

We find that these equations give two different expressions for B/A,

$$\frac{B}{A} = \frac{-\omega^2 + 2\alpha^2}{\alpha^2}, \tag{A.18}$$
$$\frac{B}{A} = \frac{\beta^2}{-\omega^2 + 2\beta^2}. \tag{A.19}$$

Therefore,

$$\frac{-\omega^2 + 2\alpha^2}{\alpha^2} = \frac{\beta^2}{-\omega^2 + 2\beta^2}.$$

This gives the two values of ω^2,

$$\omega^2 = \alpha^2 + \beta^2 \pm \sqrt{\alpha^4 + \beta^4 - \alpha^2\beta^2}.$$

Therefore, the angular frequencies of the two normal modes are

$$\omega_1 = \sqrt{\alpha^2 + \beta^2 - \sqrt{\alpha^4 + \beta^4 - \alpha^2\beta^2}},$$
$$\omega_2 = \sqrt{\alpha^2 + \beta^2 + \sqrt{\alpha^4 + \beta^4 - \alpha^2\beta^2}}.$$

Putting $\omega = \omega_1$ and $\omega = \omega_2$ into Eq. A.18 or A.19 gives the ratio of the mode amplitudes for the two normal modes. For mode 1 we get

$$\left(\frac{B}{A}\right)_1 = \frac{-\omega_1^2 + 2\alpha^2}{\alpha^2} = \frac{\alpha^2 - \beta^2 + \sqrt{\alpha^4 + \beta^4 - \alpha^2\beta^2}}{\alpha^2}.$$

For mode 2 we get

$$\left(\frac{B}{A}\right)_2 = \frac{-\omega_2^2 + 2\alpha^2}{\alpha^2} = \frac{\alpha^2 - \beta^2 - \sqrt{\alpha^4 + \beta^4 - \alpha^2\beta^2}}{\alpha^2}.$$

(b) Draw modes for the case $m_B = 2m_A$. When $m_B = 2m_A$, $\alpha^2 = 2\beta^2$. We use this to find that

$$\omega_1 = \beta\left(3 - \sqrt{3}\right)^{1/2}, \quad \omega_2 = \beta\left(3 + \sqrt{3}\right)^{1/2}.$$

$$\left(\frac{B}{A}\right)_1 = \frac{1 + \sqrt{3}}{2},$$
$$\left(\frac{B}{A}\right)_2 = \frac{1 - \sqrt{3}}{2}.$$

In mode 1, B has an amplitude that is $(1 + \sqrt{3})/2$ times that of the amplitude of A and has the same phase. In mode 2, m_B has an amplitude that is $(\sqrt{3} - 1)/2$ times that of the amplitude of A and in the opposite direction.

3.5 Figure A.17 shows the physical situation for this problem. Let y_A be the change in the length of the four springs supporting M, in addition to the change that already exists when in equilibrium with gravity. Let y_B be the change in the length of the spring from which mass m is hanging, in addition to the change that already exists at equilibrium.

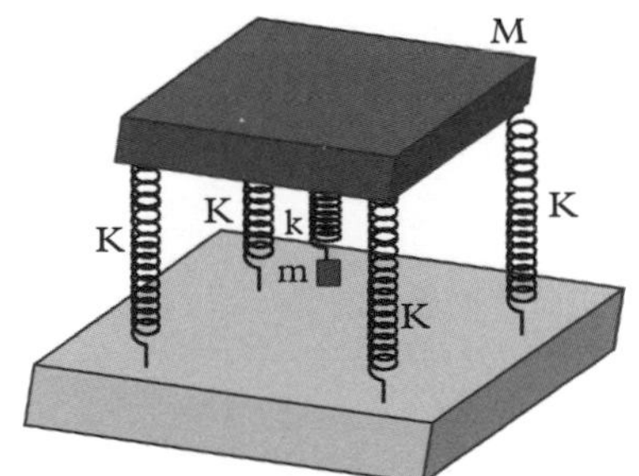

Fig. A.17 *Exercise 3.5.*

(a) The equations of motion of the two blocks will be

$$M\frac{d^2y_A}{dt^2} = -4Ky_A + k(y_B - y_A),$$
$$m\frac{d^2y_B}{dt^2} = -k(y_B - y_A).$$

(b) Let us solve this for the special situation when

$$\frac{K}{M} = \frac{k}{M} = \frac{k}{m} \equiv \omega_0^2.$$

The equations of motion simplify to

$$\frac{d^2y_A}{dt^2} = -\omega_0^2(5y_A - y_B),$$
$$\frac{d^2y_B}{dt^2} = -\omega_0^2(y_B - y_A).$$

Let $y_A = A\cos\omega t$ and $y_B = B\cos\omega t$ to help find the normal modes. The equations of motion become

$$(5\omega_0^2 - \omega^2)A = \omega_0^2 B,$$
$$(\omega_0^2 - \omega^2)B = \omega_0^2 A.$$

Solving them yields the following normal mode frequencies:

$$\omega_1 = \left(3 - \sqrt{5}\right)^{1/2}\omega_0, \quad \omega_2 = \left(3 + \sqrt{5}\right)^{1/2}\omega_0,$$

and the ratio of the amplitudes in the two modes to be

$$\left(\frac{B}{A}\right)_1 = 2 + \sqrt{5}, \quad \left(\frac{B}{A}\right)_2 = 2 - \sqrt{5}.$$

(c) Sketch is left as an exercise.

3.6 The equations of motion of the two-variable system are given to be

$$\frac{d^2x_A}{dt^2} = -x_A + \frac{2}{3}x_B,$$

$$\frac{d^2x_B}{dt^2} = -x_B + \frac{3}{8}x_A.$$

(a) With $x_A = A\cos\omega t$ and $x_B = B\cos\omega t$ in the given equations of motion we get

$$\begin{pmatrix} -1 + \omega^2 & \frac{2}{3} \\ \frac{3}{8} & -1 + \omega^2 \end{pmatrix}\begin{pmatrix} A \\ B \end{pmatrix} = 0. \tag{A.20}$$

The characteristic equation will be

$$\begin{vmatrix} -1+\omega^2 & \dfrac{2}{3} \\ \dfrac{3}{8} & -1+\omega^2 \end{vmatrix} = 0.$$

Expanding the determinant we get the following equation for ω^2:

$$(\omega^2 - 1)^2 = \frac{1}{4}.$$

Therefore, the eigenvalues are

$$\omega^2 = \frac{1}{2}, \frac{3}{2}.$$

Keeping only the positive roots we get the following angular frequencies of the two normal modes:

$$\omega_1 = \sqrt{\frac{1}{2}}, \quad \omega_2 = \sqrt{\frac{3}{2}}.$$

When we put $\omega = \omega_1$ in Eq. A.20 we get B/A for this mode,

$$\begin{pmatrix} -1+\dfrac{1}{2} & \dfrac{2}{3} \\ \dfrac{3}{8} & -1+\dfrac{1}{2} \end{pmatrix} \begin{pmatrix} A \\ B \end{pmatrix} = 0.$$

Therefore,

$$-\frac{1}{2}A + \frac{2}{3}B = 0 \implies \left(\frac{B}{A}\right)_1 = \frac{3}{4}.$$

Similarly, we put $\omega = \omega_2$ in Eq. A.20 and we get B/A for this mode,

$$\begin{pmatrix} -1+\dfrac{3}{2} & \dfrac{2}{3} \\ \dfrac{3}{8} & -1+\dfrac{3}{2} \end{pmatrix} \begin{pmatrix} A \\ B \end{pmatrix} = 0.$$

Therefore,

$$\frac{1}{2}A + \frac{2}{3}B = 0 \implies \left(\frac{B}{A}\right)_2 = -\frac{3}{4}.$$

The un-normalized eigenvectors are

$$\begin{pmatrix} A \\ B \end{pmatrix}_1 = \begin{pmatrix} 1 \\ 3/4 \end{pmatrix}, \quad \begin{pmatrix} A \\ B \end{pmatrix}_2 = \begin{pmatrix} 1 \\ -3/4 \end{pmatrix}.$$

To normalize we need to make the vector length 1. This gives

$$\begin{pmatrix} A \\ B \end{pmatrix}_1 = \begin{pmatrix} 4/5 \\ 3/5 \end{pmatrix}, \quad \begin{pmatrix} A \\ B \end{pmatrix}_2 = \begin{pmatrix} 4/5 \\ -3/5 \end{pmatrix}.$$

(b) To find the solution when the oscillators are started in an arbitrary state given by $x_A(0) = 1$, $v_1(0) = 0$, $x_B(0) = 0$, $v_2(0) = 1$, we start with the general solution,

$$x_A = C_1 \cos(\omega_1 t) + D_1 \sin(\omega_1 t) + C_2 \cos(\omega_2 t) + D_2 \sin(\omega_2 t),$$
$$x_B = \frac{3}{4}\left[C_1 \cos(\omega_1 t) + D_1 \sin(\omega_1 t)\right] - \frac{3}{4}\left[C_2 \cos(\omega_2 t) + D_2 \sin(\omega_2 t)\right].$$

The corresponding velocities will be

$$v_A = \omega_1[-C_1 \sin(\omega_1 t) + D_1 \cos(\omega_1 t)] + \omega_2[-C_2 \sin(\omega_2 t) + D_2 \cos(\omega_2 t)],$$
$$v_B = \frac{3}{4}\omega_1[-C_1 \sin(\omega_1 t) + D_1 \cos(\omega_1 t)] - \frac{3}{4}\omega_2[-C_2 \sin(\omega_2 t) + D_2 \cos(\omega_2 t)].$$

Using the initial conditions gives

$$C_1 + C_2 = 1,$$
$$\frac{3}{4}C_1 - \frac{3}{4}C_2 = 0,$$
$$\omega_1 D_1 + \omega_2 D_2 = 0,$$
$$\frac{3}{4}\omega_1 D_1 - \frac{3}{4}\omega_2 D_2 = 1.$$

These equations can be solved to obtain C_1, C_2, D_1, and D_2,

$$C_1 = C_2 = \frac{1}{2}, \quad D_1 = \frac{2\sqrt{2}}{3}, \quad D_2 = -\frac{2\sqrt{2}}{3\sqrt{3}}.$$

Therefore, the displacements are

$$x_A = \frac{1}{2}\cos\left(t/\sqrt{2}\right) + \frac{2\sqrt{2}}{3}\sin\left(t/\sqrt{2}\right) + \frac{1}{2}\cos\left(\sqrt{3/2t}\right) - \frac{2\sqrt{2}}{3\sqrt{3}}\sin\left(\sqrt{3/2t}\right),$$
$$x_B = \frac{3}{8}\cos\left(t/\sqrt{2}\right) + \frac{\sqrt{2}}{2}\sin\left(t/\sqrt{2}\right) - \frac{3}{8}\cos\left(\sqrt{3/2t}\right) + \frac{\sqrt{2}}{2\sqrt{3}}\sin\left(\sqrt{3/2t}\right).$$

3.7 Figure A.18 shows the system at an arbitrary time.

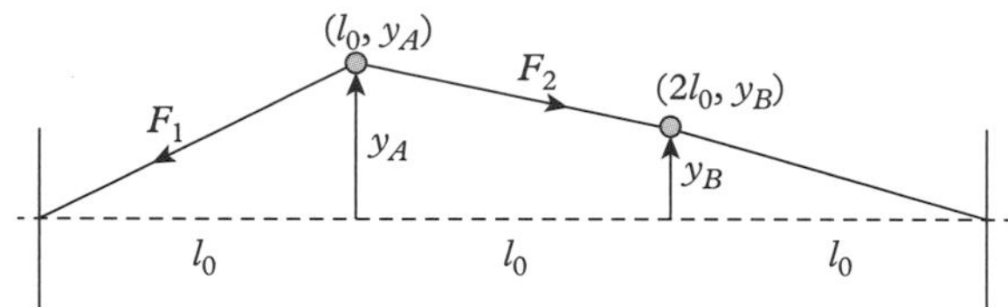

Fig. A.18 *Exercise 3.7.*

The forces F_1 on m_A have the following magnitudes:

$$F_1 = k\left[\sqrt{l_0^2 + y_A^2} - l_{01}\right],$$

where l_{01} is the length of the unstretched spring and l_0 is the length when it is in the horizontal direction. We make the approximation

$$y_A \ll l_0.$$

This simplifies F_1 to

$$F_1 \approx k(l_0 - l_{01}).$$

We need the y-component of this force. From the figure, the y-component will be

$$F_{1y} = -F_1 \frac{y_A}{\sqrt{l_0^2 + y_A^2}} \approx -F_1 \frac{y_A}{l_0} = -\frac{k(l_0 - l_{01})}{l_0} y_A.$$

Writing this as

$$F_{1y} = -k_{\text{eff}} y_A,$$

we find the following expression for the effective spring constant given in Eq. 3.94:

$$k_{\text{eff}} = \frac{k(l_0 - l_{01})}{l_0}.$$

To obtain the k_{eff} of the middle spring, you can follow a similar procedure.

3.8 Three identical pendulums of mass m and length l are coupled by two identical springs with spring constant k, as shown in Fig. A.19. As shown in the figure we will use the x-displacements, x_A, x_B, and x_C of the bobs from their corresponding equilibria as dynamical variables.

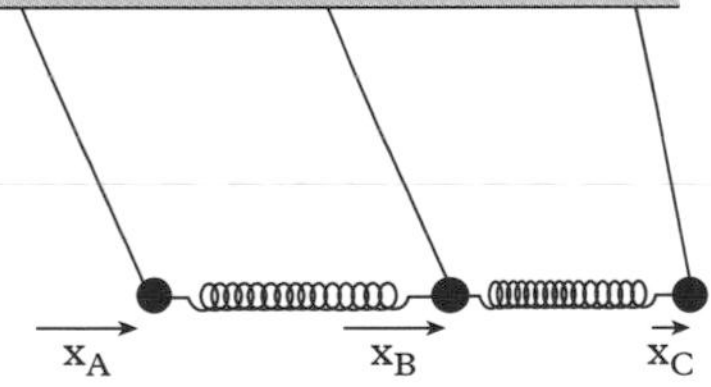

Fig. A.19 *Exercise 3.8.*

(a) From the x-components of forces on each bob we find the following equations of motion:

$$m\frac{d^2x_A}{dt^2} = -\frac{mg}{l}x_A - k(x_A - x_B),$$
$$m\frac{d^2x_B}{dt^2} = -\frac{mg}{l}x_B - k(x_B - x_A) - k(x_B - x_C),$$
$$m\frac{d^2x_C}{dt^2} = -\frac{mg}{l}x_C - k(x_C - x_B).$$

(b) Let us introduce the following symbols for brevity in writing:

$$\omega_0^2 = \frac{g}{l}, \quad \omega_c^2 = \frac{k}{m}.$$

Also, to look for the normal modes we assume the oscillating solutions

$$x_A = A\cos(\omega t), \quad x_B = B\cos(\omega t), \quad x_C = C\cos(\omega t).$$

The resulting equations can be summarized in the following matrix equation:

$$\begin{pmatrix} \omega^2 - \omega_0^2 - \omega_c^2 & \omega_c^2 & 0 \\ \omega_c^2 & \omega^2 - \omega_0^2 - 2\omega_c^2 & \omega_c^2 \\ 0 & \omega_c^2 & \omega^2 - \omega_0^2 - \omega_c^2 \end{pmatrix} \begin{pmatrix} A \\ B \\ C \end{pmatrix} = 0. \tag{A.21}$$

Therefore, the characteristic equation will be given by

$$\begin{vmatrix} \omega^2 - \omega_0^2 - \omega_c^2 & \omega_c^2 & 0 \\ \omega_c^2 & \omega^2 - \omega_0^2 - 2\omega_c^2 & \omega_c^2 \\ 0 & \omega_c^2 & \omega^2 - \omega_0^2 - \omega_c^2 \end{vmatrix} = 0.$$

This gives the following three roots for ω^2:

$$\omega_1^2 = \omega_0^2,$$
$$\omega_1^2 = \omega_0^2 + \omega_c^2,$$
$$\omega_2^2 = \omega_0^2 + 3\omega_c^2.$$

Note: When you expand the determinant it is best not to expand every term. Let us write

$$a = \omega^2 - \omega_0^2 - \omega_c^2,$$

then you would have

$$a[a(a - \omega_c^2) - \omega_c^4] - a\omega_c^4 = 0.$$

We can factor out a to get

$$a[a^2 - a\omega_c^2 - 2\omega_c^4] = 0.$$

Now, the quantities in the square bracket can be factored also to yield

$$a(a + \omega_c^2)(a - 2\omega_c^2) = 0,$$

which is

$$(\omega^2 - \omega_0^2)(\omega^2 - \omega_0^2 - \omega_c^2)(\omega^2 - \omega_0^2 - 3\omega_c^2) = 0.$$

Therefore, the normal mode frequencies are

$$\omega_1 = \omega_0, \quad \omega_2 = \sqrt{\omega_0^2 + \omega_c^2}, \quad \omega_3 = \sqrt{\omega_0^2 + 3\omega_c^2}.$$

When we had two modes, there was only one amplitude ratio, B/A, but here we have three modes, therefore, we will need two amplitude ratios for each mode, which we will take to be B/A and C/A. The three equations in the matrix equation, Eq. A.21, are

$$\begin{aligned}
&(\omega^2-\omega_0^2-\omega_c^2)A+\omega_c^2 B=0,\\
&\omega_c^2 A+(\omega^2-\omega_0^2-2\omega_c^2)B+\omega_c^2 C=0,\\
&\omega_c^2 B+(\omega^2-\omega_0^2-\omega_c^2)C=0.
\end{aligned}$$

In case of mode 1, $\omega=\omega_1=\omega_0$, these equations become

$$\begin{aligned}
&-\omega_c^2 A_1+\omega_c^2 B_1=0,\\
&\omega_c^2 A_1-2\omega_c^2 B_1+\omega_c^2 C_1=0,\\
&\omega_c^2 B_1-\omega_c^2 C_1=0.
\end{aligned}$$

Therefore,

$$\left(\frac{B}{A}\right)_1=1,\quad \left(\frac{C}{A}\right)_1=1.$$

In this mode all bobs move in the same direction with the same displacement such that the coupling springs do not exert any force on them. This explains the mode frequency being the frequency of the single pendulums. Similarly, in the case of mode 2, $\omega=\omega_2=\sqrt{\omega_0^2+\omega_c^2}$, we get

$$\begin{aligned}
&\omega_c^2 B_2=0,\\
&\omega_c^2 A_2-\omega_c^2 B_2+\omega_c^2 C_2=0,\\
&\omega_c^2 B_2=0.
\end{aligned}$$

Therefore,

$$\left(\frac{B}{A}\right)_2=0,\quad \left(\frac{C}{A}\right)_2=-1.$$

Note that in mode 2, bob B remains at rest, only bobs A and C move in opposite directions with equal amplitudes. Finally, in the case of mode 3, $\omega=\omega_3=\sqrt{\omega_0^2+3\omega_c^2}$, we get

$$\begin{aligned}
&2\omega_c^2 A_3+\omega_c^2 B_3=0,\\
&\omega_c^2 A_3+\omega_c^2 B_3+\omega_c^2 C_3=0,\\
&\omega_c^2 B_3+2\omega_c^2 C_3=0.
\end{aligned}$$

Therefore,

$$\left(\frac{B}{A}\right)_3=-2,\quad \left(\frac{C}{A}\right)_3=1.$$

(c) Figure A.20 shows a sketch of the three modes based on the ratio of the amplitudes for the corresponding mode. In each case, I chose $A=1$ and determined B and C based on A. In mode 1, the three masses oscillate independently and that is why the frequency of this mode does not depend on the spring constant.

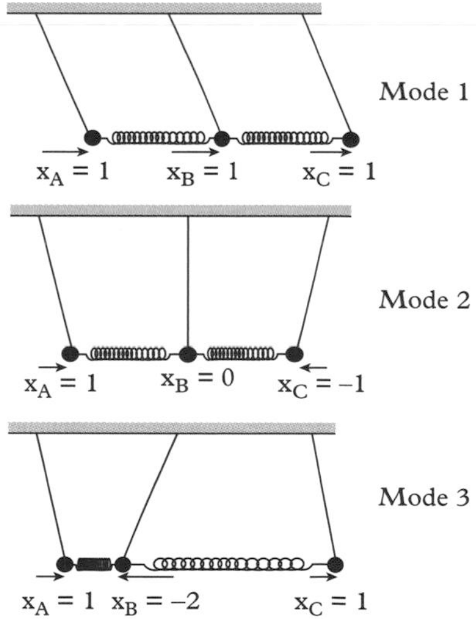

Fig. A.20 *Exercise 3.8.*

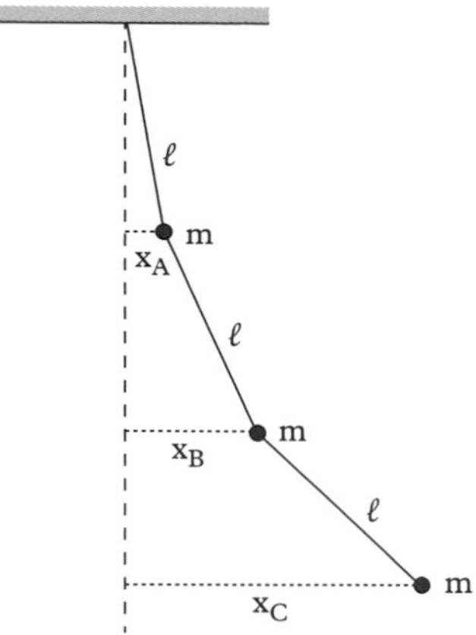

Fig. A.21 *Exercise 3.9. Displacements are exaggerated for clarity. In the text we work with small displacements.*

3.9 We choose the x-coordinates of the pendulum bobs from the vertical line passing through the suspension point as the dynamical variables, as shown in Fig. A.21. Alternatively, one could work with the angles also.

(a) It is easy to show that the tensions in the three strings will be $T = mg$ at the lowest one, $T = 2mg$ for the middle, and $T = 3mg$ in the top, since y-acceleration of the bobs will be negligible in the small-angle approximation. Then, the x-components of the equations of motion of the three bobs will be

$$m\frac{d^2x_A}{dt^2} = -3mg\frac{x_A}{l} + 2mg\frac{(x_B - x_A)}{l},$$
$$m\frac{d^2x_B}{dt^2} = -2mg\frac{(x_B - x_A)}{l} + mg\frac{(x_C - x_B)}{l},$$
$$m\frac{d^2x_C}{dt^2} = -mg\frac{(x_C - x_B)}{l}.$$

(b) For a simpler notation, let

$$\omega_0^2 = \frac{g}{l}.$$

We find normal modes as usual by seeking a solution in which all masses oscillate with the same frequency, and set

$$x_A = A\cos(\omega t), \quad x_B = B\cos(\omega t), \quad x_C = C\cos(\omega t)$$

in the equations of motion. This gives the following matrix equation for the vector A, B, C):

$$\begin{pmatrix} \omega^2 - 5\omega_0^2 & 2\omega_0^2 & 0 \\ 2\omega_0^2 & \omega^2 - 3\omega_0^2 & \omega_0^2 \\ 0 & \omega_0^2 & \omega^2 - \omega_0^2 \end{pmatrix}\begin{pmatrix} A \\ B \\ C \end{pmatrix} = 0. \tag{A.22}$$

The characteristic equation will be obtained by setting the determinant of the square matrix to zero,

$$\begin{vmatrix} \omega^2 - 5\omega_0^2 & 2\omega_0^2 & 0 \\ 2\omega_0^2 & \omega^2 - 3\omega_0^2 & \omega_0^2 \\ 0 & \omega_0^2 & \omega^2 - \omega_0^2 \end{vmatrix} = 0.$$

Expanding this we get a cubic equation for ω^2,

$$(\omega^2)^3 - 9\omega_0^2(\omega^2)^2 + 18\omega_0^2(\omega^2) - 6\omega_0^6 = 0.$$

Rather than write the exact answer, we can work out an approximate answer by a graphical method. Let us call $\lambda = \omega^2/\omega_0^2$, then this equation becomes

$$f(\lambda) = \lambda^3 - 9\lambda^2 + 18\lambda - 6 = 0.$$

Plot f as a function of λ and read off the points where f is zero,

$$\lambda_1 = 0.42, \quad \lambda_2 = 2.3, \quad \lambda_3 = 6.3.$$

These correspond to the following mode frequencies:

$$\omega_1 = 0.65\omega_0, \quad \omega_2 = 1.52\omega_0, \quad \omega_3 = 2.51\omega_0.$$

The mode amplitude ratios in the three modes can be worked out by using these mode frequencies in Eq. A.22, as we have done in other problems. The results are

$$\left(\frac{B}{A}\right)_1 = 2.29, \qquad \left(\frac{C}{A}\right)_1 = 3.92;$$

$$\left(\frac{B}{A}\right)_2 = 1.35, \qquad \left(\frac{C}{A}\right)_2 = -1.1;$$

$$\left(\frac{B}{A}\right)_3 = -0.65, \qquad \left(\frac{C}{A}\right)_3 = 0.15.$$

(c) Find B and C in each mode for $A = 1$ and then sketch the configurations of the system for the three normal modes. Figure A.22 shows the three modes.

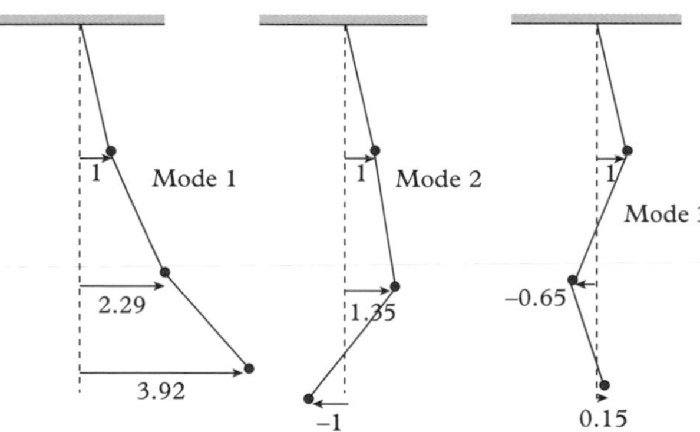

Fig. A.22 *Exercise 3.9. In mode 1, all three bobs move in the same direction so that there is no node in the motion; in mode 2, the top two move one way and the bottom one moves the other way so that there is one node between the middle and bottom; and in mode 3, the top and bottom move one way and the middle moves the other way so that there are two nodes.*

3.10 Figure A.23 shows the dynamical variables for the three masses. For each mass we refer to its displacement with respect to its equilibrium position.

(a) The equations of motion of the three masses can be written by examining the forces on them,

$$m\frac{d^2x_A}{dt^2} = -kx_A + k(x_B - x_A),$$

$$m\frac{d^2x_B}{dt^2} = -k(x_B - x_A) + k(x_C - x_B),$$

$$m\frac{d^2x_C}{dt^2} = -k(x_C - x_B) - kx_C.$$

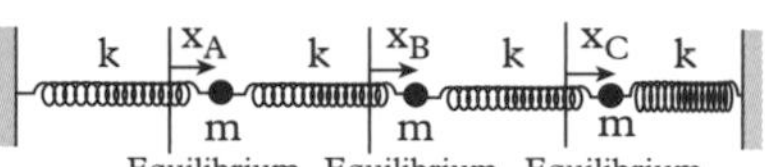

Fig. A.23 *Exercise 3.10.*

(b) Set $\frac{k}{m} = \omega_0^2$. To look for normal modes we seek an oscillatory solution for x_A, x_B, and x_C. We set

$$x_A = A\cos(\omega t),\ \ x_B = B\cos(\omega t),\ \ x_C = C\cos(\omega t)$$

in the equations of motion. This gives the following matrix equation for the vector $(A,\ B,\ C)$.

$$\begin{pmatrix} \lambda-2 & 1 & 0 \\ 1 & \lambda-2 & 1 \\ 0 & 1 & \lambda-2 \end{pmatrix} \begin{pmatrix} A \\ B \\ C \end{pmatrix} = 0, \tag{A.23}$$

where

$$\lambda = \frac{\omega^2}{\omega_0^2},$$

The characteristic equation for this matrix equation is

$$\begin{vmatrix} \lambda-2 & 1 & 0 \\ 1 & \lambda-2 & 1 \\ 0 & 1 & \lambda-2 \end{vmatrix} = 0.$$

Expanding this we get

$$(\lambda-2)\left(\lambda^2-4\lambda+2\right) = 0.$$

Therefore,

$$\lambda = 2,\ 2-\sqrt{2}, 2+\sqrt{2}.$$

Writing these as increasing frequencies we get

$$\omega_1 = \left(2-\sqrt{2}\right)^{1/2}\omega_0,\ \ \omega_2 = \sqrt{2}\,\omega_0,\ \ \omega_3 = \left(2+\sqrt{2}\right)^{1/2}\omega_0.$$

The three equations in the matrix equation, Eq. A.23, are

$$(\lambda-2)A + B = 0,$$
$$A + (\lambda-2)B + C = 0,$$
$$B + (\lambda-2)C = 0.$$

In the case of mode 1, $\lambda = \omega_1^2 = 2-\sqrt{2}$, these equations become

$$-\sqrt{2}A_1 + B_1 = 0,$$
$$A_1 - \sqrt{2}B_1 + C_1 = 0,$$
$$B_1 - \sqrt{2}C_1 = 0.$$

Therefore,

$$\left(\frac{B}{A}\right)_1 = \sqrt{2}, \quad \left(\frac{C}{A}\right)_1 = 1.$$

Similarly, in the case of mode 2, $\lambda = \omega_2^2 = 2$, we get

$$\begin{aligned} B_2 &= 0, \\ A_2 + C_2 &= 0, \\ B_2 &= 0. \end{aligned}$$

Therefore,

$$\left(\frac{B}{A}\right)_2 = 0, \quad \left(\frac{C}{A}\right)_2 = -1.$$

Note that in mode 2, bob B remains at rest, only bobs A and C move in opposite directions. Finally, in the case of mode 3, $\lambda = \omega_3^2 = 2 + \sqrt{2}$, we get

$$\begin{aligned} \sqrt{2}A_3 + B_3 &= 0, \\ A_3 + \sqrt{2}B_3 + C_3 &- 0, \\ B_3 + \sqrt{2}C_3 &= 0. \end{aligned}$$

Therefore,

$$\left(\frac{B}{A}\right)_3 = -\sqrt{2}, \quad \left(\frac{C}{A}\right)_3 = 1.$$

(c) The sketch can be drawn based on the numerical values given for the mode amplitude ratios.

3.11 (a) $+1, (1,0); -1, (0,-1)$.
(b) $+1, (1,1); -1, (1,-1)$.
(c) $+4, (1,1); -1, (3,-2)$.
(d) $0, (1,1,1); +1, (1,0,-1); +3, (1,-2,1)$.

Chapter 4

4.1 (a) Figure A.24 shows bead j and the adjoining beads along with their displacements ψ from equilibrium. The bead to the left has a change of length given by $|\psi_j - \psi_{j-1}|$ and the bead to the right, $|\psi_{j+1} - \psi_j|$.

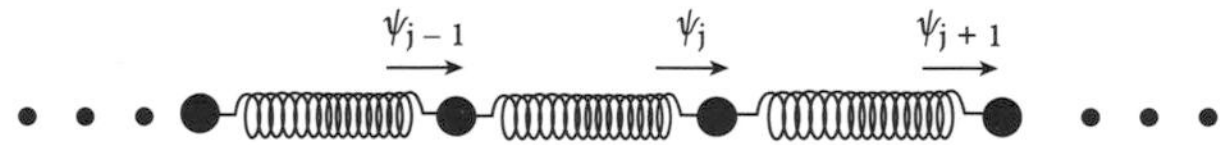

Fig. A.24 *Exercise 4.1.*

Taking into account the direction of these forces we find that the equation of motion of bead j somewhere in the system (excepting the ends) will be

$$m\frac{d^2\psi_j}{dt^2} = -k(\psi_j - \psi_{j-1}) + k(\psi_{j-1} - \psi_j). \tag{A.24}$$

The equations of motion of the beads at the ends will be different since they are not connected to beads on both sides,

$$m\frac{d^2\psi_1}{dt^2} = -k\psi_1 + k(\psi_2 - \psi_1), \tag{A.25}$$

$$m\frac{d^2\psi_N}{dt^2} = -k(\psi_N - \psi_{N-1}) - k\psi_N. \tag{A.26}$$

We can write all these equations in one equation if we introduce ψ_0 and ψ_{N+1} for the supports on the left and the right, which will be zero,

$$m\frac{d^2\psi_j}{dt^2} = -k(\psi_j - \psi_{j-1}) + k(\psi_{j+1} - \psi_j),$$

$$j = 0, 1, 2, \ldots, N+1, \tag{A.27}$$

$$\psi_0 = 0, \tag{A.28}$$

$$\psi_{N+1} = 0. \tag{A.29}$$

(b) Equations A.27–A.29 can be solved to determine the normal modes. As before, we require that masses oscillate with the same frequency when in a normal mode by setting the following for ψj:

$$\psi_i = A_j \cos(\omega t), \quad A_0 = 0, \quad A_{N+1} = 0.$$

With this form for the displacements, Eq. A.27 turns into

$$-\omega^2 A_j = \omega_0^2 A_{j-1} - 2\omega_0^2 A_j + \omega_0^2 A_{j+1},$$

where for brevity we have set

$$\omega_0^2 = \frac{k}{m}.$$

Rearranging we find

$$A_{j-1} + A_{j+1} = \left(\frac{2\omega_0^2 - \omega^2}{\omega_0^2}\right) A_j. \tag{A.30}$$

The discrete equation A.30 has a solution of the form $\sin(j\alpha)$ and $\cos(j\alpha)$. Adding two of the sines gives a result that is proportional to a sine in the middle index. For example,

$$\sin[(j-1)\alpha] + \sin[(j+1)\alpha] = (2\cos\alpha)\sin(j\alpha). \tag{A.31}$$

A similar equation holds for cosines. But since we want $\psi_0 = 0$, we will have the sine series. If we take $A_j = \sin(j\alpha)$, the discrete equation A.30 will be automatically satisfied by a proper choice of α, to be obtained below. To satisfy the boundary at $j = N + 1$ we would need

$$A_{N+1} = 0 \text{ if } \sin[(N+1)\alpha] = 0.$$

That means α can take only discrete values given by

$$(N+1)\alpha = n\pi, \quad n = 0, \pm 1, \pm 2, \ldots.$$

We will see below that $n = 1, 2, 3, \ldots, N$ only.

Let us indicate these by attaching an index n to α,

$$\alpha_n = \frac{n\pi}{N+1}. \tag{A.32}$$

Plugging $A_j = \sin(j\alpha_n)$ into Eq A.30 we get

$$\sin[(j-1)\alpha_n] + \sin[(j+1)\alpha_n] = \left(\frac{2\omega_0^2 - \omega^2}{\omega_0^2}\right) \sin(j\alpha_n).$$

Comparing this to Eq A.31 shows that $2\cos(\alpha)$ must be related to ω and ω_0 by

$$2\cos\alpha_n = \frac{2\omega_0^2 - \omega^2}{\omega_0^2},$$

which we can solve for frequency ω,

$$\omega^2 = 2\omega_0^2(1 - \cos\alpha_n) = 4\omega_0^2 \sin^2\left(\frac{\alpha_n}{2}\right).$$

This will give discrete values for the frequency, which are the normal mode frequencies here, which we will indicate by attaching the index n. Keeping only the positive roots we get

$$\omega_n = 2\omega_0 \left|\sin\left(\frac{\alpha_n}{2}\right)\right| = 2\omega_0 \left|\sin\left(\frac{n\pi}{2(N+1)}\right)\right|. \tag{A.33}$$

Let us now look at the values of n in this equation that correspond to the modes. First, note that for $n = 0,\ N+1, 2(N+1), \ldots,\ \omega = 0$, which would not be a vibration. Therefore, we will not allow these values for n. Now, note that negative and positive values of n correspond to the same frequencies and the same mode amplitudes. In the case of amplitudes, the n and $-n$ will give the same nodes and antinodes. Therefore,

$$n = 1, 2, 3, \ldots, N, N+2, N+3, \ldots, 2N+1, \text{ etc.}$$

However, even these are too many as I will show now. We can now show that we only have N modes, which can be taken to be $n = 1, 2, \ldots, N$. The mode amplitude for a particular particle j in a particular mode n is given by

$$A_j^n = \sin(j\alpha_n) = \sin\left(\frac{jn\pi}{N+1}\right).$$

Consider the amplitude for this bead j in mode $n+N+2$,

$$A_j^{n+N+2} = \sin\left[\frac{j(n+N+2)\pi}{N+1}\right]$$
$$= \sin\left[\frac{j(n+2(N+1)-N)\pi}{N+1}\right]$$
$$= \sin\left[\frac{j(n-N)\pi}{N+1}\right]$$
$$= A_j^{n-N} = -A_j^{N-n}. \tag{A.34}$$

Since the negative sign is independent of j, the amplitude of every bead in mode $n+N+2$ is just the negative of its amplitude in mode $N-n$, which would not change the mode profile. This shows that modes $n+N+2$, $n-N$, and $N-n$ are the same. Using these equivalencies, all indices with n outside of $n = 1, 2, 3, \ldots, N$ can be mapped into this range. Hence, we have only N modes. Some examples are

mode $N+2$ same as mode N (set $n=0$ in Eq. A.34),
mode $N+3$ same as mode $N-1$ (set $n=1$ in Eq. A.34),
mode $N+4$ same as mode $N-2$ (set $n=2$ in Eq. A.34),
mode $2N+1$ same as mode 1 (set $n=N-1$ in Eq. A.34).

(c) The normal modes for the $N=2$ case will have the following frequencies:

$$\omega_1 = 2\omega_0 \left|\sin\left(\frac{\pi}{2\times 3}\right)\right| = \omega_0,$$
$$\omega_2 = 2\omega_0 \left|\sin\left(\frac{2\pi}{2\times 3}\right)\right| = \sqrt{3}\,\omega_0.$$

Mode amplitude ratios will be

$$(A_2/A_1)_1 = \frac{\sin(2\alpha_1)}{\sin(\alpha_1)} = \frac{\sin(2\pi/3)}{\sin(\pi/3)} = 1,$$
$$(A_2/A_1)_2 = \frac{\sin(2\alpha_2)}{\sin(\alpha_2)} = \frac{\sin(4\pi/3)}{\sin(2\pi/3)} = -1.$$

(d) You should get sine waves like structures.

4.2 This problem is an application of the calculations in the last problem. Let mass M of the spring be distributed into N beads of mass $m = M/N$ each, to be distributed at equal intervals of separation l and the massive spring with spring constant k_s be replaced ny $N+1$ massless springs of spring constant k connecting successive beads at a distance $l = L/(N+1)$,

$$m = \frac{M}{N}, \quad l = \frac{L}{N+1}.$$

Since the original spring with spring constant k_s is being replaced with $N + 1$ springs connected in series with spring constant k each, the relation between k and k_s will be

$$\frac{1}{k} + \frac{1}{k} + \cdots + \frac{1}{k} = \frac{1}{k_s},$$

where on the left are $N + 1$ terms or $N + 1$ springs. This gives

$$k = (N + 1)k_s.$$

From the treatment of N beads in the last problem we know that the normal modes of this system will have frequencies

$$\omega_n = 2\omega_0 \left| \sin\left[\frac{n\pi}{2(N+1)}\right] \right|, \quad n = 1, 2, 3, \ldots, N,$$

where

$$\omega_0 = \sqrt{\frac{k}{m}}.$$

Now, expressing these in k_s and M of the full spring we get

$$\omega_n = 2\sqrt{\frac{N(N+1)k_s}{M}} \left| \sin\left[\frac{n\pi}{2(N+1)}\right] \right|.$$

To get the result for the continuous spring we need to take the limit $N \to \infty$ with $Nl = L$ and $Nm = M$ fixed. Let us also focus on only the low frequencies, $n \ll N$. For these modes the sine can be approximated to linear values,

$$\sin\left[\frac{n\pi}{2(N+1)}\right] \longrightarrow \frac{n\pi}{2N},$$

and the constant ω_0 becomes

$$\omega_0 = \sqrt{\frac{N(N+1)k_s}{M}} \longrightarrow N\sqrt{\frac{k_s}{M}}.$$

Therefore, the normal modes of the massive spring have the following frequencies:

$$\omega_n = n\pi\sqrt{\frac{k_s}{M}}, \quad n = 1, 2, 3, \ldots \tag{A.35}$$

The lowest frequency is $\pi\sqrt{k_s/M}$.

For the $N \to \infty$ limit, the amplitudes can be similarly obtained. For a definite N, the amplitude of the jth bead in the mode n takes the form

$$\psi_j^{(n)} = \sin\left(\frac{jn\pi}{N+1}\right).$$

Divide and multiply the argument of the sine by l, then we see that jl is the x-coordinate of the jth bead, $x_j = jl$, and the denominator is the length L of the spring,

$$\psi_j^{(n)} = \sin\left(\frac{jln\pi}{(N+1)l}\right) = \sin\left(\frac{x_j n\pi}{L}\right).$$

Taking the limit $N \to \infty$ makes the beads become continuously distributed, with x_j taking on all real values x between 0 and L, and we get the mode function $\psi_n(x)$,

$$\psi_n(x) = \sin\left(\frac{n\pi x}{L}\right). \tag{A.36}$$

4.3 (a) Consider the stretching by $d\psi$ of element dx of the spring when not in equilibrium, as shown in the figure. The spring constant of this tiny spring is not k_s, which is the spring constant of the complete spring of length L. The spring constant of the tiny spring can be obtained by thinking of the full spring as built out of tiny springs in series. This will give $\frac{L}{dx}$ tiny springs in series. Therefore, the spring constant of the tiny spring will be

$$k_{\text{eff}} = \frac{L}{dx}k_s.$$

Using this spring constant we note that the stretch $d\psi$ will cause the tension at the coordinate x to be

$$T(x,t) = \frac{L}{dx}k_s d\psi = k_s L\frac{d\psi}{dx}.$$

Since this is at a fixed instant in time, we can write this as a partial derivative also,

$$T(x,t) = k_s L\frac{\partial\psi}{\partial x}.$$

(b) The force on the element towards the fixed support will be $T(x,t)$ and the force towards the block will be $T(x+dx,t)$. Since the mass in the element will be $dm = \frac{m}{L}dx$, we will get the following equation of motion:

$$\frac{m}{L}dx\frac{\partial^2\psi}{\partial t^2} = T(x+dx,t) - T(x,t),$$

where the sign on the right side is based on the choice of positive coordinate direction. Dividing both sides by $m\frac{dx}{L}$ we obtain

$$\frac{\partial^2\psi}{\partial t^2} = \frac{L}{m}\frac{dT}{dx} = \frac{L}{m}k_s L\frac{\partial^2\psi}{\partial x^2}.$$

Therefore,

$$\frac{\partial^2\psi}{\partial t^2} = \frac{k_s L^2}{m}\frac{\partial^2\psi}{\partial x^2}. \tag{A.37}$$

This is classical wave equation with speed

$$v = \sqrt{\frac{k_s L^2}{m}}.$$

(c) At $x = 0$, the spring is attached to a fixed support. Therefore, there will be no displacement there, giving the following boundary condition at $x = 0$:

$$\psi(0, t) = 0.$$

(d) The restoring force on the block at $x = L$ is the tension T(at $x = L$). The displacement of the block is given by the function $\psi(L, t)$. Therefore, we get the equation of motion to be

$$M\frac{\partial^2\psi(L,t)}{\partial t^2} = -T(L,t),$$

which gives the following condition at the boundary at $x = L$:

$$M\frac{\partial^2\psi(L,t)}{\partial t^2} = -k_s L\left(\frac{\partial\psi}{\partial x}\right)_{x=L}. \tag{A.38}$$

(e) (i) The boundary condition for $\psi(x, t) = [A\cos(kx) + B\sin(kx)]\cos(\omega t)$ at $x = 0$ would give

$$A\cos(\omega t) = 0,$$

which can be satisfied for all t only if $A = 0$. Therefore, the solution will have only a sine function of x,

$$\psi(x,t) = B\sin(kx)\cos(\omega t).$$

(ii) Now, we plug $\psi(x, t) = B\sin(kx)\cos(\omega t)$ into Eqs. A.37 and A.38, with the following result:

$$-\omega^2 = -\frac{k_s L^2}{m}k^2, \tag{A.39}$$

$$-M\omega^2\sin(kL) = -k_s Lk\cos(kL). \tag{A.40}$$

From Eq.A.39 we get

$$\omega = kL\sqrt{k_s/m},$$

which we can use in Eq. A.40 to obtain

$$M\omega\tan\left(\omega/\sqrt{k_s/m}\right) = \frac{k_s}{\sqrt{k_s/m}}.$$

We introduce

$$\omega_0 = \sqrt{\frac{k_s}{M}}, \quad \epsilon = \frac{m}{M},$$

to write this result as

$$\tan\left(\sqrt{\epsilon}\frac{\omega}{\omega_0}\right) = \sqrt{\epsilon}\frac{\omega_0}{\omega}. \tag{A.41}$$

(f) When $\theta \to 0$ we get $\tan\theta \approx \theta$ in the linear approximation. In this approximation, Eq. A.41 becomes

$$\sqrt{\epsilon}\,\frac{\omega}{\omega_0} = \sqrt{\epsilon}\,\frac{\omega_0}{\omega},$$

which gives

$$\omega^2 = \omega_0^2 \quad \rightarrow \quad \omega = \omega_0.$$

(g) Left for student.

(h) Let us write the argument of the tangent as

$$\theta \equiv \sqrt{\epsilon}\,\frac{\omega}{\omega_0}.$$

Then, for small ϵ we will also have small θ. Now, we write the inverse of each side in Eq. A.41,

$$\frac{1}{\tan(\theta)} = \frac{\theta}{\epsilon},$$

and use the expansion of $1/\tan(x)$ for small x to get

$$\frac{1}{\theta} - \frac{1}{3}\theta \approx \frac{\theta}{\epsilon}.$$

Solving for θ we get

$$\theta^2 = \frac{3\epsilon}{\epsilon + 3}.$$

Writing this back in terms of ω, k_s, m, and M you get the required result,

$$\omega = \sqrt{\frac{k_s}{M + \frac{1}{3}m}}.$$

4.4 (a) The figures for the three lowest modes here are the same as the modes of air in a tube with one end open and the other end closed, shown in the text.

(b) From the figures for the modes we obtain the following for the wavelengths of the modes:

$$\frac{1}{4}\lambda_1 = L, \quad \frac{3}{4}\lambda_2 = L, \quad \frac{5}{4}\lambda_3 = L.$$

Therefore, the corresponding wave numbers will be

$$k_1 = \frac{2\pi}{\lambda_1} = \frac{\pi}{2L},$$
$$k_2 = \frac{2\pi}{\lambda_2} = \frac{3\pi}{2L},$$
$$k_3 = \frac{2\pi}{\lambda_3} = \frac{5\pi}{2L}.$$

(c) $k_n = \frac{(2n-1)\pi}{2L}$, $n = 1, 2, 3, \ldots$.

(d) $\psi_n(x) = \sin(k_n x)$ with the k_n as given above.

(e) $\omega_n^2 = (k_s L^2/m) k_n^2$, or, $\omega_n = \sqrt{k_s L^2/m}\, k_n$.

4.5 The normal modes of N beads have been worked out in the text. For three beads the normal modes are

$$\text{Mode 1: } \omega_1 = 2\omega_0 \sin(\pi/8) = 0.766\, \omega_0, \quad \omega_0 = \sqrt{T/ma},$$
$$A_1^{(1)} = \sin(\pi/4) = \frac{1}{\sqrt{2}},\ A_2^{(2)} = 1,\ A_3^{(2)} = \frac{1}{\sqrt{2}},$$

$$\text{Mode 2: } \omega_2 = 2\omega_0 \sin(\pi/4) = 1.414\, \omega_0,$$
$$A_1^{(2)} = \sin(\pi/2) = 1,\ A_2^{(1)} = 0,\ A_3^{(1)} = -1,$$

$$\text{Mode 3: } \omega_3 = 2\omega_0 \sin(3\pi/8) = 1.848,$$
$$A_1^{(3)} = \sin(3\pi/4) = \frac{1}{\sqrt{2}},\ A_2^{(3)} = -1,\ A_3^{(3)} = \frac{1}{\sqrt{2}}.$$

The general solution will be

$$y_1(t) = \sum_{j=1}^{3} \left(C_j \cos(\omega_j t) + S_j \sin(\omega_j t)\right), \tag{A.42}$$

$$y_2(t) = \sum_{j=1}^{3} \left[b_{2j}\left(C_j \cos(\omega_j t) + S_j \sin(\omega_j t)\right)\right], \tag{A.43}$$

$$y_3(t) - \sum_{j=1}^{3} \left[b_{3j}\left(C_j \cos(\omega_j t) + S_j \sin(\omega_j t)\right)\right], \tag{A.44}$$

where the mode amplitude ratios are

$$b_{21} = \frac{A_1^{(2)}}{A_1^{(1)}} = \sqrt{2}; \quad b_{22} = \frac{A_2^{(2)}}{A_1^{(2)}} = 0; \quad b_{23} = \frac{A_2^{(3)}}{A_1^{(3)}} = -\sqrt{2};$$

and similarly, for others. Now, we use the initial conditions given in the problem,

$$y_1(0) = 0 = y_2(0) = y_3(0), \quad v_1(0) = 0 = v_3(0), \quad v_2(0) = v_0.$$

These give

$$C_1 + C_2 + C_3 = 0, \ \sqrt{2}C_1 - \sqrt{2}C_3 = 0, \ C_1 - C_2 + C_3 = 0,$$
$$\omega_1 S_1 + \omega_2 S_2 + \omega_3 S_3 = 0, \ \sqrt{2}\omega_1 S_1 - \sqrt{2}\omega_3 S_3 = v_0, \ \omega_1 S_1 - \omega_2 S_2 + \omega_3 S_3 = 0.$$

Solving them we get

$$C_1 = C_2 = C_3 = 0, S_2 = 0, S_1 = \frac{v_0}{2\sqrt{2}\omega_1}, \ S_3 = -\frac{v_0}{2\sqrt{2}\omega_3}.$$

Putting these in the general solution we get the vibrations of each bead to be

$$y_1(t) = \frac{v_0}{2\sqrt{2}\omega_1}\sin(\omega_1 t) - \frac{v_0}{2\sqrt{2}\omega_3}\sin(\omega_3 t),$$

$$y_2(t) = \frac{v_0}{2\omega_1}\sin(\omega_1 t) + \frac{v_0}{2\omega_3}\sin(\omega_3 t),$$

$$y_3(t) = \frac{v_0}{2\sqrt{2}\omega_1}\sin(\omega_1 t) - \frac{v_0}{2\sqrt{2}\omega_3}\sin(\omega_3 t).$$

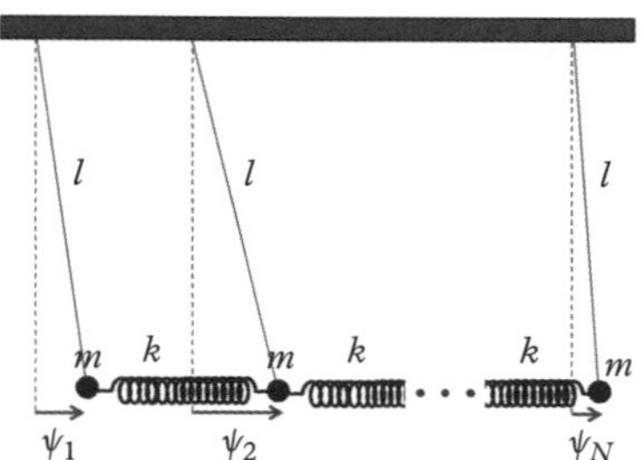

Fig. A.25 *Exercise 4.6.*

4.6 The system for this problem is shown in Fig. A.25.

(a) The motion of the bob at position j is coupled to the coordinates of the bobs at positions $j-1$ and $j+1$ through the coupling spring. Therefore, the equation of motion of the bob at position j will be

$$m\frac{d^2\psi_j}{dt^2} = -mg\frac{\psi_j}{l} - k\left(\psi_j - \psi_{j-1}\right) + k\left(\psi_{j+1} - \psi_j\right). \tag{A.45}$$

(b) As stated in the problem, let

$$\psi_j(t) = \cos(\omega t)\,[A\cos(Kja) + B\sin(Kja)],$$

where ω is the frequency and the K the wave number. Using this in Eq. A.45 and canceling the common time part we obtain

$$\begin{aligned}-\omega^2\,[A\cos(Kja) + B\sin(Kja)] =& -(\omega_0^2 + 2\omega_c^2)\,[A\cos(Kja) + B\sin(Kja)]\\ &+ \omega_c^2\,[A\cos(K(j-1)a) + B\sin(K(j-1)a)\\ &+A\cos(K(j+1)a) + B\sin(K(j+1)a)],\end{aligned}$$

where

$$\omega_0^2 = \frac{g}{l},\quad \omega_c^2 = \frac{k}{m}.$$

The cosines of $j-1$ and $j+1$ terms will combine separately from the sines and will give rise to a common factor on both sides which leads to the following relation:

$$-\omega^2 = -(\omega_0^2 + 2\omega_c^2) + 2\omega_c^2\cos(Ka).$$

Rewriting this we obtain the desired dispersion relation,

$$\omega^2 = \omega_0^2 + 4\omega_c^2\sin^2\left(\frac{Ka}{2}\right). \tag{A.46}$$

(c) With $K = 0$ in the dispersion relation, Eq. A.46, we obtain the lowest frequency to be

$$\omega_{\text{lowest}} = \omega_0 = \sqrt{\frac{g}{l}}.$$

In this mode none of the springs are participating as if the pendulums are independent of each other.

4.7 (a) The function $f(x) = A\sin(2\pi x/L)$ is already one of the sine terms in the Fourier series. So, no further work is needed,

$$B_1 = A,$$
$$B_n = 0, \quad n \neq 1,$$
$$A_n = 0, \quad \forall\, n.$$

(b) The period of the given function is L. Therefore, we will get

$$\begin{aligned} A_n &= \frac{2}{L}\int_0^L A\sin\left(\frac{\pi x}{L}\right)\cos\left(\frac{2\pi nx}{L}\right)dx \\ &= -\frac{4A}{\pi(4n^2-1)}. \end{aligned}$$

$$\begin{aligned} B_n &= \frac{2}{L}\int_0^L A\sin\left(\frac{\pi x}{L}\right)\sin\left(\frac{2\pi nx}{L}\right)dx \\ &= 0. \end{aligned}$$

(c) To find the Fourier series of $f(x) = x,\ 0 \leq x \leq L,\ f(x+L) = f(x)$, we will calculate the coefficients,

$$A_0 = \frac{2}{L}\int_0^L x\, dx = L,$$

$$A_n = \frac{2}{L}\int_0^L x\,\cos(2\pi nx/L)dx = 0,$$

$$B_n = \frac{2}{L}\int_0^L x\,\sin(2\pi nx/L)dx = -\frac{L}{n\pi}.$$

Therefore, we have

$$x = \frac{L}{2} - \frac{L}{\pi}\sum_{n=1,2,...}^{\infty}\frac{1}{n}\sin\left(\frac{2\pi nx}{L}\right).$$

(d) The Fourier coefficients in this case are:

$$A_0 = \frac{2}{L}\int_0^L x(L-x)\,dx = \frac{L^2}{3},$$

$$A_n = \frac{2}{L}\int_0^L x(L-x)\,\cos(2\pi nx/L)dx = -\frac{L^2}{n^2\pi^2},$$

$$B_n = \frac{2}{L}\int_0^L x(L-x)\,\sin(2\pi nx/L)dx = 0.$$

Therefore, we have

$$x(L-x) = \frac{L^2}{6} - \frac{L^2}{\pi^2}\sum_{n=1,2,...}^{\infty}\frac{1}{n^2}\cos\left(\frac{2\pi nx}{L}\right).$$

(e)

$$f(x) = \frac{2A}{\pi} + \frac{A}{2}\sin\left(\frac{2\pi x}{L}\right) + \frac{2A}{\pi}\sum_{n=2,4,6,\ldots}^{\infty}\frac{1}{1-n^2}\cos\left(\frac{2\pi nx}{L}\right).$$

4.8 (a) Let $f(x)$ be a periodic function of period L with the Fourier series

$$f(x) = \frac{A_0}{2} + \sum_{n=1}^{\infty} A_n \cos(k_n x) + \sum_{n=1}^{\infty} B_n \sin(k_n x), \tag{A.47}$$

where $k_n = 2\pi n/L$. Since the function is an even function we have

$$f(-x) = f(x).$$

Expressing both sides of this equation using the expansion in Eq. A.47 we get

$$\frac{A_0}{2} + \sum_{n=1}^{\infty} A_n \cos(-k_n x) + \sum_{n=1}^{\infty} B_n \sin(-k_n x)$$
$$= \frac{A_0}{2} + \sum_{n=1}^{\infty} A_n \cos(k_n x) + \sum_{n=1}^{\infty} B_n \sin(k_n x).$$

Since

$$\cos(-x) = \cos(x), \text{ and } \sin(-x) = -\sin(x),$$

the A_0 and A_n terms cancel out from both sides and give

$$2\sum_{n=1}^{\infty} B_n \sin(k_n x) = 0,$$

which easily gives

$$B_n = 0, \text{ for all } n.$$

That is, an even function has only A_0 and A_n non-zero, and therefore, has a cosine series only.

(b) Hint: Do similarly to part (a).

4.9 Let us use symbol A for the amplitude,

$$A = 2 \times 10^{-2} \text{ cm}.$$

The wave number 7π m^{-1} is one of the wave numbers for the normal modes,

$$k_n = \frac{n\pi}{L},$$

with $L = 1$ m. That is, the vibration is started in the $n = 7$ mode. Since the vibration is started in the $n = 7$ mode at rest, the subsequent vibrations will be in this mode only. Since the mode starts with zero speed for all elements of the string, the phase constant in the cosine term will be zero. Therefore, the subsequent motion will be given by

$$\psi(x,t) = A\cos(\omega_7 t)\sin(7\pi x),$$

where

$$\omega_7 = vk_7.$$

With

$$v = \sqrt{\frac{T}{\mu}} = \sqrt{\frac{10\ \text{N}}{0.1\ \text{Kg/m}}} = 10\ \text{m/s},$$

we get $\omega_7 = 70\pi\ \text{sec}^{-1}$. Therefore,

$$\psi(x,t) = A\sin(7\pi x)\cos(70\pi t).$$

4.10 The vibration is started in a superposition of modes with $n = 4$ and $n = 6$,

$$\psi(x, t=0) = 2\times 10^{-2}\ \text{cm}\ \sin(4\pi x) + 3\times 10^{-2}\ \text{cm}\ \sin(6\pi x).$$

Each mode will oscillate at its own frequency. The mode $n = 4$ will oscillate at ω_4 and the mode $n = 6$ will oscillate at ω_6. Since

$$v = \sqrt{\frac{40}{0.1}} = 20\ \text{m/s},$$

the frequencies are

$$\omega_4 = 4\pi v = 80\pi\ \text{sec}^{-1},\quad \omega_6 = 6\pi v = 120\pi\ \text{sec}^{-1}.$$

Since the vibration started with zero velocity we will have only the cosine term for the time part. Thus, the motion at subsequent times will be

$$\psi(x,t) = 2\times 10^{-2}\ \text{cm}\cos(80\pi t)\sin(4\pi x) + 3\times 10^{-2}\ \text{cm}\cos(120\pi t)\sin(6\pi x),$$

4.11 (a) The function representing the given shape in Fig. A.26 is

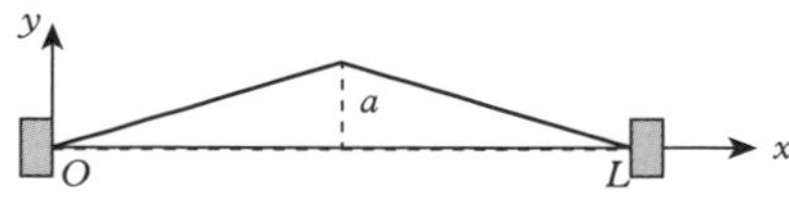

Fig. A.26 *Exercise 4.11.*

$$\psi(x) = \begin{cases} \dfrac{2a}{L}x & 0 \le x < \dfrac{L}{2} \\ -\dfrac{2a}{L}(x-L) & \dfrac{L}{2} \le x < L. \end{cases}$$

(b) We saw above that if we know the given shape in terms of a superposition of the normal modes we can evolve each mode independently. Thus, our first task is to find the series representation of the given shape in terms of the normal modes of the system. Recall that the normal modes of a string fixed at the ends are (with ends at $x = 0$ and $x = L$)

$$\psi_n(x) = \sin\left(\frac{n\pi x}{L}\right).$$

We want to know the coefficients that go with each mode in the following expansion:

$$\psi(x) = \sum_{n=1,2,...} A_n \sin\left(\frac{n\pi x}{L}\right). \tag{A.48}$$

Note that this is an expansion in normal modes of the system, which is similar to the Fourier expansion, but it is not a Fourier series since a Fourier series requires the function to be periodic. If you consider copying the given function outside of $0 \le x \le L$ and create a periodic function, then you can work with period $2L$ and you would get the same result as we will find here if you choose the inverted function for the range $L \le x \le 2L$ and then copy the new function of period $2L$.

To find the coefficients A_n in Eq. A.48, note that

$$\int_0^L \sin\left(\frac{n\pi x}{L}\right) \sin\left(\frac{m\pi x}{L}\right) dx = \begin{cases} 0 & m \neq n \\ \dfrac{L}{2} & m = n. \end{cases}$$

Now we multiply both sides of Eq. A.48 by $\sin\left(\frac{m\pi x}{L}\right)$ and integrate from $x = 0$ to L. This gives

$$\frac{L}{2} A_n = \frac{4aL}{n^2\pi^2} \sin\left(\frac{n\pi}{2}\right).$$

Therefore, the initial state expanded in modes has the following representation:

$$\psi(x) = \sum_{n=1,2,...} \frac{8a}{n^2\pi^2} \sin\left(\frac{n\pi}{2}\right) \sin\left(\frac{n\pi x}{L}\right).$$

(c) Figure A.27 shows the four plots. You can see that the function is well represented even with just three terms. The corners require many more terms.

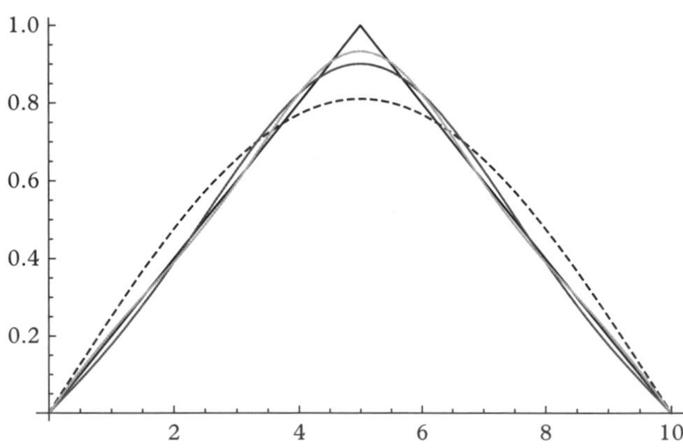

Fig. A.27 *Exercise 4.11(c).*

(d) Now, evolving this in time is very easy since each mode will evolve with its own frequency. Since, the initial velocity was zero, the phase constant is zero, which gives the following for $\psi(x, t)$:

$$\psi(x, t) = \sum_{n=1,2,...} \frac{8a}{n^2\pi^2} \sin\left(\frac{n\pi}{2}\right) \sin\left(\frac{n\pi x}{L}\right) \sin\left(\frac{n\pi vt}{L}\right),$$

where $v = \sqrt{T/\mu}$.

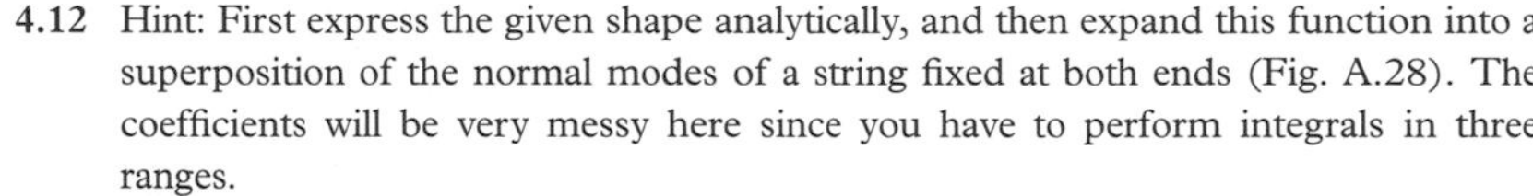

4.12 Hint: First express the given shape analytically, and then expand this function into a superposition of the normal modes of a string fixed at both ends (Fig. A.28). The coefficients will be very messy here since you have to perform integrals in three ranges.

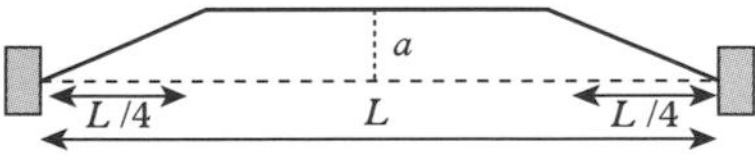

Fig. A.28 *Exercise 4.12.*

4.13 Figure A.29 shows a string hammered in the middle. This gives the following initial condition on the vertical displacement ψ of the string:

$$\psi(x, 0) = 0,$$

$$\left(\frac{\partial\psi(x, t)}{\partial t}\right)_{t=0} = v_0 L\delta\left(x - \frac{L}{2}\right).$$

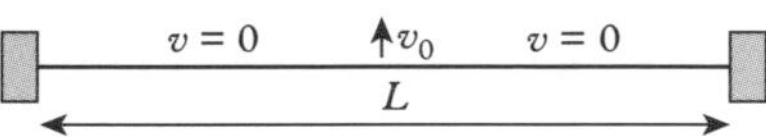

Fig. A.29 *Exercise 4.13.*

We will start with the general solution and determine the coefficients that satisfy the initial conditions.

(a) The general solution of vibrating strings fixed at $x = 0$ and $x = L$ is given by

$$\psi(x, t) = \sum_n [A_n \cos(\omega_n t) + B_n \sin(\omega_n t)] \sin(k_n x),$$

where $k_n = n\pi/L$ and $\omega_n = vk_n$ with $v = \sqrt{T/\mu}$. The initial condition on ψ gives us

$$\sum_n A_n \sin(k_n x) = 0,$$

which immediately gives

$$A_n = 0, \text{ for all } n.$$

Now, ψ simplifies to

$$\psi(x, t) = \sum_n B_n \sin(\omega_n t) \sin(k_n x).$$

The initial condition on velocity gives

$$\sum_n \omega_n B_n \sin(k_n x) = v_0 L\delta\left(x - \frac{L}{2}\right).$$

Now, multiplying both sides by $\sin(k_m x)$ and integrating from $x = 0$ to $x = L$ gives

$$B_m = \frac{2v_0}{\omega_m}\sin(k_m L/2).$$

Therefore, the subsequent motion of the hammered string is

$$\psi(x,t) = \sum_n \frac{2v_0}{\omega_n}\sin(k_n L/2)\sin(\omega_n t)\sin(k_n x).$$

(b) Note that the frequencies are multiples of a fundamental,

$$\omega_n = n\frac{\pi v}{L} = n\omega_1.$$

The longest period for any term in the sum to return to the original state would be that associated with ω_1. Since that term is not zero, that will also correspond to the largest time when the string's velocity and position will return to the initial. In each period, the string becomes horizontal twice. Therefore,

$$\Delta t = \frac{T_1}{2} = \frac{2\pi}{2\omega_1} = \frac{L}{v}.$$

4.14 Kundt's tube.

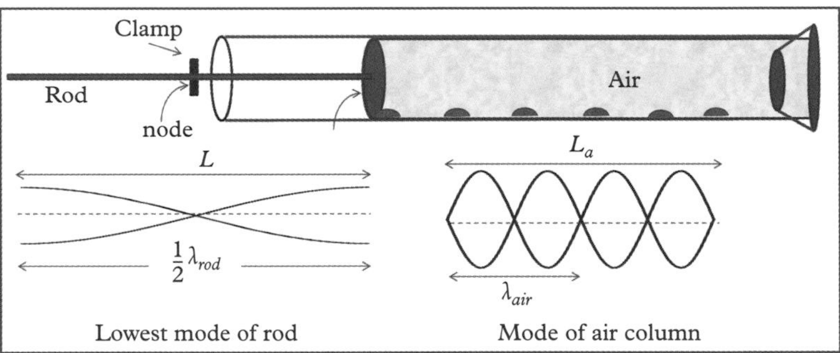

Fig. A.30 *Exercise 4.14. Kundt's Apparatus.*

(a) In Kundt's tube the heaps occur at the nodes of particle displacement. Hence the distance between heaps is equal to half a wavelength. There will be five wavelengths in an 11-heap situation. These wavelengths are in air. Let us denote them by λ_a. Therefore,

$$5\lambda_a = 40 \text{ cm}, \quad \lambda_a = 8 \text{ cm}.$$

From this data and the given velocity of sound in air (let us denote that by v_a) we can find the frequency of the wave,

$$f = \frac{v_a}{\lambda_a} = \frac{34,000 \text{ cm/s}}{8 \text{ cm}} = 4,250 \text{ Hz}.$$

The frequency will be the same for the vibrating rod. However, the wavelength λ_r of the standing wave in the rod is equal to twice the length of the rod for the fundamental mode in the rod,

$$\lambda_r = 2 \times 0.8 \text{ m}, \quad \lambda_r = 1.6 \text{ m}.$$

With the frequency the same as in air and the wavelength 1.6 m, we find the speed v_r of sound in the metal rod will be

$$v_r = f\lambda_r = 4,250 \times 1.6 = 6,800 \text{ m/s}.$$

(b) This part is the same exercise but in symbols,

$$v_r = \frac{\lambda_r}{\lambda_0} v_s = \frac{2L}{\lambda_0} v_s.$$

4.15 Drawing waves in the flute with pressure nodes where there is an opening gives the answers immediately. From the sketches in Fig. A.31, in which nodes are pressure nodes, we arrive at the following answers. (i) $\lambda = 2 \times 40$ cm = 80 cm. Therefore, the frequency for that sound will be $f = v/\lambda = 431.25$ Hz. (ii) $f = 862.5$ Hz. (iii) $f = 1725$ Hz.

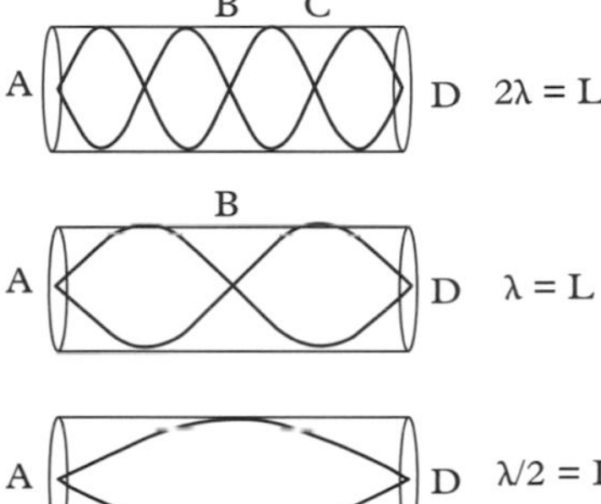

Fig. A.31 *Exercise 4.15. Modes in a simple flute.*

Chapter 5

5.1 The equation of motion we wish to solve is

$$m\frac{d^2x}{dt^2} + b\frac{dx}{dt} + kx = F_0 \sin(\omega t).$$

We will use the identity $\sin\theta = \cos(\theta - \pi/2)$ to write the driving force as a cosine,

$$m\frac{d^2x}{dt^2} + b\frac{dx}{dt} + kx = F_0 \cos(\omega t - \pi/2).$$

Let us write the right side as a real part of a complex exponential,

$$m\frac{d^2x}{dt^2} + b\frac{dx}{dt} + kx = \text{Re}\left[F_0\, e^{i(\omega t - \pi/2)}\right].$$

Let us divide both sides by m and introduce the parameters ω_0 and Γ,

$$\omega_0 = \sqrt{k/m}, \quad \Gamma = b/m.$$

We can now think of x as the real part of a complex displacement z and leave reference to the real part out of our equations,

$$m\frac{d^2z}{dt^2} + \Gamma\frac{dz}{dt} + \omega_0^2 x = \frac{F_0}{m}\, e^{i(\omega t-\pi/2)}. \tag{A.49}$$

We need a steady state solution of this equation. Let

$$z = z_0e^{(\omega t-\delta)}, \text{ with } z_0 \text{ real.}$$

Using this in Eq. A.49 and canceling out $e^{i\omega t}$ we get

$$-\omega^2 z_0 + i\omega\Gamma z_0 + \omega_0^2 z_0 = \frac{F_0}{m}\, e^{i(\delta-\pi/2)}. \tag{A.50}$$

The left side has the following polar representation:

$$\text{Left side } = z_0\sqrt{(\omega_0^2-\omega^2)^2+\omega^2\Gamma^2}\; e^{i\tan^{-1}\left[\omega\Gamma/(\omega_0^2-\omega^2)\right]}.$$

Using this in Eq. A.50, we can equate the amplitude and phase on the two sides to get

$$z_0 = \frac{F_0}{m}\frac{1}{\sqrt{(\omega_0^2-\omega^2)^2+\omega^2\Gamma^2}}, \tag{A.51}$$

$$\delta = \frac{\pi}{2} + \tan^{-1}\left(\omega\Gamma/(\omega_0^2-\omega^2)\right). \tag{A.52}$$

Now, we use

$$x(t) - \text{Re}[z],$$

to get

$$x(t) = x_0\cos(\omega t-\delta),$$

with

$$x_0 = \frac{F_0}{m}\frac{1}{\sqrt{(\omega_0^2-\omega^2)^2+\omega^2\Gamma^2}}, \tag{A.53}$$

$$\delta = \frac{\pi}{2} + \tan^{-1}\left(\frac{\omega\Gamma}{\omega_0^2-\omega^2}\right). \tag{A.54}$$

5.2 Since the system is linear, each driving force will drive displacement x independently of the other. Let x_c be the steady state solution when only $F_c\cos(\omega t)$ acts and x_s the steady state solution when only $F_s\sin(\omega t)$ acts. Then the steady state solution when both forces act will be

$$x = x_c + x_s,$$

with

$$x_c = x_{0c}\cos(\omega t-\delta_c),$$
$$x_s = x_{0s}\cos(\omega t-\delta_s),$$

where

$$x_{0c} = \frac{F_c/m}{\sqrt{(\omega_0^2 - \omega^2)^2 + \omega^2\Gamma^2}}, \quad x_{0s} = \frac{F_s/m}{\sqrt{(\omega_0^2 - \omega^2)^2 + \omega^2\Gamma^2}}$$

$$\delta_c = \tan^{-1}\left(\frac{\omega\Gamma}{\omega_0^2 - \omega^2}\right), \quad \delta_s = \frac{\pi}{2} + \tan^{-1}\left(\frac{\omega\Gamma}{\omega_0^2 - \omega^2}\right).$$

5.3 (a) The equation of motion of the damped driven oscillator with the given parameters is

$$\frac{d^2x}{dt^2} + \frac{dx}{dt} + 20^2 x = 1 \ \cos(10t),$$

where x is in m, t in sec, and ω is in sec^{-1}.

(b) In the steady state, the oscillator will oscillate with the driving frequency,

$$x = A\cos(\omega t - \delta),$$

with amplitude A given by

$$A = \frac{1}{\sqrt{(\omega_0^2 - \omega^2)^2 + \Gamma^2\omega^2}}$$
$$= \frac{1}{\sqrt{(20^2 - 10^2)^2 + 1 \times 10^2}} = 3.33 \times 10^{-3}\ \text{m}.$$

(c) The phase lag is

$$\delta = \tan^{-1}\left(\frac{\omega\Gamma}{\omega_0^2 - \omega^2}\right) = \tan^{-1}\left(\frac{10}{300}\right) \approx \frac{1}{30}\ \text{rad}.$$

(d) The average power,

$$\langle P\rangle = \frac{1}{2}\frac{F_0^2\omega}{m}\left[\frac{\Gamma\omega}{(\omega_0^2 - \omega^2)^2 + \Gamma^2\omega^2}\right].$$

Putting in the numbers we get

$$\langle P\rangle = 2.5 \times 10^{-3}\ \text{J/s}.$$

5.4 (a) The equation of motion can easily be obtained by considering the rotation of the bob about an axis through the suspension point,

$$ml^2\frac{d^2\theta}{dt^2} = -mgl\theta - bl^2\frac{d\theta}{dt} + F_0 l\cos(\omega t).$$

Divide by ml^2 and introduce $\omega_0^2 = g/l$ and $\Gamma = b/m$ and write this equation as

$$\frac{d^2\theta}{dt^2} + \Gamma\frac{d\theta}{dt} + \omega_0^2\theta = \frac{F_0}{ml}\cos(\omega t). \qquad (A.55)$$

(b) Equation A.55 is analogous to the equation of motion of a driven harmonic oscillator. Using the analogy we can immediately write the elastic and absorptive amplitudes,

$$A_{el} = \frac{F_0}{ml}\left[\frac{\omega_0^2-\omega^2}{(\omega_0^2-\omega^2)^2+\Gamma^2\omega^2}\right],$$

$$A_{ab} = \frac{F_0}{ml}\left[\frac{\Gamma\omega}{(\omega_0^2-\omega^2)^2+\Gamma^2\omega^2}\right].$$

(c) By analogy we get the time-averaged power to be

$$\langle P\rangle = \frac{F_0^2}{2m\Gamma}\left[\frac{\Gamma^2\omega^2}{(\omega_0^2-\omega^2)^2+\Gamma^2\omega^2}\right].$$

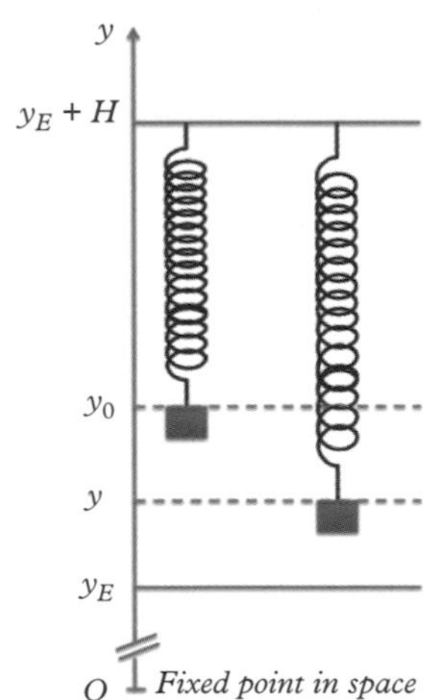

Fig. A.32 *Exercise 5.5.*

5.5 (a) We will use the coordinates shown in Figure A.32 to deduce the equation of motion of the block.

Let us first note that the change in length of the spring is given by the difference in the coordinates y and y_E,

$$\Delta l = (y_E + H) - (y + l_0).$$

The y-components of the restoring force on the block will be

$$F_1 = k[\Delta l].$$

The damping force on the block will be proportional to the rate at which the block is moving with respect to the Earth,

$$F_2 = b\frac{d}{dt}[\Delta l].$$

Using the standard substitution

$$\omega_0^2 = \sqrt{\frac{k}{m}},\quad \Gamma = \frac{b}{m},$$

we get the following for the equation of motion of the y-coordinate of the block with respect to the fixed frame:

$$\frac{d^2y}{dt^2} = \omega_0^2\Delta l + \Gamma\frac{d}{dt}(\Delta l). \tag{A.56}$$

Let us write this equation in terms of the negative of the change in length of the spring, which we will denote by the Greek letter ψ for brevity,

$$\psi = -\Delta l = (y + l_0) - (y_E + H).$$

The acceleration of the ψ variable will be

$$\frac{d^2\psi}{dt^2} = \frac{d^2y}{dt^2} - \frac{d^2y_E}{dt^2}.$$

Substituting for d^2y/dt^2 in Eq. A.56 we get

$$\frac{d^2\psi}{dt^2} + \frac{d^2y_E}{dt^2} = -\omega_0^2\psi - \Gamma\frac{d\psi}{dt}.$$

Rearranging, we find that ψ will obey the following equation of motion:

$$\frac{d^2\psi}{dt^2} + \omega_0^2\psi + \Gamma\frac{d\psi}{dt} = -\frac{d^2y_E}{dt^2}.$$

(b) For the steady state solution we note that the acceleration of the surface of the Earth provides the driving force

$$F = -m\frac{d^2y_E}{dt^2} = \left(m\omega^2 A_E\right)\cos(\omega t).$$

This is the same form as we have worked out in the text for a driven harmonic oscillator; thus, we will replace F_0 in our formulas by $(m\omega^2 A_E)$. This gives the following for the stationary state for the variable ψ:

$$\psi = \psi_0\cos(\omega t - \delta),$$
$$\psi_0 = \frac{\omega^2 A_E}{\sqrt{(\omega_0^2-\omega^2)^2+\omega^2\Gamma^2}}, \quad \delta = \tan^{-1}\left(\frac{\omega\Gamma}{\omega_0^2-\omega^2}\right).$$

(c) The resonance of ψ can be used to determine ω of the earthquake vibrations.

5.6 (a) The motion of block 1 with a rigid rod in this problem is analogous to the motion of Earth in the previous problem. Let

$$\psi = x_2 - x_1.$$

Then, we will arrive at the following equation of motion, obtained by setting $\Gamma = 0$ in the previous problem,

$$\frac{d^2\psi}{dt^2} + \omega_0^2\psi = -\frac{d^2x_1}{dt^2},$$

where $\omega_0^2 = k/m_2$. With the given $x_1(t)$ we get

$$\frac{d^2\psi}{dt^2} + \omega_0^2\psi = \omega^2\eta_0\sin(\omega t).$$

(b) The steady state solution will be given by a similar formula as in the last problem, but with the phase off by $\pi/2$ due to the sine in the driving force in place of a cosine and $\Gamma = 0$,

$$\psi = \psi_0\cos(\omega t - \delta),$$
$$\psi_0 = \frac{\omega^2\eta_0}{\sqrt{(\omega_0^2-\omega^2)^2}}, \quad \delta = \frac{\pi}{2}.$$

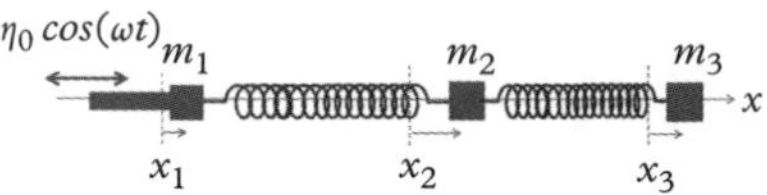

Fig. A.33 *Exercise 5.7.*

5.7 (a) Let x_2 and x_3 be the displacement of blocks m_2 and m_3 from their corresponding equilibrium positions, as shown in Fig. A.33.

The equations of motion of blocks m_2 and m_3 are

$$\frac{d^2x_2}{dt^2} + \Gamma\frac{dx_2}{dt} = -\omega_0^2(x_2 - x_1) + \omega_0^2(x_3 - x_2),$$
$$\frac{d^2x_3}{dt^2} + \Gamma\frac{dx_3}{dt} = -\omega_0^2(x_3 - x_2),$$

where $\omega_0^2 = k/m$, and $\Gamma = b/m$, and

$$x_1(t) = \eta_0 \cos(\omega t).$$

Note that the motion of $x_1(t)$ provides the driving force here. First set $x_1 = 0$ and $\Gamma = 0$ in these equations and find the normal modes by assuming

$$x_2 = A\cos(\lambda t), \quad x_3 = B\cos(\lambda t),$$

where we are using a new symbol λ for frequency since ω is used for the driving frequency. We will find that A, B, and λ must obey

$$(-\lambda^2 + 2\omega_0^2)A = \omega_0^2 B,$$
$$(-\lambda^2 + \omega_0^2)B = \omega_0^2 A.$$

Therefore,

$$\frac{B}{A} = \frac{-\lambda^2 + 2\omega_0^2}{\omega_0^2} = \frac{\omega_0^2}{-\lambda^2 + \omega_0^2}.$$

Therefore, the eigenvalues and mode ratios are

$$\lambda_1 = \left(\frac{3-\sqrt{5}}{2}\right)^{1/2}\omega_0, \quad \left(\frac{B}{A}\right)_1 = \frac{1+\sqrt{5}}{2},$$
$$\lambda_2 = \left(\frac{3+\sqrt{5}}{2}\right)^{1/2}\omega_0, \quad \left(\frac{B}{A}\right)_1 = \frac{1-\sqrt{5}}{2}.$$

Using these we can construct the two normal coordinates,

$$y_1 = x_2 + \left(\frac{B}{A}\right)_1 x_3 = x_2 + \left(\frac{1+\sqrt{5}}{2}\right) x_3, \tag{A.57}$$

$$y_2 = x_2 + \left(\frac{B}{A}\right)_2 x_3 = x_2 + \left(\frac{1-\sqrt{5}}{2}\right) x_3. \tag{A.58}$$

Now, by taking two derivatives of this equation and using the equations of motion of x_2 and x_3 you can show that y_1 and y_2 are each a separate driven oscillator,

$$\frac{d^2y_1}{dt^2} + \Gamma\frac{dy_1}{dt} + \lambda_1^2 y_1 = \omega_0^2 x_1 = \omega_0^2\eta_0\cos(\omega t),$$
$$\frac{d^2y_2}{dt^2} + \Gamma\frac{dy_2}{dt} + \lambda_2^2 y_2 = \omega_0^2 x_1 = \omega_0^2\eta_0\cos(\omega t).$$

Therefore, the steady state solutions of y_1 and y_2 will be

$$y_1 = y_{10}\cos(\omega t - \delta_1),$$
$$y_2 = y_{10}\cos(\omega t - \delta_2),$$

with

$$y_{10} = \frac{\omega^2\eta_0}{\sqrt{(\omega_1^2-\omega^2)^2+\omega^2\Gamma^2}},\quad \delta_1 = \tan^{-1}\left(\frac{\omega\Gamma}{\omega_1^2-\omega^2}\right),$$

$$y_{20} = \frac{\omega^2\eta_0}{\sqrt{(\omega_2^2-\omega^2)^2+\omega^2\Gamma^2}},\quad \delta_2 = \tan^{-1}\left(\frac{\omega\Gamma}{\omega_2^2-\omega^2}\right).$$

When we put these expressions for y_1 and y_2 into Eqs. A.57 and A.58 we can find x_2 and x_3 in the steady state by

$$x_2 = \frac{(1-\sqrt{5})y_1 + (1+\sqrt{5})y_2}{2},\quad x_3 = \frac{y_1 - y_2}{\sqrt{5}}.$$

(b) When ω is near the normal mode frequency ω_1, the amplitude y_{10} will be large compared to y_{20}. Consequently, the motion of the two blocks will be dominated by mode 1.

5.8 We use the x-coordinate from the fixed vertical line in Fig. A.34 to write the equation of motion in an inertial frame.

The equation of motion in the small-angle approximation will be

$$m\frac{d^2x}{dt^2} = -\frac{mg}{l}(x-\eta).$$

We can write this in terms of the x-displacement with respect to the suspension point,

$$\psi = x - \eta,$$

giving

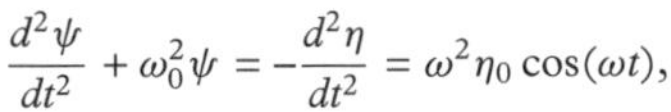

$$\frac{d^2\psi}{dt^2} + \omega_0^2\psi = -\frac{d^2\eta}{dt^2} = \omega^2\eta_0\cos(\omega t),$$

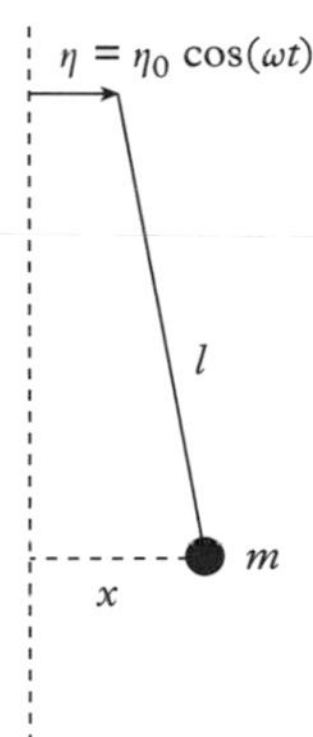

Fig. A.34 *Exercise 5.8.*

where $\omega_0^2 = \frac{g}{l}$. The steady state solution of the equation of motion for ψ will be

$$\psi = \psi_0\cos(\omega t - \delta),$$

$$\psi_0 = \frac{\omega^2\eta_0}{\sqrt{(\omega_0^2-\omega^2)^2}},\quad \delta = 0.$$

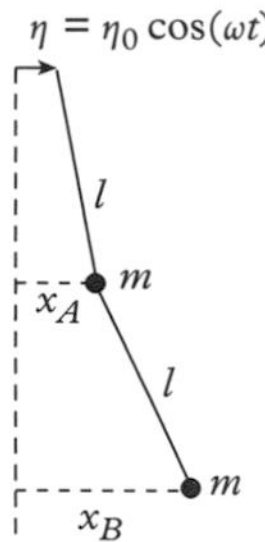

Fig. A.35 *Exercise 5.9.*

5.9 The system in Fig. A.35 is a driven coupled system. As before, we will solve it by first finding the normal modes and then rewriting the equation of motion in normal coordinates. Each normal coordinate will obey a separate driven oscillator equation, which can be solved. From the normal coordinates we will be able to deduce the original coordinates.

The equations of motion for the x-coordinates, in the small-angle approximation, with respect to an inertial frame which is the vertical line in the figure will be

$$m\frac{d^2x_A}{dt^2} = -\frac{2mg}{l}(x_A - \eta) + \frac{mg}{l}(x_B - x_A),$$
$$m\frac{d^2x_B}{dt^2} = -\frac{mg}{l}(x_B - x_A).$$

Let us introduce coordinates to measure the x-distance from the instantaneous position of the top suspension point,

$$\psi_A = x_A - \eta, \quad \psi_B = x_B - \eta. \tag{A.59}$$

Let us also introduce the following constant:

$$\omega_0^2 = \frac{g}{l}.$$

Then, we have the following equations of motion:

$$\frac{d^2\psi_A}{dt^2} + 2\omega_0^2\psi_A - \omega_0^2(\psi_B - \psi_A) = -\frac{d^2\eta}{dt^2}, \tag{A.60}$$
$$\frac{d^2\psi_B}{dt^2} + \omega_0^2(\psi_B - \psi_A) = -\frac{d^2\eta}{dt^2}. \tag{A.61}$$

To solve this for ψ_A and ψ_B we will find the normal modes of ψ_A and ψ_B. For that we need only look to the undamped undriven equations,

$$\frac{d^2\psi_A}{dt^2} + 2\omega_0^2\psi_A - \omega_0^2(\psi_B - \psi_A) = 0,$$
$$\frac{d^2\psi_B}{dt^2} + \omega_0^2(\psi_B - \psi_A) = 0.$$

We set the oscillating solutions for y_A and y_B,

$$\psi_A = Ae^{i\lambda t}, \quad \psi_B = Be^{i\lambda t},$$

in these equations to get

$$\left(-\lambda^2 + 3\omega_0^2\right)A = \omega_0^2 B,$$
$$\left(-\lambda^2 + \omega_0^2\right)B = \omega_0^2 A.$$

Therefore,

$$\frac{-\lambda^2 + 3\omega_0^2}{\omega_0^2} = \frac{\omega_0^2}{-\lambda^2 + \omega_0^2}.$$

This gives the two normal mode frequencies to be

$$\lambda_1 = \left(2 - \sqrt{2}\right)^{1/2} \omega_0, \quad \lambda_2 = \left(2 + \sqrt{2}\right)^{1/2} \omega_0,$$

and the corresponding amplitude ratios,

$$\left(\frac{B}{A}\right)_1 = 1 + \sqrt{2}, \quad \left(\frac{B}{A}\right)_2 = 1 - \sqrt{2}.$$

Now, we construct the normal coordinates. Let us denote the normal coordinates by ψ_1 and ψ_2,

$$\psi_1 = \psi_A + \left(\frac{B}{A}\right)_1 \psi_B = \psi_A + (1 + \sqrt{2})\psi_B, \tag{A.62}$$

$$\psi_2 = \psi_A + \left(\frac{B}{A}\right)_2 \psi_B = \psi_A + (1 - \sqrt{2})\psi_B. \tag{A.63}$$

Now, by taking derivatives of ψ_1 and ψ_2 and using Eqs. A.60 and A.61 we can show that

$$\frac{d^2\psi_1}{dt^2} + \lambda_1^2\psi_1 = -(2 + \sqrt{2})\frac{d^2\eta}{dt^2} = (2 + \sqrt{2})\omega^2\eta_0 \cos(\omega t), \tag{A.64}$$

$$\frac{d^2\psi_2}{dt^2} + \lambda_2^2\psi_2 = -(2 - \sqrt{2})\frac{d^2\eta}{dt^2} = (2 - \sqrt{2})\omega^2\eta_0 \cos(\omega t). \tag{A.65}$$

The steady state of the normal coordinates ψ_1 and ψ_2 can be obtained immediately by analogy to the driven one-dimensional harmonic oscillator,

$$\psi_1 = \psi_{10} \cos(\omega t - \delta_1), \quad \psi_2 = \psi_{20} \cos(\omega t - \delta_2), \tag{A.66}$$

with

$$\psi_{10} = \frac{(2 + \sqrt{2})\omega^2\eta_0}{\sqrt{(\lambda_1^2 - \omega^2)^2}}, \quad \delta = 0,$$

$$\psi_{20} = \frac{(2 - \sqrt{2})\omega^2\eta_0}{\sqrt{(\lambda_2^2 - \omega^2)^2}}, \quad \delta = 0.$$

Equation A.66 can be used in Eqs. A.62 and A.63 to get ψ_A and ψ_B,

$$\psi_A = \frac{(1 + \sqrt{2})\psi_2 - (1 - \sqrt{2})\psi_1}{2\sqrt{2}}, \quad \psi_B = \frac{\psi_1 - \psi_2}{2\sqrt{2}}.$$

Now, we can use these in Eq. A.59 to obtain x_A and x_B to be

$$x_A = \psi_A + \eta, \quad x_B = \psi_B + \eta.$$

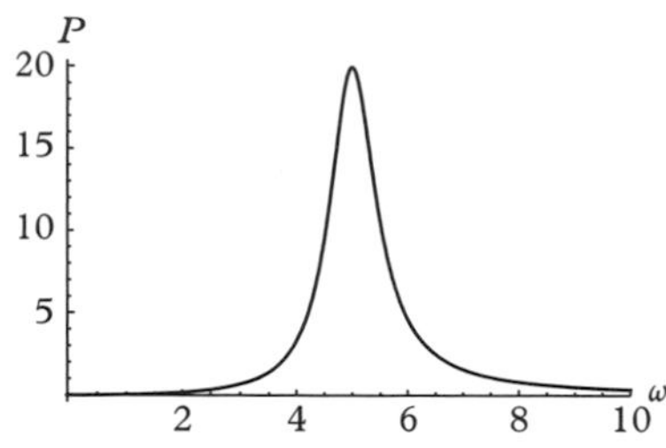

Fig. A.36 *Exercise 5.10.*

5.10 (a) See Fig. A.36.

(b) By taking the derivative of P with respect to ω and setting it to zero we find that the maximum of power occurs at $\omega = 5\ \text{sec}^{-1}$.

(c) The maximum value of power is 20 W. By setting the expression of power to 10 W we get

$$\frac{20\omega^2}{(\omega^2 - 25)^2 + \omega^2} = 10.$$

Solving this for ω^2 we get two values,

$$\omega_{1,2}^2 = \frac{1}{2}\left(51 \pm \sqrt{51^2 - 4 \times 625}\right) = 20.5,\ 30.5,$$

which give the two frequencies,

$$\omega_1 = 4.5\ \text{sec}^{-1},\ \ \omega_2 = 5.5\ \text{sec}^{-1}.$$

Therefore, the width of the peak is

$$\Delta\omega_{1/2} = 5.5 - 4.5 = 1.0\ \text{sec}^{-1}.$$

(d) The time constant is obtained by the inverse of the half width,

$$\tau = \frac{1}{1.0\ \text{sec}^{-1}} = 1.0\ \text{sec}.$$

5.11 (a) Since the hammer hits in the middle, the middle will not be a node of the excited modes since that point will be moving. Since the fundamental mode of the string does not have a node at the middle point, the lowest frequency mode excited will be the fundamental mode. The wavelength for this mode is equal to $2L$, as we have seen in systems fixed at both ends,

$$\lambda = 2L.$$

The wave number for this mode will be

$$k_{\text{lowest}} = \frac{2\pi}{\lambda} = \frac{\pi}{L} = \pi\ \text{m}^{-1}.$$

(b) The wave velocity on the taut string is

$$v = \sqrt{\frac{T}{\mu}} = \sqrt{\frac{200\ \text{N}}{0.1\ \text{kg/m}}} = 44.7\ \text{m/s}.$$

Therefore, the frequency of the fundamental mode

$$\omega_{\text{lowest}} = vk_{\text{lowest}} = 44.7\pi\ \text{sec}^{-1} = 140.5\ \text{sec}^{-1}.$$

5.12 (a) The normal mode frequencies for a sound wave in a tube opened at one end and closed at the other are given by

$$f_n = \frac{v}{\lambda_n} = \frac{(2n+1)v}{4L},\quad n = 0,\ 1,\ 2,\ 3, \ldots.$$

Using the length $L = 70$ cm and speed $v = 34,000$ cm/s in this equation we get

$$f_n = 121.43(2n + 1) \text{ Hz}.$$

The range of frequencies to be excited will be from 300 Hz to 3,000 Hz. Therefore, the lowest frequency mode to be excited will be

$$121.43(2n_{\text{low}} + 1) = 300 \implies n_{\text{low}} = 1,$$

where I have rounded to the nearest integer.

(b) The mode number for the upper end of the frequency range will be

$$121.43(2n_{\text{high}} + 1) = 3000 \implies n_{\text{high}} = 11.$$

Therefore, we will have modes $n = 1$ to $n = 11$ excited. The actual frequencies that will be excited are

$$f_1 = 364 \text{ Hz}, \quad f_2 = 607 \text{ Hz}, \quad f_3 = 850 \text{ Hz},$$
$$f_4 = 1093 \text{ Hz}, \quad f_5 = 1336 \text{ Hz}, \quad f_6 = 1579 \text{ Hz},$$
$$f_7 = 1821 \text{ Hz}, \quad f_8 = 2024 \text{ Hz}, \quad f_9 = 2307 \text{ Hz},$$
$$f_{10} = 2550 \text{ Hz}, \quad f_{11} = 2793 \text{ Hz}.$$

5.13 (a) Let us calculate ω_0 and Γ to see what type of damping is in the system,

$$\omega_0 = \sqrt{\frac{k}{m}} = 31.6 \text{ sec}^{-1}; \quad \Gamma = \frac{b}{m} = 20 \text{ sec}^{-1}.$$

Since $\omega_0 > \Gamma/2$, we have an under-damped oscillator. Therefore, the complete displacement of the oscillator will be given by

$$x(t) = e^{-\frac{\Gamma}{2}t}\left[C \cos(\omega_1 t) + S \sin(\omega_1 t)\right] + A \cos(\omega t - \delta), \tag{A.67}$$

where

$$\omega_1 = \sqrt{\omega_0^2 - \frac{\Gamma^2}{4}} = 30 \text{ sec}^{-1},$$

$$A = \frac{F_0/m}{\sqrt{(\omega_0^2 - \omega^2)^2 + \Gamma^2\omega^2}} = \frac{10/1}{\sqrt{(31.6^2 - (200\pi)^2)^2 + 20^2(200\pi)^2}}$$
$$= 2.5 \times 10^{-5} \text{ m}$$

$$\delta = \tan^{-1}\left(\frac{\omega\Gamma}{\omega_0^2 - \omega^2}\right) = \tan^{-1}\left(\frac{200\pi \times 20}{31.6^2 - (200\pi)^2}\right) = -0.032 \text{ rad}.$$

To find the constants C and S we use the initial conditions on $x(t)$,

$$x(0) = C + A\cos\delta = 0,$$
$$v(0) = \omega_1 S - \frac{\Gamma}{2}C + \omega A \sin\delta = 0.$$

Now, we solve these two to get C and S,

$$C = -A\cos\delta = -2.5 \times 10^{-5}\ \text{m},$$

$$S = \frac{\frac{\Gamma}{2}C - \omega A\sin\delta}{\omega_1} = 2.5 \times 10^{-5}\ \text{m}.$$

(b) The steady state part is the $A\cos(\omega t - \delta)$.

(c) The criterion in time is when $t \gg 2/\Gamma = 0.1$ sec so that the complementary solution becomes negligibly small due to the exponential decaying multiplying factor.

5.14 (a) Since the system is linear we can add up the effect caused by each force,

$$x = x_1 + x_2,$$

with

$$x_1 = A_1\cos(\omega_1 t - \delta_1),\quad x_2 = A_2\cos(\omega_2 t - \delta_2),$$

where (with $\omega_0^2 = k/m$ and $\Gamma = b/m$),

$$A_1 = \frac{F_0/m}{\sqrt{(\omega_1^2-\omega_0^2)^2 + \Gamma^2\omega_1^2}},\qquad \delta_1 = \tan^{-1}\left(\frac{\Gamma\omega_1}{\omega_1^2-\omega_0^2}\right),$$

$$A_2 = \frac{F_0/m}{\sqrt{(\omega_2^2-\omega_0^2)^2 + \Gamma^2\omega_2^2}},\qquad \delta_2 = \tan^{-1}\left(\frac{\Gamma\omega_2}{\omega_2^2-\omega_0^2}\right).$$

(b) The instantaneous power will simply be the net power of the net force, which will have cross-terms as well,

$$\begin{aligned} P(t) = F(t)v(t) &= [F_0\cos(\omega_1 t) + F_0\cos(\omega_2 t)]\left(\frac{dx_1}{dt} + \frac{dx_2}{dt}\right), \\ &= F_0\cos(\omega_1 t)\left(\frac{dx_1}{dt} + \frac{dx_2}{dt}\right) + F_0\cos(\omega_2 t)\left(\frac{dx_1}{dt} + \frac{dx_2}{dt}\right). \end{aligned}$$

(c) To find the average power, we need to average the instantaneous power over time. We note that if the two frequencies are different then, we will get

$$\langle\cos(\omega_1 t)\cos(\omega_2 t)\rangle = \frac{1}{2}\langle\cos(\omega_1+\omega_2)t\rangle + \frac{1}{2}\langle\cos(\omega_1-\omega_2)t\rangle = 0,$$

and when the two cosines have the same frequency,

$$\langle\cos(\omega_1 t)\cos(\omega_1 t)\rangle = \frac{1}{2}\langle 1 + \cos(2\omega_1 t)\rangle = \frac{1}{2},$$

and similarly for ω_2. Therefore,

$$\langle P\rangle = \frac{F_0 A_1\omega_1}{2}\sin\delta_1 + \frac{F_0 A_2\omega_2}{2}\sin\delta_1.$$

5.15 (a) The steady state solution will just be equal to the sum from two driving forces, as shown above, except now, the two terms have different F_0.

(b) The answer can be generalized to

$$x(t) = \sum_{i=1}^{\infty} A_i \cos(\omega_i t - \delta_i),$$

where

$$A_i = \frac{F_i/m}{\sqrt{(\omega_i^2 - \omega_0^2)^2 + \Gamma^2\omega_i^2}}, \quad \delta_i = \tan^{-1}\left(\frac{\Gamma\omega_i}{\omega_i^2 - \omega_0^2}\right).$$

5.16 (a) The multiples of the fundamental frequency are

$$f_n = \frac{n}{T}, \quad \omega_n = 2\pi f_n.$$

We need the coefficients in

$$f(t) = \frac{A_0}{2} + \sum_{n=1,2,...} A_n \cos\left(\frac{2\pi nt}{T}\right) + \sum_{n=1,2,...} B_n \sin\left(\frac{2\pi nt}{T}\right),$$

where

$$A_0 = \frac{2}{T}\int_0^T f(t)dt, \quad A_n = \frac{2}{T}\int_0^T f(t)\cos\left(\frac{2\pi nt}{T}\right)dt,$$
$$B_n = \frac{2}{T}\int_0^T f(t)\sin\left(\frac{2\pi nt}{T}\right)dt.$$

Using the given function in these integrals we get

$$A_0 = 1, \quad A_n = 0,$$
$$B_n = \frac{2}{T}\int_{T/2}^T \sin\left(\frac{2\pi nt}{T}\right)dt = \begin{cases} -\dfrac{2}{\pi n} & (n \text{ odd}) \\ 0 & (n \text{ even}). \end{cases}$$

Therefore we have

$$f(t) = \frac{1}{2} - \frac{2}{\pi}\left[\sin\left(\frac{2\pi t}{T}\right) + \frac{1}{3}\sin\left(\frac{6\pi t}{T}\right) + \frac{1}{5}\sin\left(\frac{10\pi t}{T}\right) + \cdots\right].$$

(b) The steady state will be the sum of contributions from each component of the driving force,

$$x = x_0 + \sum_{n=1,2,...} x_{n0}\cos(\omega_n t - \delta_n),$$

where

$$\omega_n = 2\pi f_n = \frac{2\pi n}{T}.$$

For the constant force $F = \frac{1}{2}$, the displacement will be moved by a constant amount (as you can easily check by setting $x = x_0$ and $F = A_0/2$ in the equation of motion),

$$x_0 = \frac{1}{2m\omega_0^2},$$

and the sine driving forces will give rise to the following expressions for the amplitude and phase lags:

$$x_{n0} = \frac{\omega_n^2 B_n/m}{\sqrt{(\omega_0^2 - \omega_n^2)^2 + \Gamma^2\omega_n^2}},$$

$$\delta_n = \frac{\pi}{2} + \tan^{-1}\left(\frac{\Gamma\omega_n}{\omega_0^2 - \omega_n^2}\right).$$

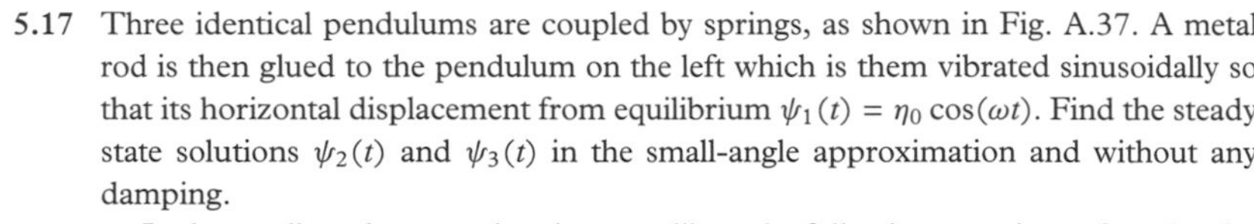

5.17 Three identical pendulums are coupled by springs, as shown in Fig. A.37. A metal rod is then glued to the pendulum on the left which is them vibrated sinusoidally so that its horizontal displacement from equilibrium $\psi_1(t) = \eta_0\cos(\omega t)$. Find the steady state solutions $\psi_2(t)$ and $\psi_3(t)$ in the small-angle approximation and without any damping.

In the small-angle approximation we will get the following equations of motion for ψ_2 and ψ_3:

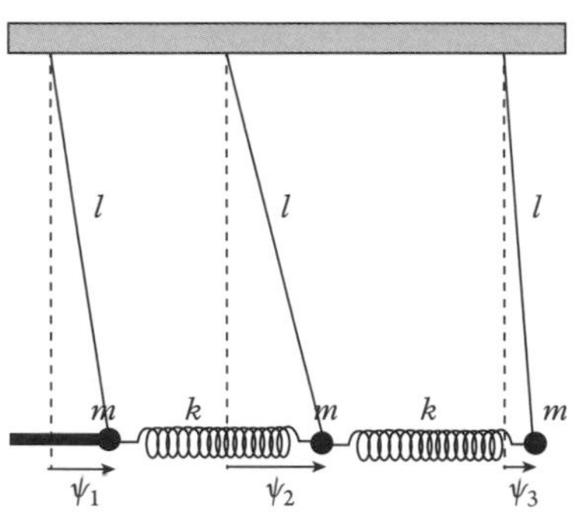

Fig. A.37 *Exercise 5.17.*

$$m\frac{d^2\psi_2}{dt^2} = -k(\psi_2 - \psi_1) + k(\psi_3 - \psi_2) - \frac{mg}{l}\psi_2,$$

$$m\frac{d^2\psi_3}{dt^2} = -k(\psi_3 - \psi_2) - \frac{mg}{l}\psi_3,$$

with ψ_1 moved rigidly to follow

$$\psi_1 = \eta_0\cos(\omega t).$$

Let us introduce

$$\omega_0^2 = \frac{g}{l},\quad \omega_c^2 = \frac{k}{m}.$$

We can find the steady state by working with normal coordinates as we have done above. Alternatively, we can find the steady state solution by demanding that the two driven masses oscillate at the same frequency as the driving force. We will carry out the latter method in this problem. Thus, let us write the displacements in exponential notation,

$$\psi_1 = \eta_0\, e^{i\omega t},\ \psi_2 = C_2\, e^{i\omega t},\ \psi_3 = C_3\, e^{i\omega t},$$

from which the actual solution can be deduced by taking the real part. Using these in the equations of motion we get

$$-\omega^2 C_2 = -\omega_c^2(C_2 - \eta_0) + \omega_c^2(C_3 - C_2) - \omega_0^2 C_2,$$

$$-\omega^2 C_3 = -\omega_c^2(C_3 - C_2) - \omega_0^2 C_3.$$

Let us organize these equations as follows:

$$\left(-\omega^2 + \omega_0^2 + 2\omega_c^2\right) C_2 - \omega_c^2 C_3 = \omega_c^2 \eta_0,$$
$$-\omega_c^2 C_2 + \left(-\omega^2 + \omega_0^2 + \omega_c^2\right) C_3 = 0.$$

These are two equations in two unknowns C_2 and C_3, which can be solved by standard methods. The answer we obtain is

$$C_2 = \frac{(-\omega^2 + \omega_0^2 + \omega_c^2)\omega_c^2 \eta_0}{(-\omega^2 + \omega_0^2 + 2\omega_c^2)(-\omega^2 + \omega_0^2 + \omega_c^2) - \omega_c^4},$$
$$C_3 = \frac{\omega_c^4 \eta_0}{(-\omega^2 + \omega_0^2 + 2\omega_c^2)(-\omega^2 + \omega_0^2 + \omega_c^2) - \omega_c^4}.$$

If you worked out the normal modes of this system by setting $\eta = 0$ in the equations of motion, you would find that the normal frequencies are

$$\omega_1 = \sqrt{\omega_0^2 + \left(\frac{3 - \sqrt{5}}{2}\right)\omega_c^2}; \; \omega_2 = \sqrt{\omega_0^2 + \left(\frac{3 + \sqrt{5}}{2}\right)\omega_c^2}.$$

The denominators in the constants C_2 and C_3 can be expressed more succinctly in terms of the normal mode frequencies as

$$C_2 = \frac{\omega_c^2(-\omega^2 + \omega_0^2 + \omega_c^2)}{(\omega^2 - \omega_1^2)(\omega^2 - \omega_2^2)}\eta_0,$$

$$C_3 = \frac{\omega_c^4}{(\omega^2 - \omega_1^2)(\omega^2 - \omega_2^2)}\eta_0.$$

5.18 (a) Since $f_R = 20$ kHz and $\Delta f = 10$ Hz we have

$$Q = \frac{\omega_R}{\Gamma} = \frac{2\pi f_R}{2\pi\,\Delta f_{1/2}} = \frac{20,000\text{ Hz}}{10\text{ Hz}} = 2,000.$$

(b) Doubling the resistance R of the circuit will double the value of Γ. Since $\Delta f \sim \Gamma$, the width $\Delta f_{1/2}$ will also double.

(c) The resonance frequency depends on the capacitance C as inverse square root,

$$f_R \sim \frac{1}{\sqrt{C}}.$$

Therefore, doubling C will reduce f_R by a factor of $\sqrt{2}$,

$$f_R = \frac{20\text{ kHz}}{\sqrt{2}} = 14.14\text{ kHz}.$$

The peak width does not depend on C.

(d) The resonance frequency and width of the resonance peak depend on L differently,

$$f_R \sim \frac{1}{\sqrt{L}}, \quad \Delta f_{1/2} \sim \frac{1}{L}.$$

Therefore, doubling L will decrease f_R by a factor of $\sqrt{2}$ and halve the $\Delta f_{1/2}$,

$$f_R = \frac{20\text{ kHz}}{\sqrt{2}} = 14.14\text{ kHz}; \quad \Delta f_{1/2} = \frac{10\text{ Hz}}{2} = 5\text{ Hz}.$$

(e) At resonance frequency, the average power will be related to the peak voltage by

$$\langle P \rangle \approx \frac{V_0^2}{2R}.$$

Therefore,

$$V_0 = \sqrt{2R\langle P \rangle} = \sqrt{2 \times 10\ \Omega \times 30\text{ W}} = 24.5\text{ Volt}.$$

5.19 Let q_4, q_5, and q_3 denote the charges on the three capacitors when the current in their branches in Fig. A.38 are I_4, I_5, and I_3 respectively.

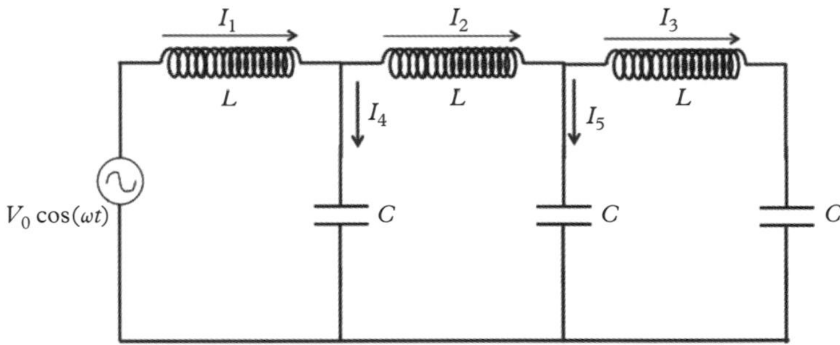

Fig. A.38 *Exercise 5.19.*

With the convention that current in the capacitor is flowing into the positive end at instant t, we get the following loop equations in the three loops:

$$V_0\cos(\omega t) - L\frac{dI_1}{dt} - \frac{q_4}{C} = 0,$$

$$\frac{q_4}{C} - L\frac{dI_2}{dt} - \frac{q_5}{C} = 0,$$

$$\frac{q_5}{C} - L\frac{dI_3}{dt} - \frac{q_3}{C} = 0.$$

Now, taking derivatives of these equations with respect to t and using the current conservation equations at the nodes, we can eliminate I_4 and I_5. Let us also introduce

$$\omega_0^2 = \frac{1}{LC}.$$

After simplifications we arrive at

$$\frac{d^2 I_1}{dt^2} + \omega_0^2 I_1 - \omega_0^2 I_2 = -\frac{\omega V_0}{L}\sin(\omega t),$$

$$\frac{d^2 I_2}{dt^2} + 2\omega_0^2 I_2 - \omega_0^2 I_1 - \omega_0^2 I_3 = 0,$$

$$\frac{d^2 I_3}{dt^2} + 2\omega_0^2 I_3 - \omega_0^2 I_2 = 0.$$

Since we do not have dissipative terms, the currents will also be oscillating as sines of the same frequency as the oscillating source. Let

$$I_1 = A_1 \sin(\omega t),$$

$$I_2 = A_2 \sin(\omega t),$$

$$I_3 = A_3 \sin(\omega t).$$

Then, the amplitudes of the current obey

$$-\omega^2 A_1 + \omega_0^2 A_1 - \omega_0^2 A_2 = -\frac{\omega V_0}{L},$$

$$-\omega^2 A_2 + 2\omega_0^2 A_2 - \omega_0^2 A_1 - \omega_0^2 A_3 = 0,$$

$$-\omega^2 A_3 + 2\omega_0^2 A_3 - \omega_0^2 A_2 = 0.$$

These are three equations in three unknowns, which is left as an exercise for the student. I got really complicated expressions for A's,

$$A_1 = \frac{-\omega^4 + 4\omega^2\omega_0^2 - 3\omega_0^4}{\text{Det}}\frac{\omega V_0}{L},$$

$$A_2 = \frac{\omega_0^2\left(\omega^2 - 2\omega_0^2\right)}{\text{Det}}\frac{\omega V_0}{L},$$

$$A_2 = \frac{\omega_0^4}{\text{Det}}\frac{\omega V_0}{L},$$

where

$$\text{Det} = -\omega^6 + 5\omega^4\omega_0^2 - 6\omega^2\omega_0^4 + \omega_0^6.$$

Note that the denominator in A_1 and the other amplitudes can be shown to be the product $(\omega^2-\omega_1^2)\,(\omega^2-\omega_2^2)\,(\omega^2-\omega_3^2)$, where ω_1, ω_2, and ω_3 are the normal frequencies of the system.

Chapter 6

6.1 The following expression of a traveling wave on a string is given in the problem,

$$y(x, t) = [0.05 \text{ m}] \cos(2\pi x - 200\pi t).$$

(a) Amplitude of the wave, $A = 0.05$ m.

(b) (i) The displacement of the particle at $x = 0$ will be

$$y(0, t) = [0.05 \text{ m}] \cos(200\pi t).$$

This is a simple harmonic motion with amplitude 0.05 m. (ii) The displacement of the particle at $x = 0.1$ m will be

$$y(0, t) = [0.05 \text{ m}] \cos(0.2\pi - 200\pi t).$$

This is also a simple harmonic motion with amplitude 0.05 m.

(c) Wave number, $k = 2\pi \text{ m}^{-1}$.

(d) Angular frequency, $\omega = 200\pi \text{ sec}^{-1}$, and frequency, $f = \omega/2\pi = 100$ Hz.

(e) Wavelength, $\lambda = 2\pi/k$, with $k = 2\pi \text{ m}^{-1}$. Therefore, $\lambda = 1$ m.

(f) Period, $T = 1/f$ with $f = 100$ Hz. Therefore,

$$T = \frac{1}{100 \text{ Hz}} = 0.01 \text{ sec}.$$

(g) Phase velocity, $v_p = \omega/k = 100$ m/s.

(h) To find the speed of the particle we first find the displacement of the article at an arbitrary instant t,

$$y(1 \text{ m}, t) = [0.05 \text{ m}] \cos(2\pi - 200\pi t) - [0.05 \text{ m}] \cos(200\pi t).$$

Taking the time derivative we get

$$v = \frac{dy}{dt} = -[0.05 \times 200\pi \text{ m/s}] \sin(200\pi t).$$

Therefore, the speed at $t = 2.5 \times 10^{-3}$ sec will be

$$v = |-[0.05 \times 200\pi \text{ m/s}] \sin(200\pi/600)| = 27.2 \text{ m/s}.$$

(i) Maximum speed of any particle of the string will occur when the sine in the expression for the velocity has value 1. When that happens, the speed will be $v_{max} = 0.05 \times 200\pi = 31.4$ m/s. This should be compared with the wave speed, which is given by the phase velocity here, which is $v_p = 100$ m/s.

(j) Minimum speed will be zero since particles will come to rest at their turning points.

(k) From the phase velocity, we can get T since $v_p = \sqrt{T/\mu}$.
Therefore,

$$T = \mu v_p^2 = 0.01 \text{ kg/m} \times (100 \text{ m/s})^2 = 100 \text{ N}.$$

6.2 (a) Let us calculate k and ω first,

$$k = \frac{2\pi}{\lambda} = 4\pi \text{ m}^{-1},$$

$$\omega = kv_p = k\sqrt{T/\mu} = 400\pi \text{ sec}^{-1}$$

Since the wave is traveling towards the positive z-axis with amplitude 4 cm, we will have

$$y(z,t) = [4\text{ cm}]\cos(4\pi\text{ m}^{-1}z - 400\pi\text{ sec}^{-1}t + \phi),$$

where ϕ is an undetermined phase constant. We can fix the value of ϕ if we know the amplitude of the wave at some point at a particular time, for instance, at $t = 0$ at $z = 0$.

(b) The sign of the k term will change in this case,

$$y(z,t) = [4\text{ cm}]\cos(-4\pi\text{ m}^{-1}z - 400\pi\text{ sec}^{-1}t + \phi).$$

6.3 (a) Since z and t terms have the same sign, the wave would be traveling towards the negative z-axis. Also, from $\varphi(z,t) = 0.8\pi z + 28\pi t$, you get $dz/dt = -35$ cm/s.

(b) From k and ω we can find the v_p of the wave by

$$k = 0.8\pi\text{ cm}^{-1},\ \omega = 28\pi\text{ sec}^{-1},$$

$$v_p = \frac{\omega}{k} = \frac{28\pi}{0.8\pi} = 35\text{ cm/s}.$$

(c) From v_p and mass density μ we can get T by,

$$T = \mu v_p^2 = 2.0\text{ kg/m} \times (0.35\text{ m/s})^2 - 0.245\text{ N}.$$

(d) The average power

$$\langle P\rangle = \frac{1}{2}Z\omega^2A^2 = \frac{1}{2}\sqrt{T\mu}\omega^2A^2.$$

Putting in the numerical values we obtain

$$\langle P\rangle = \frac{1}{2}\sqrt{0.245\text{ N} \times 2\text{ kg/m}}(28\pi\text{ s}^{-1})^2(0.02\text{ m})^2 = 1.1\text{ W}.$$

6.4 (a) From the given displacement wave function we get the velocity of the oscillating particles of the string in units of cm/s by

$$\frac{\partial y}{\partial t} = -1200\pi\sin(200\pi x + 400\pi t).$$

Evaluating this for $x = 0$ gives the displacement velocity of the particle at $x = 0$, which can be evaluated for $t = 0.001$ sec to obtain the velocity at that instant,

$$\left|\frac{dy}{dt}\right|_{x=0,t=0.001} = +1200\pi\,|\sin(0.4\pi)| = 3585\text{ cm/s}.$$

(b) The phase difference will be obtained from the difference in the argument of the cosine at the two instants,

$$\Delta\varphi = 400\pi\,\Delta t = 400\pi\,[\text{sec}^{-1}] \times 0.002\text{ sec} = 0.8\pi\text{ rad}.$$

(c) The phase difference will be obtained from the difference in the argument of the cosine at the two points in space,

$$\Delta\varphi = 200\pi\,[\text{cm}^{-1}]\Delta x = 200\pi\text{ rad}.$$

6.5 (a) From the tension and mass density we note that the wave velocity will be

$$v_p = \sqrt{T/\mu}.$$

The wave towards $x = +\infty$ will be

$$y_+(x,t) = A_+ \cos\left(\omega t - \frac{\omega}{v}x\right),$$

and the wave towards $x = -\infty$ will be

$$y_-(x,t) = A_- \cos\left(\omega t + \frac{\omega}{v}x\right),$$

where A_+ and A_- are the wave amplitudes. Now, the wave at $x = 0$ has a unique displacement whether you come from the right or from the left of the origin. That is, we must have the following boundary condition at $x = 0$.

$$y_+(0,t) = y_-(0,t) = y_0 \cos(\omega t).$$

Therefore, the two amplitudes will be equal to y_0,

$$A_+ = A_- = y_0.$$

Thus, the two waves are

$$y_+(x,t) = y_0 \cos\left(\omega t - \frac{\omega}{v}x\right),$$

$$y_-(x,t) = y_0 \cos\left(\omega t + \frac{\omega}{v}x\right).$$

(b) (i) The force at any point of the string comes from the net y-component of the tension at that point. At $x = 0$, the tension components can be obtained by using the wave amplitudes on the two sides of $x = 0$,

$$F_y = -T\left(\frac{\partial y_+}{\partial x} - \frac{\partial y_-}{\partial x}\right)_{x=0}.$$

This gives

$$F_y = 2Ty_0\frac{\omega}{v_p}\sin(\omega t).$$

(ii) To find the average power, we can evaluate the time average of the instantaneous power at $x = 0$,

$$\langle P\rangle = \left\langle F_y \frac{\partial y}{\partial t}\right\rangle.$$

This gives

$$\langle P\rangle = 2Ty_0^2\frac{\omega^2}{v_p}\left\langle \sin^2(\omega t)\right\rangle = Ty_0^2\frac{\omega^2}{v_p}.$$

6.6 (a) In the standing waves the sine or cosine of space is multiplied by a sine or cosine of the time part. Expanding the given traveling wave immediately gives the two standing waves whose sum is the traveling wave,

$$A\cos(kx-\omega t) = A\cos(kx)\cos(\omega t) + A\sin(kx)\sin(\omega t).$$

(b) Same method as above.

$$\begin{aligned}\psi(x,t) &= A\cos(kx-\omega t) + B\cos(kx+\omega t),\\ &= A\cos(kx)\cos(\omega t) + A\sin(kx)\sin(\omega t)\\ &\quad + B\cos(kx)\cos(\omega t) - B\sin(kx)\sin(\omega t),\\ &= (A+B)\cos(kx)\cos(\omega t) + (A-B)\sin(kx)\sin(\omega t),\\ &= C\cos(kx)\cos(\omega t) + D\sin(kx)\sin(\omega t).\end{aligned}$$

6.7 (a) Figure A.39 shows the circuit element between $x = (j-1)a$ and $x = ja$. The charging of the capacitor during an interval Δt will give

$$Ca\Delta V = \Delta q = \Delta I \Delta t = [I(x = ja, t) - I(x = (j-1)a, t)]\,\Delta t.$$

Dividing both sides by $a\Delta t$ and taking the limit $\Delta \to 0$ and $a \to 0$ with $x = ja$ fixed gives

$$C\frac{\partial V(x,t)}{\partial t} = -\frac{\partial I(x,t)}{\partial x} \quad (1)$$

(b) The voltage across the inductor will give

$$La\frac{\partial I(x = ja, t)}{\partial t} = -[V(x = ja, t) - V(x = (j-1)a, t)]\,.$$

Dividing by a both sides and taking the limit $a \to 0$ such that $j \to \infty$ in such a way that x remains constant we get the following relation of quantities at $x = x$ at time $t = t$:

$$L\frac{\partial I(x,t)}{\partial t} = -\frac{\partial V(x,t)}{\partial x} \quad (2)$$

(c) Taking the time derivative of (2) and using (1) to substitute for $\partial V/\partial t$ gives

$$L\frac{\partial^2 I(x,t)}{\partial t^2} = \frac{1}{C}\frac{\partial^2 I(x,t)}{\partial x^2},$$

which gives the required wave equation for the current wave,

$$\frac{\partial^2 I(x,t)}{\partial t^2} = \frac{1}{LC}\frac{\partial^2 I(x,t)}{\partial x^2},$$

with wave speed

$$v_p = \frac{1}{\sqrt{LC}}.$$

The wave equation for the voltage wave is obtained in a similar way.

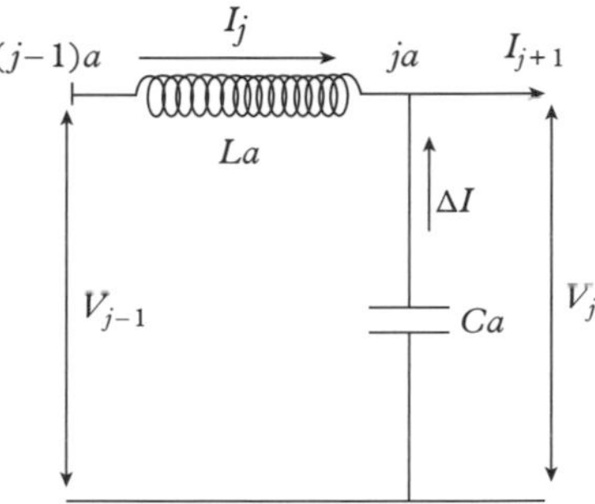

Fig. A.39 *Exercise 6.7.*

(d) Putting the wave solutions for I and V into either (1) or (2) gives the required relation between I_0 and V_0. In (1) we get

$$C\omega V_0 \sin(kx - \omega t) = kI_0 \sin(kx - \omega t),$$

in which we can cancel the sine part to get

$$C\omega V_0 = kI_0.$$

Now, we know that the wave velocity is also equal to ω/k. Therefore,

$$I_0 = Cv_p V_0 = \sqrt{C/L}\, V_0.$$

(e) The ratio V_0/I_0 gives the characteristic impedance Z,

$$Z = \sqrt{L/C}.$$

(f) The average power will be the time average of IV,

$$\langle P \rangle = \langle IV \rangle = \frac{1}{2} I_0 V_0.$$

6.8 Let us collect the numerical values we will need in this problem.

$$\begin{aligned}
&\text{Current amplitude, } I_0 = 10 \text{ mA},\\
&\text{Angular frequency, } \omega = 2\pi f = 2\pi \times 10^8 \text{ sec}^{-1},\\
&\text{Phase velocity, } v_p = 1 \times 10^8 \text{ m/s},\\
&\text{Wave number, } k = \frac{\omega}{v_p} = 2\pi \text{ m}^{-1}.
\end{aligned}$$

(a) $I(z,t) = I_0 \cos(kz - \omega t + \phi)$ gives

$$I(z,t) = 10 \text{ mA} \cos(2\pi \text{ m}^{-1} z - 2\pi \times 10^8 \text{ sec}^{-1} t + \phi),$$

where the phase constant ϕ is undetermined.

(b) The phase change is the change in the argument of the cosine in the wave for $x = 0$ and $\Delta t = 2 \times 10^{-9}$ sec,

$$\Delta\varphi = 2\pi \times 10^8 \text{ sec}^{-1} \Delta t = 0.4\pi \text{ rad}.$$

(c) We set the phase change to 0.5 rad to get the distance Δz,

$$k\Delta z = 0.5 \text{ rad}.$$

Therefore,

$$\Delta z = \frac{0.5 \text{ rad}}{2\pi \text{ m}^{-1}} = \frac{1}{4\pi} \text{ m}.$$

Note that this answer is not unique since the wave is a harmonic wave and will repeat in space. This distance is the minimum distance.

(d) The speed of a wave on a transmission line is given by $1/\sqrt{LC}$, which will equal the phase velocity,

$$L = \frac{1}{v_p^2 C} = \frac{1}{(1 \times 10^8 \text{ m/s})^2 \times 10^{-7} \text{ F/m}} = 10^{-9} \text{ H/m}.$$

(e) The characteristic impedance of a transmission line is given by

$$Z = \sqrt{\frac{L}{C}}.$$

Therefore, we get

$$Z = \sqrt{\frac{10^{-9} \text{ H/m}}{10^{-7} \text{ F/m}}} = 0.1 \ \Omega.$$

6.9 Figure A.40 shows the displacements around the pendulum i.

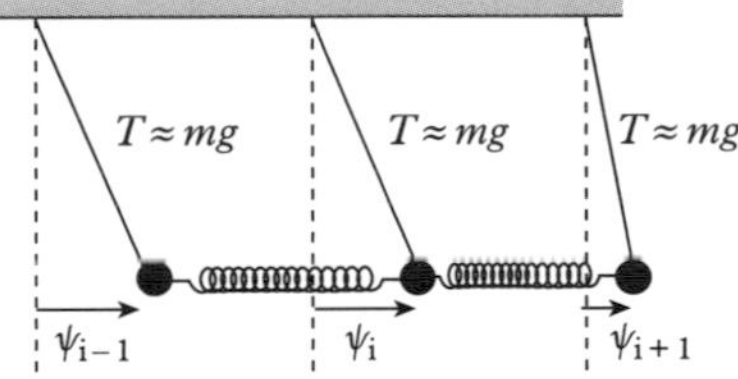

Fig. A.40 *Exercise 6.9.*

(a) The horizontal forces on the pendulum bob i will be from the two springs on the two sides and the tension in the string to which it is attached. With the x-axis pointed along the springs, we get

$$F_x[\text{on } i] = -\frac{mg}{l}\,\psi_i + K\,(\psi_{i+1} - 2\psi_i + \psi_{i-1}),$$

where the capital K is used for the spring constant. Therefore, the equations of motion will be

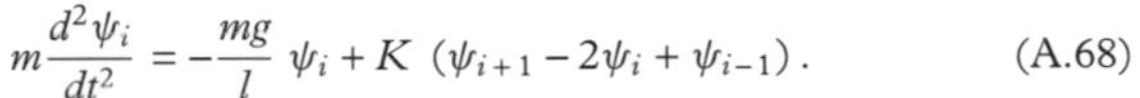

$$m\frac{d^2\psi_i}{dt^2} = -\frac{mg}{l}\,\psi_i + K\,(\psi_{i+1} - 2\psi_i + \psi_{i-1}). \qquad \text{(A.68)}$$

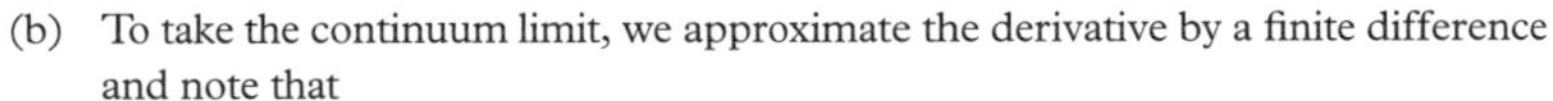

(b) To take the continuum limit, we approximate the derivative by a finite difference and note that

$$\frac{\psi_{i+1} - \psi_i}{a} \longrightarrow \frac{\partial \psi}{\partial x}.$$

The second derivatives will come from three terms,

$$\frac{\psi_{i+1} - 2\psi_i + \psi_{i-1}}{a^2} \longrightarrow \frac{\partial^2 \psi}{\partial x^2}.$$

Now, we divide both sides of Eq. A.68 by m and rewrite the discrete terms as

$$\frac{d^2\psi_i}{dt^2} = -\frac{g}{l}\,\psi_i + a^2\frac{K}{m}\,\frac{\psi_{i+1} - 2\psi_i + \psi_{i-1}}{a^2},$$

which in the limit $a \to 0$ gives

$$\frac{\partial^2 \psi}{\partial t^2} = -\frac{g}{l}\,\psi + a^2\,\frac{K}{m}\,\frac{\partial^2 \psi}{\partial x^2}. \qquad \text{(A.69)}$$

(c) To find the dispersion relation between frequency ω and wave number k (Note: small letter k) we assume a harmonic wave solution for ψ,

$$\psi(x,t) = \psi_0 e^{i(kx-\omega t)}.$$

Using this solution in the wave equation, Eq. A.69, we get the dispersion relation after canceling the common factors,

$$\omega^2 = \frac{g}{l} + \left(\frac{a^2K}{m}\right) k^2. \tag{A.70}$$

(d) The phase velocity $v_p = \omega/k$ but group velocity is $v_g = d\omega/dk$. These can be obtained from the expression for the dispersion relation, Eq. A.70,

$$v_p = \frac{\omega}{k} = \sqrt{\frac{g}{lk^2} + \left(\frac{a^2K}{m}\right)},$$
$$v_g = \frac{d\omega}{dk}.$$

To find $d\omega/dk$ we start from the expression Eq. A.70 for ω^2,

$$2\omega d\omega = 2\left(\frac{a^2K}{m}\right) kdk.$$

Therefore, the group velocity

$$v_g = \frac{d\omega}{dk} = \frac{k}{\omega}\left(\frac{a^2K}{m}\right).$$

(e) The driving of oscillations occurs by keeping the displacement of the bob at $x = 0$, as given by

$$\psi(0,t) = A\cos(\omega t).$$

The harmonic wave motion in the system is the steady state of the driven system subject to this harmonic driving displacement. The harmonic wave will match this boundary condition at $x = 0$. Therefore, the traveling wave on the coupled pendulums will be

$$\psi(x,t) = A\cos(kx - \omega t),$$

with

$$k = \sqrt{\left(\omega^2 - \frac{g}{l}\right) / \left(\frac{a^2K}{m}\right)}.$$

6.10 (a) The lowest mode will have nodes at both ends, which will give the longest wavelength to be $2L$. Therefore, the wave number of this mode will be

$$k = \frac{2\pi}{\lambda} = \frac{\pi}{L},$$

and with the speed of the wave $v_p = \sqrt{T/\mu}$, where $\mu = M/L$, the frequency of this mode will be

$$\omega = kv_p = \frac{\pi}{L}\sqrt{\frac{LT}{M}}.$$

Since the wave must vanish at the ends, which are at $x = 0$ and $x = L$, the space part will be a sine function. Therefore, the wave function for the standing wave for this mode will be

$$\psi(x,t) = A\sin(kx)\cos(\omega t + \phi).$$

(b) The kinetic energy in a small segment of length Δx and mass $\Delta m = (M/L)\Delta x$ will be

$$\Delta K = \frac{1}{2}\Delta m\left(\frac{\partial\psi}{\partial t}\right)^2 = \frac{1}{2}\frac{M}{L}\Delta x\left(-\frac{\pi v_p}{L}\right)^2 A^2\sin^2(\pi x/L)\sin^2(\omega t + \phi).$$

The time-averaged kinetic energy per unit length will vary along the position (the x-coordinate of points) on the string as given by

$$\left\langle\frac{\Delta K}{\Delta x}\right\rangle = \frac{M}{4L}\left(\frac{\pi v_p}{L}\right)^2 A^2\sin^2(\pi x/L).$$

(c) The potential energy in a length Δx at the coordinate x will be obtained by the potential energy in the wave,

$$\Delta U = \frac{1}{2}T\left(\frac{\partial\psi}{\partial x}\right)^2\Delta x.$$

Using the ψ above, and time averaging we get

$$\left\langle\frac{\Delta U}{\Delta x}\right\rangle = \frac{1}{4}Tk^2A^2\sin^2(\pi x/L).$$

Expressing $T = \mu v_p^2$ we get

$$\left\langle\frac{\Delta U}{\Delta x}\right\rangle = \frac{M}{4L}\left(\frac{\pi v_p}{L}\right)^2 A^2\sin^2(\pi x/L).$$

6.11 The wave is given as a superposition of two different wave numbers,

$$k_1 = 10\pi\ \text{cm}^{-1},\quad k_2 = 15\pi\ \text{cm}^{-1}.$$

With speed of the wave, $v = 50$ cm/s, the frequencies of the two component waves are

$$\omega_1 = k_1 v = 500\pi\ \text{sec}^{-1},\quad \omega_2 = k_2 v = 750\pi\ \text{sec}^{-1}.$$

Since the wave is moving towards the positive x-axis we can write the wave function as

$$y(x,t) = 2\cos(10\pi x - 500\pi t) + 3\cos(15\pi x - 750\pi t).$$

6.12 The wave is given as a superposition of two different frequencies,

$$\omega_1 = 10\pi \text{ sec}^{-1}, \quad \omega_2 = 15\pi \text{ sec}^{-1}.$$

With speed of the wave, $v = 50$ cm/s, the wave numbers of the two component waves are

$$k_1 = \omega_1/v = \frac{\pi}{5} \text{ cm}^{-1}, \quad k_2 = \omega_2/v = \frac{3\pi}{10} \text{ cm}^{-1}.$$

Since the wave is moving towards the positive x-axis we can write the wave function as

$$y(x,t) = 2\cos\left(10\pi t - \frac{\pi}{5}x\right) + 3\cos\left(15\pi t - \frac{3\pi}{10}x\right).$$

6.13 (a) The given function has a period of $\lambda = a$. We can write the analytic expression of the function in one period explicitly as

$$y(x) = \frac{b}{a}x, \quad 0 \le x < a.$$

The other parts of the function can be obtained by

$$y(x+a) = y(x).$$

(b) The coefficients of the Fourier series for this function are evaluated as follows:

$$A_0 = \frac{2}{a}\int_0^a y(x)dx = b,$$

$$A_n = \frac{2}{a}\int_0^a y(x)\cos\left(\frac{2n\pi x}{a}\right)dx = 0,$$

$$B_n = \frac{2}{a}\int_0^a y(x)\sin\left(\frac{2n\pi x}{a}\right)dx = -\frac{b}{n\pi}.$$

Therefore,

$$y(x) = \frac{b}{2} - \frac{b}{\pi}\sum_{n=1,2,\cdots}\frac{1}{n}\sin\left(\frac{2n\pi x}{a}\right).$$

6.14 (a) The given pulse is

$$y(0,t) = \begin{cases} 0 & t < 0, \\ b & 0 \le t \le \tau, \\ 0 & t > \tau. \end{cases}$$

Since the wave is non-dispersive, it will keep its shape while moving with speed v. The front of the pulse was generated at time $t = 0$. Therefore, the front will have moved a distance $\Delta x = vt$ in time t. The back of the pulse was generated at time $t = \tau$. At the instant where time $t = t$, the back had only a duration $t - \tau$ to move. That means, the back would have moved a distance $\Delta x = v(t - \tau)$,

$$y(x,t) = \begin{cases} 0 & x < v(t-\tau), \\ b & v(t-\tau) \le x \le vt, \\ 0 & x > vt. \end{cases}$$

(b) If the dispersion relation is

$$\omega^2 = a^2k^4,$$

then the frequency ω is not a linear function of k. This would give a wave number-dependent (equivalently, frequency-dependent) phase velocity,

$$v_p = \frac{\omega}{k} = ak = \sqrt{a\omega}.$$

That means, each harmonic wave will move at its own speed. Therefore, different Fourier components of the given pulse will move at different speeds. As a result, the pulse will not remain a square pulse, but rather will spread out.

Fourier representation of the given pulse can be obtained by finding the coefficients A and B in

$$y(0,t) = \int_0^\infty A(\omega)\cos(\omega t)d\omega + \int_0^\infty B(\omega)\sin(\omega t)d\omega, \tag{A.71}$$

with

$$A(\omega) = \frac{1}{\pi}\int_{-\infty}^\infty y(0,t)\cos(\omega t)dt = \frac{b}{\pi\omega}\sin(\omega\tau),$$

$$B(\omega) = \frac{1}{\pi}\int_{-\infty}^\infty y(0,t)\sin(\omega t)dt = \frac{b}{\pi\omega}\left(1-\cos(\omega\tau)\right).$$

The wave at x at time t will be obtained by substituting $t - x/v_p = t - x/\sqrt{a\omega}$ for t in Eq. A.71 for a wave moving towards the positive x-axis to obtain a formal expression for the wave function,

$$y(x,t) = \int_0^\infty A(\omega)\cos\left(\omega\left[t - \frac{x}{\sqrt{a\omega}}\right]\right)d\omega + \int_0^\infty B(\omega)\sin\left(\omega\left[t - \frac{x}{\sqrt{a\omega}}\right]\right)d\omega.$$

6.15 (a) For a non-dispersive system, the wave will maintain its shape. The wave at $t = 0$ is given by the following periodic function:

$$y(0,t) = \frac{b}{a}t, \quad 0 \le t < a; \quad y(0,t+a) = y(0,t).$$

Let the wave velocity be denoted by v. Then, a wave traveling towards $x = +\infty$ will be

$$y_+(x,t) = \frac{b}{a}\left(t - \frac{x}{v}\right),$$

and the wave towards $x = -\infty$ will be

$$y_-(x,t) = \frac{b}{a}\left(t + \frac{x}{v}\right).$$

(b) The instantaneous power output of the wave generator at the origin will be twice the formula for power when the wave moves only towards the positive x-axis,

$$P(t) = 2Z\left(\frac{dy}{dt}\right)^2 = 2\sqrt{\mu T}\frac{b^2}{a^2},\quad 0 \le t < a,$$

which is independent of t. Therefore, the time-averaged power is

$$\langle P \rangle = 2\sqrt{\mu T}\frac{b^2}{a^2}.$$

6.16 (a) To find the wave profile in the x-space we perform the Fourier integrals in

$$\psi(x) = \int_0^\infty A(k)\cos(kx)dk + \int_0^\infty B(k)\sin(kx)dk.$$

For the given $A(k)$ and $B(k)$ we get

$$\psi(x) = \int_{k_0}^{k_0+\Delta k}\cos(kx)dk = \frac{\sin[(k_0+\Delta k)x]}{x} - \frac{\sin(k_0 x)}{x}.$$

(b) Since the speed of a wave is independent of frequency, we would just translate the wave in time at speed v. Therefore, to obtain $\psi(x,t)$ from $\psi(x)$ at $t = 0$, we will replace x by $x - vt$,

$$\psi(x,t) = \frac{\sin[(k_0+\Delta k)(x-vt)]}{x-vt} + \frac{\sin[k_0(x-vt)]}{x-vt}.$$

(c) The plots are left as exercises for the student. Plots show that as $\Delta k \to 0$, $\Delta x \to \infty$ in space, and vice versa. This example confirms the relation $\Delta x \Delta k \sim 1$, i.e. the spreading in k-space and x-space are inversely related.

6.17 Lorentzian Wave.

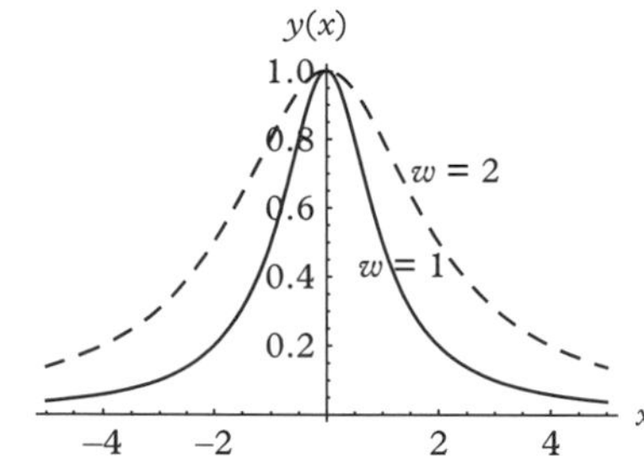

Fig. A.41 *Exercise 6.17.*

(a) Figure A.41 shows the plots for $w = 1$ and $w = 2$. Clearly, the parameter w is related to the width of the function y; the larger is the w, the wider is the y function.

(b) First we find the height of $y(x, 0)$, which occurs at $x = 0$,

$$y_{peak} = \frac{bw^2}{w^2} = b.$$

Now we set $y(x)$ to $b/2$ to find the values of x at which y has half the maximum value,

$$\frac{bw^2}{x^2 + w^2} = \frac{b}{2}.$$

This gives two values of x,

$$x_1 = -w,\ x_2 = w.$$

The width is the difference of these,

$$\Delta x = x_2 - x_1 = 2w.$$

Therefore, the width of $y(x)$ at half height is $2w$.

(c) We get velocity of the particle at x from $y(x, t)$ since y is the displacement of the particle at x. We will denote this velocity by v_y to distinguish it from the velocity of the wave, which is denoted by v here,

$$v_y = \frac{\partial y}{\partial t} = \frac{-bw^2}{\left[(x - vt)^2 + w^2\right]^2}\,[2(x - vt)(-v)]\,,$$

which upon simplifying gives

$$v_y = \frac{2bvw^2 (x - vt)}{\left[(x - vt)^2 + w^2\right]^2}.$$

(d) Figure A.42 shows the three plots. We see that in the $x < 0$ part, when the wave moves to the right, the y-values drop, giving a negative y-component of the velocity in this part. On the other hand, in the $x > 0$ part, the y-displacement rises and therefore, v_y is positive here.

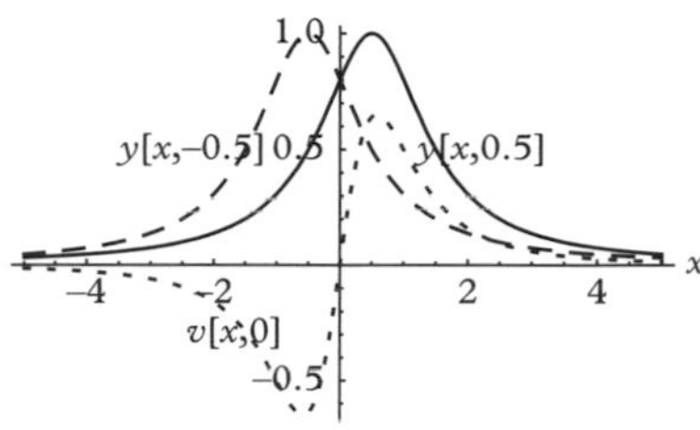

Fig. A.42 *Exercise 6.17.*

(e) Kinetic energy per unit length and potential energy per unit length will be equal in this system, as you can show from the general expressions applied to the given $y(x, t)$,

$$\frac{KE}{\Delta x} = \frac{1}{2}\mu\left(\frac{\partial y}{\partial t}\right)^2, \quad \frac{PE}{\Delta x} = \frac{1}{2}T\left(\frac{\partial y}{\partial x}\right)^2.$$

Now, using $T = \mu v^2$ and carrying out dy/dt and dy/dx we can show that

$$\frac{\partial y}{\partial t} = v\frac{\partial y}{\partial x}.$$

Therefore,

$$\frac{KE}{\Delta x} = \frac{PE}{\Delta x} = \frac{1}{2}\mu\left(\frac{\partial y}{\partial t}\right)^2.$$

Using this, the total energy per unit length will be

$$\mathcal{E} \equiv \frac{E}{\Delta x} = \mu\left[\frac{2bvw^2 (x - vt)}{\left[(x - vt)^2 + w^2\right]^2}\right]^2.$$

Evaluating this at $t = 0$ we get

$$\mathcal{E}(x, 0) = \mu\left[\frac{2bvw^2 x}{\left(x^2 + w^2\right)^2}\right]^2.$$

Integrating this over the entire string, $-\infty < x < \infty$ gives

$$E = \int_{-\infty}^{+\infty} \mathcal{E}(x, 0)\,dx = \frac{\pi b^2}{4w}\,\mu v^2.$$

6.18 (a) The function $f(x)$ below is centered about $x = x_0$,

$$f(x) = \frac{bw^2}{(x - x_0)^2 + w^2}.$$

We can show this by changing variable $z = x - x_0$. That will give $f(z)$ which is a symmetric function of z at the peak $z = 0$. Thus, the given function $y(x, t)$ at $t = 0$ has two peaks, one at $x = a$ and the other at $x = -a$. To examine the direction of movement of each pulse, work out $\partial y/\partial t$.

(b) To get the velocity at $t = 0$ we start with first taking the derivative of y with respect to t and setting $t = 0$,

$$v_y(x, 0) = \left.\frac{dy}{dt}\right|_{t=0}.$$

I leave the sketch to the student. You will find that the two peaks move towards $x = 0$ with wave speed v.

(c) When the two peaks reach $x = 0$, they overlap completely after time $t = \tau$ from when they were $2a$ apart. We get time τ from the distance a and speed v that each peak travels,

$$\tau = \frac{a}{v}.$$

(d) Momentarily, the wave appears to vanish, but since acceleration is not zero, the two pulses re-emerge after they collide at $x = 0$.

6.19 The non-relativistic Doppler effect for a moving detector is

$$f = \left(\frac{v \pm v_D}{v}\right) f_0.$$

(a)

$$f = \left(\frac{343 \text{ m/s} + 10 \text{ m/s}}{343 \text{ m/s}}\right) \times 400 \text{ Hz} = 412 \text{ Hz}.$$

(b)

$$f = \left(\frac{343 \text{ m/s} - 10 \text{ m/s}}{343 \text{ m/s}}\right) \times 400 \text{ Hz} = 388 \text{ Hz}.$$

6.20 The non-relativistic Doppler effect for a moving source is

$$f = \left(\frac{v}{v \pm v_S}\right) f_0$$

(a)

$$f = \left(\frac{343 \text{ m/s}}{343 \text{ m/s} - 10 \text{ m/s}}\right) \times 400 \text{ Hz} = 412 \text{ Hz}.$$

(b)

$$f = \left(\frac{343 \text{ m/s}}{343 \text{ m/s} + 10 \text{ m/s}}\right) \times 400 \text{ Hz} = 389 \text{ Hz}.$$

6.21 Let v_s be the speed of the fire engine, v the speed of sound in air, f_1 the frequency you hear when the fire engine is approaching and f_2 when the engine is receding while the engine produces the frequency f_0. From the non-relativistic Doppler effect we will have the following relations:

$$f_1 = \left(\frac{v}{v - v_s}\right) f_0, \quad (1)$$

$$f_2 = \left(\frac{v}{v + v_s}\right) f_0. \quad (2)$$

Dividing, we get

$$\frac{v + v_s}{v - v_s} = \frac{f_1}{f_2}.$$

Let us denote the ratio of frequencies by a,

$$\frac{f_1}{f_2} \equiv a.$$

Then, we have the following equation:

$$\frac{v + v_s}{v - v_s} = a.$$

We can solve this for v_s,

$$v_s = \left(\frac{a-1}{a+1}\right) v.$$

Now, putting in the given values,

$$a = 1650 \text{ Hz}/1550 \text{ Hz} = \frac{165}{155}, \quad v = 343 \text{ m/s},$$

we get

$$v_s = 10.7 \text{ m/s}.$$

We can put this in either (1) or (2) to obtain

$$f_0 = \left(\frac{343 - 10.7}{343}\right) \times 1650 \text{ Hz} = 1598.5 \text{ Hz}.$$

6.22 Let f_0 be the frequency of sound emitted by the humming bird. Let f' be the frequency in a frame at rest with the ground and f be the frequency you hear. then, we have the following relations:

$$f' = \left(\frac{v}{v - v_s}\right) f_0 \text{ (Source moving)},$$

$$f = \left(\frac{v + v_D}{v}\right) f' \text{ (Detector moving)}.$$

Therefore,

$$f_0 = \left(\frac{v - v_s}{v + v_D}\right) f.$$

Putting in the numerical values we obtain $f_0 = 413$ Hz.

6.23 Relativistic Doppler shift gives the following formula:

$$f = \left(\frac{1 - v/c}{1 + v/c}\right)^{1/2} f_0.$$

Here,

$$f_0 = \frac{c}{\lambda_0}, \; f = \frac{c}{\lambda}.$$

Therefore, the relation for wavelengths is

$$\left(\frac{\lambda_0}{\lambda}\right)^2 = \frac{1 - v/c}{1 + v/c}.$$

Let us denote the ratio of the wavelengths by x,

$$x = \frac{\lambda}{\lambda_0}.$$

We can solve this for v/c which gives

$$\frac{v}{c} = \frac{x^2 - 1}{x^2 + 1}.$$

From the given data we have

$$x = \frac{486.6 \text{ nm}}{121.6 \text{ nm}} = 4.$$

Therefore,

$$v = \left(\frac{16 - 1}{16 + 1}\right) c = \frac{15}{17} \times 3 \times 10^8 \text{ m/s} = 2.65 \times 10^8 \text{ m/s}.$$

6.24 (a) Let us write $\epsilon = v/c$ in the relativistic Doppler effect formula,

$$f = g(\epsilon)\, f_0,$$

where

$$g(\epsilon) = \sqrt{\frac{1 - \epsilon}{1 + \epsilon}}.$$

We can expand $g(\epsilon)$ around $\epsilon = 0$ in a McLauren series. This gives the following in the quadratic approximation:

$$\begin{aligned} g(\epsilon) &= \sqrt{\frac{1 - \epsilon}{1 + \epsilon}} = \frac{\sqrt{1 - \epsilon^2}}{1 + \epsilon} \\ &= \left(1 - \frac{1}{2}\epsilon^2 + \cdots\right)\left(1 - \epsilon + \epsilon^2 + \cdots\right) \\ &= 1 - \epsilon + \frac{1}{2}\epsilon^2. \end{aligned}$$

Keeping only the linear term in v/c we get the following approximate formula:

$$f + \left(1 - \frac{v}{c}\right) f_0.$$

(b) Numerical value of v is

$$v = 100 \text{ mph} = 44.4 \text{ m/s}.$$

Therefore, the change in frequency will be

$$\Delta f = -\frac{v}{c} f_0 = -\frac{44.4 \text{ m/s}}{3 \times 10^8 \text{ m/s}} \times 5 \times 10^9 \text{ Hz} = -741 \text{ Hz}.$$

6.25 We have $\lambda_0 = 21$ cm and $v = 0.9c$.

(a) The formula for the relativistic effect for recessional speed v in terms of wavelength gives

$$\lambda = \lambda_0 \sqrt{\frac{1 + v/c}{1 - v/c}} = 21 \text{ cm} \sqrt{\frac{1 + 0.9}{1 - 0.9}} = 91.5 \text{ cm}.$$

(b) The shift in wavelength

$$\Delta\lambda = \lambda - \lambda_0 = 70.5 \text{ cm}.$$

The z is given by the fractional shift in wavelength,

$$z = \frac{\Delta\lambda}{\lambda_0} = \frac{70.5 \text{ cm}}{21 \text{ cm}} = 3.36.$$

6.26 The red-shift z of a galaxy is defined by $z = \Delta\lambda/\lambda_0$, where $\Delta\lambda$ is the change in the wavelength and λ_0 is the emitted wavelength. A quasar is found to have a red-shift of 20. Supposing the quasar is moving away from Earth, what is the speed of the quasar with respect to Earth?

Let us rewrite the relativistic Doppler shift formula in terms of z in place of λ/λ_0 with

$$z + 1 = \frac{\lambda}{\lambda_0}.$$

That is, for a receding source and detector with relative speed v we will have,

$$z + 1 = \sqrt{\frac{1 + v/c}{1 - v/c}}.$$

Solving this for v/c we get

$$\frac{v}{c} = \frac{z^2 + 2z}{z^2 + 2z + 2}.$$

For $z = 20$ this gives

$$\frac{v}{c} = \frac{20^2 + 2 \times 20}{20^2 + 2 \times 20 + 2} = \frac{220}{221} = 0.995.$$

Therefore, $v = 0.995c$.

Chapter 7

7.1 This question is about reading properties contained in a wave function,

$$\psi(x, y, z, t) = 0.05\cos(2\pi x + \pi y - 2\pi z - 600\pi t),$$

where x, y, and z are in cm, t is in seconds, and ψ is in Pa. Determine the following quantities.

(a) Angular frequency, $\omega = 600\pi$ sec^{-1}. We can also state this as $\omega = 600\pi$ rad/sec. The frequency, $f = \omega/2\pi = 300$ Hz.

(b) The Cartesian components of the wave number vector are given in the wave function as factors that are multiplying x, y, and z in the argument of the cosine,

$$k_x = 2\pi \text{ cm}^{-1}, \quad k_y = \pi \text{ cm}^{-1}, \quad k_z = -2\pi \text{ cm}^{-1}.$$

Therefore, the propagation vector is the wave number vector,

$$\vec{k} = k_x\,\hat{x} + k_y\,\hat{y} + k_z\,\hat{z} = 2\pi\,\hat{x} + \pi\,\hat{y} - 2\pi\,\hat{z}, \ \left(\text{units: cm}^{-1}\right)$$

(c) The wave number k is the magnitude of this vector,

$$k = \sqrt{k_x^2 + k_y^2 + k_z^2} = 3\pi \text{ cm}^{-1}.$$

The wavelength is obtained from the wave number by using their relation,

$$\lambda = \frac{2\pi}{k} = \frac{2}{3} \text{ cm}.$$

(d) When we write ωt with a negative sign, the direction of wave motion is the same as the direction of the propagation vector, which is in the direction of the following unit vector,

$$\hat{u} = \frac{2}{3}\,\hat{x} + \frac{1}{3}\,\hat{y} - \frac{2}{3}\,\hat{z}.$$

(e) Wave velocity here refers to the phase velocity of the harmonic wave, which is obtained by the ratio of ω and k,

$$v_p = \frac{\omega}{k} = \frac{600\pi \text{ sec}^{-1}}{3\pi \text{ cm}^{-1}} = 200 \text{ cm/sec}.$$

(f) The amplitude of the wave is the factor multiplying the sinusoidal part,

$$A = 0.05 \text{ Pa}.$$

(g) The phase refers to the argument of the sinusoidal part. Here, the phase φ is

$$\varphi = 2\pi x + \pi y - 2\pi z - 600\pi t.$$

The change in the phase at fixed x, y, and z will be

$$\Delta\varphi = -600\pi\,\Delta t = -600\pi \text{ sec}^{-1} \times 3 \times 10^{-3} \text{ sec} = -1.8\pi \text{ rad}.$$

(h) For a fixed t, the phase change will be

$$\Delta\varphi = 2\pi\,\Delta x + \pi\,\Delta y - 2\pi\,\Delta z = 2\pi \times 1 + 0 - 2\pi \times (-1) = 4\pi,$$

which is equivalent to $\Delta\varphi = 0$ since $\Delta\varphi$ is modulo 2π radians due to the fact that $\cos(\theta) = \cos(\theta + 2\pi n)$ with n an integer.

7.2 (a) From the given expression of the harmonic wave we get

$$\omega = 2\pi \times 10^{15}\ \text{sec}^{-1} \implies f = 10^{15}\ \text{Hz}.$$

This will give wave number and wavelength to be

$$k = \frac{\omega}{v_p} = \frac{2\pi \times 10^{15}\ \text{sec}^{-1}}{3 \times 10^{8}\ \text{m/s}} = \frac{2\pi}{3} \times 10^{7}\ \text{m}^{-1},$$

$$\lambda = \frac{2\pi}{k} = 3 \times 10^{-7}\ \text{m}.$$

(b) When we set the phase of the cosine to zero we get

$$2\pi \times 10^{15}\ t + ky + \frac{\pi}{2} = 0.$$

In time duration Δt the wave will travel a distance Δy given by

$$2\pi \times 10^{15}\ \Delta t + k\Delta y = 0.$$

Therefore, the velocity has only the y-component given by

$$v_p = \frac{\Delta y}{\Delta t} = -3 \times 10^{8}\ \text{m/s}.$$

Since the y-component of the wave velocity is negative, the wave would travel towards the negative y-axis.

(c) At the origin, the electric field is given to be

$$\vec{E}(0,0,0,t) = \hat{i}\,3 \times 10^{5}\ \text{N/C}\ \cos\left(2\pi \times 10^{15}\ t + \frac{\pi}{2}\right),$$
$$= -\hat{i}\,3 \times 10^{5}\ \text{N/C}\ \sin\left(2\pi \times 10^{15}\ t\right),$$

which at $t = 0$ is

$$\vec{E}(0,0,0,t) = 0.$$

The magnetic vector will also be zero at the origin at $t = 0$.

(d) At $t = 0.25 \times 10^{-15}$ sec, the electric field at the origin will be

$$\vec{E}(0, 0, 0, 0.25 \times 10^{-15}\ \text{sec}) = -\hat{i}\, 3 \times 10^5\,(\text{N/C}).$$

Since the wave is traveling towards the negative y-axis and the electric field points towards the negative x-axis, the magnetic field will be towards the negative z-axis,

$$\vec{B}(0, 0, 0, t = 0.25 \times 10^{-15}\ \text{sec}) = -\hat{z}\,\frac{3 \times 10^5\ \text{N/C}}{3 \times 10^8\ \text{m/s}} = -\hat{z}\, 10^{-3}\ \text{T}.$$

(e) Hint: Use $\vec{B} = \hat{z}B_0 \cos\left(2\pi \times 10^{15} t + ky + \frac{1}{2}\right)$.

(f) Hint: The times are simple factors of the period.

7.3 Let us first find the propagation vector and the angular frequency,

$$\vec{k} = k_x\hat{x} + k_y\hat{y} = k\left(\frac{-\hat{x} + \hat{y}}{\sqrt{2}}\right),$$

with $k = 1000\ \text{cm}^{-1}$. The $\vec{k} \cdot \vec{r}$ will be

$$\vec{k} \cdot \vec{r} = -\frac{k}{\sqrt{2}}x + \frac{k}{\sqrt{2}}y = -\frac{1000}{\sqrt{2}}x + \frac{1000}{\sqrt{2}}y.$$

Since $f = 2$ MHz, we get

$$\omega = 2\pi f = 4\pi \times 10^6\ \text{sec}^{-1}.$$

With amplitude $A = 5\ \mu\text{m}$ we get the following expression for the wave:

$$\psi(x, y, z, t) = 5\ \mu\text{m}\ \cos\left(-\frac{1000}{\sqrt{2}}x + \frac{1000}{\sqrt{2}}y - 4\pi \times 10^6 t\right),$$

where x, y are in cm and t in sec.

7.4 (a) This is a simple application of definition. The wave energy is spread over a spherical surface area $4\pi R^2$ with R the radius of the surface,

$$I = \frac{P}{A} = \frac{10\ \text{W}}{4\pi \times (1\ \text{m})^2} = 0.8\ \text{W/m}^2.$$

(b) The intensity of an isotropic source drops as the square of the distance from the source. Therefore,

$$I = I_{\text{part (a)}} \times \left[\frac{(1\ \text{m})^2}{(2\ \text{m})^2}\right] = 0.2\ \text{W/m}^2.$$

7.5 (a) The wave number can be found by its relation to the wavelength,

$$k = \frac{2\pi}{\lambda} = \frac{2\pi}{50\ \text{cm}} = 0.126\ \text{cm}^{-1}.$$

(b) Since intensity goes as the square of the wave amplitude, $I \sim |\psi|^2$, we get the ratio of intensities from the ratio of the square of the amplitudes since the common factor will cancel out,

$$\frac{I_{200}}{I_{100}} = \frac{|\psi(r = 200\text{ cm})|^2}{|\psi(r = 100\text{ cm})|^2} = \left(\frac{100}{200}\right)^2 = \frac{1}{4}.$$

(c) In your plots notice that intensity drops off much faster than the amplitude. For simplicity you can use $I = 1$ unit when $r = \lambda$.

7.6 The intensity of the wave will come from time-averaging the square of the wave function,

$$I \sim \langle \psi^2 \rangle.$$

Suppressing the constant factor, this gives three terms,

$$I = A_1^2 + A_2^2 + 2A_1A_2 \langle \cos(k_1x - \omega t)\cos(k_2x - 2\omega t)\rangle,$$

where $k_1 = \omega/v_p$ and $k_2 = 2\omega/v_p$. Expanding the cosines we find that the time averaging is done in four terms, each of which gives zero because,

$$\langle \cos(\omega t)\cos(2\omega t)\rangle = 0,\ \langle \cos(\omega t)\sin(2\omega t)\rangle = 0,$$
$$\langle \sin(\omega t)\cos(2\omega t)\rangle = 0,\ \langle \sin(\omega t)\sin(2\omega t)\rangle = 0.$$

Therefore, the intensity will be just the sum of the intensities of the individual waves,

$$I = A_1^2 + A_2^2 = I_1 + I_2.$$

7.7 (a) Figure A.43 shows a plot of the wave profile for the case $A = 1$ across the pipe along the x-axis.

(b) It is easy to verify that the given wave function satisfies the classical wave equation by simply plugging in the given expression to the classical wave equation. The frequency will be 5.0 M Hz.

(c) Note the following identities:

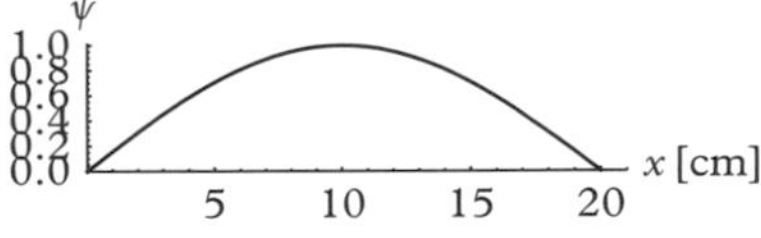

Fig. A.43 *Exercise 7.7.*

$$2\sin(A)\cos(B) = \sin(A + B) + \sin(A - B),\ \ \sin(-A) = -\sin(A).$$

We use these identities to write the mixed wave function into the sum of the traveling waves,

$$\begin{aligned}\psi(x, y, z, t) &= A\sin\left(\frac{\pi x}{20\text{ cm}}\right)\cos\left(2\pi z/30\text{ m} - 2\pi ft\right),\\ &= \frac{A}{2}\sin\left(\frac{\pi x}{20\text{ cm}} + \frac{2\pi z}{30\text{ m}} - 2\pi ft\right)\\ &\quad - \frac{A}{2}\sin\left(-\frac{\pi x}{20\text{ cm}} + \frac{2\pi z}{30\text{ m}} - 2\pi ft\right).\end{aligned}$$

(d) The wave profile shows that the amplitude of oscillations is greatest when $x = a/2$, i.e. in the middle of the cross-section. Therefore, the intensity will be strongest there.

(e) The wave function is zero at the walls at $x = 0$ and $x = a$. The intensity will be zero, i.e. the least at these walls.

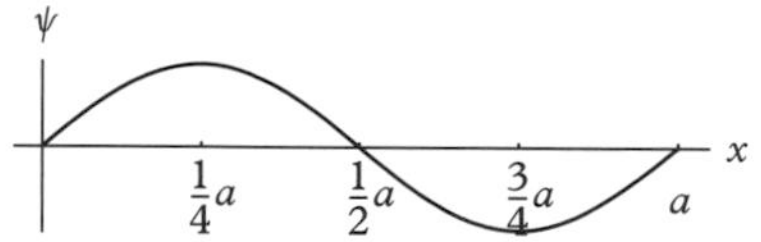

Fig. A.44 *Exercise 7.8.*

7.8 (a) Figure A.44 shows a plot of the wave profile for the case $A = 1$ across the pipe along the x-axis. A similar plot is along the y-axis.

(b) (i) Using the harmonic wave given in the problem into the classical wave equation we get the following relation between ω and k of the wave:

$$\omega^2 = v^2 \left[\left(\frac{2\pi}{a}\right)^2 + \left(\frac{2\pi}{b}\right)^2 + k^2 \right],$$

where v is the wave speed.

(ii) The minimum ω will occur when $k = 0$. This gives the smallest frequency that can propagate to be

$$\omega_{\min} = v\sqrt{\left(\frac{2\pi}{a}\right)^2 + \left(\frac{2\pi}{b}\right)^2}.$$

(c) Using the trigonometric identity between the product of a cosine and sine by the sum of two sines we get

$$\begin{aligned}\psi(x, y, z, t) &= \frac{A}{2} \sin\left(\frac{2\pi x}{a} + k_z z - \omega t\right) \sin\left(\frac{2\pi y}{b}\right) \\ &\quad - \frac{A}{2} \sin\left(-\frac{2\pi x}{a} + k_z z - \omega t\right) \sin\left(\frac{2\pi y}{b}\right).\end{aligned}$$

Now, we use the following identity to further transform this equation:

$$2 \sin\theta_1 \sin\theta_2 = \cos(\theta_1 - \theta_2) - \cos(\theta_1 + \theta_2).$$

We get

$$\begin{aligned}\psi(x, y, z, t) &= \frac{A}{4} \cos\left(\frac{2\pi x}{a} - \frac{2\pi y}{b} + k_z z - \omega t\right) \\ &\quad - \frac{A}{4} \cos\left(\frac{2\pi x}{a} + \frac{2\pi y}{b} + k_z z - \omega t\right) \\ &\quad - \frac{A}{4} \cos\left(-\frac{2\pi x}{a} - \frac{2\pi y}{b} + k_z z - \omega t\right) \\ &\quad + \frac{A}{4} \cos\left(-\frac{2\pi x}{a} + \frac{2\pi y}{b} + k_z z - \omega t\right).\end{aligned}$$

(d) From the wave profile across the cross-section of the pipe, the intensity will be strongest at $x = a/4$ and $x = 3a/4$ along the x-axis and at $y = b/4$ and $y = 3a/4$ along the y-axis.

(e) From the wave profile across the cross-section of the pipe, the intensity will be zero at the walls at $x = 0$ and $x = a$ and in the center at $x = \frac{a}{2}$ along the x-axis and at $y = 0$ and $y = b$ and in the center at $y = \frac{b}{2}$ along the y-axis.

7.9 The oscillation in local volume change can be obtained by multiplying the oscillation of displacement by

$$\delta V = A\delta x.$$

We have shown in the text that δx can be expressed in terms of pressure incremental amplitude p_m by

$$\delta x = \frac{kp_m}{\omega^2 \rho_0} e^{i(kx-\omega t+\pi/2)}.$$

Therefore, the flow rate will be

$$\frac{d\delta V}{dt} = A\frac{d\delta x}{dt} = -i\frac{Akp_m}{\omega\rho_0} e^{ikx-\omega t+\pi/2}.$$

From this we can obtain the characteristic impedance by

$$Z = \frac{p_m}{|d\delta V/dt|} = \frac{\omega\rho_0}{Ak} = \frac{\rho_0 v}{A}.$$

7.10 (a) The wave number is given in the wave function,

$$k = 10\ \text{m}^{-1}.$$

Using the wave speed $v = 350$ m/s, we get the angular frequency,

$$\omega = kv = 3,500\ \text{sec}^{-1}.$$

The wavelength and frequency will be

$$\lambda = \frac{2\pi}{k} = 0.2\pi\ \text{m},\quad f = \frac{\omega}{2\pi} = 557\ \text{Hz}.$$

(b) Intensity of sound wave is given by

$$I = \frac{p_m^2}{2Z} = \frac{p_m^2}{2\rho_0 v}.$$

Using the numerical values we get

$$I = \frac{(10\ \text{Pa})^2}{2 \times 1.225\ \text{kg/m}^3 \times 350\ \text{m/s}} = 0.12\ \text{W/m}^2.$$

(c) Adding $\pi/2$ to the phase of the pressure wave will give the phase of the displacement wave and the magnitude of the amplitude of the displacement wave will be

$$x_m = \frac{k}{\omega^2\rho_0} p_m = \frac{10}{3500^2 \times 1.225} \times 10 = 6.66 \times 10^{-6}\ \text{m}.$$

Therefore,

$$\delta x = 6.66 \times 10^{-6}\ \text{m}\ \cos(10z + \omega t + \pi/2).$$

7.11 (a) We have the following relation between impedance Z, the density ρ_0, and wave speed:

$$Z = v\rho_0, \implies v = \frac{Z}{\rho_0}.$$

Putting in the numerical values we get

$$v = \frac{1.5 \times 10^6\ \text{kg/m}^2.\text{s}}{1000\ \text{kg/m}^3} = 1.5 \times 10^3\ \text{m/s}.$$

(b) Hint: Follow the solution of the previous problem.

(c) Hint: Follow the solution of the previous problem.

(d) Hint: Follow the solution of the previous problem.

7.12 (a) When we compare the given pressure wave with the plane wave expression we find that the amplitude A/r here is analogous to p_m in the plane wave equation. A calculation along the line of the plane wave calculation will lead to

$$I = \frac{1}{2}\frac{1}{\rho_0 v}\frac{A^2}{r^2}.$$

(b) Intensity is power per unit area. When we multiply the intensity at all points of the sphere of radius $r = a$ by the area of the sphere we will get the total power radiated,

$$P_{\text{total}} = I(r = a) \times 4\pi a^2 = \frac{2\pi A^2}{\rho_0 v}.$$

7.13 (a) Figure A.45 shows the expansion of gas in the neck of the bottle. In this part we will find the force on the gas in the neck part.

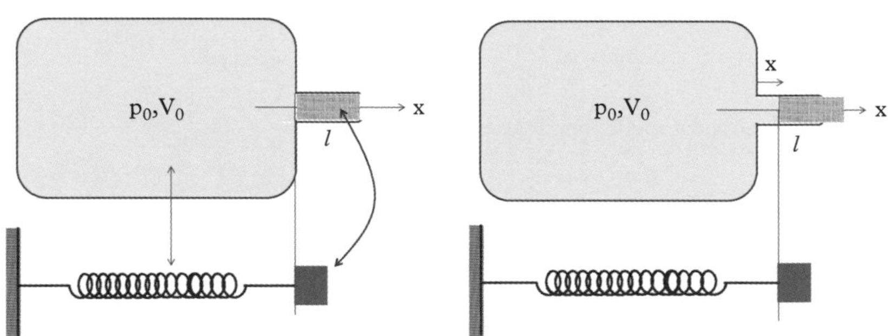

Fig. A.45 *Exercise 7.13.*

When the gas expands into the neck, the volume changes from V_0 to $V_0 + Ax$, where $Ax \ll V_0$. Let the pressure change from p_0 to $p_0 + \Delta p$. Since the process is adiabatic we will have

$$(p_0 + \Delta p)(V_0 + Ax)^\gamma = p_0 V_0^\gamma.$$

Expanding to one power in change, i.e. dropping the ΔpAx term, we get

$$V_0^\gamma \Delta p + \gamma p_0 V_0^{\gamma-1} Ax = 0.$$

Therefore,

$$\Delta p = -\frac{\gamma p_0 A}{V_0}\, x.$$

This would give the force along the x-axis,

$$F_x = \Delta p A = -\left(\frac{\gamma p_0 A^2}{V_0}\right) x. \quad (1)$$

(b) We can estimate the mass m of the air in the neck upon which the force F_x acts,

$$m = (Al)\rho_0.$$

With force (1) on this mass we get the equation of motion,

$$m\frac{d^2x}{dt^2} = F_x = -\left(\frac{\gamma p_0 A}{V_0}\right) x,$$

which is an equation of a harmonic oscillator with angular frequency,

$$\omega = \sqrt{\frac{\gamma p_0 A^2}{V_0} \div (Al\rho_0)}.$$

The frequency of oscillations,

$$f = \frac{1}{2\pi}\sqrt{\frac{\gamma p_0 A}{V_0 \rho_0 l}}. \quad (2)$$

(c) Using numerical values in (2) we get

$$f = \frac{1}{2\pi}\sqrt{\frac{(5/3) \times 1.013 \times 10^5 \text{ Pa} \times 8 \times 10^{-4} \text{ m}^2}{10^{-3} \text{ m}^3 \times 1 \text{ kg/m}^3 \times 0.04 \text{ m}}} = 292 \text{ Hz}.$$

7.14 The displacement ψ of the particle of air at $x = 0^+$ and $x = 0^-$ is given by the functions shown in Fig. A.46.

(a) The inverse Fourier transforms give $A(\omega)$ and $B(\omega)$,

$$A(\omega) = \frac{1}{\pi}\int_{-\infty}^{\infty} \psi(0, t) \cos(\omega t) dt,$$

$$B(\omega) = \frac{1}{\pi}\int_{-\infty}^{\infty} \psi(0, t) \sin(\omega t) dt.$$

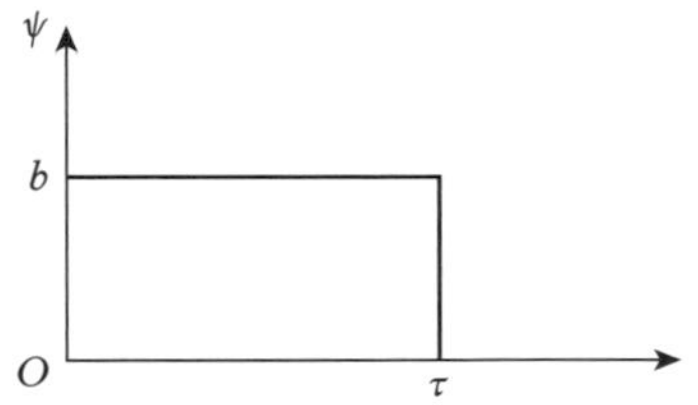

Fig. A.46 *Exercise 7.14.*

We can carry out these integrals by using the function in Fig. A.46,

$$A(\omega) = \frac{b}{\pi}\int_0^{\tau} \cos(\omega t) dt = \frac{b}{\pi\omega} \sin(\omega\tau),$$

$$B(\omega) = \frac{b}{\pi}\int_0^{\tau} \sin(\omega t) dt = \frac{b}{\pi\omega} [1 - \cos(\omega\tau)].$$

(b) The intensity of the wave in a particular frequency ω is proportional to the sum of the squares of A and B,

$$I(\omega) \propto |A(\omega)|^2 + |B(\omega)|^2,$$
$$= \frac{b^2}{\pi^2\omega^2}\left[\sin^2(\omega\tau) + (1 - \cos(\omega\tau))^2\right],$$
$$= 2\left(\frac{b\tau}{\pi}\right)^2 \frac{1}{\omega^2\tau^2}\,[1 - \cos(\omega\tau)],$$
$$= \left(\frac{b\tau}{\pi}\right)^2 \left[\frac{\sin(\omega\tau/2)}{\omega\tau/2}\right]^2.$$

Therefore,

$$I(\omega) \propto \left[\frac{\sin(\omega\tau/2)}{\omega\tau/2}\right]^2.$$

(c) Two waves will be generated, one moving to the right and the other to the left. Since the medium is non-dispersive, the speed will be independent of the frequency and hence the square shape of the wave will be maintained. Thus, the waves will be

$$\psi_+(x,t) = \begin{cases} 0 & 0 < x < vt, \\ b & vt \le x \le v(t+\tau), \\ 0 & x > v(t+\tau). \end{cases}$$

$$\psi_-(x,t) = \begin{cases} 0 & -vt < x < 0, \\ b & -v(t+\tau) \le x \le -vt, \\ 0 & x < -v(t+\tau). \end{cases}$$

7.15 (a) Let us rewrite Maxwell's equations in a linear medium with $\rho_f = \vec{J}_f = 0$,

$$\vec{\nabla}\cdot\vec{D} = 0, \tag{A.72}$$
$$\vec{\nabla}\times\vec{E} = -\frac{\partial\vec{B}}{\partial t}, \tag{A.73}$$
$$\vec{\nabla}\cdot\vec{B} = 0, \tag{A.74}$$
$$\vec{\nabla}\times\vec{H} = \frac{\partial\vec{D}}{\partial t}. \tag{A.75}$$

Taking the curl of both sides of Eq. A.73 gives the following on the left side:

$$\vec{\nabla}\times\vec{\nabla}\times\vec{E} = \vec{\nabla}(\vec{\nabla}\cdot\vec{E}) - \nabla^2\vec{E} = -\nabla^2\vec{E},$$

and on the right side,

$$-\vec{\nabla}\times\frac{\partial\vec{B}}{\partial t} = -\mu_r\mu_0\frac{\partial}{\partial t}(\vec{\nabla}\times\vec{H}) = -\mu_r\mu_0\epsilon_r\epsilon_0\frac{\partial^2\vec{E}}{\partial t^2}.$$

Therefore,

$$\nabla^2 \vec{E} = \mu_r \mu_0 \epsilon_r \epsilon_0 \frac{\partial^2 \vec{E}}{\partial t^2},$$

which is a classical wave equation with the square of the wave speed v given by

$$v^2 = \frac{1}{\mu_r \mu_0 \epsilon_r \epsilon_0}.$$

Recall that the speed of light in a vacuum is

$$c = \frac{1}{\sqrt{\epsilon_0 \mu_0}}.$$

Therefore, the speed of an electromagnetic wave in a linear homogeneous medium is

$$v = \frac{c}{\sqrt{\epsilon_r \mu_r}}.$$

(b) Using plane monochromatic wave solutions for $\vec{E}$ and $\vec{H}$,

$$\vec{E} = \vec{E}_0\, e^{i(kx-\omega t)}, \quad \vec{H} = \vec{H}_0\, e^{i(kx-\omega t)},$$

in Eq. A.73 leads to

$$kE_0 = \omega \mu_r \mu_0 H_0.$$

Therefore, the impedance will be

$$Z = \frac{|E|}{|H|} = \frac{E_0}{H_0} = \frac{\omega \mu_r \mu_0}{k}.$$

Writing $\omega/k = v$, the wave speed we get,

$$Z = v \mu_r \mu_0.$$

Now, v in a medium is

$$v = \frac{1}{\sqrt{\mu_r \mu_0 \epsilon_r \epsilon_0}}.$$

Therefore, the impedance can be simplified to

$$Z = \sqrt{\frac{\mu_r \mu_0}{\epsilon_r \epsilon_0}} = Z_0 \sqrt{\frac{\mu_r}{\epsilon_r}},$$

where $Z_0 = \sqrt{\mu_0/\epsilon_0}$.

7.16 (a) Note that the given electric field wave is moving towards the positive z-axis. Since $\vec{E} \times \vec{B}$ points in the direction of travel and since the electric field vector is pointed towards the x-axis, the magnetic field vector will be pointed towards the y-axis. The magnitude of the magnetic field vector is $B_0 = E_0/c$, where c is the speed of light. Therefore, the associated magnetic field wave will be

$$\vec{B} = \hat{y}\,\frac{E_0}{c}\cos(kz - \omega t).$$

(b) The Poynting vector $\vec{S}$ is given by the cross-product of the electric and magnetic fields,

$$\vec{S} = \frac{1}{\mu_0}\vec{E} \times \vec{B} = \hat{z}\,\frac{1}{\mu_0}\frac{E_0^2}{c}\cos^2(kz - \omega t).$$

(c) The energy density in the electric field follows from the definition,

$$u_E = \frac{1}{2}\epsilon_0 \vec{E} \cdot \vec{E} = \frac{1}{2}\epsilon_0 E_0^2 \cos^2(kz - \omega t).$$

Similarly, the energy density in the magnetic field wave is

$$u_B = \frac{1}{2\mu_0}\vec{B} \cdot \vec{B} = \frac{1}{2\mu_0}\frac{E_0^2}{c^2}\cos^2(kz - \omega t).$$

Sine $c = 1/\sqrt{\mu_0 \epsilon_0}$, this gives

$$u_B = \frac{1}{2}\epsilon_0 E_0^2 \cos^2(kz - \omega t).$$

Therefore,

$$u_E = u_B.$$

(d) First, note that $u = u_E + u_B$ will just be two times u_E or u_B. Let us write the electric field form for calculations here,

$$u = \epsilon_0 E_0^2 \cos^2(kz - \omega t).$$

Taking the magnitude of the Poynting vector found above we see that

$$|\vec{S}| = \frac{1}{\mu_0}\frac{E_0^2}{c}\cos^2(kz - \omega t).$$

Now, let us write $1/\mu_0$ as $c^2\epsilon_0$. then we get

$$|\vec{S}| = c\left[\epsilon_0 E_0^2 \cos^2(kz - \omega t)\right].$$

Comparing this to u we immediately get

$$|\vec{S}| = cu.$$

(e) Let us write the Poynting vector in terms of H_0 by writing $E_0 = cB_0 = \mu_0 c H_0$,

$$\vec{S} = \hat{z}\,\mu_0 c H_0^2 \cos^2(kz - \omega t).$$

The time-averaging of the square of the cosine gives $\frac{1}{2}$,

$$\langle \cos^2(kz - \omega t) \rangle = \frac{1}{2}.$$

Therefore,

$$\langle \vec{S} \rangle = \hat{z}\frac{1}{2}\mu_0 c H_0^2.$$

The factor $\mu_0 c$ is actually the impedance Z of a vacuum,

$$\mu_0 c = \mu_0 \frac{1}{\sqrt{\mu_0 \epsilon_0}} = \sqrt{\frac{\mu_0}{\epsilon_0}} = Z.$$

Therefore,

$$\langle \vec{S} \rangle = \hat{z}\frac{1}{2} Z H_0^2.$$

7.17 (a) Figure A.47 shows that light in area A actually strikes an area $A/\cos\theta$ if the ray strikes at angle θ. Therefore, pressure will be obtained by dividing the rate of momentum change by area $A/\cos\theta$,

$$\mathcal{P} = \frac{|dP_z/dt|}{A/\cos\theta},$$

where P_z is the z-component of the momentum in the electromagnetic wave. Here we take the z-axis normal to the surface. Recall that the momentum density of electromagnetic fields is

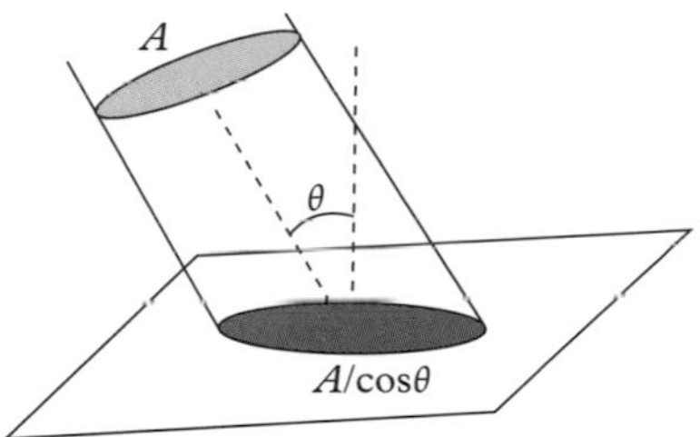

Fig. A.47 *Exercise 7.17.*

$$p = \frac{u}{c},$$

where u is the electromagnetic energy density. The momentum absorbed in time Δt will be the momentum of the wave in a tube of area A and length $c\Delta t$. Therefore,

$$|\Delta P_z| = (p\cos\theta) A c \Delta t = u A \Delta t \cos\theta.$$

Hence, the radiation pressure on the surface will be

$$\mathcal{P} = \frac{|dP_z/dt|}{A/\cos\theta} = u\cos^2\theta. \quad (1)$$

(b) The averaging will be done over a spherical surface above the flat surface. Therefore, the averaging in the angle θ will be done by multiplying the function by $\sin\theta$ first and then integrating from $\theta = 0$ to $\theta = \pi/2$,

$$\langle f(\theta) \rangle = \frac{1}{\int_0^{\pi/2} \sin\theta\, d\theta} \int_0^{\pi/2} f(\theta) \sin\theta\, d\theta.$$

Thus, for $f(\theta) = u\cos^2\theta$, we get,

$$\mathcal{P} = \frac{1}{\int_0^{\pi/2} \sin\theta\, d\theta} \int_0^{\pi/2} \cos^2\theta \sin\theta\, d\theta\, u = \frac{1}{3}u. \quad (2)$$

(c) If the surface were perfectly reflecting, the momentum change would be twice as much as for the perfectly absorbing. That would give twice as much as given in (1) and (2) above.

7.18 Since the radiation is absorbed, the pressure $\mathcal{P}$ applied would be

$$\mathcal{P} = u = \frac{1}{2}\epsilon_0 E_0^2,$$

where E_0 is the electric field amplitude. Let A be the area of cross-section of the particle. Then, the force applied to the dust particle

$$F = \mathcal{P}A.$$

Since the particle of mass m comes to rest from initial velocity v_0 in some time Δt, we must have the following from Newton's second law of motion:

$$F = m\frac{v_0 - 0}{\Delta t}.$$

Hence,

$$\frac{1}{2}\epsilon_0 E_0^2 A = m\frac{v_0}{\Delta t}.$$

Solving for E_0 we get

$$E_0 = \sqrt{\frac{2mv_0}{\epsilon_0 A \Delta t}}.$$

Putting in the numerical values gives $E_0 = 7.35 \times 10^6$ N/C.

7.19 Now, we have a two-dimensional vector situation. Let the direction of the force be the positive x-axis and the direction of the initial velocity be in the second quadrant of the xy-plane. Then, we have

$$a_x = \frac{F}{m} = \frac{v_0 - 0}{\Delta t} = \frac{0.30\ \text{m/s}}{1\ \text{s}} = 0.3\ \text{m/s}^2,\ \ a_y = 0$$
$$v_{0x} = -0.3\cos 30^\circ = -0.26\ \text{m/s},\ \ v_{0y} = 0.3\sin 30^\circ = 0.15\ \text{m/s}.$$

Since we are working with time-averaged quantities, the acceleration is constant in time. Therefore,

$$v_x = v_{0x} + a_x t = -0.26\ \text{m/s} + 0.3\ \text{m/s}^2 \times 2\ \text{s} = 0.34\ \text{m/s},$$
$$v_y = v_{0y} = 0.15\ \text{m/s}.$$

7.20 (a) We suppose that the light is emitted isotropically from the Sun in all directions equally. Then the intensity of light from the Sun will drop off quadratically with distance. This will give the following relation between the intensity at the surface of the Sun and a distance r away from the Sun:

$$R_S^2 I_S = r^2 I_r,$$

where R_S is the radius of the Sun, I_S the intensity at the surface of the Sun and similarly for other symbols. Taking r to be the distance from the Sun to the Earth we get

$$I_S = \left(\frac{r}{R_S}\right)^2 I_r = \left(\frac{149{,}600{,}000\ \text{km}}{695{,}800\ \text{km}}\right)^2 1.38\ \frac{\text{kW}}{\text{m}^2} = 64{,}000\ \frac{\text{kW}}{\text{m}^2}.$$

(b) Let us find the force F_r by the radiation pressure,

$$F_r = \mathcal{P}A = \frac{I_r}{c} \times \pi R_E^2.$$

where A is the area of cross-section of Earth, whose radius is denoted here by R_E. Putting in the numerical values gives

$$F_r = \frac{1.38 \times 10^3 \text{ W/m}^2}{3 \times 10^8 \text{ m/s}} \times \pi (6,371,000 \text{ m})^2 = 5.87 \times 10^8 \text{ N}.$$

The gravitational force on Earth will be

$$F_G = G\frac{M_S M_E}{d^2} = 6.67 \times 10^{-11} \frac{1.989 \times 10^{30} \times 5.972 \times 10^{24}}{149,600,000,000^2} = 3.54 \times 10^{22} \text{ N}$$

Thus, F_G is about 6.1×10^{13} times as great!

7.21 (a) Equating the mg to the radiation pressure will give

$$\frac{I}{c}\pi R^2 = mg.$$

Therefore,

$$I = \frac{mgc}{\pi R^2} = \frac{20 \times 10^{-9} \text{ kg} \times 9.81 \text{ m/}^2 \times 3 \times 10^8 \text{ m/s}}{\pi (2 \times 10^{-3} \text{ m})^2} = 4.7 \times 10^6 \frac{\text{W}}{\text{m}^2}.$$

(b) If completely reflected, the radiation pressure will be twice as much for the same intensity. Therefore, we need half as much intensity, i.e. $I = 2.34 \times 10^6 \text{ W/m}^2$.

7.22 The intensity of light at the location of the astronaut will be 16 times greater than the intensity on Earth since the intensity drops off quadratically with distance from the Sun,

$$I = 16 I_E.$$

Since the radiation pressure is from perfectly reflecting light which is perpendicularly incident on the reflecting surface of area A we will have

$$\frac{2I}{c}A = Ma.$$

Therefore, the acceleration of the astronaut due to the radiation pressure alone will be

$$a = \frac{32 I_E A}{Mc}.$$

For $M = 100$ kg, $A = 100 \text{ m}^2$, and $I_E = 1,400 \text{ W/m}^2$ we get

$$a = 1.5 \times 10^{-4} \text{ m/s}^2.$$

7.23 This exercise is a direct application of $I = I_0 \cos^2 \theta$,

$$I_1 = I_0 \cos^2 30^\circ = \frac{3}{4} I_0,$$

$$I_2 = I_1 \cos^2 60^\circ = \frac{1}{4} I_1 = \frac{3}{16} I_0,$$

$$I_3 = I_2 \cos^2 30^\circ = \frac{3}{4} I_2 = \frac{9}{64} I_0.$$

7.24 Let the rotating polarizer be aligned with the axis of polarizer 1 at $t = 0$; then the angle between the axes of polarizer 1 and the rotating polarizer will be

$$\theta = \omega t.$$

Therefore,

$$I_1 = I_0 \cos^2(\omega t),$$

the angle between the rotating polarizer and polarizer 2 is also θ,

$$I_2 = I_1 \cos^2(\omega t) = I_0 \cos^4(\omega t).$$

7.25 Figure A.48 shows the three cases. (a) and (b) are linearly polarized and (c) and (d) are circularly polarized.

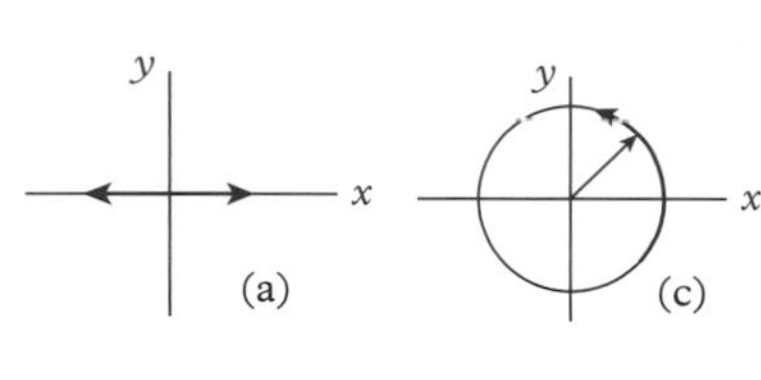

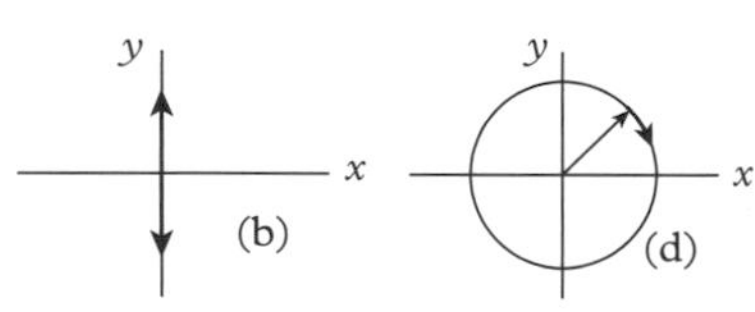

Fig. A.48 *Exercise 7.25.*

7.26 (a) Let us write $\pm i$ multiplying the unit vector $\hat{j}$ in exponential notation,

$$i = e^{i\pi/2}, \quad -i = e^{-i\pi/2}.$$

Using these, the real parts of the waves are easily found to be

$$E_{1x} = E_0 \cos(kz - \omega t), \quad E_{1y} = E_0 \cos(kz - \omega t + \pi/2) = -E_0 \sin(kz - \omega t),$$

$$E_{2x} = E_0 \cos(kz - \omega t), \quad E_{2y} = E_0 \cos(kz - \omega t - \pi/2) = E_0 \sin(kz - \omega t).$$

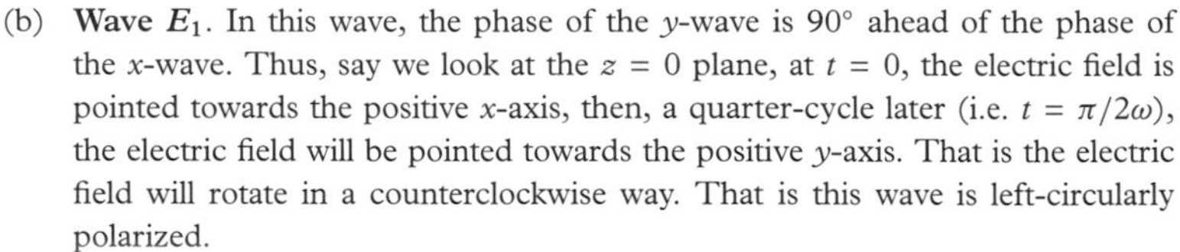

(b) **Wave E_1**. In this wave, the phase of the y-wave is 90° ahead of the phase of the x-wave. Thus, say we look at the $z = 0$ plane, at $t = 0$, the electric field is pointed towards the positive x-axis, then, a quarter-cycle later (i.e. $t = \pi/2\omega$), the electric field will be pointed towards the positive y-axis. That is the electric field will rotate in a counterclockwise way. That is this wave is left-circularly polarized.
Wave E_2. This wave will be clockwise or right-circularly polarized.

(c) To find the associated magnetic field, you can think of the given wave as a superposition of two waves, one having polarization along the x-axis and the other having polarization along the y-axis,

$$\vec{E}_1 = \vec{E}_{1a} + \vec{E}_{1b},$$

where

$$\vec{E}_{1a} = \hat{x}E_0 \cos(kz - \omega t),$$
$$\vec{E}_{1b} = \hat{y}E_0 \cos(kz - \omega t + \pi/2).$$

Now, the magnetic fields associated with these component waves are easy to work out,

$$\vec{B}_{1a} = \hat{y}\frac{E_0}{c} \cos(kz-\omega t) \longrightarrow \hat{y}\frac{E_0}{c} e^{i(kz-\omega t)},$$
$$\vec{B}_{1b} = -\hat{x}\frac{E_0}{c} \cos(kz-\omega t+\pi/2) \longrightarrow -\hat{x}\frac{E_0}{c} e^{i\left(kz-\omega t+\frac{\pi}{2}\right)}.$$

Therefore, the associated magnetic field of $\vec{E}_1$ will be

$$\vec{B}_1 = \vec{B}_{1a} + \vec{B}_{1b}, \tag{A.76}$$
$$= \left(-i\hat{x} + \hat{y}\right)\frac{E_0}{c} e^{i(kz-\omega t)}. \tag{A.77}$$

7.27 The electric field in the incident wave is

$$\vec{E}_{in} = E_0 \left[\hat{x}\cos\left(2\pi ft - \frac{2\pi}{\lambda}z\right) + \hat{y}\sin\left(2\pi ft - \frac{2\pi}{\lambda}z\right)\right].$$

The outgoing electric field will be linearly polarized in the direction of the linear polarizer. The projection of this electric field on the axis will give the amplitude of the electric field vector,

$$E_{out}(z,t) = \vec{E}_{in}\cdot\left(\frac{\hat{x}}{\sqrt{2}} + \frac{\hat{y}}{\sqrt{2}}\right) \tag{A.78}$$
$$= \frac{E_0}{\sqrt{2}}\left[\cos\left(2\pi ft - \frac{2\pi}{\lambda}z\right) + \sin\left(2\pi ft - \frac{2\pi}{\lambda}z\right)\right] \tag{A.79}$$

The factor $1/\sqrt{2}$ will give an intensity half that of the incident wave. The electric field vector would be

$$\vec{E}_{out}(z,t) = E_{out}(z,t)\left(\frac{\hat{x}}{\sqrt{2}} + \frac{\hat{y}}{\sqrt{2}}\right).$$

7.28 Let θ be the angle between the first polarizer and the inserted polarizer. The angle between the inserted polarizer and the second polarizer will be $90° - \theta$. Therefore, in order to get a fraction a as the output intensity we must have

$$\frac{I_0}{2}\cos^2\theta\cos^2(90° - \theta) = aI_0.$$

This gives

$$\sin^2(2\theta) = 8a.$$

Therefore,

$$\theta = \frac{\sin^{-1}\left(\sqrt{8a}\right)}{2}.$$

Putting in $a = 0.05$ we get

$$\theta = \frac{\sin^{-1}\left(\sqrt{0.4}\right)}{2} = 19.6^\circ.$$

7.29 Let I_0 be the intensity of the unpolarized light. Let the axis of the rotating polarizer at $t = 0$ be aligned with the first polarizer. Then, the angle of the first polarizer with the rotating polarizer at instant t will be $\theta = 2\pi ft$. Therefore, the emergent intensity will be

$$I = \frac{I_0}{2} \cos^2(2\pi ft)\cos^2(\pi/2 - 2\pi ft).$$

Simplifying we get

$$I = \frac{1}{8} I_0 \sin^2(4\pi ft).$$

7.30 (a) We will use the de Broglie equation to obtain the wavelength,

$$p = 50 \text{ kg} \times 10 \text{ m/s} = 500 \text{ kg.m/s}.$$

Therefore,

$$\lambda = \frac{h}{p} = \frac{6.626 \times 10^{-34} \text{ J.s}}{500 \text{ kg.m/s}} = 1.325 \times 10^{-36} \text{ m}.$$

(b) (i) 7.3×10^{-5} m, (ii) 7.3×10^{-8} m, (iii) 7.3×10^{-12} m.

(c) (i) 3.97×10^{-8} m, (ii) 3.97×10^{-11} m, (iii) 3.97×10^{-15} m.

7.31 An electron's position in the hydrogen atom is known with an uncertainty of 10^{-10} m.

(a) Heisenberg uncertainty relation for position and momentum uncertainties states that $\Delta x \Delta p \geq \frac{h}{4\pi}$,

$$\Delta p \geq \frac{h}{4\pi\, \Delta x} = \frac{6.626 \times 10^{-34} \text{ J.s}}{4 \times \pi \times 10^{-10} \text{ m}} = 5.273 \times 10^{-25} \text{ kg.m/s}.$$

(b) The uncertainty in speed can be obtained from Δp,

$$\Delta v = \frac{\Delta p}{m_e} \geq 5.79 \times 10^5 \text{ m/s}.$$

7.32 The probability will be based on the profile of a wave for the lowest state, which is given by

$$\psi_0(x) = \sqrt{\frac{2}{a}} \sin\left(\frac{\pi x}{a}\right).$$

(a) The probability that the electron is in the right half of the well should be half based on the symmetry of the wave function about $x = a/2$. We can also verify it by direct calculation,

$$P = \int_{a/2}^{2} |\psi_0(x)|^2 dx \Big/ \int_{2}^{2} |\psi_0(x)|^2 dx = \frac{1}{2}.$$

The detail is left as an exercise. Hint: you will need a double angle trig formula.

(b) The probability of being in the quarter of the well is not $\frac{1}{4}$. You will have to really work out the integrals this time,

$$\int_{3a/4}^{a} |\psi_0|^2 dx = \frac{1}{2}\left(\frac{1}{2} - \frac{1}{\pi}\right).$$

(c) The probability here is

$$\frac{1}{5} + \frac{1}{2\pi} \sin\left(\frac{\pi}{5}\right).$$

7.33 Hint: see the solution above.

Chapter 8

8.1 (a) Figure A.49 shows the main parts of the problem. In region I ($x < 0$) the displacement of the string will be the sum of the two waves, ψ_{in} and ψ_{re} and in region II ($x > 0$) the displacement will just be the wave ψ_{tr},

$$\psi_1(x,t) = \psi_{in}(x,t) + \psi_{re}(x,t), \quad x < 0 \tag{A.80}$$

$$\psi_2(x,t) = \psi_{tr}(x,t), \quad x > 0. \tag{A.81}$$

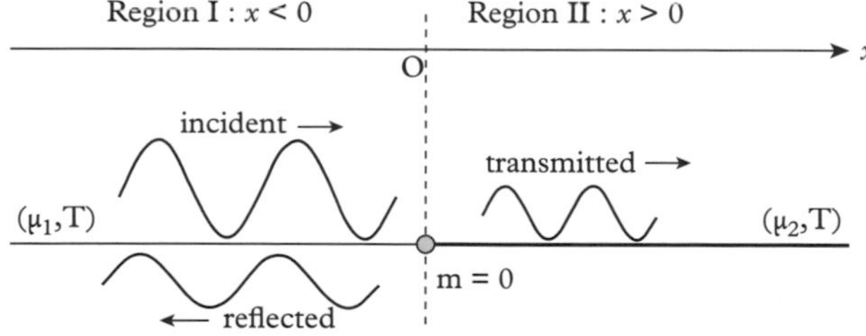

Fig. A.49 *Exercise 8.1.*

When $\mu_1 \neq \mu_2$ the wave speeds would be different in the two strings, as is clear from the dependence of wave speed on mass density and tension,

$$v_1 = \sqrt{T/\mu_1}, \quad v_2 = \sqrt{T/\mu_2}.$$

Therefore, the wave numbers of the waves on the two strings will be different even though they will have the same frequency. We will denote them by k_1 and k_2,

$$k_1 = \frac{\omega}{v_1}, \quad k_2 = \frac{\omega}{v_2}.$$

Therefore, the analytic expressions of the three waves are

$$\psi_{in}(x,t) = A\cos(k_1 x - \omega t), \tag{A.82}$$

$$\psi_{re}(x,t) = B\cos(-k_1 x - \omega t + \phi), \tag{A.83}$$

$$\psi_{tr}(x,t) = C\cos(-k_1 x - \omega t + \delta), \tag{A.84}$$

where I have added the phase shifts of the reflected wave with respect to the incident wave at $x = 0$ and $t = 0$. Therefore, the displacements of the two strings will be

$$\psi_1(x,t) = A\cos(k_1 x - \omega t) + B\cos(-k_1 x - \omega t + \phi), \quad x < 0 \tag{A.85}$$

$$\psi_2(x,t) = C\cos(-k_1 x - \omega t + \delta), \quad x > 0 \tag{A.86}$$

It is more convenient to perform the calculation using complex exponential notation, and therefore, we will write these displacements as

$$\psi_1(x,t) = A\,e^{i(k_1 x - \omega t)} + B\,e^{i(k_1 x - \omega t + \phi)}, \quad x < 0 \tag{A.87}$$

$$\psi_2(x,t) = C\,e^{i(k_1 x - \omega t + \delta)}. \quad x > 0 \tag{A.88}$$

These displacements will cause the motion of the knot at $x = 0$ and must satisfy boundary conditions there. Since the knot is assumed to be massless, we will get the continuity of the wave function as well as the derivative with respect to space coordinate across $x = 0$,

$$\psi_1(0,t) = \psi_2(0,t), \tag{A.89}$$

$$\left(\frac{\partial \psi_1}{\partial x}\right)_{x=0} = \left(\frac{\partial \psi_2}{\partial x}\right)_{x=0}. \tag{A.90}$$

Therefore,

$$A + Be^{i\phi} = C\,e^{i\delta}, \tag{A.91}$$

$$k_1\left(A - Be^{i\phi}\right) = k_2 C\,e^{i\delta}. \tag{A.92}$$

Solving these yields

$$\frac{B}{A}e^{i\phi} = \frac{k_1 - k_2}{k_1 + k_2}, \tag{A.93}$$

$$\frac{C}{A}e^{i\delta} = \frac{2k_1}{k_1 + k_2}. \tag{A.94}$$

Since k_1 and k_2 are real and positive, we can extract the amplitude ratios and the phases by observing that $e^{i\pi} = -1$,

$$\frac{B}{A} = \frac{|k_1 - k_2|}{k_1 + k_2}, \quad \phi = \begin{cases} 0 & k_1 \geq k_2 \\ \pi & k_1 < k_2 \end{cases}, \tag{A.95}$$

$$\frac{C}{A} = \frac{2k_1}{k_1 + k_2}, \quad \delta = 0. \tag{A.96}$$

We can write these in terms of the two speeds by noting that $k = \omega/v$ in each string,

$$\frac{B}{A} = \frac{|v_1 - v_2|}{v_1 + v_2}, \quad \phi = \begin{cases} 0 & v_1 \leq v_2 \\ \pi & v_1 > v_2 \end{cases}, \tag{A.97}$$

$$\frac{C}{A} = \frac{2v_2}{v_1 + v_2}, \quad \delta = 0. \tag{A.98}$$

Since $v = \sqrt{T/\mu}$ in each medium, we can also express these in terms of μ_1 and μ_2, and simplify with $\alpha = \mu_2/\mu_1$,

$$\frac{B}{A} = \frac{|1 - \sqrt{\alpha}|}{1 + \sqrt{\alpha}}, \quad \phi = \begin{cases} 0 & \alpha \leq 1 \\ \pi & \alpha > 1 \end{cases}, \tag{A.99}$$

$$\frac{C}{A} = \frac{2}{1 + \sqrt{\alpha}}, \quad \delta = 0. \tag{A.100}$$

(b) The time-average of power in a harmonic wave of amplitude A and frequency ω on a string of tension T and wave speed v is given by

$$\langle P \rangle = \frac{1}{2}\frac{T}{v}\omega^2 A^2.$$

Applying this to the three waves we get

$$\langle P_{in} \rangle = \frac{1}{2}\frac{T}{v_1}\omega^2 A^2, \tag{A.101}$$

$$\langle P_{re} \rangle = \frac{1}{2}\frac{T}{v_1}\omega^2 B^2, \tag{A.102}$$

$$\langle P_{tr} \rangle = \frac{1}{2}\frac{T}{v_2}\omega^2 C^2. \tag{A.103}$$

Now, let us verify that the power in the reflected and transmitted waves add up to the power in the incident wave,

$$\begin{aligned}\langle P_{re}\rangle + \langle P_{tr}\rangle &= \frac{1}{2}T\omega^2\left(\frac{1}{v_1}B^2 + \frac{1}{v_2}C^2\right)\\ &= \frac{1}{2}T\omega^2 A^2\left[\frac{1}{v_1}\frac{(v_1 - v_2)^2}{(v_1+v_2)^2} + \frac{1}{v_2}\frac{4v_2^2}{(v_1+v_2)^2}\right]\\ &= \frac{1}{2}T\omega^2 A^2\frac{1}{v_1}\frac{(v_1+v_2)^2}{(v_1+v_2)^2}\\ &= \frac{1}{2}T\omega^2 A^2\frac{1}{v_1} = \langle P_{in}\rangle.\end{aligned}$$

Therefore, on average, the energy arriving at the knot per unit time is carried away from the knot as the same amount.

(c) The average power of the incident wave is independent of the value of α. When we look at the dependence of B/A and C/A on α we find the following:

(i) When $\alpha = 1$, we have

$$\frac{B}{A} = 0,\ \ \frac{C}{A} = 1,\ \ \delta = 0,\ \ \phi = 0.$$

This is the case when the two strings are the same. The incident wave just moves on without any reflection.

(ii) When $\alpha = \infty$, we have

$$\frac{B}{A} = 1,\ \ \frac{C}{A} = 0,\ \ \delta = 0,\ \ \phi = \pi.$$

This is the case when string 1 is just tied to a fixed support. The reflected wave is 180° out of phase with respect to the incident wave.

(d) We can use either $\psi_1(0,t)$ or $\psi_2(0,t)$ to obtain the displacement of the knot. Using ψ_2 we get

$$\psi(t) = \psi_2(0,t) = C\cos(-\omega t + \delta) = \frac{2v_2 A}{v_1 + v_2}\cos(\omega t).$$

8.2 (a) The equation of motion of an arbitrary bob, say, labeled n, will be

$$m\frac{d^2\psi_n}{dt^2} = -\frac{mg}{l}\psi_n - K(\psi_n - \psi_{n-1}) + K(\psi_{n+1} - \psi_n),$$

where $K = K_L$ for a bob at $x < 0$ and $K = K_R$ for a bob at $x > 0$. This equation can be rewritten as

$$\frac{d^2\psi_n}{dt^2} = -\frac{g}{l}\psi_n - \frac{Ka^2}{m}\left(\frac{\psi_{n+1} - 2\psi_n + \psi_{n-1}}{a^2}\right).$$

Taking $a \to 0$ with $na = x$, the x-coordinate, we get

$$\frac{\partial^2\psi(x,t)}{\partial t^2} = -\frac{g}{l}\psi(x,t) - \frac{Ka^2}{m}\frac{\partial^2\psi(x,t)}{\partial x^2}. \tag{A.104}$$

(b) From the continuity of the displacement we note that one condition will be

$$\psi_L(0,t) = \psi_R(0,t).$$

When we integrate Eq. A.104 from $x = -\epsilon$ to $x = +\epsilon$, where $\epsilon \ll 1$ and take the limit $\epsilon \to 0$ we would get

$$K_L \left.\frac{\partial \psi_L(x,t)}{\partial x}\right|_{x=0} = K_R \left.\frac{\partial \psi_R(x,t)}{\partial x}\right|_{x=0}.$$

(c) Let us first find the dispersion relation based on the wave equation in part (a) so that we can know whether wave numbers would be different in $x < 0$ and $x > 0$ regions. By using $\psi(x,t) = A\cos(kx - \omega t)$ we find that

$$\omega^2 = \frac{g}{l} + \left(\frac{a^2K}{m}\right)k^2.$$

Clearly, k will depend on whether $K = K_L$ or $K = K_R$. Let us denote them as k_L and k_R respectively,

$$k_L = \sqrt{\frac{m}{a^2K_L}\left(\omega^2 - \frac{g}{l}\right)},$$
$$k_R = \sqrt{\frac{m}{a^2K_R}\left(\omega^2 - \frac{g}{l}\right)}.$$

These show that the wave numbers in $x < 0$ and $x > 0$ regions are related by

$$\frac{k_L}{k_R} = \sqrt{\frac{K_R}{K_L}}.$$

The displacements on the $x < 0$ and $x > 0$, i.e. ψ_L and ψ_R, in terms of incident, reflected, and transmitted waves will be

$$\psi_L(x,t) = \psi_{in} + \psi_{re} = A\cos(k_Lx - \omega t) + B\cos(-k_Lx - \omega t + \phi),$$
$$\psi_R(x,t) = \psi_{tr} = C\cos(k_Rx - \omega t + \delta),$$

or, in complex notation,

$$\psi_L(x,t) = A\,e^{i(k_Lx - \omega t)} + Be^{i\phi}\,e^{i(-k_Lx - \omega t)},$$
$$\psi_R(x,t) = Ce^{i\delta}\,e^{i(k_Rx - \omega t)},$$

where I have used $e^{-i\omega t}$ for the time part for each exponential. It is easier to conduct the calculation in complex exponential notation rather than cosines. The boundary conditions yield

$$A + Be^{i\phi} = Ce^{i\delta},$$
$$K_Lk_L\left(A - Be^{i\phi}\right) = K_Rk_RCe^{i\delta}.$$

Let us introduce the following for calculation purposes:

$$\rho = \frac{B}{A}e^{i\phi}, \quad \tau = \frac{C}{A}e^{i\delta}.$$

Then,

$$1 + \rho = \tau,$$
$$K_L k_L (1 - \rho) = K_R k_R \tau.$$

Therefore,

$$\rho = \frac{\sqrt{K_L} - \sqrt{K_R}}{\sqrt{K_L} + \sqrt{K_R}},$$

$$\tau = \frac{2\sqrt{K_L}}{\sqrt{K_L} + \sqrt{K_R}}.$$

Therefore,

$$\frac{B}{A} = \frac{|\sqrt{K_L} - \sqrt{K_R}|}{\sqrt{K_L} + \sqrt{K_R}}, \quad \phi = \begin{cases} 0 & K_L \geq K_R \\ \pi & K_L < K_R. \end{cases} \tag{A.105}$$

$$\frac{C}{A} = \frac{2\sqrt{K_L}}{\sqrt{K_L} + \sqrt{K_R}}, \quad \delta = 0. \tag{A.106}$$

8.3 Figure A.50 shows that there are two types of oscillators in the system in the two regions $x < 0$ and $x > 0$.

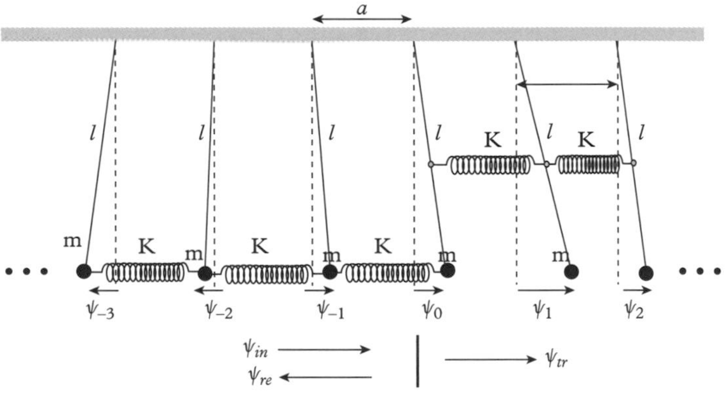

Fig. A.50 *Exercise 8.3.*

(a) To deduce the wave equation for this problem we will start with the equations of motion for an arbitrary bob in the two regions,

$$x < 0: \quad m\frac{d^2\psi_n}{dt^2} = -\frac{mg}{l}\psi_n + K\left(\psi_{n+1} - 2\psi_n + \psi_{n-1}\right),$$

$$x > 0: \quad m\frac{d^2\psi_n}{dt^2} = -\frac{mg}{l}\psi_n + K\left(\frac{\psi_{n+1} - 2\psi_n + \psi_{n-1}}{2}\right).$$

The distance between bobs is a. We divide the last term in the two expressions by a^2 and take $a \to 0$ limit to obtain the wave equation for pendulums in the continuum limit,

$$x < 0: \quad \frac{\partial^2 \psi(x,t)}{\partial t^2} = -\frac{g}{l}\psi 9x, t) + \frac{a^2 K}{m}\frac{\partial^2 \psi(x,t)}{\partial x^2},$$

$$x > 0: \quad \frac{\partial^2 \psi(x,t)}{\partial t^2} = -\frac{g}{l}\psi 9x, t) + \frac{a^2 K}{2m}\frac{\partial^2 \psi(x,t)}{\partial x^2}.$$

If we try the harmonic wave solution, $\psi = Ae^{ikx-i\omega t)}$, in these equations we get a different wave number on the two sides of $x = 0$ for the same frequency,

$$x < 0: \quad k_L = \sqrt{\frac{m}{a^2 K}\left(\omega^2 - \frac{g}{l}\right)},$$

$$x > 0: \quad k_R = \sqrt{\frac{2m}{a^2 K}\left(\omega^2 - \frac{g}{l}\right)} = \sqrt{2}k_L,$$

where the wave number k is now labeled as k_L and k_R.

(b) Let us denote the displacement on the left side by psi_L and that on the right side by ψ_R. In terms of the incident, reflected and transmitted waves these displacements will be

$$\psi_L = \psi_{in} + \psi_{re} \quad (x < 0),$$
$$\psi_R = \psi_{tr} \quad (x < 0).$$

To find the reflection and transmission coefficients we make use of the following boundary conditions based on the wave equation:

$$\psi_L(0,t) = \psi_R(0,t),$$
$$K_L\left(\frac{\partial \psi_L}{\partial x}\right)_{x=0} = K_R\left(\frac{\partial \psi_R}{\partial x}\right)_{x=0},$$

where

$$K_L = K, \quad K_R = \frac{K}{2}.$$

Fitting the boundary conditions to the given wave functions, written in complex notation, we get

$$A + Be^{i\phi} = Ce^{i\delta},$$
$$Kk_L\left(A - Be^{i\phi}\right) = \frac{K}{2}k_R Ce^{i\delta}.$$

With $k_R = \sqrt{2}k_L$ we get

$$A + Be^{i\phi} = Ce^{i\delta},$$
$$A - Be^{i\phi} = \frac{1}{\sqrt{2}}Ce^{i\delta}.$$

Therefore,

$$\frac{B}{A} = \frac{\sqrt{2}-1}{\sqrt{2}+1}, \quad \phi = 0,$$
$$\frac{C}{A} = \frac{2\sqrt{2}}{\sqrt{2}+1}, \quad \delta = 0.$$

8.4 (a) We will apply the following relation:

$$\frac{\partial I}{\partial x} = -C\frac{\partial V}{\partial t} \quad (1)$$

to the three waves separately. Recall the following additional relation in a transmission line,

$$\frac{\omega}{k} = v(\text{wave}) = \frac{1}{\sqrt{LC}}.$$

This becomes the following at a point on the negative x-axis in the figure:

$$\frac{\omega}{k_1} = v_1(\text{wave}) = \frac{1}{\sqrt{L_1 C_1}},$$

where L_1 and C_1 are inductance and capacitance per unit length of the cable on the left. At a point on the positive x-axis in the figure we have

$$\frac{\omega}{k_2} = v_2(\text{wave}) = \frac{1}{\sqrt{L_2 C_2}},$$

where L_2 and C_2 are inductance and capacitance per unit length of the cable on the right.

(i) With $I = I_{in}$ and $V = V_{in}$ we find the following at a point where $x < 0$:

$$k_1 I_0 e^{i\theta} = -C_1 \omega V_0,$$

where k_1 is the wave number in this region and C_1 the capacitance per unit length at that point. The ratio

$$I_0 e^{i\theta} = -\frac{C_1\omega}{k_1} V_0. \quad (2)$$

We can write the constants on the right side as

$$\frac{C_1\omega}{k_1} = C_1 v_1 = \frac{C_1}{\sqrt{L_1 C_1}} = \sqrt{\frac{C_1}{L_1}} = \frac{1}{Z_1}.$$

Hence, if we write $-1 = e^{i\pi}$ in (2) we get the desired result,

$$I_0 = \frac{V_0}{Z_1}, \quad \theta = \pi.$$

(ii) When we use $I = I_{re}$ and $V = V_{re}$ in (1) we get

$$I_{0R} = \frac{V_{0R}}{Z_1}, \quad \phi_r' = \phi_r.$$

(iii) When we use $I = I_{tr}$ and $V = V_{tr}$ in (1) we get

$$I_{0T} = \frac{V_{0T}}{Z_2}, \quad \phi_t' = \phi_t + \pi.$$

(b) The boundary conditions in this problem are the continuity of voltage and current across the boundary at $x = 0$. Let $V_1(x, t)$, and $I_1(x, t)$ be the voltage and current at an arbitrary point on transmission line 1 in the region $x < 0$, and $V_2(x, t)$, and $I_2(x, t)$ be the voltage and current at an arbitrary point on transmission line 2 in the region $x > 0$. Note the voltage and current would be equal to the sum of the incident and reflected wave on 1,

$$V_1(x, t) = V_{in}(x, t) + V_{re}(x, t), \quad I_1(x, t) = I_{in}(x, t) + I_{re}(x, t),$$
$$V_2(x, t) = V_{tr}(x, t), \quad I_2(x, t) = I_{tr}(x, t).$$

The boundary conditions would then be

$$V_1(0, t) = V_2(0, t), \quad I_1(0, t) = I_2(0, t).$$

These conditions give us the following relations:

$$V_{in}(0, t) + V_{re}(0, t) = V_{tr}(0, t),$$
$$I_{in}(0, t) + I_{re}(0, t) = I_{tr}(0, t).$$

Using the results from above we can rewrite these equations as

$$V_0 + V_{0R}e^{i\phi_r} = V_{0T}e^{i\phi_t},$$
$$-\frac{V_0}{Z_1} + \frac{V_{0R}e^{i\phi_r}}{Z_1} = -\frac{V_{0T}e^{i\phi_t}}{Z_2}.$$

Solving these we obtain

$$V_{0R}e^{i\phi_r} = \frac{Z_2 - Z_1}{Z_1 + Z_2},$$
$$V_{0T}e^{i\phi_t} = \frac{2Z_2}{Z_1 + Z_2}.$$

Reading off the amplitude and phase, we get

$$V_{0R} = \frac{|Z_1 - Z_2|}{Z_1 + Z_2}, \quad \phi_r = \begin{cases} 0 & Z_2 > Z_1, \\ \pi & Z_1 > Z_2. \end{cases}.$$

$$V_{0T} = \frac{|2Z_2}{Z_1 + Z_2}, \quad \phi_t = 0.$$

(c) To verify the conservation of energy, we note that the time-averaged power for any sinusoidal wave is

$$\langle P \rangle = \frac{1}{2}\frac{V_0^2}{Z}.$$

Therefore, the power values in the three waves are

$$\langle P_{in} \rangle = \frac{1}{2}\frac{V_0^2}{Z_1},$$

$$\langle P_{re} \rangle = \frac{1}{2}\frac{V_{0R}^2}{Z_1} = \frac{1}{2}\frac{(Z_2 - Z_1)^2}{(Z_1 + Z_2)^2}\frac{V_0^2}{Z_1},$$

$$\langle P_{tr} \rangle = \frac{1}{2}\frac{V_{0T}^2}{Z_2} = \frac{1}{2}\frac{4Z_2 Z_1}{(Z_1 + Z_2)^2}\frac{V_0^2}{Z_1}.$$

It is easily seen that

$$\langle P_{re} \rangle + \langle P_{tr} \rangle = \langle P_{in} \rangle.$$

8.5 (a) The shorted connection at $x = 0$ means we will have

$$V_{in}(0, t) + V_{re}(0, t) = 0.$$

Therefore, in this situation the reflection coefficient of the voltage wave will be

$$\rho_v = -1.$$

The relation of current to voltage gives

$$I_0 = \frac{V_0}{Z_0}, \quad \rho_i I_0 = -\frac{\rho_v V_0}{Z_0},$$

where Z_0 is the impedance of the line and ρ_i is the reflection coefficient of the current wave. Therefore,

$$\rho_i = -\rho_v = 1.$$

That means $I_{in} = I_{re}$, and net current at $x = 0$, will be $2I_{in}(0, t)$.

(b) In the case of the open circuit we will have the net current zero,

$$I_{in}(0, t) + I_{re}(0, t) = 0.$$

Therefore, in this situation the reflection coefficient of the voltage wave will be

$$\rho_i = -1.$$

Since $\rho_v = -\rho_i$, we will have $\rho_v = 1$. Since $V_{re} = V_{in}$, the voltage at $x = 0$ will be $2V_{in}(0, t)$.

(c) (i) Writing Ohm's law for the load we get

$$I_0 + \rho_i I_0 = \frac{V_0 + \rho_v V_0}{Z_0}.$$

Since $I_0 = V_0/Z_0$ we will get

$$\rho_i = \rho_v.$$

Since $\rho_i = -\rho_v$ from the equation relating the reflected voltage and current waves, we must have

$$\rho_i = \rho_v = 0.$$

That is, no return wave. The current and voltage in the line are all in the incident wave. (ii) The power delivered to the load will be the power of the incident wave,

$$\langle P \rangle = \frac{1}{2} \frac{V_0^2}{Z_0}.$$

8.6 (a) The fundamental mode for one end closed and the other end open will have the wavelength

$$\lambda = 4L,$$

since a quarter of the wave will fit in the tube that has the correct boundary condition. Therefore, the frequency of the excited mode will be

$$f = \frac{v}{\lambda} = \frac{v}{4L}.$$

(b) The shape of the fundamental mode is one quarter of the sine function. If we place the origin at the closed end and point the positive x-axis towards the open end the mode will have the following expression:

$$\psi(x, t) = A \sin(kx) \cos(\omega t),$$

where

$$k = \frac{2\pi}{\lambda} = \frac{\pi}{2L}, \quad \omega = 2\pi f = \frac{\pi v}{2L}.$$

We can write the product of the sine and cosine as the sum of two sines,

$$\psi(x, t) = \frac{A}{2} \sin(kx - \omega t) - \frac{A}{2} \sin(-kx - \omega t).$$

The first term is the wave moving towards the positive x-axis, which is the reflected wave and the second term is the wave moving towards the negative x-axis, which is the wave that is incident on the closed end. Note the relative sign of the two terms, which indicates the two waves have opposite phase.

8.7 (a) Let the air/water interface be at $x = 0$ with air in the $x < 0$ region and water in the $x > 0$ region. Let ω be the angular frequency of the waves and k_1 the wave number when the wave is in air and k_2 the wave number when in water. Let v_1 and v_2 be the speed of sound in air and water respectively. Let us write the incident, reflected, and transmitted waves as follows:

$$\text{In air: } \psi_{in} = \psi_0 e^{ik_1 x - i\omega t}, \quad \psi_{re} = R\psi_0 e^{ik_1 x - i\omega t + i\phi_r},$$
$$\text{In water: } \psi_{tr} = T\psi_0 e^{ik_2 x - i\omega t + i\phi_t}.$$

The displacement in air is, of course, the sum of the two waves in the air. Now, applying the boundary conditions we find

$$1 + Re^{i\phi_r} = Te^{i\phi_t},$$
$$1 - Re^{i\phi_r} = \frac{\rho_2 v_2^2 k_2}{\rho_1 v_1^2 k_1} Te^{i\phi_t}.$$

Let us denote the bunch of constants on the right side with one symbol,

$$\alpha = \frac{\rho_2 v_2^2 k_2}{\rho_1 v_1^2 k_1} = \frac{\rho_2 v_2}{\rho_1 v_1},$$

where I have made use of the fact that $\omega = k_1 v_1 = k_2 v_2$ and canceled out the common terms. Now, we can solve for the reflection and transmission coefficients as

$$Re^{i\phi_r} = \frac{1-\alpha}{1+\alpha},$$
$$Te^{i\phi_t} = \frac{2}{1+\alpha}.$$

Therefore,

$$R = \frac{|1-\alpha|}{1+\alpha}, \quad \phi_r = \pi \text{ if } \alpha > 1 \text{ else } \phi_r = 0,$$
$$T = \frac{2}{1+\alpha}, \quad \phi_t = 0.$$

(b) Intensity of sound is related to the particle displacement amplitude A of a harmonic wave $\psi = A\cos(kx - \omega t)$ as follows,

$$I = \frac{1}{2}\rho v \omega^2 A^2.$$

The intensity is the average power per unit cross-sectional area. Therefore, we will have the following expressions for the intensities of the three waves, two in air and one in water:

$$I_{in} = \frac{1}{2}\rho_1 v_1 \omega^2 \psi_0^2, \quad I_{re} = \frac{1}{2}\rho_1 v_1 \omega^2 R^2 \psi_0^2,$$
$$I_{tr} = \frac{1}{2}\rho_2 v_2 \omega^2 T^2 \psi_0^2.$$

To convert these into power we will multiply by the area of cross-section perpendicular to the direction in which the wave is traveling.

(c) If we divide the intensity of the transmitted wave by the intensity of the incident wave we will get the fraction of the energy that goes into water,

$$\text{Fraction of energy} = \frac{\rho_2 v_2}{\rho_1 v_1} T^2 = \alpha T^2.$$

This can be written in α as

$$\text{Fraction of energy} = \frac{4\alpha}{(1+\alpha)^2}.$$

From the given numerical values we have

$$\alpha = \frac{1000 \text{ kg/m}^3 \times 1500 \text{ m/s}}{1.3 \text{ kg/m}^3 \times 350 \text{ m/s}} = 3,297.$$

Therefore,

$$\text{Fraction of energy} = \frac{4 \times 3,297}{3,298^2} = 0.0012.$$

8.8 From the definition of Brewster's angle we get

$$\theta_B = \tan^{-1}\left(\frac{n_2}{n_1}\right) = \tan^{-1}\left(\frac{3/2}{4/3}\right) = 48.4°.$$

8.9 (a) Since the speed of an electromagnetic wave in a medium is $v = c/n$, we get the wavelengths in the two media as follows:

$$\lambda_1 = \frac{c}{n_1 f} = \frac{3 \times 10^8 \text{ m/s}}{1.5 \times 4.5 \times 10^{14} \text{ Hz}} = 4.44 \times 10^{-7} \text{ m},$$
$$\lambda_2 = \frac{c}{n_2 f} = \frac{3 \times 10^8 \text{ m/s}}{1.0 \times 4.5 \times 10^{14} \text{ Hz}} = 6.67 \times 10^{-7} \text{ m}.$$

(b) Here the situation is transverse electric (TE). Therefore, we will use the TE formulas for the reflection and transmission coefficients,

$$\rho_\perp = \frac{n_1 \cos\theta_1 - n_2 \cos\theta_2}{n_1 \cos\theta_1 + n_2 \cos\theta_2},$$
$$\tau_\perp = \frac{2n_1 \cos\theta_1}{n_1 \cos\theta_1 + n_2 \cos\theta_2}.$$

We need θ_2, which can be obtained from Snell's law,

$$n_2 \sin\theta_2 = n_1 \sin\theta_1, \quad \theta_2 = \sin^{-1}\left(\frac{n_1 \sin\theta_1}{n_2}\right) = \sin^{-1}(0.75) = 48.59°.$$

Therefore, we get the following numerical values of the coefficients:

$$\rho_\perp = 0.325, \quad \tau_\perp = 1.325.$$

(c) The energy can be obtained from the intensity formula for harmonic electromagnetic waves by multiplying the area projected on the plane of incidence. Thus we have the following for average power arriving at the plane of incidence and leaving the plane of incidence:

$$\langle P_{in}\rangle = \frac{1}{2}\epsilon_1 v_1 E_0^2 A\cos\theta_1,$$
$$\langle P_{re}\rangle = \frac{1}{2}\epsilon_1 v_1 \rho_\perp^2 E_0^2 A\cos\theta_1,$$
$$\langle P_{tr}\rangle = \frac{1}{2}\epsilon_2 v_2 \tau_\perp^2 E_0^2 A\cos\theta_2,$$

where $\epsilon_{1,2}$ and $v_{1,2}$ are the electric susceptibility and speed of the wave in the two media, and R and T are the reflection and transmission coefficients. To find the fraction of energy entering the second medium, we just need the ratio of $\langle P_{tr}\rangle$ to $\langle P_{in}\rangle$,

$$\begin{aligned}\frac{\langle P_{tr}\rangle}{\langle P_{in}\rangle} &= \frac{\epsilon_2 v_2 \cos\theta_2}{\epsilon_1 v_1 \cos\theta_1}\tau_\perp^2,\\ &= \frac{n_2\cos\theta_2}{n_1\cos\theta_1}\tau_\perp^2,\\ &= \frac{1\times\cos 48.59^\circ}{1.5\times\cos 30^\circ}1.325^2 = 0.894.\end{aligned}$$

That is, 89.4% of energy is transmitted.

8.10 (a) Since the speed of an electromagnetic wave in a medium is $v = c/n$, we get the wavelengths in the two media as follows:

$$\lambda_1 = \frac{c}{n_1 f} = \frac{3\times 10^8\ \text{m/s}}{1.5\times 4.5\times 10^{14}\ \text{Hz}} = 4.44\times 10^{-7}\ \text{m},$$
$$\lambda_2 = \frac{c}{n_2 f} = \frac{3\times 10^8\ \text{m/s}}{1.3\times 4.5\times 10^{14}\ \text{Hz}} = 5.128\times 10^{-7}\ \text{m}.$$

(b) From Snell's law we get

$$\theta_2 = \sin^{-1}\left(\frac{1.5\sin 30^\circ}{1.3}\right) = 35.23^\circ.$$

Here the situation is transverse magnetic (TM). Therefore, we will use the TM formulas for the reflection and transmission coefficients,

$$\rho_\parallel = \frac{n_2\cos\theta_1 - n_1\cos\theta_2}{n_2\cos\theta_1 + n_1\cos\theta_2} = -0.0423,$$
$$\tau_\parallel = \frac{2n_1\cos\theta_1}{n_1\cos\theta_1 + n_2\cos\theta_2} = 1.105.$$

(c) The fraction of energy transmitted will be

$$\frac{P_{tr}}{P_{in}} = \frac{n_2\cos\theta_2}{n_1\cos\theta_1}\tau_\parallel^2 = 0.998.$$

That is, 99.8% of energy is transmitted.

Chapter 9

9.1 Let l_1 be the distance from S_1 to D and l_2 be the distance from S_2 to D. Let y be the y-coordinate of S_2. Then, we have

$$l_2 = \sqrt{l_1^2 + y^2}.$$

The phase difference will be

$$\Delta\phi = k\,(l_2 - l_1) = \frac{2\pi}{\lambda}\,(l_2 - l_1).$$

For numerical values, we have $l_1 = 100$ cm, $\lambda = 4$ cm, and y different for different parts.

(a) When $y = 1$ cm, we get $\Delta\phi = 0.0079$ rad or $\Delta\phi = 0.45°$.

(b) When $y = 3$ cm, we get $\Delta\phi = 0.071$ rad or $\Delta\phi = 4.05°$.

(c) When $y = 5$ cm, we get $\Delta\phi = 0.20$ rad or $\Delta\phi = 11°$.

(d) When $y = 7$ cm, we get $\Delta\phi = 0.38$ rad or $\Delta\phi = 22°$.

9.2 The coherence time τ_c is related to the spread Δf of frequencies in the wave,

$$\tau_c = \frac{1}{\Delta f}.$$

The coherence length l_c is related to the distance traveled in the coherence time,

$$l_c = v\tau_c.$$

(a) The spread in frequency is $\Delta f = 10$ kHz. Therefore, coherence time is

$$\tau_c = \frac{1}{\Delta f} = \frac{1}{10\ \text{kHz}} = \frac{1}{10}\ \text{msec}.$$

Since the wave travels at the speed of light, we get

$$l_c = c\tau_c = 3 \times 10^8\ \text{m/s} \times \frac{1}{10}\ \text{msec} = 3 \times 10^4\ \text{m}.$$

(b) The spread in frequency is $\Delta f = 5$ Hz. Therefore, coherence time is

$$\tau_c = \frac{1}{\Delta f} = \frac{1}{5\ \text{Hz}} = \frac{1}{5}\ \text{sec}.$$

Since the wave travels at the speed of sound, which we will take to be 340 m/s we get

$$l_c = c\tau_c = 340\ \text{m/s} \times \frac{1}{5}\ \text{sec} = 68\ \text{m}.$$

(c) To find the spread in frequency from the spread in wavelength we use the following relation:

$$f = \frac{v}{\lambda}.$$

Taking the differential of this relation gives

$$|\Delta f| = \frac{v}{\lambda^2}|\Delta\lambda|.$$

The range of wavelength can be written as the average and the spread, $\lambda \pm \Delta\lambda =$ 550 nm $\pm$ 150 nm. Therefore,

$$|\Delta f| = \frac{3 \times 10^8 \text{ m/s}}{(550 \text{ nm})^2} \times 150 \text{ nm} = 1.49 \times 10^{14} \text{ Hz}.$$

Therefore, coherence time is

$$\tau_c = \frac{1}{\Delta f} = \frac{1}{1.49 \times 10^{14} \text{ Hz}} = 6.7 \times 10^{-15} \text{ sec}.$$

Since the wave travels at the speed of light, we get

$$l_c = c\tau_c = 3 \times 10^8 \text{ m/s} \times 6.7 \times 10^{-15} \text{ sec} = 2.0 \times 10^{-6} \text{ m}.$$

(d) We apply the formulas given above to obtain

$$|\Delta f| = \frac{3 \times 10^8 \text{ m/s}}{(590 \text{ nm})^2} \times 0.0004 \text{ nm} = 3.45 \times 10^8 \text{ Hz},$$
$$\tau_c = \frac{1}{\Delta f} = \frac{1}{3.45 \times 10^8 \text{ Hz}} = 2.9 \times 10^{-9} \text{ sec},$$
$$l_c = c\tau_c = 3 \times 10^8 \text{ m/s} \times 2.9 \times 10^{-9} \text{ sec} = 0.87 \text{ m}.$$

9.3 (a) First let us find the positions of the constructive interferences. Here they occur in the directions given by the angle θ that obeys

$$d \sin\theta = m\lambda, \quad m = 0, \pm 1, \pm 2, \ldots.$$

This gives the following angles for three values of m:

$$m = 0 \implies \theta_0 = 0.$$
$$m = 1 \implies \theta_1 = \sin^{-1}\left(\frac{\lambda}{d}\right) = \sin^{-1}\left(\frac{0.6328\ \mu\text{m}}{10\ \mu\text{m}}\right) = +3.63°.$$
$$m = -1 \implies \theta_2 = \sin^{-1}\left(\frac{-\lambda}{d}\right) = -3.63°.$$

The position on the screen can be given by the y-coordinate from the center beam,

$$y_0 = 0,$$
$$y_1 = L\tan\theta_1 = 1.5\ \text{m}\tan(3.63°) = 9.5\ \text{cm},$$
$$y_{-1} = -9.5\ \text{cm}.$$

Here, the destructive interferences occur at the angle given by

$$d\sin\theta = \left(m + \frac{1}{2}\right)\lambda,\quad m = 0, \pm 1, \pm 2, \ldots.$$

We will find the angles for $m = 0$ and $m = -1$,

$$m = 0 \implies \theta_0' = \sin^{-1}\left(\frac{\lambda}{2d}\right) = +1.81°.$$
$$m = -1 \implies \theta_1' = \sin^{-1}\left(\frac{-\lambda}{2d}\right) = -1.81°.$$

The position on the screen can be given by the y-coordinate from the center beam,

$$y_0' = L\tan\theta_0' = 1.5\ \text{m}\tan(1.81°) = 4.8\ \text{cm},$$
$$y_1' = L\tan\theta_1' = -4.8\ \text{cm}.$$

(b) In the constructive interference formula, $d\sin\theta = n\lambda$ we set the maximum value of $\sin\theta = 1$ to obtain the theoretical maximum order,

$$n_{\text{max}} = \frac{d}{\lambda} = 15.$$

9.4 The wave function at the detector will be a sum of the three waves,

$$\psi = \psi_1 + \psi_2 + \psi_3.$$

As Fig. A.51 shows, let l_0 be the distance traveled by the central ray. Then, the distances traveled by the other two waves will differ from this distance by

$$\Delta l = \pm\frac{d}{2}\sin\theta.$$

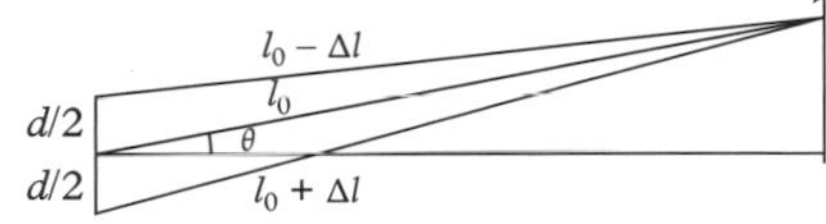

Fig. A.51 *Exercise 9.4.*

Writing the waves as complex exponentials, we will get ψ as follows:

$$\psi = Ae^{-\iota\omega t}\left(e^{ikl_0} + e^{ikl_0 + ik\Delta l} + e^{ikl_0 - ik\Delta l}\right),$$

where ω is the angular frequency and k is the wave number which is related to the wavelength λ as usual,

$$k = \frac{2\pi}{\lambda}.$$

The intensity will be proportional to the absolute value of the square of the wave function ψ. This will give

$$|\psi|^2 = A^2 \left| 1 + 2\cos\left(\frac{\pi d}{\lambda}\sin\theta\right)\right|^2. \tag{A.107}$$

The maximum of $|\psi|^2$ will give the direction θ in which we have the constructive interference. The maximum will occur when the cosine has the value +1,

$$\cos\left(\frac{\pi d}{\lambda}\sin\theta\right) = +1.$$

Therefore, the directions for destructive interference are

$$\frac{\pi d}{\lambda}\sin\theta = 0, 2\pi, -2\pi, 4\pi, -4\pi, \ldots$$

That is,

$$d\sin\theta_m = m\lambda, \quad m \text{ even integer.}$$

The minimum of $|\psi|^2$ will give the direction θ in which we have the destructive interference. The minimum will occur when the cosine has the value −1 or 0.

$$\text{When} \qquad \cos\left(\frac{\pi d}{\lambda}\sin\theta\right) = 0.$$

Therefore, the directions for destructive interference are

$$\frac{\pi d}{\lambda}\sin\theta = \pm\frac{\pi}{2}, \pm\frac{3\pi}{2}, \pm\frac{5\pi}{2}, \ldots$$

That is,

$$d\sin\theta_m = m\frac{\lambda}{2}, \quad m \text{ odd integer.}$$

When cos in Eq. A.107 is −1, we get $d\sin\theta_m = m\lambda$, m odd.

9.5 (a) From the frequency and the speed of the wave we get the wavelength,

$$\lambda = \frac{v}{f} = \frac{340 \text{ m/s}}{2000 \text{ Hz}} = 17 \text{ cm.}$$

With $d = 66$ cm we get

$$\frac{\lambda}{d} = \frac{17 \text{ cm}}{66 \text{ cm}} = 0.258.$$

The constructive interference will occur when

$$\sin\theta_n = n\frac{\lambda}{d} = 0.258\, n.$$

Therefore, for $n = 0, \pm 1, \pm 2, \pm 3, \pm 4$ we get

$$\theta_0 = 0, \quad \theta_1 = \pm\sin^{-1}(0.258) = \pm 15°,$$
$$\theta_2 = \pm\sin^{-1}(2 \times 0.258) = \pm 31°, \quad \theta_3 = \pm\sin^{-1}(3 \times 0.258) = \pm 51°.$$

When $n = 4$ we get the condition $\sin\theta_4 = 4 \times 0.258 = 1.032 > 1$, which does not have a real number solution.

(b) The intensity at the screen is given by

$$I = I_{ref}\cos^2\left(\frac{\pi d \sin\theta}{\lambda}\right).$$

To find the angle when the intensity will be half that at the center we set up the following equations:

$$\frac{I_{ref}}{2} = I_{ref}\cos^2\left(\frac{\pi d \sin\theta}{\lambda}\right),$$

which yields

$$\frac{\pi d \sin\theta}{\lambda} = \pm\cos^{-1}\left(\frac{1}{\sqrt{2}}\right) = \pm\frac{\pi}{4}.$$

Solving for $\sin\theta$ we get

$$\sin\theta = \pm\frac{\lambda}{4d} = \pm 0.0645.$$

Therefore, the angles are

$$\theta = \pm 3.7°,$$

from which we get the angular width to be

$$\Delta\theta_{1/2} = 7.4°.$$

9.6 Let ϕ_1 and ϕ_2 be the phase changes when the wave travels in a vacuum and when it travels through the glass slide. The wavelength in the vacuum is λ_0 but in glass of refractive index n it is

$$\lambda = \frac{\lambda_0}{n}.$$

We know that for each wavelength of travel, the phase change is 2π rad. Therefore, we will get

$$\phi_1 = 2\pi\,\frac{L}{\lambda_0},$$
$$\phi_2 = 2\pi\,\frac{(L-d)}{\lambda_0} + 2\pi\,\frac{nd}{\lambda_0}.$$

Therefore, the phase difference at the end will be

$$\Delta\phi = \phi_2 - \phi_1 = 2\pi\,\frac{(n-1)d}{\lambda_0}.$$

9.7 (a) The phase difference between waves comes from the optical path difference, not just the path difference. We will calculate the phase difference due to path difference and due to medium difference separately and then add them,

$$\Delta\phi = \Delta\phi_{\text{path}} + \Delta\phi_{\text{medium}}.$$

The phase difference due to the path difference has been worked out in the text,

$$\Delta\phi_{\text{path}} = 2\pi\,\frac{d\sin\theta}{\lambda_0}.$$

The change due to the glass plate will be

$$\Delta\phi_{\text{medium}} = 2\pi\,\frac{(n-1)}{\lambda_0}\,\Delta l_{\text{glass}}.$$

Since the beam goes through the glass plate of thickness h at an angle θ, the path in glass will have the geometric distance given by

$$\Delta l_{\text{glass}} = \frac{h}{\cos\theta}.$$

Therefore, the net phase difference will be

$$\Delta\phi = 2\pi\,\frac{d\sin\theta}{\lambda_0} + 2\pi\,\frac{h(n-1)}{\lambda_0\cos\theta}.$$

(b) Constructive interference will occur when $\Delta\phi$ is an integral multiple of 2π,

$$\Delta\phi = 2\pi m, \quad m = \text{integer}.$$

This becomes

$$2\pi\,\frac{d\sin\theta_m}{\lambda_0} + 2\pi\,\frac{h(n-1)}{\lambda_0\cos\theta_m} = 2\pi m.$$

For a small angle, let us approximate

$$\sin\theta \approx \theta, \quad \cos\theta = 1,$$

to simplify to

$$\theta_m = m\frac{\lambda_0}{d} - \frac{(n-1)h}{d}, \quad \theta_m \ll 1.$$

(c) Now,

$$n = 1.5, \quad \frac{h}{d} = 2, \quad \frac{\lambda_0}{d} = \frac{1}{2}.$$

With these values we get the following for the order m:

$$\theta_m = \frac{m}{2} - 2.$$

This relation shows that when $m = 4$, there is an interference maximum in the $\theta = 0$ direction, and we do not need to look for the case where this may not be the case. Therefore, the interference order for center direction, $\theta = 0$ will be

$$m = 4.$$

9.8 (a) The variation in intensity is due to the interference of the two waves.

(b) Figure A.52 shows five places where the car would encounter the strongest signals due to constructive interference at those locations.

Let us write the constructive interference condition in terms of the frequency f as

$$d \sin\theta_m = m\lambda = m\frac{c}{f}, \quad m \text{ integer}.$$

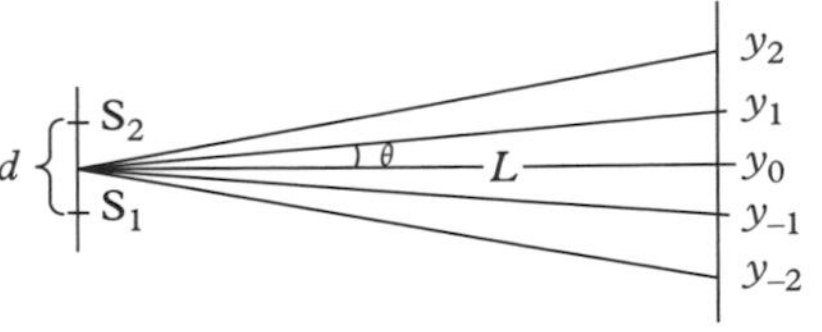

Fig. A.52 *Exercise 9.8.*

The y-coordinates on the road will be

$$y_m = L\tan\theta_m \approx L\sin\theta_m = m\frac{Lc}{df}.$$

Thus,

$$m = 0 \implies y_0 = 0,$$

$$m = 1 \implies y_1 = \frac{Lc}{df} = \frac{20,000 \text{ m} \times 3 \times 10^8 \text{ m/s}}{1000 \text{ m} \times 4 \times 10^6 \text{ Hz}} = 1500 \text{ m},$$

$$m = 2 \implies y_1 = 2\frac{Lc}{df} = 3000 \text{ m},$$

$$m = -1 \implies y_{-1} = -1500 \text{ m},$$

$$m = -2 \implies y_{-2} = -3000 \text{ m}.$$

9.9 (a) From the geometry of two rays emanating from the source, one (ray SP) going directly to the detector at P and the other (ray SBP) first being reflected at B and then meeting ray SP at point P, as shown in Fig. A.53, gives the following path difference between the two paths:

$$\Delta l = 2D\sin\theta.$$

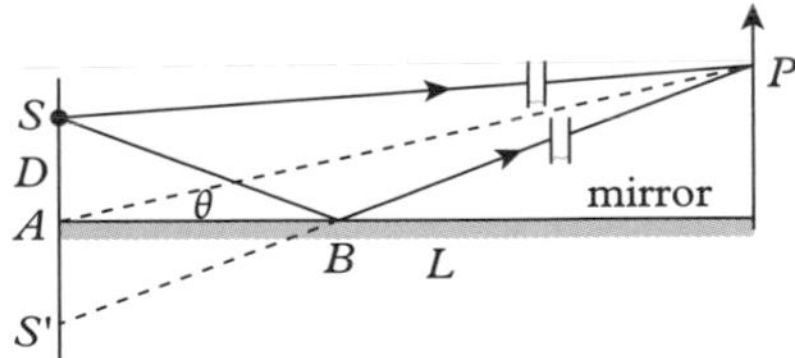

Fig. A.53 *Exercise 9.9.*

But the phase difference between the two waves SP and SBP also has an additional phase difference due to reflection at B, which adds π to the phase of SBP. Therefore, the net phase difference between the two paths is

$$\Delta\phi = 2\pi\frac{\Delta l}{\lambda} + \pi = 2\pi\frac{2D\sin\theta}{\lambda} + \pi.$$

For constructive interference this phase change must be an integral multiple of 2π radians,

$$\text{Constructive: } \Delta\phi = 2\pi m, \quad m \text{ integer},$$

and the destructive interference,

$$\text{Destructive: } \Delta\phi = \pi m, \quad m \text{ odd integer}.$$

Therefore,

$$\text{Constructive: } 2D\sin\theta = m\frac{\lambda}{2}, \quad m \text{ odd integer,}$$
$$\text{Destructive: } 2D\sin\theta = m\lambda, \quad m \text{ integer.}$$

(b) The y-coordinate of the detector is related to θ as follows:

$$y = L\tan\theta \approx L\sin\theta \implies \sin\theta = \frac{y}{L}.$$

Therefore, in terms of the y-coordinate of the detector, we will have the following conditions:

$$\text{Constructive: } y = m\frac{\lambda L}{4D}, \quad m \text{ odd integer,}$$
$$\text{Destructive: } y = m\frac{\lambda L}{2D}, \quad m \text{ integer.}$$

Therefore, we will have the following bright and dark places:

Constructive:

$$y_1 = 1 \times \frac{589 \text{ nm} \times 2 \text{ m}}{4 \times 2\ \mu\text{m}} = 14.7 \text{ cm},$$
$$y_2 = 3 \times 14.7 \text{ cm} = 44.2 \text{ cm},$$

Destructive:

$$y_1 = 1 \times \frac{589 \text{ nm} \times 2 \text{ m}}{2 \times 2\ \mu\text{m}} = 29.4 \text{ cm},$$
$$y_2 = 2 \times 29.4 \text{ cm} = 58.9 \text{ cm}.$$

9.10 (a) From Snell's law at the air/oil interface, as shown in Fig. A.54(a) we get

$$\theta_2 = \sin^{-1}\left(\frac{n_1 \sin\theta_1}{n_2}\right) = \sin^{-1}\left(\frac{1.0 \sin 60^\circ}{1.2}\right) = 46.2^\circ.$$

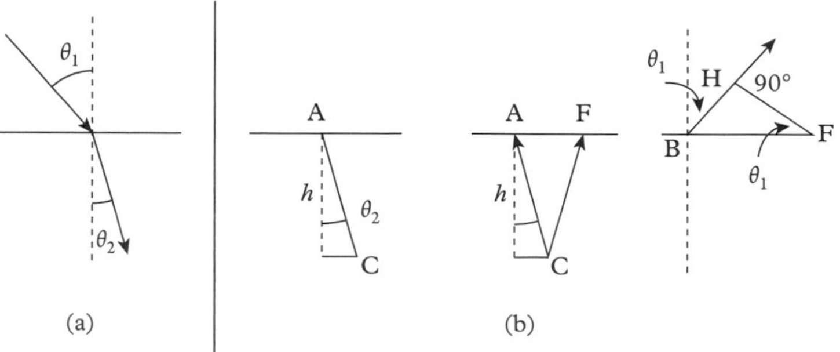

Fig. A.54 *Exercise 9.10.*

(b) Referring to figures in Fig. A.54(b) we find the following:

$$AC = \frac{h}{\cos\theta_2} = \frac{150\ \text{nm}}{\cos 46.2^\circ} = 216.7\ \text{nm},$$
$$AF = 2h\tan\theta_2 = 2 \times 150\ \text{nm} \times \tan 46.2^\circ = 312.8\ \text{nm},$$
$$BH = BF\sin\theta_1 = AF\sin\theta_1 = 2h\tan\theta_2\sin\theta_1 = 270.9\ \text{nm}.$$

(c) We can readily find the phase difference by employing the following rules:

- For a path difference Δl, the phase difference is

$$\Delta\phi = \frac{2\pi\,\Delta l}{\lambda}.$$

- For a reflection of light when $n_1 < n_2$,

$$\Delta\phi = \pi\,.$$

(i) Since A to B is a reflection in which $n_1(= 1.0) < n_2(= 1.2)$ we get a phase change of π radians,

$$\phi_B - \phi_A = \pi\,.$$

(ii) Since C to D is a reflection in which $n_1(= 1.2) < n_2(1.33)$ we get a phase change of π radians,

$$\phi_D - \phi_C = \pi\,.$$

(iii) The distance between E and F is infinitesimal with no reflection. Therefore

$$\phi_F - \phi_E = 0.$$

(iv) In the path AH we have the flip for A to B and then path travel from B to H,

$$\phi_H - \phi_A = \pi + 2\pi\frac{BH}{\lambda_0}.$$

Using the numerical values we get

$$\phi_H - \phi_A = \pi + 2\pi\frac{270.9\ \text{nm}}{550\ \text{nm}} = 1.985\pi\ \text{rad}.$$

(v) The path A to F has one phase changing reflection at C. With $AC = CF$ we get

$$\phi_F - \phi_A = \pi + 2\pi\frac{2AC}{\lambda_2}.$$

Putting in the numerical values we get

$$\phi_F - \phi_A = \pi + 2\pi\frac{2 \times 216.7\ \text{nm}}{550\ \text{nm}/1.22} = 2.923\pi\ \text{rad}.$$

(vi) The phase difference between H and F can be obtained by comparing the phase changes from A to H and from A to F,

$$\phi_F - \phi_H = (\phi_F - \phi_A) - (\phi_H - \phi_A) = 2.923\pi - 1.985\pi = 0.938\pi.$$

(d) Here we want a general formula for any h, λ, and angles θ_1 and θ_2. Using the general formulas we have written for the phase changes we have

$$\phi_F - \phi_H = (\phi_F - \phi_A) - (\phi_H - \phi_A) = \pi + 2\pi\frac{2AC}{\lambda_2} - \left(\pi + 2\pi\frac{BH}{\lambda_0}\right),$$

which simplifies to

$$\phi_F - \phi_H = \frac{4\pi h}{\lambda_0 \cos\theta_1}(n_{\text{oil}} - \sin\theta_1 \sin\theta_2),$$

where n_{oil} is the refractive index of oil. Equating this to integer multiples of 2π will give the constructive interference condition,

$$\frac{4\pi h}{\lambda_0 \cos\theta_2}(n_{\text{oil}} - \sin\theta_1 \sin\theta_2) = 2\pi m, \quad (m \text{ integer}). \tag{A.108}$$

Now, note that θ_1 and θ_2 are related by

$$n_{\text{air}} \sin\theta_1 = n_{\text{oil}} \sin\theta_2.$$

With $n_{\text{air}} = 1$, this gives

$$\sin\theta_1 = n_{\text{oil}} \sin\theta_2.$$

Therefore, we will get

$$\frac{(n_{\text{oil}} - \sin\theta_1 \sin\theta_2)}{\cos\theta_2} = n_{\text{oil}} \cos\theta_2.$$

Using this in Eq. A.108, we find the condition for constructive interference,

$$\cos\theta_2 = m\frac{\lambda_0}{2hn_{\text{oil}}} \Longrightarrow \sin\theta_1 = \sqrt{n_{\text{oil}}^2 - \frac{m\lambda_0}{2h}}.$$

9.11 (a) The phase difference between the wave reflected from the front face and that from the bottom of the air wedge will be

$$\phi = 2\frac{hx}{l}\frac{2\pi}{\lambda_0} + \pi,$$

where x is the horizontal distance from the corner and the phase change π comes about due to reflection from the bottom of the air wedge. For constructive interference this phase change must equal an integral multiple of 2π.

$$2\frac{hx}{l}\frac{2\pi}{\lambda_0} + \pi = 2\pi m, \quad m = 0, \pm 1, \pm 2, \ldots.$$

Therefore, for some particular integer $m = M$,

$$2\frac{hx}{l}\frac{2\pi}{\lambda_0} + \pi = 2\pi M. \tag{A.109}$$

When we move towards the wider gap we will get the interference at order $M+1$ since the path length will now be longer. Successively moving we will observe $M+2$, $M+3$ etc. Question is asking Δx when we will see $M+1$,

$$2\frac{h(x+\Delta x)}{l}\frac{2\pi}{\lambda_0} + \pi = 2\pi(M+1). \tag{A.110}$$

Subtracting Eq. A.109 from Eq. A.110 we get

$$\Delta x = \frac{\lambda_0 l}{2h}.$$

(b) Using the numerical values we get

$$\Delta x = \frac{550\text{ nm} \times 25\text{ cm}}{2 \times 100\ \mu\text{m}} = 0.69\text{ mm}.$$

9.12 When you move one mirror of the Michelson interferometer by a distance D you change the difference in path in the two arms by $2D$, since you change the path to the mirror and from the mirror. Therefore, when a mirror is moved D we will get the number N to be

$$N = \frac{2D}{\lambda}.$$

Therefore,

$$D = N \times \frac{1}{2}\lambda.$$

Putting in the numerical values yields

$$D = 8 \times \frac{1}{2} \times 632.8\text{ nm} = 2.531\ \mu\text{m}.$$

9.13 Let n be the refractive index when the container is filled with the gas. Since the light will pass a distance of $2d$ through the gas (d in each direction), the change in phase due to filling the gas will be

$$\Delta\phi = 2\pi\,\frac{2d(n-1)}{\lambda}.$$

For 1220 fringes to have passed, this phase change must have gone through 1220 cycles of 2π. That is, this phase change will be 1220 times 2π radians,

$$2\pi\,\frac{2d(n-1)}{\lambda} = 1220 \times 2\pi.$$

Therefore,

$$n = 1 + 610\frac{\lambda}{d} = 1 + 1.93 \times 10^{-2} = 1.02.$$

Chapter 10

10.1 (a) Since the slit is horizontal, the horizontal bands of the diffraction pattern will be spread out vertically.

(b) The diffraction minima from a single slit occur when $\beta = m\pi$ for $m = \pm 1, \pm 2, \ldots$. In terms of the slit width b and the direction θ from the slit this condition becomes

$$\frac{\pi b \sin\theta_m}{\lambda} = m\pi, \implies b\sin\theta_m = m\lambda, \quad m = \pm 1, \pm 2, \cdots.$$

Using this formula we obtain the following values for θ_1, θ_2 and θ_3:

$$\theta_1 = \sin^{-1}\left(\frac{\lambda}{b}\right) = \sin^{-1}\left(\frac{530\text{ nm}}{1500\text{ nm}}\right) = 0.36\text{ rad} = 21°,$$

$$\theta_2 = \sin^{-1}\left(\frac{2\lambda}{b}\right) = \sin^{-1}\left(\frac{2\times 530\text{ nm}}{1500\text{ nm}}\right) = 0.78\text{ rad} = 45°,$$

$$\theta_3 = \sin^{-1}\left(\frac{3\lambda}{b}\right) = \sin^{-1}\left(\frac{3\times 530\text{ nm}}{1500\text{ nm}}\right) = \sin^{-1}(>1), \quad \text{does not exist.}$$

Therefore, the positions of the first- and second-order bands on the screen will be given by the y-coordinates,

$$y_{\pm 1} = L\tan(\theta_{\pm 1}) = (1.2\text{ m})\tan(\pm 21°) = \pm 46\text{ cm},$$
$$y_{\pm 2} = L\tan(\theta_{\pm 2}) = (1.2\text{ m})\tan(\pm 45°) = \pm 120\text{ cm}.$$

(c) Maxima when $\beta = 0, \pm 1.4303\,\pi, \pm 2.4590\,\pi, \ldots$. Therefore, the bright spots will be in the following directions:

$$\theta_0 = 0°, \quad \text{central bright spot,}$$

$$\theta_1 = \sin^{-1}\left(\frac{1.4303\,\lambda}{b}\right) = 30°,$$

$$\theta_2 = \sin^{-1}\left(\frac{2.4590\,\lambda}{b}\right) = 60°.$$

The y-coordinates of these bright spots are

$$y_0 = 0, \quad y_{\pm 1} = 69.3\text{ cm}, \quad y_{\pm 2} = 210\text{ cm}.$$

(d) The width of the central bright spot will be between the y_{-1} and y_{+1} of the diffraction minima worked out in part (a),

$$\Delta y_{\text{central}} = y_{+1} - y_{-1} = 46\text{ cm} - (-46\text{ cm}) = 92\text{ cm}.$$

(e) The $m = 1$ bright spot will be between y_{+1} and y_{+2} of the diffraction minima. Therefore, the width of the $m = 1$ bright spot will be

$$\Delta y_{\text{m}=1} = [y_{+2} - y_{+1}]_{\text{min}} = 1.2\text{ m} - (46\text{ cm}) = 74\text{ cm}.$$

(f) The ratio of the $m = 1$ peak to $m = 0$ peak is found to be

$$\frac{I_{m=1}}{I_{m=0}} = \frac{I(\theta = 30°)}{I(\theta = 0)}.$$

Using the formula for the intensity in a single-slit diffraction this gives

$$\frac{I_{m=1}}{I_{m=0}} = \left[\frac{\sin\beta_1}{\beta_1}\right]^2,$$

where

$$\beta_1 = \frac{\pi b}{\lambda}\sin(30°) = 4.45 \text{ rad}.$$

Therefore,

$$\frac{I_{m=1}}{I_{m=0}} = \left[\frac{\sin(4.45 \text{ rad})}{4.45 \text{ rad}}\right]^2 = 0.047.$$

10.2 (a) We will find that vertically oriented lines are spread out horizontally.

(b) The diffraction minima will occur when

$$\sin\theta_m = m\frac{\lambda}{b}.$$

This gives the x-coordinates of the minima bands to be

$$x_m = L\tan\theta_m = \frac{L\sin\theta_m}{\sqrt{1-\sin^2\theta_m}}.$$

Therefore, x-coordinates of $m = \pm 1$ and $m = \pm 2$ bands are

$$x_{\pm 1} = \frac{L\lambda/b}{\sqrt{1-\lambda^2/b^2}}, \quad x_{\pm 2} = \frac{2L\lambda/b}{\sqrt{1-4\lambda^2/b^2}}.$$

Numerical values:

$$\frac{\lambda}{b} = \frac{0.630 \text{ µm}}{1.5 \text{ µm}} = 0.42, \quad \frac{L\lambda}{b} = 50.4 \text{ cm}.$$

These give

$$x_{\pm 1} = \frac{50.4 \text{ cm}}{\sqrt{1-0.42^2}} = 55.5 \text{ cm}, \quad x_{\pm 2} = \frac{2 \times 50.4 \text{ cm}}{\sqrt{1-4\times 0.42^2}} = 186 \text{ cm}$$

(c) Beside the central maximum for $\theta = 0$, we will get two maxima corresponding to $m = \pm 1$,

$$\sin\theta_{\pm 1} = \pm\left(1.4303\frac{\lambda}{b}\right) = \pm 0.60.$$

Therefore,

$$\theta_{\pm} = \pm 36.9^\circ.$$

This gives

$$y_{\pm 1} = L \tan \theta_{\pm 1} = \pm 90 \text{ cm}.$$

Note that the constructive diffraction condition for $m = \pm 2$ is

$$\sin \theta_{\pm 2} = \pm \left(2.4590 \frac{\lambda}{b} \right) = 1.03278,$$

which is greater than 1, meaning we do not have a real solution.

(d) The angular width of the central peak will be

$$\Delta\theta = \frac{2\lambda}{b} = 2 \times 0.42 \text{ rad} = 0.84 \text{ rad} = 48^\circ.$$

10.3 (a) The width of the central peak will be the distance between $m = -1$ and $m = +1$ diffraction minima, which occur in directions given by

$$\sin \theta_{\pm 1} = \pm \frac{\lambda}{b}.$$

For small angles the width of the central peak will be

$$\Delta\theta = \theta_{+1} - \theta_{-1} = \frac{2\lambda}{b}.$$

Now, the distance Δy on the screen between the two minima will be

$$\Delta y = L \tan \Delta\theta \approx L\Delta\theta = \frac{2L\lambda}{b}.$$

Therefore, the slit width b will be

$$b = \frac{2L\lambda}{\Delta y}.$$

Now, putting in numbers we get

$$b = \frac{2 \times 1.6 \text{ m} \times 632.8 \text{ nm}}{5 \text{ cm}} = 40.5 \text{ µm}.$$

(b) With $b = 40.5$ µm and $\lambda = 540$ nm we will get the width on the screen to be

$$\Delta y = \frac{2L\lambda}{b} = \frac{2 \times 1.6 \text{ m} \times 540 \text{ nm}}{40.5 \text{ µm}} = 4.3 \text{ cm}.$$

10.4 (a) We plot the functions x^2 and $\sin(x)$ versus x on the same graph paper. The intersections of the two graphs correspond to the solution of the equation $x^2 = \sin(x)$. The solution is shown in Fig. A.55. They are $x = 0$ and $x = 0.875$. To obtain the more accurate number, I plotted (not shown) within a tighter range of 0.8 to 0.9.

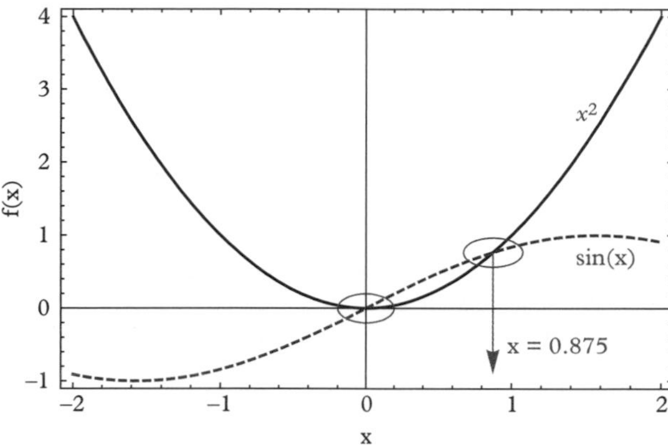

Fig. A.55 *Exercise 10.4.*

(b) Do a similar plot. Here you will obtain an infinite number of solutions since $1/x$ and $\tan(x)$ will cross at infinitely many points.

10.5 (a) We will first find the angular width of the central peak due to diffraction. Then, we will determine how many interference peaks will lie within that angular width.

Since the central diffraction peak is bounded by the $m = \pm 1$ minima, the angular width will be

$$\Delta\theta = 2\theta_1 = 2 \times \sin^{-1}\left(\frac{\lambda}{b}\right) = 2 \times \sin^{-1}\left(\frac{0.450\ \mu\text{m}}{2\ \mu\text{m}}\right) = 26^\circ.$$

We now need to find interference maxima between $\theta = -13^\circ$ and $\theta = +13^\circ$. The condition for interference maxima is

$$a \sin\theta_m = m\lambda, \quad m = 0, \pm 1, \pm 2, \ldots.$$

Let us first calculate the value of m when we require $\theta_m = 13^\circ$.

$$6\ \mu\text{m} \sin 13^\circ = m \times 0.450\ \mu\text{m}, \implies m = 3.$$

That means $m = 3$ constructive interference occurs at the point where the diffraction minimum also occurs. Therefore, the $m = 3$ interference will not be observed. Hence, we have $m = 0, \pm 1, \pm 2$ within the central diffraction maximum. That is, there will be five interference peaks within the central peak.

(b) If the slit width is doubled, the angular width of the central diffraction peak will be

$$\Delta\theta = 2 \times \sin^{-1}\left(\frac{0.450\ \mu\text{m}}{4\ \mu\text{m}}\right) = 13°.$$

So, the question is how many interference maxima are there between $\theta = -6.5°$ and $\theta = +6.5°$? Using $\theta_m = 6.5°$ we find

$$(6\ \mu\text{m})\ \sin 6.5° = m \times 0.450\ \mu\text{m}, \implies m = 1.5.$$

Therefore, there will only be three interference peaks in the central maximum now for $m = 0, \pm 1$.

(c) Here the central maximum of diffraction is still 26° wide, but since the slits are separated more, the interference maxima will be closer together. Using $\theta = 13°$ in the interference condition gives the maximum $m = 6$. Therefore, there will be 11 interference peaks corresponding to $m = 0, \pm 1, \pm 2, \pm 3, \pm 4, \pm 5$.

(d) Hint: Follow the given calculation in (a).

(e) The $m = 0$ of the interference constructive peak occurs at $\theta = 0$ and $m = 1$ occurs at $\theta = \sin^{-1}(\lambda/a) = 4.3°$. The peak from interference is diminished due to the diffraction effect. Therefore, we need only the β part,

$$\beta(4.3°) = \frac{\pi \times 2\ \mu\text{m} \sin 4.3°}{0.45\ \mu\text{m}} = 1.0469\ \text{rad}.$$

Therefore, the intensity of the $m = 1$ peak will be

$$I(4.3°)/I(0) = \left(\frac{\sin(1.0469\ \text{rad})}{1.0469\ \text{rad}}\right)^2 = 0.68.$$

(f) The ratio will depend on the wavelength of light through the dependence of β on the wavelength of light.

(g) The ratio will depend on the width b of the slits through the dependence of β on these parameters.

10.6 From the number of interference fringes in the central maximum when we use 430 nm light we determine the width of the central maximum. First note that the maximum order of the interference peak to fit in the half-width of the central max will obey

$$a \sin\theta_{1/2} = \left(\frac{m_{\text{max}} + 1}{2}\right)\lambda,$$

where $\theta_{1/2}$ is the half-width of the central max of diffraction. From 11 fringes we get $m_{\text{max}} = 5$. Therefore,

$$\theta_{1/2} = \sin^{-1}\left(\frac{6 \times 0.430\ \mu m}{2 \times 5.0\ \mu m}\right) = 15°.$$

This gives us the condition of a minimum for the $m = 1$ diffraction minimum when we use 430 nm light,

$$b \sin\theta_{1/2} = \lambda, \implies b = \frac{0.430\ \mu\text{m}}{\sin 15°} = 1.7\ \mu\text{m}.$$

Thus, by using the data for 430 nm light we have found the width of the slits. Now, we apply these to the red light. For the red light the central maximum half-width will be different,

$$\theta_{1/2}(\text{red}) = \sin^{-1}\left(\frac{\lambda}{b}\right) = \sin^{-1}\left(\frac{0.6328\ \mu\text{m}}{1.7\ \mu\text{m}}\right) = 22°.$$

This gives

$$m_{\max}(\text{red}) = \text{integer part of}\left[\frac{a \sin 22°}{\lambda}\right] = 2.$$

Therefore, there will be five fringes in the central max for light of wavelength 632.8 nm.

10.7 Let D be the diameter of the lens and f its focal length. Then, treating the lens as a circular aperture the passage of the beam through the lens will cause diffraction and the central bright spot will have the following radius at the focal plane:

$$R = 1.22\frac{f\lambda}{D} = 1.22\frac{20\ \text{cm} \times 550\ \text{nm}}{8\ \text{cm}} = 1.67\ \mu\text{m}.$$

10.8 In order for the two images to be resolvable, the angle θ in Fig. A.56 must be greater than the half-width of the central max for image I_1.

Therefore, we get the condition

$$\theta = 1.22\ \frac{\lambda}{D} = 1.22 \times \frac{540\ \text{nm}}{40 \times 2.54\ \text{cm}} = 6.48 \times 10^{-7}\ \text{rad}.$$

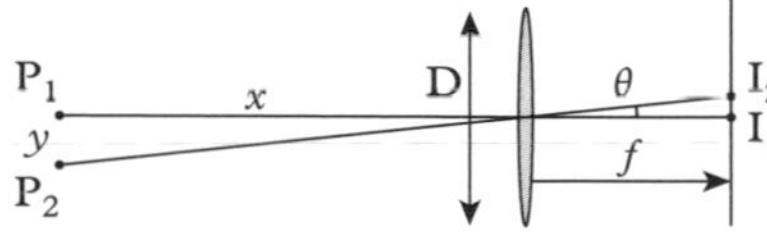

Fig. A.56 *Exercise 10.8.*

Using this angle with the distance x to the moon, we find the separation y on the moon to be

$$y \approx x\theta = 4 \times 10^8\ \text{m} \times 6.48 \times 10^{-7}\ \text{rad} = 259\ \text{m}.$$

10.9 Refer to Fig. A.56 of the last problem. We need to find x here,

$$x = \frac{y}{\theta}, \text{ and } \theta = 1.22\ \frac{\lambda}{D}.$$

Therefore,

$$x = \frac{yD}{1.22\ \lambda} = \frac{1\ \text{m} \times 4.5\ \text{mm}}{1.22 \times 580\ \text{nm}} = 6.4\ \text{km}.$$

This is too far for the eye to see. The main problem will be the low intensity of light reaching the eye.

10.10 (a) The separation between the lines of the grating will be

$$a = \frac{1}{500}\ \text{mm} = 2\ \mu\text{m}.$$

(b) The principal maxima in a diffraction grating occur at $a \sin \theta_m = m\,\lambda$ for $m = 0, \pm 1, \pm 2, \ldots$, where a is the distance between adjacent lines. Therefore, the angular direction of the $m = 1$ peaks for the two wavelengths are

$$\theta_1\,(\text{for } 590\ \text{nm}) = 17.16^\circ,\quad \theta_1\,(\text{for } 630\ \text{nm}) = 18.36^\circ$$

Therefore, the angular separation is 1.2°.

10.11 Let there be N lines per mm. Then we will have the separation between adjacent slits to be

$$a = \frac{1000}{N}\ \mu\text{m}. \tag{A.111}$$

From the given data of the second-order peak we get

$$a \sin \theta_2 = 2\,\lambda. \tag{A.112}$$

From Eqs. A.111 and A.112 we get

$$N - \frac{500 \sin \theta_2}{\lambda[\text{in } \mu\text{m}]} - 379.4.$$

10.12 The resolving power needed is

$$R = \frac{\Delta\lambda}{\lambda} = 274.5.$$

For the order $m = 1$ the condition of resolution in terms of number of diffraction lines N will be

$$N = R.$$

That is, we need 275 lines of the grating over which the beam is to be incident. Since the beam covers (400 lines/mm × 4 mm = 800) lines of the grating, the lines would be resolved in the first order.

10.13 (a) From the data the separation between successive slits is

$$a = \frac{1\ \text{mm}}{400} = 2.5\ \mu\text{m}.$$

From the condition for $m = 1$ peak,

$$\lambda = a \sin \theta = 2.5\ \mu\text{m} \sin 20^\circ = 0.855\ \mu\text{m}.$$

(b) From the given data, we have the number of grating lines upon which the light is incident, $N = 400 \times 5 = 2000$. The half-width of the peak is given by

$$\theta_{\rm hw} = \frac{\lambda}{N\,a\,\cos\,\theta} = \frac{0.855\ \mu\text{m}}{2000 \times 2.5\ \mu\text{m}\ \cos\,20^\circ} = 1.8 \times 10^{-4}\ \text{rad} = 0.01^\circ.$$

10.14 Figure A.57 shows interference peaks within the central diffraction maximum. In the small-angle approximation, the angles $\Delta\theta_i$ between successive interference peaks is related to the separation a of the slits and the wavelength λ,

$$\Delta\theta_i = \frac{\lambda}{a},$$

and the angular width of the central peak $\Delta\theta_d$ is related to the width of the slits b and the wavelength λ,

$$\Delta\theta_d = \frac{\lambda}{b}.$$

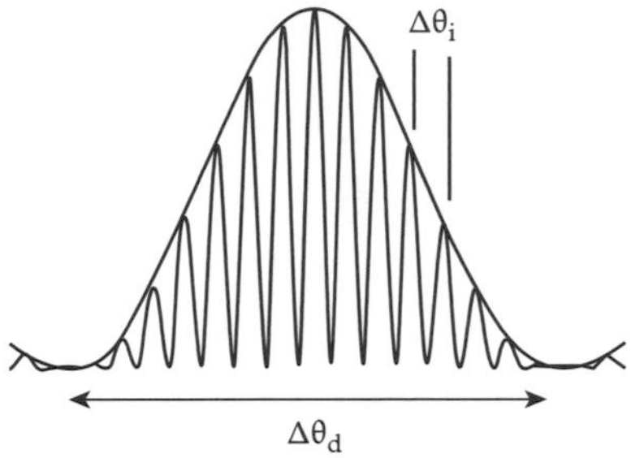

Fig. A.57 *Exercise 10.14.*

(a) Counting the interference peaks from 0 to N on the positive angle side, and -1 to $-N$ on the negative angle side, we will have $2N + 1$ interference peaks if N is the maximum order. Let us find N by dividing $\frac{1}{2}\Delta\theta_d$ by $\Delta\theta_i$,

$$N = \frac{\frac{1}{2}\Delta\theta_d}{\Delta\theta_i}.$$

Therefore,

$$N = \frac{a}{2b}.$$

Since we have $a = pb$ we get the total number of interference peaks,

$$\text{Total number on interference peaks} = 2N + 1 = 2\frac{a}{2b} + 1 = 2p + 1.$$

(b) In the small-angle approximation the direction angle at which interference peaks occur is given by

$$\theta_m = m\frac{\lambda}{a}, \quad (m\ \text{integer}),$$

and the direction angle at which the diffraction minima occur is

$$\theta_{m'} = m'\frac{\lambda}{b}, \quad (m'\ \text{integer}).$$

The missing interference peaks occur when

$$\theta_m = \theta_{m'}.$$

That is, when

$$\frac{m}{a} = \frac{m'}{b} \implies m = \frac{a}{b}m' = pm'.$$

This shows that p has to be a ratio of integers to satisfy the condition for missing order.

(i) For $p = 3$, following are some examples of the missing order.

$$\text{When } m' = \pm 1, \text{ the interference order } m = \pm 3 \text{ will be missing,}$$
$$\text{When } m' = \pm 2, \text{ the interference order } m = \pm 6 \text{ will be missing,}$$

(ii) For $p = \frac{3}{2}$, following are some examples of the missing order.

$$\text{When } m' = \pm 2, \text{ the interference order } m = \pm 3 \text{ will be missing,}$$
$$\text{When } m' = \pm 4, \text{ the interference order } m = \pm 6 \text{ will be missing,}$$

(iii) For $p = \frac{5}{3}$, following are some examples of the missing order.

$$\text{When } m' = \pm 3, \text{ the interference order } m = \pm 5 \text{ will be missing,}$$
$$\text{When } m' = \pm 6, \text{ the interference order } m = \pm 10 \text{ will be missing,}$$

(iv) When $p = \sqrt{2}$, there are no integers m and m' whose ratio will be equal to p. Therefore, there will be no missing orders.

10.15 The additional spread from the radius of 1 mm at the exit point will be given by the diffraction from a circular aperture. The angle of spread is

$$\theta = 1.22\frac{\lambda}{D} = 1.22\frac{461.9 \text{ nm}}{2 \text{ mm}} = 2.82 \times 10^{-4} \text{ rad}.$$

The radius of the spot a distance L from the exit point will be

$$R = r_{exit} + q = r_{exit} + L\theta,$$

where I have replaced $\tan\theta$ by θ, which is a good approximation for small angles, as is the case here.

(a) When $L = 1$ m, we get $R = 1.282$ mm.

(b) When $L = 1$ km, we get $R = 283$ mm.

(c) When $L = 1000$ km, we get $R = 283$ m.

(d) When $L = 400,000$ km, we get $R = 113$ m.

10.16 For two point sources of light separated by a distance x to be resolvable by the Hubble telescope, the angle θ subtended by the two light points at the Hubble must be greater than θ_1, given by the diffraction formula

$$\theta_1 = \frac{1.22\lambda}{D},$$

where D is the diameter of the light collection mechanism of Hubble. Therefore, to get x we multiply the angle by the distance L to the Hubble,

$$x \geq L\tan\theta_1 \approx \frac{1.22\lambda L}{D}.$$

Putting in the numerical values we get the minimum separation to be

$$x = \frac{1.22 \times 550 \text{ nm} \times 600 \text{ km}}{2.4 \text{ m}} = 16.8 \text{ cm}.$$

10.17 From Fig. A.58 we find that we need to compare the wavefront AD to the wavefront AB, rather than AC as was the case when the incident wave was incident normally on the slit. This gives the following Δl between the top and bottom of the slit:

$$\Delta l = BC + CD = b \sin \psi + b \sin \theta.$$

The diffraction minima would occur when this distance is an integral multiple of a wavelength, giving the following condition:

$$b \sin \psi + b \sin \theta = m\lambda, \quad (m \text{ integer}).$$

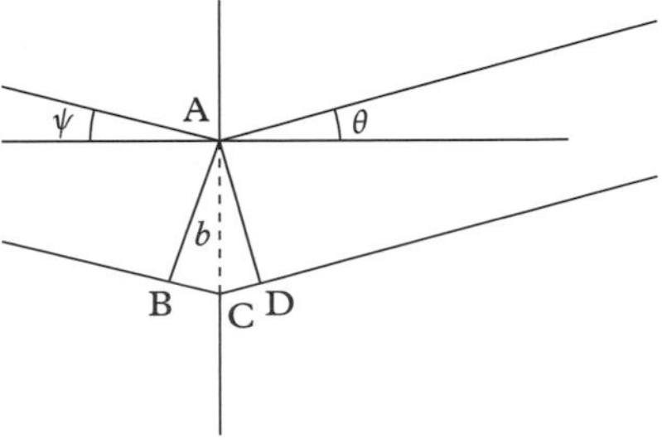

Fig. A.58 *Exercise 10.17.*

Index